GENES, CHROMOSOMES, AND NEOPLASIA

The University of Texas System Cancer Center
M. D. Anderson Hospital and Tumor Institute
33rd Annual Symposium on Fundamental Cancer Research

Published for
The University of Texas System Cancer Center
M. D. Anderson Hospital and Tumor Institute
Houston, Texas, by Raven Press, New York

The University of Texas System Cancer Center
M. D. Anderson Hospital and Tumor Institute
33rd Annual Symposium on Fundamental Cancer Research

Genes, Chromosomes, and Neoplasia

Edited by

Frances E. Arrighi, Ph.D.
Department of Cell Biology
The University of Texas System
Cancer Center
M. D. Anderson Hospital
and Tumor Institute
Houston, Texas

Potu N. Rao, Ph.D.
Department of Developmental
Therapeutics
The University of Texas System
Cancer Center
M. D. Anderson Hospital
and Tumor Institute
Houston, Texas

Elton Stubblefield, Ph.D.
Department of Cell Biology
The University of Texas System
Cancer Center
M. D. Anderson Hospital
and Tumor Institute
Houston, Texas

Raven Press ■ New York

Raven Press, 1140 Avenue of the Americas, New York, New York 10036

Library of Congress Cataloging in Publication Data

Symposium on Fundamental Cancer Research, 33d,
    Anderson Hospital and Tumor Institute, 1980.
    Genes, chromosomes, and neoplasia.

    "Published for the University of Texas System
Cancer Center, M. D. Anderson Hospital and Tumor
Institute, Houston, Texas."
    Proceedings of symposium held March 4–7, 1980 in
Houston, Texas.
    Includes bibliographical references and indexes.
    1.   Cancer—Genetic aspects—Congresses.
2.   Chromosomes—Congresses.   3.   Gene expression—
Congresses.   I.   Arrighi, Frances E.   II.   Rao, Potu N.
III.   Stubblefield, Elton.   IV.   Anderson Hospital and
Tumor Institute, Houston, Tex. V. Title.
RC268.4.S95 1981   616.9′94042   80–24702
ISBN 0–89004–532–1

This volume is a compilation of the proceedings of The University of Texas System Cancer Center M. D. Anderson Hospital and Tumor Institute 33rd Annual Symposium on Fundamental Cancer Research, held March 4–7, 1980, in Houston, Texas.

The material contained in this volume was submitted as previously unpublished material, except in the instances in which credit has been given to the source from which some of the illustrative material was derived.

Great care has been taken to maintain the accuracy of the information contained in the volume. However, the Editorial Staff and The University of Texas System Cancer Center cannot be held responsible for errors or for any consequences arising from the use of the information contained herein.

Preface

This volume, based on the 1980 symposium on Fundamental Cancer Research of M. D. Anderson Hospital and Tumor Institute, reviews our current understanding of the alterations in normal genetic material that occur prior to or as a consequence of cancer in the organism. Major sections discuss chromatin and chromosome structure, the *src* gene, gene expression, and the genetics of human cancer.

Frank Ruddle and T. C. Hsu, two pioneers who continue to advance the understanding of genetic mechanisms, set the stage by giving histories of somatic cell genetics and cytogenetics and projecting future advances. Insights available from the analysis of the structure of chromatin and chromosomes are provided by Potu Rao, Elton Stubblefield, and Albert Ting of Bert O'Malley's group.

Viral oncogene studies have contributed much to the understanding of neoplastic transformation. This monograph devotes several chapters to the *src* viral locus and its normal homologue, the *sarc* locus.

Gene expression is examined in subsequent chapters by, for example, Donald Robberson, who carefully models the expression of genetically altered mitochondrial DNA, supporting the model with electron micrographic evidence. Somatic cells, rat hepatoma cells, enzymes in human tumor cell lines, and DNA-transformed cells further the discussion of gene expression in normal and malignant cells.

Gene amplification is reviewed by Albert Levan and O. J. Miller. In his survey of 4 years' work on the SEWA mouse tumor, Levan concludes that C-bandless chromosomes and double minutes are in the same category of cytogenetic phenomena. Miller discusses methylation's role in controlling amplification.

Part of the evidence for a genetic cause for neoplastic transformation has been the finding of specific chromosome changes in cells from particular cancers. Among those discussed in this book are human leukemia, for which chromosome changes have led to the identification of new entities, lymphoma and the 14 q+ anomaly, and ovarian cystadenocarcinomas and the 14 q+ and 6q− anomalies. Experimental animal tumors are also reviewed in light of specific chromosome changes.

The monograph concludes with four discussions of particular human populations and with a summary by James Evans. Relatives of breast cancer patients, Utah Mormons, relatives of lung cancer patients, and patients with polyposis of the colon and their relatives are analyzed for the insight they provide into the relation of genetics and cancer. For example, the problem of altering a familial predisposition to cancer is covered by John Mulvihill.

This volume will be of interest to basic researchers in disciplines concerned with the cellular basis of cancer, including cell and molecular biologists, and to medical epidemiologists and human geneticists.

Editors' Foreword

"Genes, Chromosomes and Neoplasia" was the topic for the 33rd Annual Symposium on Fundamental Cancer Research, which was held in Houston March 4–7, 1980. This was a timely topic because of the increase in evidence that nonrandom chromosome changes are associated with certain tumors. The symposium was well attended, with 734 scientists participating representing 38 states and 18 foreign countries. At this symposium, Dr. T. C. Hsu became the first staff member of The University of Texas System Cancer Center M. D. Anderson Hospital and Tumor Institute to receive the Bertner Award.

As cochairpersons, we acknowledge our appreciation to members of the organizing and advisory committees for their advice and guidance.

Our appreciation and special thanks are due to Frances Goff and her staff for the many functions that they have expertly performed. We also wish to thank Forma Scientific, Scimetrics, Beckman Instruments, Inc., Dolan Scientific, and GIBCO Laboratories for providing funds for the hospitality room wherein speakers and other participants could meet and discuss topics of mutual interest. Our appreciation is also extended to the National Cancer Institute and American Cancer Society, Texas Division, without whose support this meeting could not have been held. We wish to thank the staff of the Graduate School of Biomedical Sciences of The University of Texas Health Science Center at Houston for their assistance. The efforts of the Departments of Scientific Publications and Public Information and Education are also appreciated, especially the efforts of Walter J. Pagel and Leslie Wildrick, who played important roles in the compilation and manuscript editing of this volume.

Finally we acknowledge the interest, concern, and efforts of Dr. R. Lee Clark, President Emeritus, and Dr. Charles A. LeMaistre, President, The University of Texas System Cancer Center.

Frances E. Arrighi
Potu N. Rao
Elton Stubblefield
Co-Editors

Contents

Contributors

R. Axel
Cancer Center
College of Physicians and Surgeons
Columbia University
New York, New York 10032

D. T. Bishop
Department of Medical Biophysics and
    Computing
LDS Hospital and
University of Utah Medical Center
Salt Lake City, Utah 84143

J. Michael Bishop
Department of Microbiology and Immu-
    nology
University of California
San Francisco, California 94143

J. S. Butel
Department of Virology and Epidemiology
Baylor College of Medicine
Houston, Texas 77030

D. Carmelli
Department of Medical Biophysics and
    Computing
LDS Hospital and
University of Utah Medical Center
Salt Lake City, Utah 84143

R. Lee Clark
The University of Texas System Cancer
    Center
M. D. Anderson Hospital and Tumor In-
    stitute
Houston, Texas 77030

Marc S. Collett
Department of Microbiology
University of Minnesota
Minneapolis, Minnesota 55455

Sara Courtneidge
Department of Microbiology and Immu-
    nology
University of California
San Francisco, California 94143

L. Scott Cram
Experimental Pathology Group
Los Alamos Scientific Laboratory
University of California
Los Alamos, New Mexico 87545

A. Peter Czernilofsky
Department of Microbiology and Immu-
    nology
University of California
San Francisco, California 94143

Larry L. Deaven
Health Effects Research Division
Office of Health and Environmental Re-
    search
Office of Environment
United States Department of Energy
Washington, D.C. 20545

Jean Deschatrette
Centre de Genetique Moleculaire du
    C.N.R.S.
Gif-sur-Yvette, France

Eleanor Erikson
Department of Pathology
University of Colorado Health Sciences
    Center School of Medicine
Denver, Colorado 80262

Raymond L. Erikson
Department of Pathology
University of Colorado Health Sciences
    Center School of Medicine
Denver, Colorado 80262

Bernard F. Erlanger
Cancer Center/Institute for Cancer Re-
    search
Department of Microbiology
College of Physicians and Surgeons
Columbia University
New York, New York 10032

H. J. Evans
Medical Research Council
Clinical and Population Cytogenetics Unit
Western General Hospital
Edinburgh, United Kingdom

E. Gardner
Department of Medical Biophysics and
 Computing
LDS Hospital and
University of Utah Medical Center
Salt Lake City, Utah 84143

Marion L. Gay
Department of Molecular Biology
The University of Texas System Cancer
 Center
M. D. Anderson Hospital and Tumor In-
stitute
Houston, Texas 77030

James German
Laboratory of Human Genetics
The New York Blood Center
New York, New York 10021

Thomas J. Gonda
Department of Microbiology and Immu-
nology
University of California
San Francisco, California 94143

Ramareddy V. Guntaka
Cancer Center/Institute for Cancer Re-
search
Department of Microbiology
College of Physicians and Surgeons
Columbia University
New York, New York 10032

Felix L. Haas
Division of Biology
The University of Texas System Cancer
 Center
M. D. Anderson Hospital and Tumor In-
stitute
Houston, Texas 77030

R. Hadley
Department of Medical Biophysics and
 Computing
LDS Hospital and
University of Utah Medical Center
Salt Lake City, Utah 84143

S. Hasstedt
Department of Medical Biophysics and
 Computing
LDS Hospital and
University of Utah Medical Center
Salt Lake City, Utah 84143

J. R. Hill
Department of Medical Biophysics and
 Computing
LDS Hospital and
University of Utah Medical Center
Salt Lake City, Utah 84143

Walter N. Hittelman
Department of Developmental Therapeu-
tics
The University of Texas System Cancer
 Center
M. D. Anderson Hospital and Tumor In-
stitute
Houston, Texas 77030

T. C. Hsu
Department of Cell Biology
The University of Texas System Cancer
 Center
M. D. Anderson Hospital and Tumor In-
stitute
Houston, Texas 77030

S. Hunt
Department of Medical Biophysics and
 Computing
LDS Hospital and
University of Utah Medical Center
Salt Lake City, Utah 84143

J. Jackson
Cancer Center
College of Physicians and Surgeons
Columbia University
New York, New York 10032

Richard Katz
Department of Microbiology
College of Physicians and Surgeons
Columbia University
New York, New York 10032

Alfred G. Knudson, Jr.
The Institute for Cancer Research
The Fox Chase Cancer Center
Philadelphia, Pennsylvania 19111

Paul M. Kraemer
Experimental Pathology Group
Los Alamos Scientific Laboratory
University of California
Los Alamos, New Mexico 87545

Charles A. LeMaistre
The University of Texas System Cancer
Center
M. D. Anderson Hospital and Tumor Institute
Houston, Texas 77030

Albert Levan
Institute of Genetics
University of Lund
Lund, Sweden

Göran Levan
Institute of Genetics
University of Gothenburg
Gothenburg, Sweden

Arthur D. Levinson
Department of Microbiology and Immunology
University of California
San Francisco, California 94143

Leon Levintow
Department of Microbiology and Immunology
University of California
San Francisco, California 94143

I. Lowy
Cancer Center
College of Physicians and Surgeons
Columbia University
New York, New York 10032

Paul Luciw
Department of Microbiology and Immunology
University of California
San Francisco, California 94143

J. L. Lyon
Department of Family and Community
Medicine
LDS Hospital and
University of Utah Medical Center
Salt Lake City, Utah 84143

Nils Mandahl
Institute of Genetics
University of Lund
Lund, Sweden

Kenneth B. McCredie
Department of Developmental Therapeutics
The University of Texas System Cancer
Center
M. D. Anderson Hospital and Tumor Institute
Houston, Texas 77030

Orlando J. Miller
Department of Human Genetics and Development
Department of Obstetrics and Gynecology
Cancer Center/Institute for Cancer Research
College of Physicians and Surgeons
Columbia University
New York, New York 10032

Felix Mitelman
Department of Clinical Genetics
Lund University Hospital
Lund, Sweden

Emma E. Moore
Department of Pathology
University of Colorado Health Sciences
Center School of Medicine
Denver, Colorado 80262

John J. Mulvihill
Clinical Epidemiology Branch
National Cancer Institute
Bethesda, Maryland 20205

Bert W. O'Malley
Department of Cell Biology
Baylor College of Medicine
Houston, Texas 77030

Hermann Oppermann
Department of Microbiology and Immunology
University of California
San Francisco, California 94143

M. Ostrander
Cancer Center
College of Physicians and Surgeons
Columbia University
New York, New York 10032

B. Panigrahy
Department of Veterinary Microbiology
College of Veterinary Medicine
Texas A&M University
College Station, Texas 77840

S. Pathak
Department of Cell Biology
The University of Texas System Cancer
Center
M. D. Anderson Hospital and Tumor In-
stitute
Houston, Texas 77030

A. Pellicer
Cancer Center
College of Physicians and Surgeons
Columbia University
New York, New York 10032

Anthony F. Purchio
Department of Pathology
University of Colorado Health Sciences
Center School of Medicine
Denver, Colorado 80262

Potu N. Rao
Department of Developmental Therapeu-
tics
The University of Texas System Cancer
Center
M. D. Anderson Hospital and Tumor In-
stitute
Houston, Texas 77030

James H. Ray
Laboratory of Human Genetics
The New York Blood Center
New York, New York 10021

Donald L. Robberson
Department of Molecular Biology
The University of Texas System Cancer
Center
M. D. Anderson Hospital and Tumor In-
stitute
Houston, Texas 77030

J. Roberts
Cancer Center
College of Physicians and Surgeons
Columbia University
New York, New York 10032

D. Robins
Cancer Center
College of Physicians and Surgeons
Columbia University
New York, New York 10032

Janet D. Rowley
Department of Medicine and The Frank-
lin McLean Memorial Research Insti-
tute
The University of Chicago
Chicago, Illinois 60637

Frank H. Ruddle
Department of Biology
Yale University
New Haven, Connecticut 06511

Avery A. Sandberg
Division of Medicine
Roswell Park Memorial Institute
Buffalo, New York 14263

Diana K. Sheiness
Department of Microbiology and Immu-
nology
University of California
San Francisco, California 94143

Michael J. Siciliano
Department of Medical Genetics
The University of Texas System Cancer
Center
M. D. Anderson Hospital and Tumor In-
stitute
Houston, Texas 77030

S. Silverstein
Cancer Center
College of Physicians and Surgeons
Columbia University
New York, New York 10032

G.-K. Sim
Cancer Center
College of Physicians and Surgeons
Columbia University
New York, New York 10032

Louis Siminovitch
Hospital for Sick Children
Department of Medical Genetics
University of Toronto
Toronto, Ontario, Canada

M. Skolnick
Department of Medical Biophysics and
    Computing
LDS Hospital and
University of Utah Medical Center
Salt Lake City, Utah 84143

C. R. Smart
Department of Surgery
LDS Hospital and
University of Utah Medical Center
Salt Lake City, Utah 84143

Linda Sperling
Centre de Genetique Moleculaire du
    C.N.R.S.
Gif-sur-Yvette, France

Louise C. Strong
Department of Medical Genetics
The University of Texas System Cancer
    Center
M. D. Anderson Hospital and Tumor In-
    stitute
Houston, Texas 77030

Elton Stubblefield
Department of Cell Biology
The University of Texas System Cancer
    Center
M. D. Anderson Hospital and Tumor In-
    stitute
Houston, Texas 77030

Prasad S. Sunkara
Department of Developmental Therapeu-
    tics
The University of Texas System Cancer
    Center
M. D. Anderson Hospital and Tumor In-
    stitute
Houston, Texas 77030

R. Sweet
Cancer Center
College of Physicians and Surgeons
Columbia University
New York, New York 10032

Umadevi Tantravahi
Department of Human Genetics and De-
    velopment
College of Physicians and Surgeons
Columbia University
New York, New York 10032

Albert C. Ting
Department of Cell Biology
Baylor College of Medicine
Houston, Texas 77030

J. J. Trentin
Division of Experimental Biology
Baylor College of Medicine
Houston, Texas 77030

Ming-Jer Tsai
Department of Cell Biology
Baylor College of Medicine
Houston, Texas 77030

Harold E. Varmus
Department of Microbiology and Immu-
    nology
University of California
San Francisco, California 94143

Bjorn Vennstrom
Department of Microbiology and Immu-
    nology
University of California
San Francisco, California 94143

Norio Wake
Division of Medicine
Roswell Park Memorial Institute
Buffalo, New York 14263

Mary C. Weiss
Centre de Genetique Moleculaire du
    C.N.R.S.
Gif-sur-Yvette, France

Robert S. Wells
Experimental Pathology Group
Los Alamos Scientific Laboratory
University of California
Los Alamos, New Mexico 87545

Cheryl E. Wilkins
Department of Molecular Biology
The University of Texas System Cancer Center
M. D. Anderson Hospital and Tumor Institute
Houston, Texas 77030

Dan C. Williams
Board of Regents
The University of Texas System
Austin, Texas 78701

R. R. Williams
Department of Internal Medicine
LDS Hospital and
University of Utah Medical Center
Salt Lake City, Utah 84143

B. Wold
Cancer Center
College of Physicians and Surgeons
Columbia University
New York, New York 10032

David A. Wright
Department of Medical Genetics
The University of Texas System Cancer Center
M. D. Anderson Hospital and Tumor Institute
Houston, Texas 77030

Genes, Chromosomes, and Neoplasia, edited by
Frances E. Arrighi, Potu N. Rao, and Elton Stubblefield.
Raven Press, New York © 1981.

Introductory Remarks

Charles A. LeMaistre

President, The University of Texas System Cancer Center, M. D. Anderson Hospital and Tumor Institute, Houston, Texas 77030

Welcome to the 33rd Annual Symposium on Fundamental Cancer Research sponsored each year by The University of Texas System Cancer Center, M. D. Anderson Hospital and Tumor Institute. M. D. Anderson hosts the Symposium on Fundamental Cancer Research so that scientists may exchange ideas about a particular topic of current interest.

This year's symposium, entitled "Genes, Chromosomes, and Neoplasia" was an exciting one. It comes at a time when scientists are poised on the brink of gleaning new insights into the control mechanisms of the cell. This knowledge is now accessible because of new techniques that recently have come into use in the sciences of cell biology and cytogenetics.

One of those new techniques, in particular, has not only excited scientists, but captured the imagination of the public through the news media. After only 3 to 4 years of serious study, recombinant DNA techniques already have shown themselves capable of producing hormones and other natural substances in large amounts. This ability to propagate specific DNA fragments is a genetic technique that is without previous counterpart. But recombinant DNA also shows promise of revealing the molecular events taking place in normal and abnormal cells. Recombinant DNA is making it possible to analyze the structure and organization of genes on a scale not imaginable only a few years ago.

I am sure you can imagine how such new basic research discoveries excite those involved in the treatment of cancer. At M. D. Anderson a comprehensive program of basic research undertaken since the early 1940s already has provided a solid foundation for progress in the successful treatment of many types of cancer.

But all of these modes of treatment have been developed to fight a disease that was already present, a disease whose origins and causes were a mystery. The search for new ways of halting a cancer's progression will certainly continue I hope that recent developments in cellular biology will one day enable scientists to find methods to correct cellular abnormalities, thereby eliminating diseases such as cancer before they occur.

I am sure you are all aware of just how far these recent developments have brought the science of cytogenetics. But perhaps by looking at the last symposium held by M. D. Anderson on this topic, we can see just how far this decade's developments have taken us in the understanding of cellular processes.

In 1969, M. D. Anderson's 23rd Annual Symposium was devoted to "Genetic Concepts and Neoplasia." Some of you who are here today presented papers at the symposium that pointed to possibilities that are now fact—made possible because specific technologies available today were not then known to be possible.

That symposium took place the year before Doctors T. C. Hsu and Frances Arrighi reported on their studies of chromosome banding patterns in 1970. That discovery enabled chromosomes to be banded into patterns—what they described as looking like bands on "argyle socks." That discovery also led the way to finding specific chromosome abnormalities common to specific types of cancer, particularly leukemias.

When those common rearrangements have been identified, in some cases they have been found to be diagnostic and predictive of how certain types of cancer will respond to therapy. Now, studies are being conducted to determine chromosome abnormalities in samples of solid tumors.

Other new techniques discovered in the last decade and now being perfected include gene sequencing, location of integrated viral segments, gene transfer, and gene amplification. These techniques have enabled scientists to conduct more precise experiments with heretofore unexpected results, all of which lead to a better understanding of the role of genetics in the cancer cell.

In this year's monograph, authors will be discussing some new and exciting findings. Perhaps we will look back on these reports 10 years hence with equal interest. The authors will be discussing how they have used the new techniques to come up with some interesting observations about the behavior of normal and abnormal cells. They will be looking at the cell from the microscopic structure of the chromosome itself, the chromosome at the molecular level, and the structure of specific genes. Some of the interesting discussions we can look forward to include those about gene transfer and how it has been used to probe cancer cells, the avian sarcoma virus and how it is involved in cancer in birds, double minutes and the role they play, chromosomal rearrangements in cancer cells, and the exciting area of gene amplification. The last session will be devoted to the practical application of what we've found out about cancer genetics.

I cannot close these remarks without adding my congratulations to this year's Bertner Award winner, Dr. T. C. Hsu. Dr. Hsu has been a member of our faculty since 1955 and recently was honored with an appointment to the Olga Keith Wiess Chair for Cancer Research here at M. D. Anderson.

In the short time that I have known Dr. Hsu, I have been immensely impressed by the startling contributions he has made to the entire field of cell biology. His laboratory here at M. D. Anderson continues to be a focal point of research and education in cytology and cytogenetics.

In closing, I would like to express my sincere thanks to the National Cancer Institute and the Texas Division of the American Cancer Society for their assistance in co-sponsoring this symposium. This long-standing partnership is one that we value greatly, for it indicates the importance these agencies place on the role of basic research in finding a solution to the cancer problem.

Genes, Chromosomes, and Neoplasia, edited by
Frances E. Arrighi, Potu N. Rao, and Elton Stubblefield.
Raven Press, New York © 1981.

Welcome Address

Dan C. Williams

Chairman, Board of Regents, The University of Texas System, Austin, Texas 78701

On behalf of the Board of Regents of The University of Texas System, I am pleased to welcome each of you to the 33rd Annual Symposium on Fundamental Cancer Research. The subject of this year's symposium—"Genes, Chromosomes, and Neoplasia"—is indeed a timely one because of the current attention focused on the role that recombinant DNA techniques may play in helping provide a solution to the problem of cancer.

A strong supporter of the need for rapid exchange of scientific information, The University of Texas System is proud of its long association with this series of symposia. Having grown from a regional meeting to one of international stature, the symposium had its beginning in the mid-1940s. At that time, the need for an increased emphasis upon coordination of many disciplines in cancer research was becoming more and more apparent. So in 1946, the idea was created for a symposium in which scientists could gather to discuss different facets of cancer research.

The first symposium resulting from this idea featured seven speakers from Austin, Galveston, and Houston. It was not until 1952 that the format changed to focus on a central theme reflecting prevalent areas of interest and progress to cancer scientists. Since then, a broad range of research approaches has been covered in successive symposia.

At this time, I would like to extend my warm congratulations to Dr. T. C. Hsu of M. D. Anderson Hsopital and Tumor Institute. Dr. Hsu was named the 29th recipient of the annual Bertner Award for his distinguished contributions to cancer research in the field of cell biology. Dr. Hsu is the first Anderson scientist to receive the Bertner Award, which honors the late Dr. Ernst W. Bertner. Dr. Bertner was M. D. Anderson Hospital's first acting director and first president of the Texas Medical Center. Dr. Hsu, one of M. D. Anderson's most widely respected researchers, is Professor of Biology, Chief of the Section of Cell Biology, and the holder of the Olga Keith Wiess Chair for Cancer Research.

On behalf of the Board of Regents, I also would like to congratulate Drs. Marc S. Collett of the University of Colorado Health Sciences Center and Peter T. Lomedico of Harvard University. Drs. Collett and Lomedico are co-recipients

5

of the Ninth Annual Wilson S. Stone Memorial Award and are of special interest to this year's symposium.

To the American Cancer Society and the National Cancer Institute, I express a special note of thanks and appreciation for their continued help and support of this important scientific meeting.

Once again, I extend a hearty welcome to you all. I trust that the information and knowledge shared here will aid you in further research and bring us one step closer to the ultimate conquest of cancer.

Genes, Chromosomes, and Neoplasia, edited by
Frances E. Arrighi, Potu N. Rao, and Elton Stubblefield.
Raven Press, New York © 1981.

Keynote Address: Somatic Cell Genetics—Past, Present, and Future

Frank H. Ruddle

Department of Biology, Yale University, New Haven, Connecticut 06511

It is a special pleasure to present the introductory lecture in this symposium which honors my good friend and colleague, Professor T. C. Hsu. I have known Dr. Hsu since 1955 when I was travelling west to start my graduate work at the University of California. I had already become interested in the genetics of cultured cells, and I had been influenced by Dr. Hsu's publications, which demonstrated the possibilities of mammalian somatic cell cytogenetics. He was a great inspiration to me as a student, and he has been a valued mentor and colleague over the past quarter century. T. C. Hsu's work has influenced the development of somatic cell genetics profoundly. His contribution to modern mammalian cytogenetics has served as a foundation for present-day cell genetics. It gives me great pleasure to dedicate to him this discussion of somatic cell genetics—its past, present, and future.

INTRODUCTION

Gene mapping by somatic cell genetic procedures had its conceptual origins with microbial genetics systems. It was conjectured that parasexual mechanisms might be identified that could be applied to mammalian cells in tissue culture in much the same way as for bacteria (Lederberg 1958, Pontecorvo 1958, 1962, Stern 1958). The breakthroughs in developing systems of parasexuality came with demonstration of cell hybridization by Barski and his colleagues in 1960, chromosome loss or segregation from hybrid cells by Ephrussi and Weiss (1967), the generation of appropriate genetic marker systems by Markert (1968), and cytological methods that allowed the unambiguous identification of genetically distinct chromosomes by Caspersson et al. (1970). It is interesting to note that the conceptual stage dominated the decade of the fifties, the assembly of the requisite techniques took place in the sixties, and the success of the mapping system was realized in the seventies.

The somatic cell system of genetic analysis is extremely straightforward. One simply correlates a particular segregating phenotypic trait with a specific chromosome or another biochemical marker that has been previously mapped. In this way, the relevant gene can be assigned to a particular chromosome. Moreover, by making use of segregating subsegments of chromosomes, it is possible to assign genes to subchromosome regions and also to establish the order of genes and the spacing of genes along the arms of chromosomes (Ricciuti and Ruddle 1973). In this way, genetic maps of rather low resolution can be produced (Ruddle and Creagan 1975).

It is of historical interest to review the development of the human gene map since 1911, when the first gene assignment, for color blindness to the X chromosome, was published by E. B. Wilson (1911) (Table 1). In 1921, T. Painter, a famous cytologist and faculty member of The University of Texas, and ultimately its president, produced some of the first cytological preparations of the human chromosomes, but *incorrectly* concluded that the haploid number was 24 (Painter 1923). I cannot pass over the fact that for a while in the early 1950s, T. C. Hsu also supported Painter's claim—perhaps out of patriotism to the State of Texas and loyalty to a fellow Texan. In 1956, Tjio and Levan, making use of tissue cultured cells, showed clearly that the chromosome number was 2n = 46, and by that time, the number of X-linked genes had risen to 36 and three linkage groups had been defined. No autosomal genes had yet been mapped. In 1970, Caspersson and co-workers introduced the fluorescent staining procedure. This was the last technical development required for the somatic cell genetic mapping of chromosomes, and thereafter there was a rapid increase in the number of mapped genes. The gene map now contains over 350 entries, and in many instances the chromosomal location of specific genes has been determined (McKusick and Ruddle 1977, 1980).

Progress has been rapid, some might say dramatic; but certainly much remains to be done. The number of human structural genes has been estimated as being approximately 100,000. Thus, only 0.3% have been mapped. Approximately three genes are mapped per month, so that several hundred years will be required to map all the genes of man using current methods. Moreover, the resolution of mapping is of a low order of discrimination. How then can we increase the speed of gene assignment, and how can we improve the resolution of the map?

TABLE 1. Progress in human gene mapping

| | | Autosomal | | X chromosome assignments |
Year	Total	Linkage groups	Assignments	
1911	0	0	0	1
1923	61	0	0	7
1956	356	3	0	36
1971	780	14	3	86
1975	1,049	20	71	94
1978	1,259	22	204	107
1979	1,329	23	230	110

In this paper, I shall discuss recent developments that promise to improve the accession rate of genes to the map and to improve resolution. I shall also point out some opportunities that somatic cell genetics presents for the cancer researcher and for the cell biologist. I shall emphasize (1) the use of recombinant DNA techniques to map genes, (2) how chromosome-mediated gene transfer can be used to establish linkage relationships between genes at the centiMorgan level of resolution, and (3) the use of DNA-mediated gene transfer to test the functional properties of genes.

MAPPING GENES BY DNA HYBRIDIZATION USING SOMATIC CELL HYBRIDS

Somatic cell hybrids have been extensively used to map the "constitutive" markers of mammalian cells, including enzymes (Nichols and Ruddle 1973), structural proteins (Sundar Raj et al. 1977), and cell surface antigens (Dorman et al. 1978, Aden et al. 1978, Jones et al. 1979). More recently our laboratory has been involved in mapping genes not normally expressed in hybrid cells, particularly genes that code for developmentally specific functions. In these cases, the sequence specificity of the DNA is used to distinguish donor and recipient genes. Denatured DNA extracted from somatic cell hybrids can be tested for its ability to form duplexes with a donor-specific probe molecule by conventional C_0t analysis (Britten and Kohne 1968). Using a DNA probe to human α globin, which cross-reacted very weakly with mouse cell DNA, we screened a panel of human $\times$ mouse hybrid cells for human α globin sequences and thus mapped the gene to human chromosome 16 (Deisseroth et al. 1977). Similarly, we were able to assign the β and γ globin genes to human chromosome 11 (Deisseroth et al. 1978). Neither α, nor β, nor γ globin polypeptides were expressed by these hybrids, and the map position of each gene was achieved by scoring the presence of the human gene itself by nucleic acid hybridization and correlating it with the presence of a human chromosome.

A second approach to the mapping of genes that are not normally expressed in hybrid cells depends on the specificity imparted by the distribution of restriction endonuclease sites. High molecular weight DNA purified from a hybrid cell is digested with a restriction endonuclease to produce fragments of defined size. Either the entire digestion mixture or the mixture prefractionated by reverse-phase chromatography (Tiemeier et al. 1977) is separated by agarose gel electrophoresis and transferred to nitrocellulose filters (Southern 1975). Specific DNA fragments are identified on the filters by hybridization with nucleic acid probe. This procedure is far more efficient than liquid hybridization, as a single assay requires less than 30 μg of cellular DNA. The procedure is applicable through evolution, since even closely homologous genes of different species tend to possess distinct restriction endonuclease sites. This results in different fragment patterns following blotting, and allows the homologous genes of different species to be distinguished.

We have demonstrated the effectiveness of the restriction endonuclease mapping procedure using two different systems. In both, we have used cloned recombinant DNA molecules as probes. First, we confirmed the assignment of the β globin gene to human chromosome 11 by associating the presence of β globin restriction fragments with the presence of human chromosome 11 in human × mouse hybrids (Huttner, Scangos, and Ruddle, unpublished results). Other laboratories have used this same approach to obtain information on the regional localization of the γ and β globin loci (Jeffreys et al. 1979, Gusella et al. 1979, Scott et al. 1979). Secondly, we have assigned the mouse immunoglobulin κ gene cluster to mouse chromosome 6 (Swan et al. 1979) and have assigned the mouse heavy chain gene cluster to mouse chromosome 12 (D'Eustachio et al. 1980) in analyses of mouse × Chinese hamster hybrids. It has become clear to us from these studies that any cloned DNA sequence can be mapped to a chromosomal site using this approach, including regulatory genes, introns, and structural genes. In addition, it will be possible to map genes whose protein product is not readily detectable or distinguishable between species, as long as an appropriate nucleic acid probe is available.

MAPPING TUMOR-SPECIFIC TUMOR ANTIGENS

In regard to mapping systems, I will take a few minutes to report on preliminary studies relevant to this meeting. It is well known that carcinogen-induced tumors frequently express unique tumor-specific transplantation antigens (TSTA). The function of these antigens and the genetic mechanism responsible for their variability are unknown. Dr. Dimitrina Pravtcheva, working together with Drs. Lloyd Old and Albert DeLeo at the Sloan-Kettering Institute, has used a somatic cell genetic approach to establish the map position of a specific TSTA that we call Meth A. The Meth A, BALB/c, tumor was induced by Dr. Old in 1962 using methylcholanthrene (Old et al. 1962). Dr. DeLeo has been successful more recently in producing a specific cytotoxic antiserum to the Meth A antigen (DeLeo et al. 1977). Therefore, we possess a means of identifying the antigen in cultured cells and in mouse × Chinese hamster hybrids that segregate the mouse chromosomes. Both antigen-negative and -positive hybrids can be produced. Moreover, the antigen phenotype correlates with mouse chromosome 12. In addition, further analysis permits us to assign the Meth A gene to the distal half of chromosome 12. More independent cases are required in order to complete the story, but already this early result provides useful information. It suggests that TSTAs need not be coded at the major histocompatibility locus as some have suggested. It also advances an intriguing hypothesis: there may exist a meaningful relationship between TSTA antigen variation and the molecular mechanism that accounts for immunoglobulin diversity. I believe this is a particularly cogent example of the usefulness of somatic cell genetics in regard to cancer studies, since TSTAs are limited to tumors, and thus their properties cannot be analyzed by Mendelian genetics.

REGIONAL MAPPING OF GENES USING CHROMOSOME-MEDIATED GENE TRANSFER

Chromosome-mediated gene transfer was first described by McBride and Ozer in 1973. In the past 5 years, we and others have analyzed the properties of this transfer system extensively. It is useful to summarize briefly our present understanding of the transformation process in order to define more precisely a series of questions concerning the fine structure of eukaryotic chromosomes.

Purified metaphase chromosomes from a donor cell population are precipitated with calcium phosphate and applied to recipient cells in multiplicities of 0.5 to 2 genome equivalents per cell. Under the appropriate conditions (Miller and Ruddle 1978), when conditionally auxotrophic recipient cells (e.g., HPRT⁻) are treated with chromosomes from a prototrophic donor, cells of the recipient type that now express the donor phenotype (transformants) can be recovered at frequencies as high as 2×10^{-5} per treated recipient cell. In this system, subchromosomal fragments are transferred to the recipient cell. The fragments range in size from large pieces readily detected by light microscopy to pieces carrying no detectable genetic information beyond the selected prototrophic marker itself. These fragments typically are derived from a single donor chromosome, but can also arise from the fusion of fragments derived from multiple donor chromosomes in the course of the transformation process. We have coined the term "transgenome" to describe these fragments (Miller and Ruddle 1978, Klobutcher and Ruddle 1979).

The transferred phenotype is initially expressed in an unstable fashion. That is, between 2% and 10% of the cells lose the transferred phenotype at each cell division. The loss is an all-or-none phenomenon. All detectable donor markers are retained or lost as a group. Further, the phenotypic loss appears to be due to the physical loss of the transgenome, and not a modulation phenomenon, as it is impossible to recover cells that re-express the transferred phenotype following back selection (Willecke and Ruddle 1975). These data suggest that the replication of the transgenome and its orderly partition into daughter cells at mitosis may both be defective to some extent. In all of these respects, the unstable transgenome bears a striking resemblance to an autonomous episomal element.

When such an unstable transformant is maintained in culture in selective medium, stable sublines of it may arise at a low frequency (Degnen et al. 1977, Athwal and McBride 1977, Klobutcher et al. 1980). In the presence of continued selection such a stable transformant will gradually overgrow the population. In all stable transformants so far examined, the transgenome is closely associated with a recipient cell chromosome (Fournier and Ruddle 1977, Willecke et al. 1978). Indeed, in those cases in which a donor chromosome fragment was detectable in the unstable line, it formed a morphologically distinct region of a recipient cell chromosome following stabilization (Klobutcher and Ruddle 1979). An interesting additional observation is that material can be lost from the transgenome during stabilization; this loss can be detected both morphologically (the retained fragment is measurably smaller) and genetically (previously linked donor markers are lost from the stabilized cell line) (Klobutcher and Ruddle 1979).

The chromosome-mediated gene transfer process can be used to obtain intrachromosomal, or regional, mapping data. The method makes use of transformed cell lines with visible donor chromosome fragments in conjunction with the genetic loss that occurs during stabilization for the generation of deletion maps. We have used this method to localize three genes on the long arm of human chromosome 17 (Klobutcher and Ruddle 1979). A human thymidine kinase (TK) transformed cell line that had received the intact long arm of human chromosome 17, and thus the genes for galactokinase *(GALK)* and procollagen type I *(Col I)*, which are located on this segment (Elsevier et al. 1974, Sundar Raj et al. 1977), was used in this analysis. Subclones were constructed and followed through stabilization. In each stable subclone, the chromosome fragment was found to associate with a recipient cell chromosome. More importantly, genetic information was lost during the process and was manifested by a decrease in the size of the fragment and loss of *GALK* expression in some lines. By correlating the portion of 17q retained by the stable subclones with the chromosome 17 genes expressed, it was possible to localize *TK, GALK,* and *Col I* to regions of human chromosome 17. A gene order of centromere *GALK-(TK, ProCol I)* was indicated from this analysis (Klobutcher and Ruddle 1979).

Intrachromosomal mapping data derived from chromosome-mediated gene transfer deletion maps and nonselected gene cotransfer frequencies provide a level of genetic resolution intermediate between those obtainable by somatic cell hybridization and restriction enzyme analysis. This type of mapping data will clearly be necessary to develop concise and detailed mammalian gene maps.

DNA-MEDIATED GENE TRANSFER

DNA-mediated gene transfer provides a means of analyzing the genetic organization of chromosomes at still higher levels of resolution. It also provides a means of defining conditions that promote or inhibit gene expression. The transfer of genetically active DNA into mammalian cells was established by Graham and Van der Eb (1973) using a viral transfection system. Mammalian donor genes were transferred ultimately to mammalian recipient cells using a variation of this procedure (Wigler et al. 1978). In this report, I shall limit my comments to a single DNA-mediated gene transfer system. This system makes use of the HSV-1 thymidine kinase gene inserted into the BamHI site of pBR322. This is a convenient system since large amounts of the donor gene can be obtained easily.

In the experiments to be discussed here, the donor plasmid is cleaved with Hind III to produce a linear molecule. Nanogram amounts of HSV-*TK*-plasmid are mixed with microgram amounts of carrier DNA isolated from the recipient cell (LTK⁻). The prototrophic and carrier DNAs are coprecipitated by the calcium phosphate procedure of Graham and Van der Eb. Transformants are selected by the hypoxanthine-aminopterin-thymidine (HAT) selection medium of Littlefield (1964). Transformants can be classified into two main types as in the case of chromosome-mediated gene transfer. These are the unstable and stable transformants. In addition, microcell hybrids can be produced using the stably transformed cells as donors and TK⁻ Chinese hamster cells as recipients. Thus, DNAs derived from the unstable transformant, its stable derivative, and derivative microcell populations can be compared using the Southern blotting procedure, with HSV-*TK*-pBR322 used as the labeled probe.

The results of this type of analysis have shown that the original linear plasmid material has been substantially reduced in size as a result of deletions from either end of the linear fragment (Scangos, Huttner, and Ruddle, unpublished results). Southern analysis also indicates that the plasmid becomes associated with adventitious DNA prior to or simultaneous with the formation of the unstable transgenome (Scangos, Huttner, and Ruddle, unpublished results). Preliminary experiments also suggest the adventitious DNA is in all likelihood carrier DNA (Huttner, Scangos, and Ruddle, unpublished results). The Hirt (1967) method of separation of DNAs from unstable transformants shows that unintegrated transgenomes are large, being in excess of 100–150 kb. This would mean that the HSV-*TK*-pBR322 fragment represents only 5% or less in the final unstable transgenome product.

Evidence also exists for a higher copy number of *TK* genes in the unstable transformant than for the stable derivatives. This is based on quantitative and qualitative changes in the Southern blot patterns in comparisons of DNAs from stable and unstable transformants. In addition, the specific activity of Herpes *TK*-specific activity decreases after conversion to stability. The gene copy number appears to decrease about fivefold during the transition from the unstable to stable phenotype. Thus, one can visualize the unstable transgenome as an independent, unintegrated unit whose HSV-*TK*-pBR322 component represents only approximately 5% or less of the total unstable transgenomes. This structure may have multiple copies of HSV-*TK*-pBR322 associated with it. Moreover, the entire unstable transgenome may be represented in numerous copies in the unstable cells, whereas only one unit will survive the transition to stability.

The current interpretation of the unstable transgenome likens it to a length of nondescript adventitious DNA (possibly carrier) in which is interspersed HSV-*TK*-pBR322 plasmid segments. Independently arising unstable transformants possess unstable transgenomes of this general type, but each is identifiable in terms of its unique restriction pattern. Thus, recognizably different transgenomes exist in transformed populations prior to the transition to stability. A single clonally derived transformant may possess several different transgenomes, all of which carry at least one active *TK* gene. This can be demonstrated by subcloning the transformed population. Clonally derived unstable subpopulations as a rule contain a single transgenome.

These findings now suggest a number of new questions, amenable to experimental analysis. How are the unstable transgenomes formed initially? What mechanisms mediate the ligation of plasmid DNA to adventitious DNA? Do linkage units arise in a single large molecule that is subsequently divided, or do linkage units arise independently? Must the unstable transgenome be so large in order to accommodate an intact replicon structure, or to include other entities necessary to its normal function? How is integration into the host chromosome mediated? These are some of the questions that our current understanding of the transgenome calls to mind.

FUTURE POSSIBILITIES FOR SOMATIC CELL GENETICS

It seems likely that the speed and resolution of gene mapping will rapidly increase in the immediate future. The development of restriction mapping will contribute in a significant way to an acceleration of gene map data accession. This new methodology will also make it possible to map all kinds of genetic elements, irrespective of their transcriptional or translational expression in hybrid cells.

It will also be important to increase the resolution of gene mapping. It seems quite possible that any scheme that promotes the breakage and retention through selection of specified fragments of linkage groups (see also Goss and Harris 1975) will be useful in this regard. Chromosome-mediated gene transfer would appear to possess many opportunities for regional gene mapping at intermediate levels of genetic resolution. This type of system may be quite valuable as a means of constructing mammalian linkage maps, since it can provide information on gene order intermediate between that attainable by cell hybridization and restriction mapping. Transformation with DNA permits the transfer of even smaller fragments of donor DNA. This may provide gene linkage data at still higher levels of resolution.

Gene transfer has additional important applications other than gene mapping. It serves as a means of altering the genetic constitution of cells. As genetic transformation systems become refined, it will be feasible to genetically modify cells with efficiency and precision. In this way, it will be possible to test the effects of specifically mutated genes on the biology of the recipient cells. This could provide a means of analyzing genes that specify dominant phenotypes. Recessive traits may also be amenable to study if specific integration and replacement of endogenous genes can be realized in mammalian cells. This type of experimental system may be particularly valuable in the analysis of genes that potentiate neoplasia. Moreover, experience in the experimental genetic modification of cells by means of gene transfer can be expected to set the stage for the introduction of genetic therapy methodologies.

Gene transfer systems of this sort can be expected to play an important role in the analysis of developmental mechanisms. Genes of developmental consequence can be isolated, cloned, and modified in various ways. These products can then be tested in terms of their epigenetic effects by introducing them into cells of an appropriate developmental type. The ultimate refinement of such an approach would be the genetic transformation of the germ line of experimental organisms so that the transferred gene could be studied in the context of the complete and essentially normal ontogenetic plan.

In the future, we can expect somatic cell genetics to make an ever-increasing contribution to our basic knowledge of tumor cell origin and tumor progression. We shall learn in the course of this symposium that particular chromosome rearrangements are associated with neoplastic transformation in certain epigenetically defined cell types. As the human genetic map becomes more detailed, we can test for an association between neoplasia and specific genetic loci. As the symposium unfolds, we shall also learn of genetic studies that implicate genetic factors in the origin of neoplasia. Somatic cell genetics provides a means to map these genes, and to test their mode of action. As implied above, it will also be possible to test the neoplastic effects of such genes on specific cell types by means of genetic transformation. In the future, both gene mapping and gene transfer analyses will permit us to more readily devise and test hypotheses that bear on the nature of the neoplastic cell.

ACKNOWLEDGMENT

This work was supported by a grant from the United States Public Health Service (5 R01 GM09966).

REFERENCES

Aden, D. P., R. Mausner, and B. B. Knowles. 1978. Production of HLA antibody in mice immunized with syngeneic mouse-human hybrid cells containing human chromosome 6. Nature 271:375–377.

Athwal, R., and O. W. McBride. 1977. Serial transfer of a human gene to rodent cells by sequential chromosome-mediated gene transfer. Proc. Natl. Acad. Sci. USA 74:2943–2947.

Barski, G., S. Sorieul, and F. Cornefert. 1960. Production dans de cultures in vitro de deux souches cellulaires en association, de cellules de caractere 'hybride'. C. R. Acad. Sci. Paris 251:1825–1830.

Britten, R. J., and D. E. Kohne. 1968. Repeated sequences in DNA. Science 161:529–540.

Caspersson, T., L. Zech, and C. Johansson. 1970. Analysis of the human metaphase chromosome set by aid of DNA binding fluorescent agents. Exp. Cell Res. 62:490–492.

Degnen, G. E., I. L. Miller, E. A. Adelberg, and J. M. Eisenstadt. 1977. Overexpression of an unstably inherited gene in cultured mouse cells. Proc. Natl. Acad. Sci. USA: 3956–3959.

Deisseroth, A., A. Nienhuis, J. Lawrence, R. Giles, P. Turner, and F. H. Ruddle. 1978. Chromosomal localization of the human beta globin gene to human chromosome 11 in somatic cell hybrids. Proc. Natl. Acad. Sci. USA 75:1456–1460.

Deisseroth, A., A. Nienhuis, P. Turner, R. Velez, W. F. Anderson, F. H. Ruddle, J. Lawrence, R. Creagan, and R. Kucherlapati. 1977. Localization of the human alpha globin structural gene to chromosome 16 in somatic cell hybrids by molecular hybridization assay. Cell 12:205–218.

DeLeo, A. B., H. Shiku, T. Takahashi, M. John, and L. J. Old. 1977. Cell surface antigens of chemically induced sarcomas of the mouse. I. Murine leukemia virus-related and alloantigens on cultured fibroblasts and sarcoma cells. Description of a unique antigen on BALB/c Meth A sarcoma. J. Exp. Med. 146:720–734.

D'Eustachio, P., D. Pravtcheva, K. Marcu, and F. H. Ruddle. 1980. Chromosomal localization of the structural gene cluster encoding murine immunoglobulin heavy chains. J. Exp. Med. 151:1545–1550.

Dorman, B. P., N. Shimizu and F. H. Ruddle. 1978. Genetic analysis of the human cell surface: Antigenic marker for the human X chromosome in human-mouse hybrids. Proc. Natl. Acad. Sci. USA 75:2363–2367.

Elsevier, S. M., R. Kucherlapati, E. A. Nichols, R. P. Creagan, R. E. Giles, F. H. Ruddle, K. Willecke and J. K. McDougall. 1974. Assignment of the gene for galactokinase to human chromosome 17 and its regional localization to band q 21–22. Nature 251:633–635.

Ephrussi, B., and M. C. Weiss. 1967. Regulation of the cell cycle in mammalian cells: Inferences and speculations based on observations of interspecific somatic hybrids. Dev. Biol. Suppl. I:136–169.

Fournier, R. E. K. and F. H. Ruddle. 1977. Stable association of the human transgenome and host murine chromosomes demonstrated with trispecific microcell hybrids. Proc. Natl. Acad. Sci. USA 74:3937–3941.

Goss, S. J., and H. Harris. 1975. New method for mapping genes in human chromosomes. Nature 255:680–684.

Graham, F. L., and A. J. Van der Eb. 1973. A new technique for the assay of human adenovirus 5 DNA. Virology 52:456–467.

Gusella, J., A. Varsanyi-Breiner, F.-T. Kao, C. Jones, T. Puck, C. Keys, S. Orkin, and D. Housman. 1979. Precise localization of human beta-globin gene complex on chromosome 11. Proc. Natl. Acad. Sci. USA 76:5239–5243.

Hirt, B. 1967. Selective extraction of polyoma DNA from infected mouse cell cultures. J. Mol. Biol. 26:365–369.

Jeffreys, A. L., I. W. Craig, and U. Francke. 1979. Localization of the G-gamma-, A-gamma-, delta-, and beta-globin genes on the short arm of human chromosome 11. Nature 281:606–608.

Jones, C., E. E. Moore, and D. W. Lehman. 1979. Genetic and biochemical analysis of the a-1 cell-surface antigen associated with human chromosome 11. Proc. Natl. Acad. Sci. USA 76:6491–6495.

Klobutcher, L. A., C. L. Miller, and F. H. Ruddle. 1980. Chromosome mediated gene transfer results in two classes of unstable transformants. Proc. Natl. Acad. Sci. USA 77:3610–3614.

Klobutcher, L. A., and F. H. Ruddle. 1979. Phenotype stabilization and transgenome integration in chromosome-mediated gene transfer. Nature 280:657–660.

Lederberg, J. 1958. Genetic approaches to somatic cell variation: summary comment. J. Cell. Comp. Physiol. 52(Suppl. 1):383–401.

Littlefield, J. 1964. Selection of hybrids from mating of fibroblasts in vitro and their presumed recombinants. Science 145:709–710.

Markert, C. L. 1968. The molecular basis for isozymes. Ann. N.Y. Acad. Sci. 151:14–40.

McBride, O. W., and H. L. Ozer. 1973. Transfer of genetic information by purified metaphase chromosomes. Proc. Natl. Acad. Sci. USA 70:1258–1262.

McKusick, V. A., and F. H. Ruddle. 1977. The status of the gene map of the human chromosomes. Science 196:390–405.

McKusick, V. A., and F. H. Ruddle. 1980. The status of the gene map of the human chromosomes. Science (in press).

Miller, C. A., and F. H. Ruddle. 1978. Co-transfer of human X-linked markers into murine somatic cells via isolated metaphase chromosomes. Proc. Natl. Acad. Sci. USA 75:3346–3350.

Nichols, E. A., and F. H. Ruddle. 1973. A review of enzyme polymorphism, linkage and electrophoretic conditions for mouse and somatic cell hybrids in starch gels. J. Histochem. Cytochem. 21:1066–1081.

Old, L. J., E. A. Boyse, D. A. Clark, and E. Carswell. 1962. Antigenic properties of chemically induced tumors. Ann. N.Y. Acad. Sci. 101:80–106.

Painter, T. S. 1923. Studies in mammalian spermatogenesis. II. The spermatogenesis of man. J. Exp. Zool. 37:291–336.

Pontecorvo, G. 1958. Trends in Genetic Analysis. Columbia University Press, New York.

Pontecorvo, G. 1962. Methods of microbial genetics in an approach to human genetics. Br. Med. Bull. 18:81–84.

Ricciuti, F., and F. H. Ruddle. 1973. Assignment of three gene loci (PGK, HGPRT, G6PD) to the long arm of the human X chromosome by somatic cell genetics. Genetics 74:661–678.

Ruddle, F. H., and R. P. Creagan. 1975. Parasexual approaches to the genetics of man. Annu. Rev. Genet. 9:407–486.

Scott, A. F., J. A. Phillips, and B. R. Migeon. 1979. DNA restriction endonuclease analysis for location of human beta- and delta-globin genes on chromosome 11. Proc. Natl. Acad. Sci. USA 76:4563–4565.

Southern, E. M. 1975. Detection of specific sequences among DNA fragments separated by gel electrophoresis. J. Mol. Biol. 98:503–517.

Stern, C. 1958. The nucleus and somatic cell variation. J. Cell. Comp. Physiol. 52(Suppl. 1):1–34.

Sundar Raj, C. V., R. L. Church, L. A. Klobutcher, and F. H. Ruddle. 1977. Genetics of the connective tissue proteins. Assignment of the gene for human type I procollagen to chromosome 17 by analysis of cell hybrids and microcell hybrids. Proc. Natl. Acad. Sci. USA 74:4444–4448.

Swan, D., P. D'Eustachio, L. Leinwand, J. Seidman, D. Keithley, and F. H. Ruddle. 1979. Chromosomal assignment of the mouse κ light chain genes. Proc. Natl. Acad. Sci. USA 76:2735–2739.

Tiemeier, D., S. Tilghman, and P. Leder. 1977. Purification and cloning of a mouse ribosomal gene fragment in coliphage lambda. Gene 2:173–191.

Tjio, J. H., and A. Levan. 1956. The chromosome number of man. Hereditas 42:1–6.

Wigler, M., A. Pellicer, S. Silverstein, and R. Axel. 1978. Biochemical transfer of single-copy eucaryotic genes using total cellular DNA as donor. Cell 14:725–731.

Willecke, K., R. Mierau, A. Krüger, and R. Lange. 1978. Chromosomal gene transfer of human cytosol thymidine kinase into mouse cells. Molec. Gen. Genet. 161:49–57.

Willecke, K., and F. H. Ruddle. 1975. Transfer of human gene for hypoxanthineguanine phosphoribosyltransferase via isolated human metaphase chromosomes into mouse L-cells. Proc. Natl. Acad. Sci. USA 72:1792–1796.

Wilson, E. B. 1911. The sex chromosomes. Literarisch-kritische Rundschau. Arch. Mikrosk. Anat. 77:249–271.

Genes, Chromosomes, and Neoplasia, edited by
Frances E. Arrighi, Potu N. Rao, and Elton Stubblefield.
Raven Press, New York © 1981.

Presentation of the Wilson S. Stone Memorial Award

Felix L. Haas

Division of Biology, The University of Texas System Cancer Center, M. D. Anderson Hospital and Tumor Institute, Houston, Texas 77030

It is my privilege and great honor to present the Wilson S. Stone Memorial Award for 1980. This award is presented each year by The University of Texas System Cancer Center M. D. Anderson Hospital and Tumor Institute for outstanding achievement in the biomedical sciences by a graduate student or a postdoctoral student in the United States.

First, I would like to say a few words about the man that this award commemorates, Dr. Wilson S. Stone. I knew Dr. Stone intimately as a friend, as my teacher, and as my counselor. I first became acquainted with him in 1946 when I was a junior at The University of Texas. He gave me a job as a research technician in his laboratory, and I stayed on after I had the B.A. degree and took the M.S. and Ph.D. degrees under his supervision. I had the great privilege to be his first graduate student. With the death of Dr. Stone on February 28, 1968, the University lost one of its greatest leaders and one who played a major role in the development of the science of genetics throughout the world and of modern biology in The University of Texas System.

Dr. Stone was born at Junction, Texas, attended high school at Brackenridge High School in San Antonio, and then entered Baylor. In 1926 he transferred to The University of Texas at Austin where he received the B.A., M.A., and the Ph.D. degrees. He was appointed to the Zoology faculty in 1932, and he remained with the University until his death, with the exception of three years that he served with the Air Force during World War II.

Mainly because of Dr. Stone's endeavors, the Genetics Foundation of The University of Texas was organized in 1952, and he remained a member of its Board of Directors until his death. He was chairman of the Department of Zoology at the University from 1959–1963, and during this time he attracted the interests of established biologists, biochemists, and physicists to the institution, and also induced leading scientists at other institutions to send their young graduates to The University of Texas at Austin. He initiated the plans and found the funds for the construction of the J. T. Patterson Laboratory building at The University of Texas in Austin. After serving as advisor to the Chancellor on graduate and research programs for The University of Texas System for two years, he was appointed Vice Chancellor of the system in 1965, and was serving in this capacity at the time of his death.

Dr. Stone long served as a consultant to The University of Texas M. D. Anderson Hospital and Tumor Institute. He was a close friend and counselor of the Director, Dr. Clark, and of many of the professional staff of the institution during its early formative years. He aided greatly in the planning and the development of this Annual Symposium on Fundamental Cancer Research; and by working with the staff, Dr. Clark, and Chancellor Harry Ransom, he was largely instrumental in creating a graduate program for The University of Texas units at Houston from which The University of Texas Graduate School of Biomedical Sciences finally emerged.

Nationally, Dr. Stone served on too many committees and review boards to mention. Recognition by scientists and other educators of his acuteness is shown by the large number of the nation's most significant committees to which he was appointed, including the President's Advisor for Science to President Eisenhower. In 1960, he was elected to the National Academy of Sciences.

As a person he was one of the finest, most understanding friends one could have. He always shared his research ideas with others and always acknowledged the ideas or help that anyone gave to him. Dr. Stone had a strong sense of integrity and expected the same from others, be they colleagues, collaborators, or his students. They always found him to be interested not only in their research but also in their personal problems. He inspired a permanent attitude of loyalty, which he graciously respected and always returned. The warm relationship remained after students became mature investigators and teachers in their own right; and they always continued to seek his counsel. This was the man that this award commemorates—Wilson Stuart Stone.

Ladies and gentlemen, it gives me great pleasure on behalf of the selection committee and of Dr. Stone and The University of Texas System Cancer Center to present to you the recipients of the Wilson S. Stone Memorial Award for 1980. For the first time this award honors two young scientists, Dr. Marc Collett and Dr. Peter Lomedico. Dr. Collett, of the Department of Pathology School of Medicine, The University of Colorado Health Sciences Center, receives the Stone Memorial Award for his work on the structure and function of the avian sarcoma virus–transforming gene product.

Dr. Collett carried out his predoctoral research at the University of Michigan with Professor Anthony Faras, where he received the Ph.D. degree. During the course of his work at Michigan, Dr. Collett made a number of highly significant experimental contributions to understanding the mechanism of reverse transcription in oncornavirus DNA synthesis. After an outstanding undergraduate career, he received a Damon Runyon—Walter Winchell Cancer Research Fellowship and is presently carrying out his postdoctoral research with Dr. R. L. Erikson at the University of Colorado Health Science Center. There Dr. Collett is continuing his molecular biology research on the structure and function of the avian sarcoma virus—transforming gene product. His experiments have helped to identify, purify, and characterize the transforming gene product and have shown that this product was responsible for protein phosphorylation.

One significance of this work is that for the first time we can begin to describe biochemically the pathways that lead to malignant transformation in an experimental system. Furthermore, since in normal cells there is a highly conserved protein very similar in structure and function to that of the avian sarcoma virus–transforming protein, his results are also important with regard to understanding normal cell functions. Dr. Collett's contributions greatly advance the basis for new approaches to the biochemistry of malignancy.

Dr. Lomedico of the Biological Laboratories of Harvard University receives the Wilson S. Stone Memorial Award for his work on the structure and expression of insulin genes. This work was initiated while Dr. Lomedico was a graduate student at The University of Texas Graduate School of Biomedical Sciences at Houston, and was carried out at The University of Texas System Cancer Center under the direction of Dr. Grady Saunders. This marks the first time that a Cancer Center alumnus or faculty member has received this award. At the Cancer Center, Dr. Lomedico used a cell-free system to produce the primary translation product of bovine insulin messenger RNA, and he was the first to show a 23 amino acid sequence, now called the *signal sequence,* on the amino terminal end of the proinsulin translation product. After receiving the Ph.D. degree from the Graduate School of Biomedical Sciences in 1977, Dr. Lomedico was awarded a postdoctoral fellowship and has continued his work on the insulin gene at the Harvard University Biological Laboratories with Professor Walter Gilbert. He has used recombinant DNA technology, first to establish most of the primary structure of the messenger RNA for rat preproinsulin, and then to isolate the chromosomal rat preproinsulin gene. DNA sequence analysis of this gene then allowed Dr. Lomedico to complete the primary structure determination of the preproinsulin messenger RNA, to establish the complete amino acid sequence for the pre-region of preproinsulin, and to predict that the primary gene transcript must be spliced to yield the mature messenger RNA, a prediction that has now been proved to be true. He has established a testable model for how *introns,* the intervening sequences in genetic DNA, evolve that will considerably extend our understanding of gene evolution. The work makes possible a series of research projects on dissecting key elements for the regulation of the insulin gene that are the forerunners of things to come. As Dr. Lomedico said in his seminar, "Now, correction of inborn errors of metabolism and some other classes of genetic defects in the newborn are not only possibilities, they are just around the corner."

Genes, Chromosomes, and Neoplasia, edited by
Frances E. Arrighi, Potu N. Rao, and Elton Stubblefield.
Raven Press, New York © 1981.

Presentation of the Ernst W. Bertner Memorial Award

R. Lee Clark

*President Emeritus, The University of Texas System Cancer Center, M. D. Anderson Hospital
and Tumor Institute, Houston, Texas 77030*

This is a memorable event in the 33-year history of The University of Texas System Cancer Center M. D. Anderson Hospital and Tumor Institute Annual Symposium on Fundamental Cancer Research. For the first time, one of our own investigators is being honored for his contributions to basic cell research and cancer research. We are especially proud because most of his studies have been conducted here at M. D. Anderson Hospital.

Dr. T. C. Hsu is a respected and well-liked scientist and teacher who skillfully combines humor with precise scientific language to enlighten students, audiences, and readers. His standards for academic and scientific performance are exacting for himself and others, yet the organization of his laboratory simultaneously creates an environment allowing everyone "elbow room" to progress at a pace that stimulates thoughtful investigation. The spirited cooperative exchange of ideas and information, not only among the laboratory personnel but with investigators around the world, is unique in the scientific world. The introductions written by Dr. Hsu for the *Mammalian Chromosomes Newsletter* since 1960 exemplify this active fostering of free-flowing information exchange.

Dr. Hsu was born in China and received a considerable part of his education in that country. After graduating from college, he worked in the fields of entomology and plant cytology.

He came to the United States from China in 1948 as a graduate student to work and study in Professor J. T. Patterson's laboratory at The University of Texas main campus. Dr. Patterson assigned him a graduate research project to trace the cytogenetic relationships of the species in the *virilis* group of *Drosophila.* Dr. Hsu's *virilis* polytene chromosome map is still the standard map used by geneticists who work with *Drosophila.*

Dr. Hsu was awarded his Doctor of Philosophy degree by The University of Texas in 1951. His plans to return to China to his wife and infant daughter, whom he had not yet seen, were thwarted first by the People's Revolution in China in 1949 and then the Korean War, which began in 1950. U.S. governmental policy prohibited the return of Chinese scholars to China but allowed them to seek full-time employment and to apply for permanent residence.

Professor G.M. Pomerat, a leader in the field of tissue culture studies at The University of Texas Medical Branch in Galveston, offered Dr. Hsu a position to study nuclear phenomena of cells in culture. Although Dr. Hsu had not worked with human and other mammalian systems, his interest and urgent need for employment led him to accept the position. The unpredictable finger of fate thereby directed him down the path to a series of major contributions in the field of cell biology and cytogenetics. One of the projects on which Drs. Hsu and Pomerat worked was time-lapse motion pictures of cellular activities, particularly in mitosis. These films are exceptionally beautiful and are still being shown.

Dr. Hsu's first major impact on the field of vertebrate cytogenetics was the well-known serendipitous discovery in 1952 of chromosome spreading in mitotic cells after washing in hypotonic solution, a key to modern chromosome study. Dean Chauncey D. Leake appointed him to the faculty as assistant professor in the Department of Anatomy in 1953, and Dr. Hsu was then compelled to learn human histology in order to carry his share of the teaching responsibilities.

Dr. Hsu became a citizen of the United States in 1953, and his wife and daughter joined him in the U.S. in 1954.

Teaching medical students left Dr. Hsu too little time for research, and in 1955 he accepted a position as Associate Biologist at another University of Texas unit, the young but rapidly developing M. D. Anderson Hospital, where the new facilities for the section of cytology were so full of potential (walls, floor, ceiling, and utilities only) that Dr. Hsu had the rare opportunity to build from scratch.

Initially, Dr. Hsu established three primary goals for his research: (1) to limit his investigations to the cell nucleus, (2) to apply his fundamental cell research to understand cancer cells, with a balance between basic research and applied research, (3) to avoid the siren lure to stagnation through lifelong repetition of the same type of work, by seeking and accepting change, whether anticipated or unanticipated, that might lead to progress.

He is one of those few individuals who initially devoted much time and concern over an 8-year period to create The University of Texas Graduate School of Biomedical Sciences at Houston and has helped sustain its high levels of graduate work through the subsequent years. Dr. Hsu was president of the graduate faculty of the graduate school in 1972 and 1973. He has personally supervised or sat on the advisory committees of 40 students earning their masters or doctoral degrees or working as postdoctoral fellows, and many more students in various categories have been influenced by his guidance and encouragement.

Through his work with graduate students and postdoctoral fellows, he gradually acquired his staff by inviting some of the fellows to join him as colleagues in research. During the 1960s, these young people worked with Dr. Hsu to define the ultrastructure of nuclear components with electron microscopy, and to further elucidate mammalian phylogeny and chromosome physiology.

With regard to one of the three goals established for the section of cytology,

Dr. Hsu's career-long approach to problems has been to develop new or better techniques with which to study phenomena or to solve problems, rather than to develop methods for the sake of methods.

And Dr. Hsu is refreshingly candid about the role of fortuity in many events and decisions of his scientific career. For example, as he explains in his recently published *Human and Mammalian Cytogenetics: An Historical Perspective,* two fortuitous events led to the establishment of Dr. Hsu's frozen zoo—the unexpected acquisition in 1961 of three summer high school students instead of one, who had to be kept busy, and a long-term association and friendship with the superintendent of the Houston Zoological Garden, Mr. John Werler. This remarkable collection of thousands of tissue specimens of rare animal species, from aardvark to zebra, some of which are probably destined for extinction, not only provided a valuable survey of karyotypes of closely related species of mammals but also has provided opportunities for cell research in other areas. In addition, a *Time* magazine article appearing in 1971 states, "Hsu's hope is that scientists of the future will be able to use the genetic 'codes' locked in the cells of each species to reconstitute the original animals." This prospect seems much closer to reality now than it did in 1971.

Drs. Hsu and Kurt Benirschke have worked together since 1966 to edit 10 volumes of *An Atlas of Mammalian Chromosomes,* the first published in 1967, and three volumes of *Chromosome Atlas: Fish, Amphibians, Reptiles, and Birds,* the first published in 1971. With regard to studies of mammals, one must mention the detailed molecular and cytogenetic studies on the American rodent genus *Peromyscus* by Dr. Hsu and his colleagues. Many investigators are now using these species in their laboratories.

I will briefly enumerate other "firsts" for which Dr. Hsu, often with collaborators, is responsible. In 1963, Drs. Hsu, Kit, Dubbs, and Piekarski published a paper describing the isolation of the first drug-resistant mammalian deletion-mutant cell line, known as LM(TK$^-$) cells. These mouse fibroblasts are resistant to bromodeoxyuridine and lack thymidine kinase activity. This cell line is still widely used for research.

Following on the heels of fluorescent staining of chromosomes by Caspersson's team and others, the perfection of the in situ hybridization technique of mouse cells by Pardue and Gall and independently by K. Jones, Buongiorno-Nardelli, and Amaldi, and the noting of more intense staining of centromeric or heterochromatic regions (C-bands), in 1970 Drs. Hsu and Frances Arrighi developed the first reliable C-banding technique for human chromosomes using Giemsa solution to define heterochromatin, particularly in human chromosomes 1,9,16, and Y. Because this technique was so valuable for cytogenetic studies, for a time the publication in *Cytogenetics* was one of the five most frequently cited in the world scientific literature.

Soon after, using a modification of the C-banding technique (which consisted of a longer incubation period), Drs. M. E. Drets and Margery Shaw of our institution demonstrated G-banding of human chromosomes, which made possible the identification of each pair of human chromosomes. Subsequently, a number of other stains were also used by other laboratories to distinguish individual mammalian chromosomes. Techniques for analyzing chromosome complements took a remarkable jump forward.

One of Dr. Hsu's particular interests in the early 1970s centered on repetitive DNA. Currently, diverse activities are being conducted in Dr. Hsu's laboratory. Work is nearing completion of the development of quick, economical, accurate protocols to assay for environmental chemical mutagens using cultured cells. There are also good preliminary results from in vivo animal models being used to determine the mutagenicity of chemicals.

Characterization of consistent chromosomal aberrations, specifically translocations and deletions, for each type of cancer and the study of the phenomenon of gene amplification are in preliminary phases, as are the current efforts to characterize the significance of double minutes, to analyze nucleosome structure through DNA cloning procedures, and to futher define gametogenesis.

I have given only a brief glimpse of the varied scientific activities of Dr. Hsu. As a sidelight, I would like to mention that Dr. Hsu once claimed to have the *second*-best laboratory for chromosome research in the world, but when urged to name the *best* laboratory, he was unable to come up with a single suggestion. This type of humility I particularly understand, as I have been reluctant to say that M. D. Anderson Hospital and Tumor Institute is the best comprehensive cancer center in the world, but I am also reluctant to correct anyone else who makes such a claim.

Dr. Hsu has authored or coauthored more than 200 scientific articles and approximately two dozen chapters in books and textbooks, and has delivered lectures in 32 of the 50 United States and in three foreign countries, including participation in a Nobel symposium in Sweden in 1973.

During the last 25 years, he has received grant and contract funds from the National Cancer Institute, the American Cancer Society, the Damon Runyon/Walter Winchell Memorial Fund, the National Science Foundation, the Food and Drug Administration, and from other National Institutes of Health. More recently, he has also received funds from the John S. Dunn Research Foundation.

He has been a consultant and member of the Board of Scientific Counselors of the National Institute of Environmental Health Sciences since 1973, has been on the editorial boards of numerous scientific journals, has been invited to review approximately a dozen books and textbooks on cytogenetics and cell biology, and is frequently invited to chair sessions at scientific meetings, not only because of his scientific erudition but because he is such a hard worker and because his humor enlivens any gathering he attends.

In 1979, two major events occurred at M. D. Anderson Hospital. Dr. Hsu's Section of Cell Biology, within the Department of Biology, was given full departmental status, and in December, The Univeristy of Texas System Board of Regents honored Dr. Hsu by naming him as the first recipient of the Olga

Keith Wiess Chair for Oncology. We honor Dr. Hsu again today by presenting him with the Ernst W. Bertner Memorial Award. This award, established in 1950 to honor the memory of the first acting director of the M. D. Anderson Hospital and Tumor Institute, is presented annually to a scientist or physician who has made major contributions to cancer research. We are singularly proud to present the award to Dr. Hsu, who has been a key vertebra in the backbone of this institution, the long-term staff members who are responsible for its remarkable growth, not only in size but reputation, in the science of oncology specifically and of biomedical science generally.

Genes, Chromosomes, and Neoplasia, edited by
Frances E. Arrighi, Potu N. Rao, and Elton Stubblefield.
Raven Press, New York © 1981.

The Ernst W. Bertner Memorial Award Lecture: Cytogenetics: Today and Tomorrow

T. C. Hsu

Department of Cell Biology, The University of Texas System Cancer Center, M. D. Anderson Hospital and Tumor Institute, Houston, Texas 77030

Cytogenetics is a discipline of biological science with a relatively long standing. In the earlier days, most pertinent information was obtained from studies using plant and insect materials. However, the conclusions did not strike a responsive cord, even among cytogeneticists, and few thought them applicable to man also. Most medical scientists ignored the importance of cytogenetic findings until Jerome Lejeune's discovery on trisomy 21 in 1959. It was then obvious that genetic abnormalities similar to those of *Datura, Nicotiana,* and *Drosophila* can and indeed did occur in our own families, communities, and populations.

I must emphasize that human and mammalian cytogenetics does not merely confirm what has been described in other systems. During the last 10 to 12 years, there have been many new and significant discoveries, e.g., in situ hybridization to locate DNA sequences, gene amplification, somatic cell hybridization, chromosome-mediated gene transfer, and the visualization of interphase chromosomes. Banding of somatic chromosomes was not achieved by classic cytologists. Without banding, human genetics would not have advanced to the present state and cancer cytogenetics would have remained ambiguous. However, I shall not dwell on a historic recount of progress in this field since I have recently made such a resume (Hsu 1979). Instead, I would like to describe briefly some of the contemporary accomplishments (many are being covered by other authors in this book), to speculate on what I consider important problems for the next few years, and to pose some additional questions. Of course, I have no idea whatsoever how some of these questions can be answered; many must await breakthroughs. Unfortunately, most breakthroughs come unexpectedly. But I will ask these questions anyway for whatever they are worth. I am certain that microscopic analyses of chromosomes will continue either as the dominant approach for some research programs or as an adjunct in connection with other disciplines, especially biochemical investigations.

CHROMOSOME STRUCTURE AND GENOMIC COMPOSITION

Prior to the elucidation of the basic chromatin structure, the nucleosome, biologists had studied chromosomes on two ends, namely, the microscopically observable chromosomes on one side and the chemically analyzable macromole-cules (DNA, RNA, proteins) on the other. The organization or interplay of the macromolecules between the two ends was not understood. The discovery of the nucleosome structure has been one step, at the submicroscopic level, toward the goal of solving the problem of the molecular architecture of chromo-somes. The next step is, of course, to understand the higher order of organization.

In the next few years, the controversial subject of the chromosome core (Stub-blefield and Wray 1971) or the scaffold (Paulsen and Laemmli 1977) should have a more definite conclusion. It is important to establish the validity of the scaffold structure because it vitally affects our concept of chromosome organi-zation and its modifications.

The explosive activities and new methodologies in molecular biology during recent years have enabled biologists to obtain detailed information on the com-position of certain genes in higher animals, e.g., the ovalbumin gene of the chicken. Therefore, to read the genome of any organism is *theoretically* feasible, using existing technology and improvements yet to come. When our knowledge of genomic composition is improved, we should then be able to critically examine the age-old problem of uninemy *vs.* multinemy and the difference in DNA content among many life forms. Do all the "regular" chromosomes contain a unineme structure? In amphibia, it is well known that the DNA content per cell varies enormously, indicating a redundancy of the genome. In the broad bean, *Vicia,* it is also well known that *V. faba* has a much larger DNA content and larger chromosomes than its close relative, *V. sativa;* yet the diploid number is the same for both. In fishes, the DNA content ranges from one similar to that of mammalian cells to one with only 20% of that. Yet all of them possess the genes required to develop into fishes. How is the genetic material organized? Is there some possibility of dinemy or multinemy, or is the redundancy strictly longitudinal?

As a cytologist, I have been interested in the organization and function of heterochromatin for a number of years. Our understanding of heterochromatin has advanced considerably during recent years, but many fundamental questions are still unanswered. We have learned that constitutive heterochromatin contains a high proportion of highly repetitive DNA, but highly repetitive DNA is not a unique property of heterochromatin. The degree of diversity of highly repetitive sequences within a genome and between related genomes has not been fully deter-mined. I hope some molecular biologists will help cell biologists obtain informa-tion on this subject.

Even though the problem of the anatomy of repetitive DNA is completely solved, many questions relating to heterochromatin require extensive investiga-tions. Thus far, we do not have a reasonable hypothesis regarding the function of heterochromatin. Moreover, we need to learn why heterochromatin is cytologi-cally different from euchromatin. What is the molecular organization that causes heterochromatin to condense at the time euchromatin does not? Does heterochro-matin remain condensed in all cells and at all stages except when it replicates its own DNA? We have evidence to show that it is not so. In spermatids of the mouse, for example, the centromeric areas (heterochromatin) are decondensed instead of condensed when the technique of premature chromosome condensation is applied (Figure 1). Presumably, the molecules that condense the heterochromatin area are either gone or replaced. Moreover, what causes facultative heterochromatin to condense? Is it possible to change experimentally an active gene into an inactive gene by causing it to become facultative heterochromatin?

One might consider that the formation of highly repetitive DNA is a form of gene amplification (or saltatory replication). Analyzing differences in heterochromatin content between individuals or between related species only reveals that such a variability exists, not how it is formed.

In plants, there is an interesting system that may allow experimental study of de novo generation of heterochromatin. In tobacco *(Nicotiana tabacum)*, the chromosomes have a minimum amount of heterochromatin. In *N. otophora*, five pairs of chromosomes carry a noticeable amount of heterochromatin in their long arms (Burns 1966). When the two species are crossed, some cells of the hybrids will exhibit one or more extraordinarily long heterochromatic chromosomes known as megachromosomes (Gerstel and Burns 1966). The longest megachromosome was found to be 15 times as long as the longest regular chromosomes. The origin of the megachromosome is still a mystery, but it is

FIG. 1. Prematurely condensed chromosomes of a mouse spermatid. Note extended centromeric (heterochromatic) regions forming associations. (Courtesy Ms. Helen Drwinga).

probably safe to regard it as a repeated occurrence of DNA amplification during cellular development of the hybrid. Since tobacco cells are easy to cultivate in vitro, and single cells can be grown to differentiate into full plants, this system of megachromosome formation should offer an excellent opportunity for studies on gene amplification. At present we do not know whether the megachromosomes contain highly repetitive DNA, and if they do, what the DNA sequence is. We also do not have information on whether the megachromosomes of one hybrid plant have the same DNA characteristics as those of another plant. In fact, it is possible that megachromosomes of different cells of the same plant are different.

One might also study nematodes, in which chromatin diminution (especially heterochromatic segments) takes place in meiosis. The regeneration of heterochromatin in embryogenesis may also be a good system for such study.

In a number of cases, especially in human beings, extra large C-bands can easily be explained by unequal crossing-over during meiosis. It is even possible to transfer a piece of C-band from a chromosome to a nonhomologous chromosome by mispairing during the zygotene stage since the sequences are the same. However, such mechanisms do not explain how related species have entirely changed satellite DNA. For example, all the centromeric heterochromatin pieces of the laboratory mouse *(Mus musculus)* are enriched with the same satellite DNA (Pardue and Gall 1970). In a related species, *M. caroli,* the *M. musculus* satellite DNA is replaced by other satellite sequences, yet a certain degree of homology between the two still exists (Sutton and McCallum 1972). Let us assume, for the sake of convenience, that the *M. musculus* satellite sequence is the ancestral form. A mutational event in a *M. musculus* sequence may change it into the *M. caroli* sequence. Amplification of the *M. caroli* sequence is feasible; but how can this new sequence move to other chromosomes and replace all the original sequence? Does the new sequence possess an adaptative advantage over the old sequence? To evolutionary biologists, answers to such questions may hold a key to speciation.

There are other interesting phenomena that do not have good explanations. For example, the long arm of the X chromosome and nearly the entire Y chromosome of the Chinese hamster are C-band positive. Yet in situ hybridization using highly repetitive DNA as a probe failed to demonstrate that these heterochromatic segments contain highly repetitive DNA sequences. It does not seem reasonable to consider that the huge Y chromosome of the Chinese hamster is facultative heterochromatin. Then how can this dichotomy be explained? Perhaps such a phenomenon is not limited to the Chinese hamster.

Cytogeneticists might also look into the classification of C-bands. Some C-bands (usually centromeric ones) are very resistant to alkaline treatments while others are not. In the Syrian hamster, for example, many short arms are C-band positive if the preparations are given a short alkaline treatment. Prolonged alkaline treatment will cause them to stain more or less like euchromatin. The Y chromosomes of most animals are moderately darkly stained in C-band preparations, but many authors have reported that the Y chromosomes are C–negative. Does this mean the Y chromosomes are compositionally euchromatic, the alkalinity too high, or the treatment too long? It is probably necessary to apply the in situ hybridization technique using a variety of DNA fractions as probes to study different types of heterochromatin in detail. It is safe to say that we still have a great deal to learn about the composition and behavior of heterochromatin. The recent review of John and Miklos (1979) also emphasizes this point.

DEVELOPMENTAL CYTOGENETICS

Gene activities change during cell development, cell differentiation, and cell maturation. It is highly probable that some of these changes are expressed at the chromosome level. Indeed, elegant demonstrations have been made using the polytene chromosomes, both during normal development and under experimental conditions, to show that gene activities can be observed with a microscope. A combination of molecular biology and cytological methods has produced tangible and significant information concerning chromosome physiology. The macronucleus of some protozoa is another good system for studying gene activities under the microscope. However, polytene chromosomes are specialized chromosomes limited to certain larval tissues of one group of insects, and the macronucleus can be found only in ciliated protozoa. Such superior materials are not available in the tissues of the great majority of life forms.

Nevertheless, even in higher animals, a number of systems are readily available and can be effectively exploited for studies in developmental cytogenetics, and more are expected to be found in the future. I can think of a few such systems without searching diligently.

Neoplasia

Numerous papers, both in the prebanding era and in the postbanding era, dealt with the chromosome constitution of neoplasms. The primary purpose of these investigations was an attempt to test the Boveri hypothesis, i.e., chromosome imbalance initiates cancer. The general conclusion is that the chromosome constitution of neoplastic cells is indeed abnormal, and the abnormality is not limited to numerical changes. In chronic myelogenous leukemia (CML) of man, the basic abnormality appears to be a specific (22/9) translocation.

Nevertheless, the debate continues. Specific chromosome changes in some tumors may or may not be the event that triggers the change of a normal cell to a neoplastic one. In other words, the changes may have occurred as the result of neoplastic transformation. Thus, tackling this problem by analyzing even an enormous number of cases would not convince those who doubt that chromosome alteration is an etiological factor.

In experimental animal systems, some new light has come into existence. This is particularly exemplified by studies on murine leukemias. It was discovered by Dofoku et al. (1975) that the karyotype of T-cell leukemias of the AKR strain consistently exhibited a trisomy 15. Subsequently, the same phenomenon was found in the NZB strain (Friedman et al. 1978) and in induced lymphomas in the C57BL strain (Chang et al. 1977, Wiener et al. 1978). The nonrandom abnormality was again expressed even when chromosome 15 was centrically fused with chromosome 1 or chromosome 6. The lymphoma cells showed three 1/15 or 6/15 metacentrics (Spira et al. 1979). It therefore appears that chromosome 15 is critically important for normal physiology of T-cell lymphocytes of the mouse and an excess of it would lead to uncontrolled proliferation. Even when chromosome 15 is fused with chromosome 1, the duplication of the attached genetic material must be tolerated as malignancy develops. When chromosome 15 was broken into two segments and translocated, the trisomy in lymphomas was always in the distal portion, suggesting that the gene responsible for lymphoma development is located in that segment (Wiener et al. 1979). Of even greater interest is the report of Ohno et al. (1979) on the chromosomes of mouse plasmacytomas. Chromosome 15 was again involved, but instead of producing a trisomy, it was translocated to chromosome 6 or chromosome 12. Genetic studies have suggested that the heavy chain of the *Ig* gene is located in chromosome 12, and the light chain in chromosome 6. Thus, it appears that the nonrandom chromosome changes in plasmacytomas are associated with the genetic function of the target cell and that translocations may have affected the structure or function of the genes responsible for the differentiation of normal plasma cells.

The data from experimental animal systems, therefore, strongly suggest that specific genetic changes in cells constitute the initial etiological factor for neoplasia and that such genetic changes can, in the cases of murine leukemias and plasmacytomas, be observed at the cytological level. The inference is also strong that specific chromosomal changes in human cancers (e.g., the 22/9 translocation in CML, the interstitial deletion in 13q of retinoblastomas) reflect basic genetic changes that initiate neoplastic transformation.

Of course, critical data in this area of research are just emerging, while the majority of cases of human cancers still require painstaking confirmation. The chromosome constitutions of human carcinomas and sarcomas are usually so extensively altered that many of them almost defy analysis even with superb banding preparations. Because of technical difficulties in obtaining quality cytogenetic preparations directly from biopsy specimens of solid tumors, most of the previous chromosome analyses came from long-term cell lines. However, the data are not very useful since chromosome constitutions are known to change in cell culture. In fact, even biopsy specimens from metastatic lesions (including effusions) represent very late stages of tumor development, so that a great deal of karyotypic evolution has already taken place. In CML, for example, the primary change is the 22/9 translocation. But many additional changes can be noted when the disease reaches the crisis stage. Thus, in order to analyze the chromosomes of primary lesions or at least earlier stages of tumor development, one must procure proper biopsy samples and refrain from using long-term cell lines.

Cytogeneticists must seek cooperation from clinical colleagues not simply to supply tissues but also to participate actively in planning and interpretation. The first step, of course, is to improve the technology for obtaining suitable preparations, since the mitotic rate of tumors in situ is usually exceedingly low. Dissociation of tissues probably can be accomplished by treatment with collagenase (Kusyk et al. 1979) or other mild proteolytic enzymes. The dissociated cells can be fixed directly or used for short-term cultures. Various media should be tried to increase the yield of mitosis.

The newer method for immediate cloning of dissociated tumor cells (Hamburger et al. 1978, Courtenay et al. 1978, Hamburger and Salmon 1979) may be advantageously utilized for analyzing chromosomes (Trent, in press). Colonies can be pooled to initiate vigorous growth. Even though this procedure takes several weeks, the time is shorter than that for long-term cell lines. Anyway, technical improvements are urgently needed to offer a consistent supply of metaphase figures.

One can then deal with a number of problems. First, of course, is whether specific aberrations are always associated with a specific type of neoplasm. Second, it would be of great interest to determine whether each type of neoplasm can be classified into cytological types similar to the way human and murine leukemias have been. One can also find out, by using serial biopsies from the same patients, whether the extensive chromosome aberrations take place all at once or are accumulated in many steps. With earlier lesions, it is possible to determine whether the cells of the original stemline are always in the diploid range or whether some of them start in the tetraploid range. If the latter is the case, then what is the proportion of each? It is also possible to estimate the effects of various therapeutic regimens on the progression of solid tumors. In the future, genetic and molecular biology research can be combined to study some tumor systems to find out the basic cause of cancer at the DNA level.

Gametogenesis

Gonads are very specialized organs with unique cytological phenomena in each sex. The events that can be observed by microscopists are well known, and classical cytogeneticists have effectively utilized meiosis to analyze species relationships and other subjects. However, the physiology, the structural changes in chromosomes, and the molecular events in different stages of gametogenesis are still not well understood. Indeed, even ultrastructural investigations on meiosis are not plentiful. Hotta, Stern, and their collaborators have taken advantage of the synchronized meiosis in plants to study DNA synthetic phenomena in meiotic prophase and extended their study to mammals (Chandley et al. 1977, Hotta et al. 1979), but they have merely scratched the surface.

Studies in the biochemistry of gametogenesis of higher animals have been

severely hampered by the fact that the cells of premeiotic, meiotic, and postmeiotic stages are not synchronized. However, the development of cell fractionation
and purification procedures has greatly alleviated this difficulty. Even at the
present time, it is feasible to obtain highly concentrated fractions of cells in
pachytene stage or second meiosis, spermatids, etc. (Meistrich 1977). With further improvements, many sophisticated experiments can be carried out to elucidate the biochemistry and molecular biology of meiosis.

We have known for years the fine structure of the synaptonemal complex
(SC), but isolation and purification procedures for this cell organelle are still
in the infant state. We learned that the SC has a great affinity for silver staining
in cytological preparations (Pathak and Hsu 1979, Fletcher 1979), as much
affinity as a fraction of nucleolar proteins, which can be purified and identified
(Hubbel et al. 1979). Perhaps with adaptations of the biochemical techniques
of Hubbel and others, we can isolate the proteins of the SC. If the chromosome
core or scaffold is an essential component of somatic chromosomes and not
an artifact, do pachytene chromosomes have scaffolds also? What is the relationship between the scaffold and the SC? How many protein species are in the
SC and how are they arranged? What happens to the SC proteins when the
first meiotic division is completed? Do the chromosomes of second meiosis and
spermatids have scaffolds?

We have also known for years that histones are discarded during spermiogen-
esis. Correlation between the stages of spermatid development and biochemical
changes should be determined. What is the chromosome organization in spermatids before and after the discarding of histones? Do the chromosomes still have
regular nucleosomes, modified nucleosomes, or none at all? What is the chromatin structure of the pronucleus, and if it is different from that of the regular
nuclei, how does it change when cleavage divisions ensue? We know that it is
feasible to induce prematurely condensed chromosomes from spermatozoa and
spermatids (Johnson et al. 1970, Drwinga et al. 1979). A combination of morphological, cytochemical, and biochemical investigations should yield some pertinent
information in this highly specialized but nearly universal phenomenon of cellular
development.

Gene Amplification

Some years ago gene amplication was thought to be limited to some specialized
tissues of invertebrates and lower vertebrates. In human and mammalian systems
gene amplification was suspected but there was no direct evidence. Biedler and
Spengler (1976) discovered, in Chinese hamster and human cells resistant to
methotrexate, segments of chromosomes that did not differentiate into bands
when the G-band procedure was applied. These chromosome segments are referred to as homogeneously staining regions (HSR); since biochemists had already
found that methotrexate-resistant cells had elevated dihydrofolate reductase activity, Biedler and Spengler (1976) considered it a possibility that HSR might

represent the amplified dihydrofolate reductase gene. By elegant biochemical experiments, Alt et al. (1978) conclusively demonstrated that gene amplification was a real phenomenon, and Nunberg et al. (1978) and Dolnick et al. (1979), using the in situ hydridization technique, proved that HSR indeed represents the chromosome segment with amplified genes.

The discovery of double minutes (dm) occurred early in the 1960s in human tumor cells (cf., Barker and Hsu 1979). Double minutes are small, spherical, paired, DNA-containing bodies found in metaphase. They vary in number from cell to cell, ranging from none to several hundred. The nature and the function of the dm have been a puzzle and the object of debate. Recently, the observations and experiments of Kaufman et al. (1979) on methotrexate-resistant cell lines whose resistance is unstable in the absence of the drug proved that dm are also cytological expressions of gene amplification. That HSR and dm are related structures has been observed by Balaban-Malenbaum and Gilbert (1977) in a human neuroblastoma cell line, by Levan et al. (1978) in a transplantable mouse tumor, and by Quinn et al. (1979) in a human colon carcinoma.

Thus, some tentative conclusions can be drawn: (1) the gene amplification phenomenon also exists in mammalian cells, (2) amplified genes may be cytologically detectable either as dm or HSR, and (3) dm may represent an unstable phenotype while HSR may represent a more stable phenotype of amplification. These findings, though preliminary, are exciting not only because they explain one of the mechanisms of drug resistance but also because cytologists can actually observe the structures containing the amplified genes. However, many questions have been raised by these discoveries. For example, in many tumor cells, dm or HSR have been observed. Some of the patients had undergone multiple chemotherapy. If dm or HSR in these cells also represent gene amplification, what genes are being amplified? Are some of them genes mediating resistance to certain drugs? Presumably they are different in different cases. In some patients no therapy was administered prior to biopsy, yet their cells also exhibit dm or HSR (Quinn et al. 1979). What then do the dm or HSR represent? Do the genes being amplified determine a higher degree of anaplasia, growth potential, or invasiveness? Biochemical and molecular biological technology of the present day should be able to answer these questions.

Another series of questions concerns the mechanism of gene amplification and the structure of dm and HSR. When a gene (or genes) becomes amplified, does it involve only the gene *per se* or does it involve the neighboring chromosome segments as well? What are the mechanisms for increasing the gene? In the case of dm, how does the gene leave the chromosome and become an independent entity? How many gene copies are contained in each dm? Presumably they vary because the size of dm varies, but whether the size variation coincides with the number of gene copies is not yet known. Moreover, we have no idea whether the structure of dm is exactly the same as that of a normal chromosome. Since it appears that cells with dm can lose dm but exhibit HSR, the two structures may well be interchangeable, especially from dm to HSR. We do not know whether dm has a scaffold structure. If it does not, how can dm be incorporated into a chromosome to become HSR, which undoubtedly contains the scaffold? Can HSR break down into dm?

The gene amplification phenomenon points to other questions: Do highly differentiated tissues (e.g., Sertoli cells and neurons) have amplified genes? If so, how can they be demonstrated? Is it possible, by isolating dm or by microcell techniques, to transfer dm with a known function (e.g., drug resistance) to cells without this property and thus transfer the property? One can think of numerous questions relating to this area of research in developmental cytogenetics as well as molecular biology.

Pathology of Chromosome Aberrations

In human cytogenetics, a large amount of data has been accumulated on morphological and physiological expressions of complete and partial trisomies. However, only a few types of trisomies can survive embryonic wastage. Thus, a systematic pathological study of the effects of aneuploidy in all chromosomes on embryonic development is difficult in human beings. In experimental animals, especially mice, a beautiful system is available. Investigations by A. Gropp, E. Capanna, and their collaborators on the chromosomes of feral mice in the Alps and other parts of Europe offered excellent experimental material for this problem. There are a variety of Robertsonian translocations involving different chromosomes, and breeding of these mice enabled the investigators to obtain stocks homozygous for Robertsonian fusions involving practically any two particular chromosomes.

A mouse homozygous for a Robertsonian fusion between chromosomes A and B possesses a diploid number of 38 with two metacentrics. Another mouse homozygous for a Robertsonian fusion between chromosomes A and C displays a similar gross karyotype to the one with an A/B fusion, but the fusion partners are different. Banding can identify the individual chromosomes involved. When these two individuals are crossed, the hybrid will again have two metacentrics, but these two are A/B and A/C. Meiosis of such heterozygous individuals should produce a certain proportion of gametes with both metacentrics and an equal number of gametes with no metacentrics. If such a heterozygous mouse is mated with a normal mouse (2n = 40, all acrocentrics), the zygotes from the gametes with both A/B and A/C metacentrics will be trisomic for chromosome A, while those from gametes with no metacentrics will be monosomic for chromosome A. Using stock animals carrying different Robertsonian fusions, one can experimentally produce zygotes trisomic and monosomic for any given chromosome. Systematic analyses by Gropp and his collaborators showed that monosomic embryos invariably succumb prior to implantation, but trisomies can develop to various stages or even survive after birth (e.g., tri-19). Furthermore, each trisomy seems to exhibit its own developmental errors. Such an experimental system should enable biologists to study not only the pathology of trisomy but also the biochemical differences in developing tissues and organs, which may serve as a model to explain some of the human trisomic conditions, especially when data on gene homology between human and mouse chromosomes are established. I think that this will be one of the most exciting areas of cytogenetics for the next few years.

CHROMOSOMES AND ORGANIC EVOLUTION

Application of various banding techniques to comparative cytogenetic investigations on higher vertebrates, especially mammals, has uncovered a number of interesting facts. Although some of these findings have been known for decades in more classical materials, such as plants and insects, the banding techniques on somatic metaphase chromosomes have acquired a degree of precision not achieved by plant cytologists in the past. Even at the present time, G-banding of plant and insect chromosomes is not well developed.

The classical methods of analyzing phylogenetic relationships by cytogenetic means call for obtaining hybrids between related taxa and examining either meiosis (mainly plants) or polytene chromosomes (dipterous insects). The resolution of polytene chromosomes is so good that even minor rearrangements (e.g., small paracentric inversions) can be detected with precision. However, in the vast majority of life forms, such luxury does not exist. The gross morphology of somatic chromosomes offers little encouragement, and hybridizing related forms to study gametogenesis usually yields disappointment.

Even though metaphase chromosome banding is relatively crude compared to what can be achieved with polytene chromosomes, it at least represents a significant improvement over the conventionally stained preparations. The banding patterns can be used to identify chromosome homology between related species without resorting to crosses, and many changes that have taken place in the chromosomes (rearrangements) can be identified. Perhaps the report of Mascarello et al. (1974) best illustrates this principle. These investigators compared the G-banding patterns of a number of rodent species in the families Cricetidae and Muridae. They found that chromosome homology is almost absolute between closely related species. As taxonomic relationships become more and more distant, the number of unmatched chromosomes progressively increases.

This approach is applicable to higher vertebrates, including mammals, birds, and reptiles. G-banding has not been induced in amphibia and fishes, although C-banding is feasible. Similar problems exist in the study of chromosomes of invertebrates. It is possible that the chromosome structure of some taxa is such that G-banding cannot be induced, but it is also possible that efforts to modify the procedures have not been active enough to achieve a workable protocol.

When two forms differed in the number of chromosome arms but not the diploid number, classic cytogeneticists always interpreted the difference as being the result of pericentric inversions and asymmetrical translocations. Banding techniques, primarily C-banding, drastically amended this concept. A chromosome arm can be made exclusively of heterochromatin. The best illustration can be found in the rodent genus *Peromyscus*. In nearly 30 species analyzed, the diploid number of all was found to be 48, but the number of total chromosome arms varied from 56 *(P. boylei* and *P. crinitus)* to 96 *(P. eremicus* and *P. collatus).* C-banding of somatic cells of *P. eremicus* revealed that all short arms were heterochromatic, while the G-banding patterns of the long arms (euchromatin) between *P. eremicus* and *P. crinitus* were identical. Therefore, aside from the addition (or deletion) of heterochromatin, the G-banding pattern of the two species is essentially unchanged despite the apparent difference in chromosome morphology (Pathak et al. 1973a).

Addition of heterochromatin is not limited to the formation of new chromosome arms or extension of euchromatic arms. In some mammalian taxa, e.g., *Rattus* (Yosida and Sagai 1975) and *Uromys* (Baverstock et al. 1977), supernumerary heterochromatic chromosomes similar to the B chromosomes of plants have been found in natural populations. Heterochromatin must play some important role in evolution since such a phenomenon has been found in many organisms, including plants, grasshoppers, *Drosophila,* and others. It is now common knowledge that to avoid misinterpretation in studies on karyological evolution and cytogenetics, C-banding is the necessary first step.

It is also well known that two acrocentric chromosomes can fuse in the centromeric areas to form a biarmed element. This phenomenon of centric fusion (or Robertsonian fusion) is prevalent in numerous life forms, including mammals. A karyotype of a species with an all-acrocentric complement showing a metacentric and a reduced diploid number is considered to be the result of a Robertsonian fusion. However, without banding only empirical identification of the chromosomal elements involved in the translocation could be made. In the human population, D/D, D/G, and G/G translocations were found in numerous cases prior to banding, but identification of the elements involved was not feasible. Banding techniques solved this problem. In D/D translocation, the highest frequency was found to be 13/14, but other D chromosomes can also be involved. Thus, when metacentrics resulting from Robertsonian translocation are found in two individuals, they may or may not be homologous unless there is a blood relationship between the individuals. This concept is illustrated in the work of Pathak et al. (1973b) on the climbing rat *(Tylomys).* Two species possess a large metacentric with an identical morphology. Without banding, they would be considered homologous. Therefore, it would have to be assumed that the two species came from a common ancestral stock that had this centric fusion. G-banding showed that different chromosomes were involved, indicating that they occurred as independent events. Hence, the two species did not come from the same ancestral form. Analyses of *Mus musculus* populations in Europe (Gropp et al. 1972) also emphasize this point.

The frequency of tandem translocations was not expected to be very high because it was surmised that meiotic abnormalities arise in heterozygotes. However, banding analyses of related species indicated that tandem translocations (centromere-telomere or telomere-telomere) are more frequent than we had thought. I shall cite two well-documented examples to illustrate this phenomenon.

In the cricetid rodent genus *Sigmodon* (cotton rat), centric fusion is very common. The highest diploid number is 52 *(S. hispidus)*, all acrocentic. However, one species, *S. mascotensis,* has a diploid number of 28, again all acrocentric (Zimmerman 1970). The suspicion that tandem translocations must have occurred between the karyotypes of *S. hispidus* and *S. mascotensis* was verified by banding analyses (Elder 1980). In another species, *S. arizonae,* the diploid number is reduced further to 22, with two pairs of metacentrics. Again, evidence for tandem translocation as well as Robertsonian fusion was strong. The two species with low diploid numbers apparently shared some of the arrangements, indicating a common ancestry (Elder 1980).

The most spectacular karyotypic difference between related species was found in the muntjacs. The Reeves (or Chinese) muntjac, *Muntiacus reevesi,* has a diploid number of 46, all acrocentric. The Indian muntjac, *M. muntjak,* has a diploid number of 6 in females and 7 in males. All six chromosomes are very long, and all are biarmed. It is obvious that if the two karyotypes have any relationship, tandem translocations must have played an important role. Recently unequivocal evidence was obtained to demonstrate that hybrids between the two species can be obtained with animals in captivity, and that the hybrid has 26 (♀) or 27 (♂) chromosomes, as expected. G-banding of the hybrid cells suggests that a series of tandem translocations has taken place from the *M. reevesi* karyotype to the *M. muntjak* karyotype with only some loss of heterochromatic material (Shi et al. 1980).

Despite the increase in resolution, metaphase or even prophase banding is still crude in terms of detecting minor rearrangements such as small paracentric inversions, duplications, and deletions. However, pericentric inversions can usually be identified. Just how large a role pericentric inversions play in karyotypic evolution in higher vertebrates is not yet known.

It is anticipated that data will be accumulated on the cytogenetic changes of various life forms, especially by taxonomists. However, many questions of more fundamental nature cannot be answered by an increase in catalogue. We already have sufficient examples to obtain a general picture of karyological changes in phylogeny. Perhaps it is time for us to ask some questions about the reasons these phenomena exist and how they developed. Of course I do not have the vaguest idea how to obtain the answers to these questions, but I believe that they may be useful to keep in mind.

Susceptibility to Translocations

In some rodent genera, e.g., *Peromyscus* and *Onychomys,* extensive karyological analyses failed to show a single individual with a translocation. All species of these two genera have a diploid number of 48. Karyotypic variability is mainly the result of addition or deletion of heterochromatic arms and, to a minor extent, pericentric inversions. In some species with a higher number of acrocentrics, e.g., *P. boylei,* there is plenty of opportunity for centric fusions; yet they have never been recorded. Tandem translocation is also absent. On the other hand, in related genera, such as *Sigmodon* and *Neotoma,* especially the former, both types of translocations are common among different species.

Similarly, all mouse species closely related to *Mus musculus* were found to possess a diploid number of 40, all acrocentrics. They have a wide distribution in southeastern Asia (Hsu et al. 1978). No evidence of centric fusion or tandem translocation has been obtained. Yet in the Alps, centric fusions in *Mus musculus* are rampant.

What is the underlying factor or factors that causes such a difference? Since inversions have been found between species of *Peromyscus,* chromosome breakage and restitution definitely can happen. In cell cultures, Robertsonian translocations have been induced by treatment of the cells of *P. eremicus* with mitomycin C; yet in nature, such translocations, if formed, apparently did not survive. This lack of Robertsonian fusions in some groups of animals is an interesting problem, but how to tackle it remains an open question.

Heterochromatin

It is obvious that the variability in the amount of heterochromatin among life forms is more pronounced in some taxa than others. In felids, for example, heterochromatin content and variation are minimal. In a number of rodent genera, a wide spectrum of variability has been noted. The reason for this variability and the role of heterochromatin in karyotypic evolution constitute intriguing problems for workers in cytogenetic evolution. It appears that when the amount of heterochromatin is minimal in the karyotypes, the karyotypic variability is also minimal. When the amount of heterochromatin is highly variable, karyotypic variability is likewise increased. It is possible that heterochromatin is a genetic structure that may facilitate karyotypic changes in the course of organic evolution, but such a notion lacks critical proof.

Mechanism of Tandem Translocations

In a number of cases of human cogenital anomalies, tandem fusion of two chromosomes, particularly the X chromosomes, has occurred repeatedly. It appears, from banding analysis, that in a number of cases the two X chromosomes fuse directly at the telomeres (Therman et al. 1974). The resulting compound chromosome should therefore be dicentric. Yet this long chromosome usually behaves as a monocentric element. What happens to the second centromere? The idea of centromere inactivation seems plausible, but it has no experimental support. If centromere inactivation exists and its mechanism is elucidated, it should have many ramifications on the study of chromosome structure and physiology, human genetics, and organic evolution.

What are the underlying differences between a dicentric that behaves as a dicentric and a dicentric that does not behave as a dicentric? Can the latter type of dicentric persist in nature with one of the centromeres permanently inactivated? Is there any rule determining which centromere should be inactivated? Is there any condition under which the inactivated centromere would become reactivated?

I think there are numerous questions that one should ask in relation to cytogenetic evolution. Mammalian systems, whose cells can be cultured, cloned, and hybridized relatively easily, and their chromosomes banded easily, offer superior materials for an experimental cytogenetic approach to evolutionary problems.

APPLIED CYTOGENETICS

In agriculture, cytogenetics has been employed as an effective tool for plant breeding. The improvements in mammalian and human cytogenetic technology during recent years have slowly gained momentum so that applied research in some areas is not only feasible but also profitable. I can think of several areas.

Medical Cytogenetics and Demography

Collection of data on congenital anomalies associated with chromosomal changes should be vigorously expanded for a worldwide effort in constructing an accurate picture of epidemiology and demography of genetic diseases in man (and perhaps in domestic animals also). Genetic counseling and amniocentesis naturally should receive more attention and support.

Livestock and Poultry Breeding

It has been found that in cattle, spontaneous Robertsonian fusion between chromosomes 1 and 29 is common. Animals heterozygous for this and other translocations have reduced fertility. It is highly probable that such a phenomenon also occurs in other domestic animals, farm animals, and fowl. Breeders may have to determine the karyotypes of their stocks in order to avoid difficulties.

Cancer Chemotherapy

There is a growing body of evidence suggesting that neoplasms with similar pathological characteristics may be genetically different. This notion stemmed from the responses of CML to chemotherapy. Patients respond well to the conventional chemotherapeutic regimen if the CML cells exhibit the Ph^1 chromosome, but CML patients without the Ph^1 are refractory to the same treatment. It appears, therefore, that there are two types of CML differing cytologically as well as physiologically. Recent data indicated that similar differences exist in other leukemias as well. Aside from studies on the basic genetics and physiology of these cancers, cytogenetic determination should render useful assistance to clinical endeavor.

This phase of work should certainly continue not only because of its clinical importance but also because of its fundamental significance. We hope that similar investigations can be performed in various solid tumors when cytogenetic studies on solid tumors become technically developed enough to permit critical determination of the cytogenetic pictures of each lesion.

Mutagen Assays

Utilization of chromosomes as an object for studies on mutagens started soon after the discovery that the X ray is a mutagen. Levan and other cytologists extensively investigated the effects of chemicals on plant chromosomes.

Our deteriorating environment demands effective tests for screening mutagens, carcinogens, and teratogens. Cytogenetic criteria constitute a simple, reliable, and economical assay system. There are a number of materials (*Drosophila,* grasshoppers, plants, mammalian cells in vitro and in vivo) that are suitable for such assays. At this moment, a standardized protocol has not been worked out, but its completion is just a matter of time. After that, cytogenetic assays for detrimental chemicals should make important contributions to environmental research, and, as a bonus, some data may also contribute to our understanding of chromosome physiology and drug action.

CONCLUSION

I am confident that the field of cytogenetics, especially in combination with biochemistry, will see a continuing growth phase for years to come.

EPILOGUE

I would like to comment on my personal involvement in this field of endeavor. It is inevitable that everyone sooner or later comes to face the reality of age and deterioration in physical and mental health. I am afraid that my days of active scientific research are numbered. But let no one feel sad over this seemingly depressing proclamation, for this is the normal process of life. In the body there is tissue renewal, and in the natural population there is individual renewal. A biochemist friend once told me: "One hundred years from now who is going to remember your name or my name?" He was absolutely right, of course. A Chinese proverb says, "The back waves of the Yangtze River push the front waves." There are and will be plenty of new waves in every profession.

REFERENCES

Alt, F. W., R. E. Kellum, J. R. Bertino, and R. T. Schimke. 1978. Selective multiplication of dihydrofolate reductase in methotrexate-resistant variants of cultured murine cells. J. Biol. Chem. 253:1357–1370.

Balaban-Malenbaum, G., and F. Gilbert. 1977. Double minute chromosomes and homogeneously staining regions in chromosomes of a human neuroblastoma cell line. Science 198:739–742.

Barker, P. E., and T. C. Hsu. 1979. Double minutes in human carcinoma cell lines, with special reference to breast tumors. J. Natl. Cancer Inst. 62:257–262.

Baverstock, P. R., C. H. S. Watts, and J. T. Hogarth. 1977. Chromosome evolution in Australian rodents. I. The Pseudomynae, the Hydromyinae and the *Uromys/Molomys* group. Chromosoma 61:95–125.

Biedler, J. L., and B. A. Spengler. 1976. Metaphase chromosome anomaly: Association with drug resistance and cell-specific products. Science 191:185–187.

Burns, J. A. 1966. The heterochromatin of two species of *Nicotiana*. J. Hered. 57:43–47.

Chandley, A. C., Y. Hotta, and H. Stern. 1977. Biochemical analysis of meiosis in the male mouse. I. Separation and DNA labeling of specific spermatogenic stages. Chromosoma 62:243–253.

Chang, T. D., J. L. Biedler, E. Stockert, and L. J. Old. 1977. Trisomy 15 in X-ray induced mouse leukemia. (Abstract) Proc. Am. Assoc. Cancer Res. 18:225.

Courtenay, V. D., P. J. Selby, I. E. Smith, J. Miles, and M. J. Peckham. 1978. Growth of human tumor cell colonies from biopsies using two soft-agar techniques. Br. J. Cancer 38:77–81.

Dofoku, R., J. L. Biedler, B. A. Spengler, and L. J. Old. 1975. Trisomy of chromosome 15 in spontaneous leukemia of AKR mice. Proc. Natl. Acad. Sci. USA 72:1515–1517.

Dolnick, B. J., R. J. Berenson, J. R. Bertino, R. J. Kaufman, J. H. Nunberg, and R. J. Schimke. 1979. Correlation of dihydrofolate reductase elevation with gene amplification in a homogeneously staining region in L 5178Y cells. J. Cell Biol. 83:394–402.

Drwinga, H. L., T. C. Hsu, and S. Pathak. 1979. Induction of prematurely condensed chromosomes from testicular cells of the mouse. Chromosoma 75:45–50.

Elder, F. F. B. 1980. Tandem fusion, centric fusion, and chromosomal evolution in the cotton rat, genus *Sigmodon*. Cytogenet. Cell Genet. 27:31–38.

Fletcher, J. M. 1979. Light microscopic analysis of meiotic prophase chromosomes by silver staining. Chromosoma 72:241–248.

Friedman, J. M., P. J. Fialkow, J. Bryant, A. L. Reddy, and A. C. Salo. 1978. Neoplastic behavior of chromosomally abnormal clones in New Zealand black mice. Int. J. Cancer 22:458–464.

Gerstel, D. U., and J. A. Burns. 1966. Chromosomes of unusual length in hybrids between two species of *Nicotiana*, in Chromosomes Today, C. D. Darlington and K. R. Lewis, eds. Oliver & Boyd, Edinburgh, pp. 41–56.

Gropp, A., H. Winking, L. Zech, and H. Miller. 1972. Robertsonian chromosonal variation and identification of metacentric chromosomes in feral mice. Chromosoma 39:265–288.

Hamburger, A. W., and S. E. Salmon. 1979. Primary bioassay of human tumor stem cells. Science 197:461–463.

Hamburger, A. W., S. E. Salmon, M. B. Kim, J. M. Trent, B. J. Soehnien, D. S. Alberts, and J. Schmidt. 1978. Direct cloning of human ovarian carcinoma cells in agar. Cancer Res. 38:3438–3444.

Hotta, Y., A. C. Chandley, H. Stern, A. G. Searle, and C. V. Beechey. 1979. A disruption of pachytene DNA metabolism in male mice with chromosomally derived sterility. Chromosoma 73:287–300.

Hsu, T. C. 1979. Human and Mammalian Cytogenetics: An Historical Perspective. Springer-Verlag, New York.

Hsu, T. C., A. Markvong, and J. T. Marshall. 1978. G-band patterns of six species of mice belonging to subgenus *Mus*. Cytogenet. Cell Genet. 20:304–307.

Hubbell, H. R., L. L. Rothblum, and T. C. Hsu. 1979. Identification of a silver banding protein associated with the cytological silver staining of actively transcribing nucleolar regions. Cell Biology International Reports 3:615–622.

John, B., and G. L. Miklos. 1979. Functional aspects of satellite DNA and heterochromatin. Int. Rev. Cytol. 58:1–114.

Johnson, R. T., P. N. Rao, and H. D. Hughes. 1970. Mammalian cell fusion. III. A HeLa cell inducer of premature chromosome condensation active in cells from a variety of animal species. J. Cell Physiol. 76:151–158.

Kaufman, R. T., P. C. Brown, and R. T. Schimke. 1979. Amplified dihydrofolate reductase genes in unstably methotrexate resistant cells are associated with double minute chromosomes. Proc. Natl. Acad. Sci. USA 76:5669–5673.

Kusyk, C. J., C. L. Edwards, F. E. Arrighi, and M. M. Romsdahl. 1979. Improved method for cytogenetic studies of solid tumors. J. Natl. Cancer Inst. 63:1199–1203.

Levan, A., G. Levan, and N. Mandahl. 1978. A new chromosome type replacing the double minutes in a mouse tumor. Cytogenet. Cell Genet. 20:12–23.

Mascarello, J. T., A. D. Stock, and S. Pathak. 1974. Conservatism in the arrangement of genetic material in rodents. J. Mammal. 55:695–704.

Meistrich, M. L. 1977. Separation of spermatogenic cells and nuclei from rodent testes. Methods Cell Biol. 15:15–54.

Nunberg, J. H., R. J. Kaufman, R. T. Schimke, G. Urlaub, and L. A. Chasin. 1978. Amplified dihydrofolate reductase genes are localized to a homogeneously staining region of a single chromosome in a methotrexate-resistant Chinese hamster ovary cell line. Proc. Natl. Acad. Sci. USA 75:5553–5556.

Ohno, S., M. Babonits, F. Wiener, J. Spira, G. Klein, and M. Potter. 1979. Nonrandom chromosome changes involving the Ig gene-carrying chromosomes 12 and 6 in pristane induced mouse plasmacytomas. Cell 18:1001–1007.

Pardue, M. L., and J. G. Gall. 1970. Chromosomal localization of mouse satellite DNA. Science 168:1356–1358.

Pathak, S., and T. C. Hsu. 1979. Silver-stained structures in mammalian meiotic prophase. Chromosoma 70:195–203.

Pathak, S., T. C. Hsu, and F. E. Arrighi. 1973a. The role of heterochromatin in karyotypic evolution. Cytogenet. Cell Genet. 12:315–326.

Pathak, S., T. C. Hsu, L. Shirley, and J. D. Helm. 1973b. Chromosome homology in the climbing rats, genus *Tylomys* (Rodentia, Cricetidae). Chromosoma 42:215–225.

Paulsen, J. R., and U. K. Laemmli. 1977. The structure of histone depleted metaphase chromosomes. Cell 12:817–828.

Quinn, L. A., G. E. Moore, R. J. Morgan, and L. K. Woods. 1979. Cell lines from human colon carcinoma with unusual cell products, double minutes and homogeneously staining regions. Cancer Res. 39:4914–4924.

Shi, L., Y. Ye, and D. Duan. 1980. Comparative cytogenetic studies on the red muntjac, Chinese muntjac and their F1 hybrids. Cytogenet. Cell Genet. 26:22–27.

Spira, J., F. Wiener, S. Ohno, and G. Klein. 1979. Is trisomy cause or consequence of murine T cell leukemia development? Studies on Robertsonian translocation mice. Proc. Natl. Acad. Sci. USA 76:6619–6621.

Stubblefield, T. E., and W. Wray. 1971. Architecture of the Chinese hamster metaphase chromosomes. Chromosoma 32:262–294.

Sutton, W. D., and M. McCallum. 1972. Related satellite DNA's in the genus *Mus*. J. Mol. Biol. 71:633–656.

Therman, E., G. E. Sarto, and K. Pätau. 1974. Apparently isocentric but functionally monocentric X chromosome in man. Am. J. Hum. Genet. 26:83–92.

Trent, J. M. 1980. Protocols of procedures and technique in chromosome analysis of tumor stem cell cultures in soft agar, *in* Human Tumor Cloning, Salmon and Buick, eds. Alan R. Liss & Co., New York (in press).

Wiener, F., J. Spira, S. Ohno, N. Haran-Ghera, and G. Klein. 1978. Chromosome changes (trisomy 15) in murine T-cell leukemia induced by 7, 12 dimethylbenz(a)anthracene (DMBA). Int. J. Cancer 22:447–453.

Wiener, F., J. Spira, S. Ohno, N. Haran-Ghera, and G. Klein. 1979. Non-random duplication of chromosome 15 in murine T-cell leukemias induced in mice heterozygous for translocation T(14:15)6. Int. J. Cancer 23:504–507.

Yosida, T. H., and T. Sugai. 1975. Variation of C-bands in the chromosomes of several subspecies of *Rattus rattus*. Chromosoma 50:283–300.

Zimmerman, E. G. 1970. Karyology, systematics and chromosomal evolution in the rodent genus *Sigmodon*. Michigan State University Publications Museum Biology Series. 49:389–454.

Chromatin and Chromosome Structure

Genes, Chromosomes, and Neoplasia, edited by
Frances E. Arrighi, Potu N. Rao, and Elton Stubblefield.
Raven Press, New York © 1981.

Chromosome Condensation Factors of Mammalian Cells

Potu N. Rao, Prasad S. Sunkara, and David A. Wright

Departments of Developmental Therapeutics and Medical Genetics, The University of Texas System Cancer Center, M. D. Anderson Hospital and Tumor Institute, Houston, Texas 77030

Nearly a decade ago, Johnson and Rao (1970) used the term "premature chromosome condensation" to describe the events that follow fusion between mitotic and interphase cells. Since that time, this phenomenon has been studied extensively. The purpose of this presentation is to review briefly the literature on the phenomenon of premature chromosome condensation and then discuss our latest experiments on the isolation and characterization of the factors involved in the condensation of chromosomes.

With the exception of certain flagellates and the salivary gland cells of dipterans, chromosomes in eukaryotic cells are visible for only a brief period during the cell cycle, i.e., either during mitosis or meiosis. The technique of cell fusion made it possible to visualize the chromosomes of interphase cells. Fusion between mitotic and interphase cells induced either by ultraviolet-inactivated Sendai virus or polyethylene glycol results in the premature condensation of the interphase chromatin into chromosomes (Kato and Sandberg 1968a,b,c, Takagi et al. 1969, Sandberg et al. 1970, Johnson and Rao 1970) under the influence of factors present in the mitotic cells. This phenomenon described by Johnson and Rao (1970) as premature chromosome condensation has also been referred to as "prophasing" by Matsui et al. (1972). The products of this phenomenon are the prematurely condensed chromosomes or PCC. The morphological changes observed during the induction of PCC in interphase nuclei parallel those seen during the beginning of mitosis. Immediately after fusion between mitotic and interphase cells, the interphase nucleus undergoes chromatin condensation similar to that of a prophase nucleus, which is rapidly followed by dissolution of the nuclear membrane and further condensation of chromatin into discrete chromosomes. The morphology of the PCC reflects the position of an interphase cell in the cell cycle at the time of fusion. The G_1 cells yield PCC with a single chromatid. G_2-PCC exhibit two chromatids, while PCC from S phase cells present a "pulverized" appearance (Figure 1). The cell cycle phase-specific morphology of PCC has formed the basis for the development of a new method of cell cycle analysis (Rao et al. 1977).

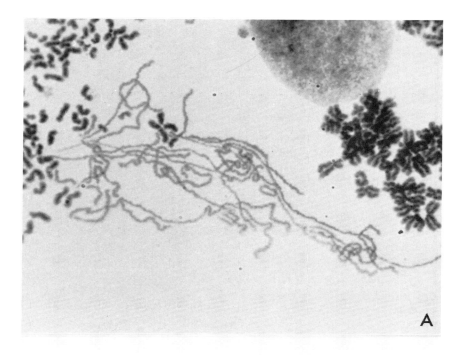

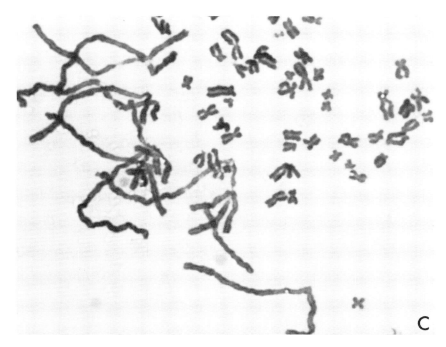

FIG. 1. Prematurely condensed chromosomes of rat kangaroo cells *(Potorous tridactylis)* cell line (Pt K₂) after fusion with mitotic HeLa cells. (A) G₁-PCC with single chromatids; (B) S-PCC with pulverized appearance; (C) G₂-PCC with two chromatids closely attached to each other. (Reprinted from Sperling and Rao 1974, with permission of Humangenetik.)

ABSENCE OF SPECIES-SPECIFICITY

The factors present in the mitotic cells which are responsible for the induction of PCC have no species barriers. The mitotic cells of human origin can, upon fusion, induce PCC in the interphase cells of other mammals, rodents, amphibians and insects, and vice versa. PCC can be induced not only in cycling tissue culture cells but also in noncycling and differentiated cells from animal tissues. This method has made it possible, for the first time, to visualize the chromosomes of differentiated cells, such as unstimulated horse or human lymphocytes, chick erythrocytes, and bovine spermatozoa (Johnson et al. 1970).

FACTORS THAT REGULATE PCC INDUCTION

The rate and frequency of PCC induction are dependent on the ratio of mitotic to interphase nuclei in the fused cell. The concentration of the mitotic factors (or the inducer molecules present in the mitotic cells) determines the rate of PCC induction. For example, when three mitotic cells are fused with one

G_1 cell, PCC is induced in all the heterokaryons within 15 minutes. But, when the ratio is reversed (1 mitotic:3 interphases), PCC induction occurs in only 40% of the heterokaryons (Johnson and Rao 1970). If the interphase to mitotic ratio is greater, the reverse of PCC phenomenon occasionally occurs, i.e., under the dominant influence of the interphase factors, membranes form around each of the mitotic chromosomes similar to the formation of micronuclei. This phenomenon has been termed "telophasing" (Ikeuchi et al. 1971, Obara et al. 1974a,b). Therefore, the induction of either PCC or telophasing depends on the balance between mitotic and interphase factors. The dosage effect of the mitotic factors on the incidence of PCC has been well established (Johnson and Rao 1970).

The addition of mitotic extracts to the fusion mixture of mitotic and interphase cells has increased the frequency of PCC induction (Figure 2). The mitotic extract in combination with $MgCl_2$ (2 mM) produced an even more significant increase in the incidence of PCC than either of them added individually.

The effects of various positively and negatively charged compounds on the PCC phenomenon have been examined (Rao and Johnson 1971). Among the different positively charged compounds tested, spermine, putrescine, and Mg^{++} were specific in promoting PCC induction, while spermidine was unique in inhibiting this event. All the negatively charged compounds, including estradiol-17β, Na_2HPO_4, EDTA, cyclic AMP, and heparin, were uniformly inhibitory (Rao and Johnson 1971). Although the addition of spermine, putrescine, or $MgCl_2$ (all at a concentration of 2×10^{-3} M) to the mitotic-interphase fusion mixture increases the incidence of PCC, these substances by themselves or in combination with mitotic extracts do not induce PCC in interphase cells or isolated nuclei.

MIGRATION OF MITOTIC PROTEINS TO PCC

When mitotic HeLa cells that were prelabeled with ^{3}H-amino acids prior to their entry into mitosis were fused with unlabeled interphase cells, label was

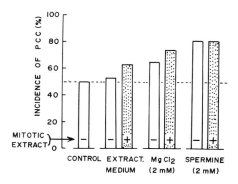

FIG. 2. Effect of mitotic extract on the induction of PCC in the mitotic-interphase fused cells. The presence or absence of the mitotic extract in the fusion mixture is indicated by + or −. The compound listed below each pair of columns indicates its presence in the fusion mixture. "Extract.Medium" = Medium used for the homogenization of mitotic cells. (Reprinted from Rao and Johnson 1971, with permission of J. Cell. Physiol.)

TABLE 1. *Migration of protein labeled with specific amino acids during S*M/I fusion*

	[³H]Arginine	[³H]Lysine	[³H]Tryptophan
Number of grains on mitotic chromosomes	19.6	47.1	25.2
Number of grains on PCC	10.0	23.0	21.5
Ratio grain count (PCC/M)	0.5	0.5	1.0

(Reprinted from Rao and Johnson 1974, with permission of Cold Spring Harbor Laboratory.)

found not only on the mitotic chromosomes but also on the PCC of the interphase cells (Rao and Johnson 1974). This observation indicates the migration and association of the labeled mitotic proteins with the PCC (Table 1). The ratio of grains on mitotic chromosomes and PCC was 1:1 when the mitotic cells were prelabeled (during G_2 phase) with tryptophan, whereas it was 2:1 if either arginine or lysine was used for labeling (Table 2). This indicates that the tryptophan-rich proteins, which are likely to be nonhistone proteins, became associated with PCC more extensively than the arginine- or lysine-rich proteins.

INDUCTION OF GERMINAL VESICLE BREAKDOWN AND CHROMOSOME CONDENSATION IN AMPHIBIAN OOCYTES BY THE MITOTIC FACTORS FROM HeLa CELLS

All attempts to induce PCC, by incubating untreated or permeabilized interphase cells or isolated nuclei with mitotic extracts, were of no avail. In our search for a suitable system to test the activity of the mitotic factors, we found the amphibian oocytes ideal. In amphibian oocytes, meiotic maturation, which could be induced by in vitro incubation with progesterone, involves breakdown of the nuclear envelope (germinal vesicle breakdown or GVBD), chromosome condensation, and progression through the first meiotic division (Masui 1967, Subtelny et al. 1968, Merriam 1972). Meiotic maturation can also be induced by injecting the immature oocytes with cytoplasmic extracts from mature oocytes

TABLE 2. *Migration of labeled mitotic protein to prematurely condensed chromosomes*

Type of fusion	Number of grains/karyotype Mitotic chromosomes		PCC
	Cells not fusing	Cells fusing	
G_1*M/I	25.0	24.5	21.2
S*M/I	67.4	44.1	34.3
G_2*M/I	69.8	70.0	61.0

(Reprinted from Rao and Johnson 1974, with permission of Cold Spring Harbor Laboratory.)

or amphibian embryos in cleavage stages (Masui and Markert 1971, Drury and Schorderet-Slatkine 1975, Wasserman and Masui 1976, Wasserman and Smith 1978). The activity of the cytoplasmic extracts, which fluctuates during the division cycle of the embryonic cells, reaches a peak during mitosis (Wasserman and Smith 1978). Also, the experiments of Gurdon (1968), who observed premature chromosome condensation following the injection of somatic cell nuclei into maturing *Xenopus* oocytes, suggest that the factors involved in chromosome condensation during meiosis may be similar to those involved in mitosis. Since meiotic maturation resembles PCC induction, particularly with regard to the GVBD and chromosome condensation, we decided to test whether mitotic factors from mammalian cells could induce maturation in *Xenopus* oocytes. If they could, then this system could be useful as a biological assay for isolation and characterization of mitotic factors from mammalian cells. The procedures for obtaining extracts from mitotic HeLa cells and the preparation of oocytes from *Xenopus* females were described earlier (Sunkara et al. 1979a).

The results of these studies indicate that the extracts from mitotic HeLa cells were very effective in inducing GVBD and chromosome condensation in *Xenopus* oocytes (Figure 3). No such activity was found in the extracts of cells synchronized in either G_1 or S. However, GVBD-inducing activity was present at minimal levels in early G_2 cells, but the activity increased rapidly as the G_2 cells progressed toward mitosis (Figure 4). Extracts from HeLa cells arrested irreversibly in G_2 by cis-acid treatment had no activity at all (Table 3).

These observations indicate that the mitotic factors from mammalian cells can induce GVBD and chromosome condensation in amphibian oocytes, and these factors have no species-specificity. Furthermore, the mitotic extracts appear to be highly efficient in inducing meiotic maturation, since a protein equivalent of 1,000 mitotic HeLa cells can cause GVBD and chromosome condensation in an oocyte that is about 500,000 times greater in volume than a HeLa cell. In other words, the mitotic factors are effective even at a dilution of 1:500 in the oocyte. This study also revealed that active mitotic factors are present in G_2 cells but at a relatively lower concentration than mitotic cells.

CHARACTERIZATION OF MITOTIC FACTORS

Since there is an excellent correlation between the PCC-inducing ability of the mitotic HeLa cells and their capacity to induce GVBD and chromosome condensation in *Xenopus* oocytes, we used the oocyte system to characterize the factors involved in the condensation of chromsomes. The experimental methods for these studies have been described elsewhere (Sunkara et al. 1979b).

Nature of Mitotic Factors

Incubation of freshly prepared extracts from mitotic HeLa cells (7 mg/ml) with RNase (1.5 units/ml) at 25°C for 1 hour had no effect on the maturation-

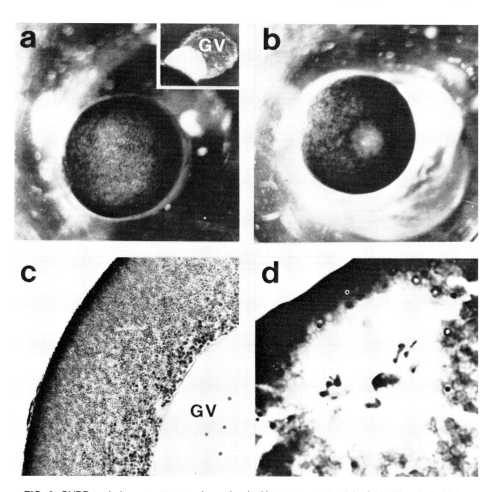

FIG. 3. GVBD and chromosome condensation in *Xenopus* oocytes injected with mitotic HeLa cell extracts. (a) Oocyte animal hemisphere 3 hours after injection with 65 nl of extraction medium. (X 26.) (Inset) Clear germinal vesicle (GV) dissected from buffer-injected living oocyte. (b) Appearance of an oocyte animal hemisphere at 3 hours after injection with mitotic HeLa cell extract (228 ng of protein in 65 nl). Note the bright spot indicating the depigmented area caused by GVBD. (c) Histological section of an oocyte 4 hours after injection with S phase HeLa cell extract (406 ng of protein in 65 nl). (Stained with Feulgen/fast green; X 236.) Note intact germinal vesicle. (d) Condensed chromosomes on meiotic spindle near oocyte surface 1.5 hours after injection with HeLa mitotic cell extract (309 ng of protein in 65 nl). (Stained with Feulgen/fast green; X 1,750.) (From Sunkara *et al.* 1979a.)

promoting activity (MPA). But a similar treatment with protease (subtilopeptidase-A from *Bacillus subtilis*) (0.2 units/ml) reduced the activity of the mitotic extract to 18% of the control (Table 4).

These studies also revealed that inhibition of protein synthesis in the oocytes

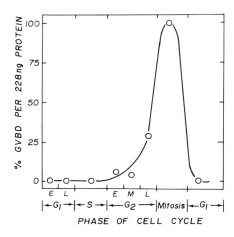

FIG. 4. Meiotic maturation-promoting activity of cell extracts during HeLa cell cycle. The data for this graph are derived from Table 3. Because 228 ng of mitotic protein induced GVBD in 100% of the cases, the percent activity for other phases of the cell cycle was normalized to that amount of protein. E, early; M, mid-; and L, late. (From Sunkara *et al.* 1979a.)

by incubation with cycloheximide for 1 hour before and 2 hours after injection had no effect on the MPA of the extracts (Table 5). These data indicate that no new protein synthesis is required for the induction of meiotic maturation by the mitotic extracts. The MPA of the mitotic factors appears to be highly sensitive to Ca^{++} (1 mM) but unaffected by Mg^{++} even at relatively high concentrations (10 mM) (Table 6). The MPA of the mitotic extracts was unaffected by incubation of the extract at $40°C$ for 10 minutes. However, a 10-minute

TABLE 3. *Maturation-promoting activity of cytoplasmic extracts of HeLa cells from different phases of the cell cycle*

Cell cycle phase	Protein injected* (ng)	Oocytes injected	Oocytes showing GVBD	Induction of GVBD (%)
Early G_1	325	23	0	0
Late G_1	293	19	0	0
S	455	22	0	0
Early G_2	351	24	2	8.3
Mid-G_2	325	29	2	6.9
Late G_2	293	22	8	36.1
G_2 arrested	358	22	0	0
Mitosis	455	15	15	100
Mitosis	228	26	26	100
Mitosis	114	16	3	18.7
Mitosis	57	16	0	0
Extracts from progesterone- stimulated oocytes	1300	15	15	100
Extraction medium	—	20	0	0

* A total volume of 65 nl of extracts was injected into each oocyte.
(From Sunkara et al. 1979a.)

TABLE 4. *Effect of RNase and protease treatments on the maturation-promoting activity of mitotic cell extracts**

Substance injected	Number of oocytes injected	Number showing GVBD	% GVBD
Mitotic extracts	11	11	100
RNase (1.5 units/ml)	11	0	0
Mitotic extracts treated with RNase (1.5 units/ml)	5	5	100
Protease (0.2 units/ml)	11	0	0
Mitotic extracts treated with protease (0.2 units/ml)	11	2	18

* Freshly prepared mitotic extracts (7 mg of protein/ml) were incubated with either RNase or protease at 25°C for 1 hour. At the end of incubation, 65 nl of the treated extracts were injected into each oocyte.
(Reprinted from Sunkara et al. 1979b, with permission of Alan R. Liss, Inc.)

TABLE 5. *Maturation-promoting activity of mitotic cell extracts in cycloheximide-treated oocytes**

Treatment	Number of oocytes	Number showing GVBD	% GVBD
Injected with mitotic extract	11	11	100
Incubated with cycloheximide (20 μg/ml) alone	8	0	0
Mitotic extracts injected into cycloheximide-treated oocytes	12	12	100

* Oocytes were incubated with cycloheximide (20 μg/ml) for 1 hour prior to injection. After injection of the mitotic extract, oocytes were further incubated with cycloheximide for another 1.75 hours, at which time they were scored for GVBD.
(Reprinted from Sunkara et al. 1979b, with permission of Alan R. Liss, Inc.)

TABLE 6. *Effect of Ca^{2+} and Mg^{2+} on maturation-promoting activity of mitotic cell extracts**

Oocytes injected with	Number of oocytes injected	Number of oocytes showing GVBD	% GVBD
Ca^{2+} and Mg^{2+}—free buffer (CMF buffer)	7	0	0
Mitotic extract in CMF buffer	9	8	89
Mitotic extract			
in CMF buffer +1 mM Mg^{2+}	8	7	88
in CMF buffer +2.5 mM Mg^{2+}	8	8	100
in CMF buffer +10 mM Mg^{2+}	13	13	100
in CMF buffer +1 mM Ca^{2+}	10	0	0

* A total volume of 65 nl of extracts made from 20×10^6 cells in 1 ml (7 mg of protein/ml) was injected into each oocyte.
(Reprinted from Sunkara et al. 1979b, with permission of Alan R. Liss, Inc.)

TABLE 7. *Effect of temperature on the maturation-promoting activity of mitotic cell extracts**

Temperature	Number of oocytes injected	Number of oocytes showing GVBD	% GVBD
0°C	10	9	90
40°C	12	10	83
50°C	12	1	8.3
60°C	6	0	0
100°C	6	0	0

* Freshly prepared mitotic extracts were incubated at different temperatures for 10 minutes. At the end of incubation, extracts were centrifuged at 30,000 × g at 4°C to remove the precipitate. A total volume of 65 nl of the supernatant was injected into each oocyte.
(Reprinted from Sunkara et al. 1979b, with permission of Alan R. Liss, Inc.)

incubation at 50°C or higher resulted in a rapid loss of the activity, as indicated by the decrease in the frequency of GVBD (Table 7).

Dialysis of the mitotic extracts against extraction medium in the pH range of 6.5 to 7.5 had no significant effect on its maturation-promoting activity. However, dialysis against medium with a pH of 8.0 resulted in a rapid loss of the activity (Table 8).

The fractionation of mitotic extracts by linear sucrose gradient centrifugation revealed the activity in a single peak (Figure 5). This fraction has an estimated sedimentation value of 4–5S.

In summary, the factors from mitotic HeLa cells that can induce GVBD and chromosome condensation in *Xenopus* oocytes are heat labile, nondialyzable, and Ca^{++}-sensitive proteins with a molecular size of about 4–5S. The characteristics of the factors isolated from cytoplasms of mature amphibian oocytes appear to be similar to those of HeLa mitotic factors, since they are also Ca^{++}-sensitive and heat-labile proteins. However, there are some differences between the mitotic factors and the oocyte factors with regard to their molecular size and the tempera-

TABLE 8. *Effect of pH of the extraction medium on the maturation-promoting activity of mitotic cell extracts**

pH of dialysis buffer	Number of oocytes injected	Number showing GVBD	% GVBD
Undialyzed	10	10	100
6.5	10	10	100
7.0	10	10	100
7.5	10	10	100
8.0	10	1	10

* Mitotic extracts were dialyzed overnight against buffers of different pH, and 65 nl of the dialyzed extracts were injected into each oocyte. The oocytes were examined for GVBD at 1.75 hours after the injection.
(Reprinted from Sunkara et al. 1979b, with permission of Alan R. Liss, Inc.)

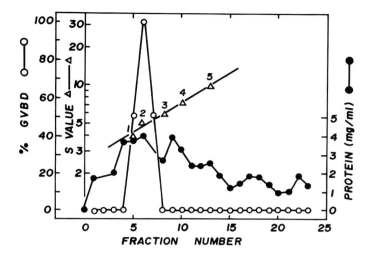

FIG. 5. Separation of mitotic factors on sucrose gradients. Mitotic extract (0.4 ml; 15 mg/ml) was layered on 5-ml linear sucrose gradient (5%–20%)and spun at 85,000 × g for 17 hours in a Beckman SW50-1 rotor. The fractions were collected from the top by a fraction collector. Each of these fractions was assayed for maturation-promoting activity and for the presence of (1) mannose phosphate isomerase (3.9S), (2) malate dehydrogenase (5S), (3) 6-phosphoglu-conate dehydrogenase (6S), (4) lactate dehydrogenase (7.3S), and (5) malic enzyme (10S) as internal standards. (Reprinted from Sunkara *et al.* 1979b, with permission of Alan R. Liss, Inc.)

tures at which they become inactivated. The oocyte factors seem to consist of three different molecular forms, i.e., 4S, 15S, and 32S (Wasserman and Masui 1976) in contrast to a single form (4S–5S) of the mitotic factors (Sunkara et al. 1979b). As expected, the amphibian oocyte factors are heat labile at a much lower temperature (within 15 minutes at 37°C and 4 hours at 25°C) than the mammalian mitotic factors (Wasserman and Masui 1976, Sunkara et al. 1979b).

These data provide convincing evidence that the factors responsible for chromosome condensation during mitosis, meiosis, and for premature chromosome condensation in the cells of a wide range of animal species (probably even plants) are surprisingly similar, if not identical.

ACKNOWLEDGMENTS

This investigation was supported in part by grants CA-11520, CA-14528, CA-23878, and CA-05831 from the National Center Institute, Department of Health, Education and Welfare.

REFERENCES

Drury, K. C., and S. Schorderet-Slatkine. 1975. Effects of cycloheximide on the "autocatalytic" nature of the maturation promoting factor (MPF) in oocytes of *Xenopus laevis.* Cell 4:269–274.

Gurdon, J. 1968. Changes in somatic cell nuclei inserted into growing and maturing amphibian oocytes. J. Embryol. Exp. Morphol. 20:401–414.

Ikeuchi, T., M. Sanbe, H. Weinfeld, and A. A. Sandberg. 1971. Induction of nuclear envelopes around metaphase chromosomes after fusion with interphase cells. J. Cell Biol. 51:104–115.

Johnson, R. T., and P. N. Rao. 1970. Mammalian cell fusion: Induction of premature chromosome condensation in interphase nuclei. Nature 226:717–722.

Johnson, R. T., P. N. Rao, and S. D. Hughes. 1970. Mammalian cell fusion. III. A HeLa cell inducer of premature chromosome condensation active in cells from a variety of animal species. J. Cell. Physiol. 76:151–158.

Kato, H. and A. A. Sandberg. 1968a. Chromosome pulverization in human cells with micronuclei. J. Natl. Cancer Inst. 40:165–179.

Kato, H., and A. A. Sandberg. 1968b. Chromosome pulverization in Chinese hamster cells induced by Sendai virus. J. Natl. Cancer Inst. 41:1117–1123.

Kato, H., and A. A. Sandberg. 1968c. Cellular phase of chromosome pulverization induced by Sendai virus. J. Natl. Cancer Inst. 41:1125–1131.

Masui, Y. 1967. Relative roles of the pituitary, follicle cells and progesterone in the induction of oocyte maturation. J. Exp. Zool. 166:365–376.

Masui, Y., and C. L. Markert. 1971. Cytoplasmic control of nuclear behavior during meiotic maturation of frog oocytes. J. Exp. Zool. 177:129–146.

Matsui, S., H. Yoshida, H. Weinfeld, and A. A. Sandberg. 1972. Induction of prophase in interphase nuclei by fusion with metaphase cells. J. Cell Biol. 54:120–132.

Merriam, R. W. 1972. On the mechanism of action in gonadotropic stimulation of oocyte maturation in Xenopus laevis. J. Exp. Zool. 180:421–426.

Obara, Y., L. S. Chai, H. Weinfeld, and A. A. Sandberg. 1974a. Synchronization of events in fused interphase-metaphase binucleate cells: Progression of the telophase-like nucleus. J. Natl. Cancer Inst. 53:247–259.

Obara, Y., L. S. Chai, H. Weinfeld, and A. A. Sandberg. 1974b. Prophasing of interphase nuclei and induction of nuclear envelopes around metaphase chromosomes in HeLa and Chinese hamster, homo- and heterokaryons. J. Cell Biol. 62:104–113.

Rao, P. N., and R. T. Johnson. 1971. Mammalian cell fusion. IV. Regulation of chromosome formation from interphase nuclei by various chemical compounds. J. Cell. Physiol. 78:217–224.

Rao, P. N., and R. T. Johnson. 1974. Regulation of cell cycle in hybrid cells, in Control of Proliferation in Animal Cells. Cold Spring Harbor Laboratory, Cold Spring Harbor, New York, pp. 785–800.

Rao, P. N., B. A. Wilson, and T. T. Puck. 1977. Premature chromosome condensation for cell cycle analysis. J. Cell. Physiol. 91:131–142.

Sandberg, A. A., T. Aya, I. Ikeuchi, and H. Weinfeld. 1970. Definition and morphologic features of chromosome pulverization: A hypothesis to explain the phenomenon. J. Natl. Cancer Inst. 45:615–623.

Sperling, K., and P. N. Rao. 1974. The phenomenon of premature chromosome condensation: Its relevance to basic and applied research. Humangenetik 23:235–258.

Subtelny, S., L. D. Smith, and R. E. Ecker. 1968. Maturation of ovarian frog eggs without ovulation. J. Exp. Zool. 168:39–48.

Sunkara, P. S., D. A. Wright, and P. N. Rao. 1979a. Mitotic factors from mammalian cells induce germinal vesicle breakdown and chromosome condensation in amphibian oocytes. Proc. Natl. Acad. Sci. USA 76:2799–2802.

Sunkara, P. S., D. A. Wright, and P. N. Rao. 1979b. Mitotic factors from mammalian cells: A preliminary characterization. J. Supramol. Struct. 11:189–195.

Takagi, N., T. Aya, H. Kato, and A. A. Sandberg. 1969. Relation of virus-induced cell fusion and chromosome pulverization to mitotic events. J. Natl. Cancer Inst. 43:335–347.

Wasserman, W. J., and Y. Masui. 1976. A cytoplasmic factor promoting maturation: Its extraction and preliminary characterization. Science 191:1266–1268.

Wasserman, W. J., and L. D. Smith. 1978. A cyclic behavior of a cytoplasmic factor controlling nuclear membrane breakdown. J. Cell Biol. 78:R15–R22.

Genes, Chromosomes, and Neoplasia, edited by
Frances E. Arrighi, Potu N. Rao, and Elton Stubblefield.
Raven Press, New York © 1981.

The Molecular Organization of Mammalian Metaphase Chromosomes

Elton Stubblefield

Department of Cell Biology, The University of Texas System Cancer Center, M. D. Anderson Hospital and Tumor Institute, Houston, Texas 77030

The encoded genetic message is in the primary structure of cellular DNA as a linear sequence of nucleotide pairs specifying protein amino acid sequences and transcription control. This is a well-established concept even though there are many details about genetic control still to be uncovered. The amount of information potentially encoded in mammalian DNA is a staggering three billion nucleotide pairs per haploid genome, although some unknown fraction of this total may have functions other than genetic information storage. The logistic problems involved in managing and replicating the approximately two meters of molecular DNA in every mammalian cell nucleus are formidable, especially so considering that the DNA molecule is helical and must unwind to replicate.

CHROMATIN FIBERS

During recent years, a part of this problem has been illuminated by the discovery of how DNA and histones interact to make chromatin. By coiling the DNA around histone clusters in a regular pattern to generate nucleosomes at closely spaced intervals, like beads on a string, and then coiling the beaded necklace and additional proteins into the thicker fiber characteristic of chromatin, the length of DNA is effectively reduced from 2 m to about 2 cm of chromatin fiber. In the human cell, this is apportioned to 46 chromosomes, giving each chromosome about 0.5 mm of chromatin, more or less, depending on the size of the chromosome. Upon duplication, then, the average human chromosome contains 1 mm of chromatin or the equivalent 10 cm of DNA. The remaining question, as yet not completely resolved, is how this 1000 μm length of chromatin is organized into an average human metaphase chromosome only 4 μm long.

CHROMOSOME CORE FIBERS

At the resolution level of the light microscope, the metaphase chromosome is a double structure consisting of two chromatids attached at the centromere.

FIG. 1. Isolated Chinese hamster chromosomes, unstained and dried by Anderson critical point procedure. The chromosome on the right was swollen in distilled water before drying to loosen the fiber arrangement. 25,000X.

Each chromatid is a rod-like structure with distinct helical organization. By using electron microscopy, we can see these features clearly (Figure 1) and at the same time readily resolve chromatin fibers as the major structural components. The chromatin appears to be arranged as hundreds of loops, somehow anchored internally. Are these loops attached internally to some kind of a core structure? This has been a major point of debate over the last decade, with one extreme denying the existence of any kind of a core structure in mitotic chromosomes (Comings and Okada 1974) and the other extreme claiming that mitotic chromosomes do indeed contain core structures of DNA and protein (Stubblefield and Wray 1971, Stubblefield 1973) or scaffolds made of protein (Paulson and Laemmli 1977).

LAMPBRUSH CHROMOSOMES

Probably the oldest evidence for the existence of a chromosome core comes from studies of meiotic chromosomes. For many years the large chromosomes of amphibian oocytes have been studied extensively (see Gall 1966). Their architecture has been described as similar to a "lampbrush," so they were called lampbrush chromosomes (Rückert 1892). The lampbrush architecture consists of a stiff linear core with many attached loops of material radiating from the core, like the brushes used to clean the soot from the glass chimneys of oil lamps. Oocyte chromosomes from many diverse organisms also have lampbrush chromosomes, but they are not as large or as easily isolated as those in amphibian oocytes. Lampbrush chromosomes are peculiar to meiosis, however, so it is possible that the axial structure clearly visible in such chromosomes is found only in this stage of meiosis and is not present in mitotic chromosomes. Ultrastructure studies of meiotic cells have shown that in most organisms an axial structure, the synaptonemal complex, appears as the homologous chromosomes pair and then vanishes as the paired chromosomes separate in later stages (Moses 1968). This axial complex is quite distinct in structure from the chromatin around it, so it is readily found in sectioned material as shown in Figure 2. It can be uniquely stained with silver (Fletcher 1979), so it is also readily visible with light microscopy.

MITOTIC CHROMOSOME CORES

Ultrastructural studies of mitotic chromosomes have always failed to reveal a distinct axial element like the synaptonemal complex in either sectioned or whole mount preparations. However, it was recently demonstrated by light microscopy that an axial core could be shown in mitotic chromosomes (Figure 3) by the same silver treatment that stains the synaptonemal complex in meiotic chromosomes (Howell and Hsu 1979). Therefore, though a distinctive morphological core is not readily visible in mitotic chromosomes, a core structure of

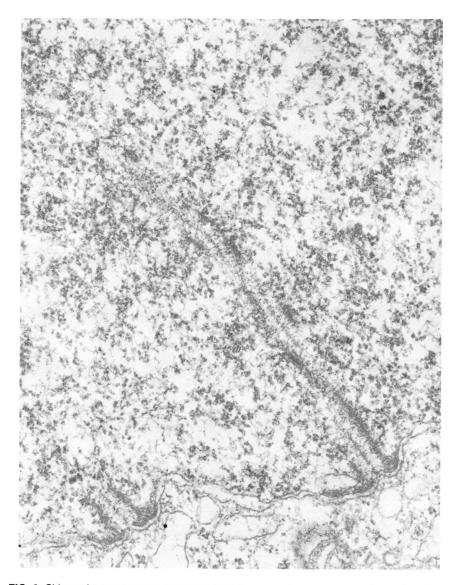

FIG. 2. Chinese hamster spermatocyte nucleus demonstrating synaptonemal complex. These tripartite structures occur at the axis of paired homologous chromosomes in meiosis. Two complexes terminate at the nuclear membrane in this section. The complex on the right is twisted. 50,000X.

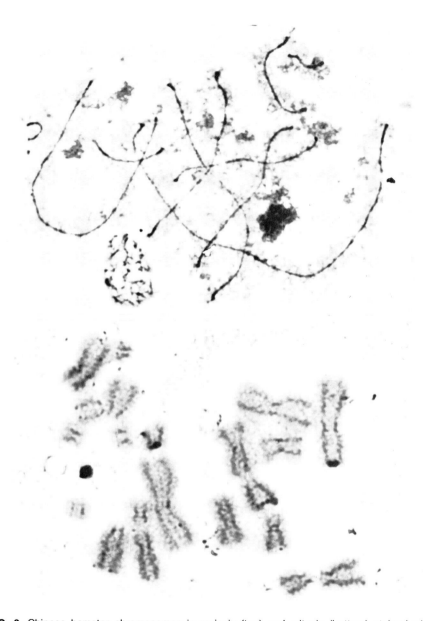

FIG. 3. Chinese hamster chromosomes in meiosis (top) and mitosis (bottom) stained with silver. The synaptonemal complex stains in the meiotic spermatocyte, and an axial core structure stains by the same procedure in the chromatids of a CHO cell in mitosis. Preparations by Dr. Sen Pathak.

similar biochemical composition is nonetheless present, presumably with a morphology very similar to the surrounding chromatin fibers that are attached to it.

The studies of Laemmli and co-workers (Paulson and Laemmli 1977, Marsden and Laemmli 1979) have also confirmed the existence of a chromosome core structure, which they called a "scaffold," and also demonstrated the looped arrangement of the chromatin in mitotic chromosomes. Using isolated human metaphase chromosomes, they first removed the histones and then spread the DNA on a liquid surface by the technique of Kleinschmidt (1968). The DNA fibers were then seen arrayed around a residual core, which retained the overall morphology of the chromosome, as in Figure 4. In one case the DNA could be shown to be organized as closed loops originating and reinserting at adjacent sites on the core. Their studies indicated that the chromatin organization in mitotic chromosomes also resembles a lampbrush configuration. Analysis of the nonhistone proteins left in the scaffold structures revealed a complex assortment of proteins and a small amount of DNA that was resistant to DNase digestion. Since the histones had been removed, no information about the possible presence of histones in the scaffold could be obtained.

Laemmli's scaffold-and-loops model was challenged by Comings and Okada (1979), who found that addition of a chelating agent to the preparation could completely eliminate the scaffold structure. However, such an observation does not prove that the scaffold is an artifact, as Comings and Okada claim, but rather it simply demonstrates that the scaffold is held together by its nonhistone protein complement and can be disrupted by removing divalent ions.

Our chromosome model with a core-and-loops architecture was based on studies of isolated Chinese hamster chromosomes (Stubblefield and Wray 1971). Extraction of isolated chromosomes with 6.0 M urea or 2.0 M NaCl removes most of the DNA and proteins but leaves a core structure resembling the chromosome in shape (Figure 5). A similar core structure also remains following brief exposure to 0.1 N NaOH (Stubblefield 1973). The extraction procedure of Ohnuki (1968) for karyotype preparations on slides also appears to stain the chromosome core as a spiral structure in each chromatid.

In the light of such evidence the opposition to a lampbrush model of chromosome structure has gradually diminished. Most workers acknowledge that some kind of a core structure can be demonstrated in mitotic chromosomes, and clearly some organizational principle is needed to gather the 1,000-μm length of chromatin into a 4-μm long chromosome. It is also widely agreed that chromosome cores contain certain of the major kinds of nonhistone proteins and a small amount of DNA. However, other nonhistone proteins are found in the chromatin loops, so nonhistone proteins are not limited to the core. Core DNA may be only the attached ends of the loops, or it may consist of repetitive DNA species limited to the core fibers, as we initially suggested (Stubblefield 1973).

One form of core fiber that we were able to isolate from metaphase cells

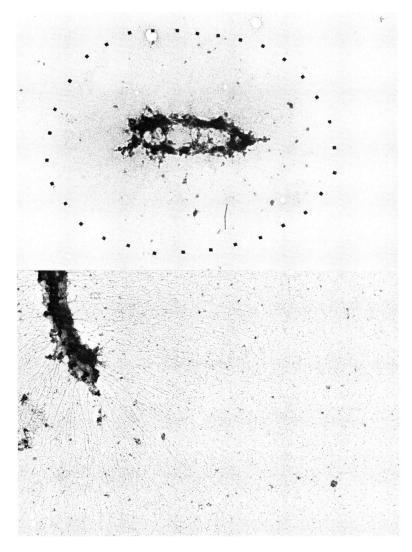

FIG. 4. Dehistonized metaphase chromosomes spread to demonstrate the DNA loops surrounding the scaffold. The upper figure shows an intact core structure of an acrocentric chromosome surrounded by the DNA fibers (out to the dotted line). 8,300X. At higher magnification the lower micrograph shows the DNA loops attached to a core scaffold. 19,000X.

using a sonication technique consisted of thin ribbons 0.4 μm wide (Stubblefield and Wray 1971). These fibers stain brightly with either ethidium bromide (Figure 6) or Hoechst 33258, suggesting that they contain DNA. Our suggestion was that the ribbons are coiled into a tight fiber that would be somewhat stiffer

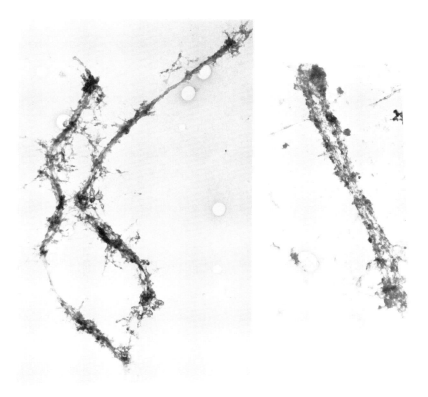

FIG. 5. Chromosome cores produced by extraction of isolated chromosomes by 6.0 M urea (left) or 2.0 M NaCl (right). 10,000X.

than the chromatin loops attached to it. Only one edge is available for epichromatin attachment in such a fiber, since the other edge of the ribbon becomes buried within the coiled structure. Such a coiled filament is an ideal candidate for a chromosome core, inasmuch as it could be expected to accomplish the observed coiling changes seen in mitotic chromosomes as they progress from long, thin prophase chromosomes to short, fat metaphase chromosomes.

EPICHROMATIN LOOPS

An average human chromosome may have about 1,000 loops of chromatin attached to its core structure. Each loop would contain from 30 to 150 μm of DNA length or the equivalent values of 90 to 450 kilobase pairs. This range is similar to the interorigin distances in replicons studied by Huberman and Riggs (1968). It may also be that each loop contains only one or a few structural genes. The pairing mechanism in meiosis seems to operate at this level of struc-

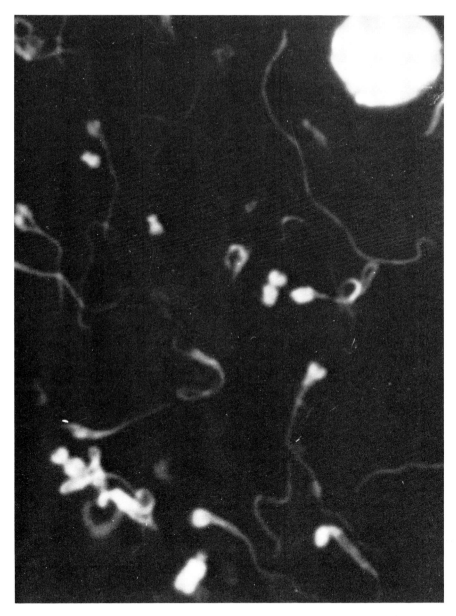

FIG. 6. Chromosome core structures, intact chromosomes, and a nucleus stained by ethidium bromide (25 μg/ml). Although the cores are largely stripped of any epichromatin DNA, they still contain some DNA that fluoresces.

ture, bringing the ends of corresponding homologous loops together in the synaptonemal complex.

CHROMOSOME BANDS

Clusters of epichromatin loops within a chromosome region seem to replicate together and share biochemical properties that create stainable bands on chromosomes (Stubblefield 1975). A variety of staining procedures have demonstrated banding in chromosomes. In several cases it has been reported that G-bands (or Q-bands) correspond to late-replicating segments of chromosomes (Pathak et al. 1973, Sharma and Dhaliwal 1974), but this does not hold for the Chinese hamster (Stubblefield 1975). An average chromosome band may include 100 loops of epichromatin extending for perhaps 0.5 μm along the chromosome. Such a band should contain as much as 10 mm of DNA, perhaps all in one length of double helix corresponding to 30,000 kilobase pairs. The smallest human chromosomes contain only a few such bands. Okada and Comings (1974) have pointed out a similarity of pattern in G-bands of mitotic chromosomes and chromomeres of meiotic chromosomes in the Chinese hamster.

It is of interest to our discussion that the classic case of banded chromosomes, the polytene chromosomes of *Drosophila melanogaster* salivary glands, may also be structurally similar to lampbrush chromosomes (Bencze and Brasch 1980). Upon treatment with alkali-urea, the bands open up into chromatin loops attached to a linear core, and the process is at least partly reversible when the alkali-urea solution is replaced by saline. Thus, although these chromosomes are clearly multistranded, they still appear to have a core-and-loops type of architecture. We think this may be the basic pattern of most eukaryotic chromosomes.

STRANDEDNESS

The existence of half chromatids has been suggested by observations of many workers over a span of 50 years in the cells of many organisms, including human cells (see Egozcue 1973 for a review). What has been usually seen by light microscopy is an apparent doubleness in each chromatid. If it were an optical artifact, it could be readily photographed, even in living cells, by phase-contrast time-lapse cinematography (Bajer 1965). We also found evidence of a half chromatid in our ultrastructure studies of isolated chromosomes (Stubblefield and Wray 1971). Even though considerable visual evidence thus exists for a bineme chromosome, there has been a great reluctance to assimilate such evidence into current concepts about chromosome structure, mainly because it seems to complicate matters unnecessarily for genetics.

If each half chromatid were originally identical in a chromosome, the pair of homologous chromosomes in a diploid cell would contain a minimum of four copies of each gene before chromosome replication. Each copy would presumably be capable of independent mutation. Four copies would provide additional genetic stability for a cell only if a mechanism exists to ensure that the replicated half chromatid does not become a whole chromatid; the two daughters of each half chromatid must segregate to different chromatids (and hence to different cells) each time. Such a segregation mechanism must be part of the replication mechanism and structure of the chromosome. If this segregation rule should be compromised, so that a half chromatid could, upon replication, become both halves of a chromatid, then a mutant gene would occupy both positions on a chromatid, displacing the normal gene and negating the advantage of four gene copies per cell. Failure of this half chromatid segregation mechanism

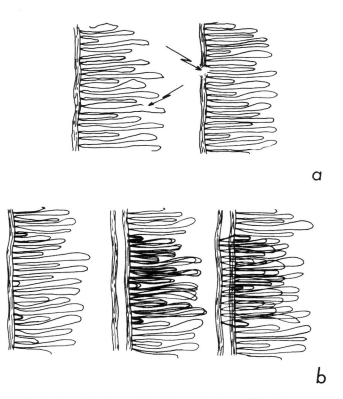

FIG. 7. a. Radiation-induced breakage of epichromatin loop DNA (left) will not produce a visible chromosome break, whereas breakage of the core fiber (right) will effectively break the chromosome. b. Diagram demonstrating epichromatin loop DNA duplication within a stainable band and chromosome core duplication separated in space and time. The new loops must attach to the new core fiber, and mistakes in attachment could easily lead to sister chromatid exchange.

could be one explanation for the apparent increase in the rate of mutation observed in certain aneuploid cell lines compared to normal diploid cells. The genetic effects of binemy in this form would be difficult to detect. These aspects of the chromosome structure problem have already been discussed (Stubblefield 1973).

CHROMOSOME BREAKAGE

A large segment of the cytogenetics research force has been studying the breakage of chromosomes by various chemical agents and radiation. Chromosome breakage, translocations, rearrangements, and sister chromatid exchanges have been the usual cytologic end points for such studies. The basic premises of this endeavor may require revision if the core-and-loops chromosome model is correct. DNA may be broken (in the loops) without any visible effect on the chromosome (Figure 7a). To create breaks or rearrangements of the chromosome would clearly involve manipulation of the core structure. Sister chromatid exchange may be more a phenomenon of loop-to-core attachment rather than a direct breakage-reunion problem (Figure 7b). As we begin to understand the

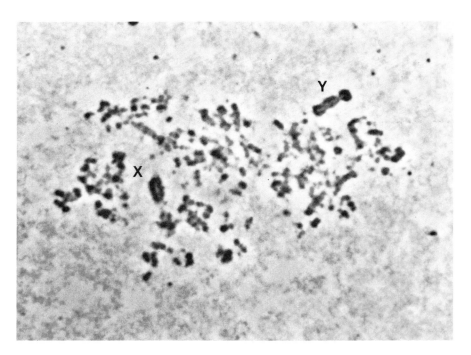

FIG. 8. Fragmented chromosomes in a male Chinese hamster metaphase cell. Regions corresponding to the Y chromosome and long arms of the X chromosome, which are late replicating, remain intact. Such configurations may result from failure in core structure duplication.

roles of the various proteins and DNA in the core structure of chromosomes, we should benefit chromosome breakage studies very much by providing a better structural model.

Finally, the phenomenon of chromosome fragmentation or pulverization (Kato and Sandberg 1966) becomes readily understandable if a chromosome core exists. Such cells occasionally appear in mitosis when the previous S phase was interrupted by drug treatment (Figure 8). The premature condensation of S phase chromatin, by fusion to a mitotic cell, also results in such a fragmented configuration, and this suggests that in such cases the chromosome core is missing or broken. Since G_1 and G_2 interphase chromatid can be condensed into visible chromosomes by fusion to mitotic cells (Johnson and Rao 1970), we suspect that chromosome cores are present or can be rapidly generated at any time in the cell cycle except the S phase. It is quite likely that the core is duplicated in S phase along with the rest of the chromosome.

ACKNOWLEDGMENTS

I wish to thank Dr. Sen Pathak of our Department of Cell Biology for providing the micrographs in Figure 3. For the free use of their electron microscope facilities, I also thank Dr. Donald Robberson and Dr. Bill R. Brinkley. This work was supported in part by grant PCM 79-05428 from the National Science Foundation.

REFERENCES

Bajer, A. 1965. Subchromatid structure of chromosomes in the living state. Chromosoma 17:291–302.

Bencze, J. L., and K. Brasch. 1980. The morphology of normal and denatured polytene chromosomes from *Drosophila melanogaster*. Cytobios 25:93–104.

Comings, D. E., and T. A. Okada. 1974. Some aspects of chromosome structure in eukaryotes. Cold Spring Harbor Symp. Quant. Biol. 38:145–153.

Comings, D. E., and T. A. Okada. 1979. Chromosome scaffolding structure—real or artifact? (Abstract) J. Cell Biol. 83:150a.

Egozcue, J. 1973. Banding patterns of uncoiled mammalian chromosomes. Nucleus 16:53–56.

Fletcher, J. M. 1979. Light microscope analysis of meiotic prophase chromosomes by silver staining. Chromosoma 72:241–248.

Gall, J. G. 1966. Techniques for the study of lampbrush chromosomes. Methods Cell Physiol. 2:37–60.

Howell, W. M., and T. C. Hsu. 1979. Chromosome core structure revealed by silver staining. Chromosoma 73:61–66.

Huberman, J. A., and A. D. Riggs. 1968. On the mechanism of DNA replication in mammalian chromosomes. J. Mol. Biol. 32:327–341.

Johnson, R. T., and P. N. Rao. 1970. Mammalian cell fusion: Induction of premature chromosome condensation in interphase nuclei. Nature 226:717–722.

Kato, H., and A. A. Sandberg. 1967. Chromosome pulverization in human binucleate cells following Colcemid treatment. J. Cell Biol. 34:35–45.

Kleinschmidt, A. K. 1968. Monolayer techniques in electron microscopy of nucleic acid molecules. Methods Enzymol. 12:361–377.

Marsden, M. P. F., and U. K. Laemmli. 1979. Metaphase chromosome structure: Evidence for a radial loop model. Cell 17:849–858.

Moses, M. J. 1968. Synaptinemal complex. Ann. Rev. Genet. 2:363–412.

Ohnuki, Y. 1968. Structure of chromosomes. I. Morphological studies of the spiral structure of human mitotic chromosomes. Chromosoma 25:402–428.

Okada, T. A., and D. E. Comings. 1974. Mechanisms of chromosome banding. III. Similarity between G-bands of mitotic chromosomes and chromomeres of meiotic chromosomes. Chromosoma 48:65–74.

Pathak, S., T. C. Hsu, and T. Utakoji. 1973. Relationships between patterns of chromosome banding and DNA synthetic sequences: A study on the chromosomes of the Seba's fruit bat, *Carollia perspicillata.* Cytogenet. Cell Genet. 12:157–164.

Paulson, J. R., and U. K. Laemmli. 1977. The structure of histone-depleted metaphase chromosomes. Cell 12:817–828.

Rückert, J. 1892. Zur Entwickelungsgeschichte des Ovarialeies bei Selachiern. Anat. Anz. 7:107–158.

Sharma, T., and M. K. Dhaliwal. 1974. Relationship between patterns of late S DNA synthesis and C- and G-banding in muntjac chromosomes. Exp. Cell Res. 87:394–397.

Stubblefield, E. 1973. The structure of mammalian chromosomes. Int. Rev. Cytol. 35:1–60.

Stubblefield, E. 1975. Analysis of the replication pattern of Chinese hamster chromosomes using 5-bromodeoxyuridine suppression of 33258 Hoechst fluorescence. Chromosoma 53:209–221.

Stubblefield, E., and W. Wray. 1971. Architecture of the Chinese hamster metaphase chromosome. Chromosoma 32:262–294.

Genes, Chromosomes, and Neoplasia, edited by
Frances E. Arrighi, Potu N. Rao, and Elton Stubblefield.
Raven Press, New York © 1981.

Analysis of Complementarity between snRNAs and Splice Junctions from Ovalbumin and Ovomucoid Genes

Albert C. Ting, Ming-Jer Tsai, and Bert W. O'Malley

Department of Cell Biology, Baylor College of Medicine, Houston, Texas 77030

Many structural genes of eukaryotic cells and viruses that code for polyadenylated messenger RNAs have been found to be interrupted by introns, or intervening sequences, which appear to be transcribed but are not found in the mature messenger RNAs (Berget et al. 1977, Breathnach et al. 1977, Chow et al. 1977, Jeffreys and Flavell 1977, Dugaiczyk et al. 1978, Tilghman et al. 1977; for review see Abelson 1979). In the chick oviduct system, large nuclear precursors of the ovalbumin and the ovomucoid genes have been found to include some, if not all, of the introns (Roop et al. 1978, Nordstrom et al. 1979). To form the mature RNA, the introns must be removed by a process called splicing. In the yeast tRNA system, the splicing reaction has been shown to consist of two steps, the endonucleolytic cleavage step and the ligation step (Peebles et al. 1979, Knapp et al. 1979). However, the recognition of intron/exon junctions remains an enigmatic problem.

The first question to be addressed in approaching this problem is whether the primary sequence contains sufficient specificity to determine the splice site. When the junction regions of introns from various genes were examined, limited sequence homology between different junction regions was used to generate consensus sequences (Catterall et al. 1978, Seif et al. 1979). However, such limited homology would require an improbably large number of enzymes if the enzyme recognition site were to depend solely on primary sequence. Unique secondary structures also were not readily apparent around splice sites (Breathnach et al. 1978, Catterall et al. 1978, Konkel et al. 1978).

The presence of adenine- and uridine-rich sequences has raised the possibility of an alignment scheme in which the polyA tail orients the splice junctions by forming a triple-stranded structure (Bina et al. 1980). Another attractive proposal involves using small nuclear RNA (snRNA) to align splice junctions. In the adenovirus-2 (Ad2) system, Murray and Holliday (1979) have found significant homology between the virus-associated RNA (VA-RNA) and the intron and exon junctions of a late Ad2-mRNA. This has led them to propose three models using segments of VA-RNA as bridging sequences to bring and hold splicing sites in juxtaposition (Murray and Holliday 1979). Another source of bridging sequences has been proposed—namely the snRNAs (Prestayko et al. 1970, Lerner et al. 1980, Rogers and Wall 1980). Several of these snRNAs (U1a, U2, U3b, U6, 4.5S) isolated from Novikoff hepatoma cell nuclei have been sequenced (Ro-Choi et al. 1972, Shibata et al. 1975, Ro-Choi and Henning 1977, Reddy et al. 1979). U1a, U2, U6, and 4.5S RNAs are all known to be localized in the extranucleolar region of the nucleus (Weintraub and Penman 1968, Ro-Choi and Busch 1974). U1a and U2 among others have been found within snRNPs precipitated with anti-RNP antibodies from patients with lupus erythematosus (Lerner et al. 1980). Therefore, these snRNAs are all possible candidates to participate in the splicing mechanism. Furthermore, from examining the primary sequence of a major snRNA (U1a), considerable sequence homology was found between the 20 bases at the 5' end of U1a and the consensus sequences of a number of splice junctions (Lerner et al. 1980, Rogers and Wall 1980). Thus, an attractive proposal would consist of U1a-RNA acting to bridge donor and acceptor splicing sites together in the framework of hnRNPs. Then, the splicing and ligating activities would act to splice out the introns and join the two structural sequences. In this report, we have addressed this proposal by performing sequence analysis and sequence complementarity tests between the established sequence of certain snRNAs and all splicing junctions of ovalbumin and ovomucoid genes.

METHODS

To analyze for complementarity, we used a program modified from that designed by Staden (1977). The snRNA sequences we used were U1a (Ro-Choi and Henning 1977), U2 (Shibata et al. 1975), U3b (Reddy et al. 1979), 4.5S (Ro-Choi et al. 1972), and U6 (4.7S RNAs) (Epstein et al., submitted for publication). Presently, the only snRNA sequences available to us are from rat hepatoma cell nuclei. However, fingerprints of HeLa, rat, and mouse U1-RNAs indicate that their respective RNA sequences are identical (Lerner et al. 1980). Recently, ovalbumin gene has been completely sequenced (Catterall et al. 1978, manuscript in preparation). The ovomucoid gene has been partially sequenced (Catterall et al., manuscript in preparation), and all the splicing junctions are available.

All sequences are numbered from the 5' end. Position 1 of the ovalbumin gene is the A nucleotide, 134 bases upstream from the 5' end of the first exon. Position 1 of the ovomucoid is the T nucleotide, 126 bases upstream from the 5' end of the first exon. Position 1 of U1a, U2, U3b, and U6 RNA is the nucleotide immediately following the cap structure. Position 1 of 4.5 RNA is the first G nucleotide of the 4.5S-RNA sequence.

As a first approximation of base-pairing stability, sequences were initially compared scoring only for Watson-Crick base pairs. Then, the comparison was made, scoring for GU base pairs as well. Since energies of stacking were found to be important in determining stability of base pairing (Tinoco et al. 1973), we have made the energy calculations for base pairing between U1a and the splice junctions of ovalbumin and ovomucoid according to the energies summarized by Salser (1977).

For all complementarity determinations, a search string of 20 bases or 16 bases was compared with every possible position on the searched sequence. When intron junctions were compared with a given sequence, a search string consisting of eight bases from the 5' end and 12 bases from the 3' end of each intron was used (Figure 1). The first eight bases from the 5' end were used because the position on the U1a-RNA complementary to the splice site is exactly eight bases away from the 5' end of U1a (Lerner et al. 1980, Rogers and Wall 1980). Complementarity of U1a-RNA to the consensus sequence remains quite good 20 bases from the 5' end of the U1a-RNA; therefore, 12 bases from the 3' region of the intron were added to the eight bases from the 5' region to form a convenient 20-base search string (see Figure 1). The exon search string consisted of eight bases from both the 5' and the 3' ends of the exon junction (Figure 2).

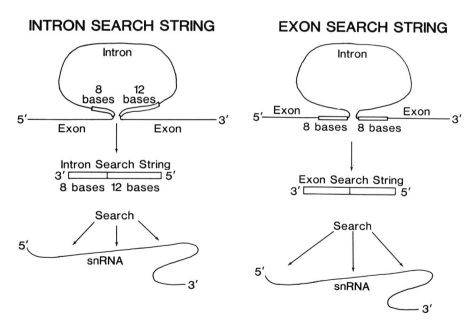

FIG. 1, left. To generate the intron search string, eight bases from the 5' end and 12 bases from the 3' end of the intron were joined at the putative splice sites. The string was oriented in 3' to 5' direction with the eight-base segment at the 3' end. This string was used to search for sequence complementarity with each position of the snRNA.

FIG. 2, right. To generate the exon search string, eight bases at the 3' end of one exon were joined to eight bases at the 5' end of the next exon. The string was oriented in 3' to 5' direction and used to search for sequence complementarity with each position of the snRNA.

As a control, 16- and 20-nucleotide sequences were randomly extracted from the ovalbumin sequence to be used as search strings. A random number generator was used to find seven random positions along the ovalbumin gene. Depending on whether the control was for introns (search string is 20 bases long) or exons (16 bases long), 20- or 16-base sequences at those positions were extracted to be used as search strings. These search strings were then used to search through all positions of U1a-RNA. Searches were made both excluding and including GU base pairs.

RESULTS

Search of U1a-RNA with Intron Junctions of Ovalbumin and Ovomucoid

Intron search strings from ovalbumin and ovomucoid RNAs were used to search through U1a-RNA. The results are shown in Table 1. Complementarity is measured in percent of bases in the search string that can form Watson-Crick base pairs with the specified regions of U1a-RNA. For example, 11 bases of intron A junction string can base pair with the 20-base sequence of U1a beginning at position 1. From this table, it is immediately apparent that the

TABLE 1. *Complementarity of intron junctions with various positions of U1a-RNA*

	Complementarity					
	70% (14/20)	65% (13/20)	60% (12/20)	55% (11/20)	50% (10/20)	45% (9/20)
Ovalbumin introns						
A				1	6	5,16,126
B			1			97
C						17,58,73,93
D			1		98,118	17,97,107,130
E				1		6,57,85,127
F			16		58,107	1,106
G				1	2	16,58,73,87,103
Ovomucoid introns						
A		1		5,41,102,103		6,85,101
B			41		85	19,22,38,70,104,118
C				1,100	16	20,58
D					1,127	20,36,41,97,118
E		1			22	2,17,107,113,114
F				1	41,97,118	87,127
G	1				16	100,130

Intron junctional search strings from ovalbumin and ovomucoid genes were scored for percent of sequence complementarity with every position in U1a-RNA. Only Watson-Crick base pairs were scored. The positions of U1a-RNA having sequences with greater than 40% complementarity to the specified introns are tabulated. The actual fractions of the 20-base sequence forming complementary base pairs are given in parentheses.

sequence at position 1 of U1a-RNA has a high degree of complementarity with most of the intron junctions. This complementarity is most easily seen with the ovalbumin introns but can also be seen consistently with the ovomucoid introns.

The data can be viewed in another form. For the sequence at each position of U1a-RNA, the base complementarity (number of base pairs formed) with each of the introns is summed. The sums are then converted to percent of complementarity cumulative over the seven introns by the following equation.

$$\% = \frac{\text{Sum of Base Complementarity}}{140 \ (\text{Maximum Base Complementarity})} \times 100$$

where Maximum Base Complementarity = 20 Possible Base Pairings per Intron × 7 Introns

The positions are then ranked with respect to decreasing percent of complementarity. The results are given in Table 2. Scorings for Watson-Crick base pairs excluding (−) and including (+) GU base pairs are both shown. Again, it is evident that only position 1 occurs consistently, ranks the highest, and has significantly higher percentage of complementarity than the second-ranked positions.

To serve as a control, the results of percent complementarity cumulative over seven randomly selected sequences from ovalbumin are also included in Table 2. When the results were compared, two observations could be made. First, position 1 of U1a is not highly favored in the control with respect to percent of complementarity. Second, position 1, when matched against the intron junctions, is seen to have a significantly higher percentage of complementarity than the highest ranked position of the control. Therefore, both the consistently high rank of position 1 and its high complementarity distinguish the 20 bases at position 1 of U1a as a candidate for the intron-bridging sequence.

Introns with Position 1 of U1a-RNA

To further investigate base pairing at position 1 of U1a-RNA, sequences of the intron junctions and the 5′ region of U1a-RNA have been aligned and drawn in the left panel of Figure 3. Complementary base pairs are indicated by vertical lines, where GC pairs are shown with triple lines, AU with double, and GU with single lines. The vertical lines dividing all the sequences are the sites of the putative splice points. With this figure, the extent and the distribution of base pairing can be visually ascertained. It can be seen that the amount of base pairing is extensive for most intron junctions. Also, the base pairing seems to be clustered, for the most part, around the putative splice point.

As shown in Table 1, introns C and F of ovalbumin and intron B of ovomucoid have 45% complementarity or less with position 1. Also, complementarity to

TABLE 2. *Percent complementarity for each position on Ula-RNA cumulative over seven introns*

Rank ⟶

Ovalbumin (−GU)							
% Complementarity	53% (74)*	42.9% (60)	40% (56)	38.6% (54)	37.1% (52)	35.7% (50)	35% (49)
Position	1	16	97	6	58,118	2,73	38,113
Ovalbumin (+GU)							
% Complementarity	66.4% (93)	63.6% (89)	60.7% (85)	57.1% (80)	56.4% (79)	55.7% (78)	55% (77)
Position	1	38	127,133	2	56,131	98	126
Ovomucoid (−GU)							
% Complementarity	54.3% (76)	43.6% (61)	40.7% (57)	37.1% (52)	35.7% (50)	35% (49)	
Position	1	41	16,118	2,6,97,100	38,85,103	5,20,57	
Ovomucoid (+GU)							
% Complementarity	72.9% (102)	63.6% (89)	59.3% (83)	58.6% (82)	57.9% (81)	56.4% (79)	55.7% (78)
Position	1	133	38	39,131	41,127	126	2,37,40,137
Control (−GU)							
% Complementarity	34.3% (48)†	32.9% (46)	32.1% (45)	30.7% (43)	30% (42)	29.3% (41)	
Position	115	16	19	83,99	22,113,121	1,10,12,64,100	
Control (+GU)							
% Complementarity	51.4% (72)	49.3% (69)	47.9% (67)	47.1% (66)	46.4% (65)	45.7% (64)	
Position	133	138	127	10,130	2,40,124,128	52,53,136	

* Actual number of base pairings for each position summed (cumulative) over seven introns.

† Seven randomly selected sequences replaced the seven intron search sequences.

For each position of Ula-RNA, the numbers of complementary base pairs formed with the introns were summed together. The resulting sums are expressed as percent of complementarity (see Results for conversion equation). The percentages are then ranked in decreasing order, and the positions of Ula are listed under their respective percentages. Scores for Watson-Crick base pairs excluding (−) and including (+) GU base pairs are both shown. Complementarity of Ula-RNA with seven intron search strings from both ovalbumin and ovomucoid is shown, along with the control of seven randomly selected ovalbumin search strings.

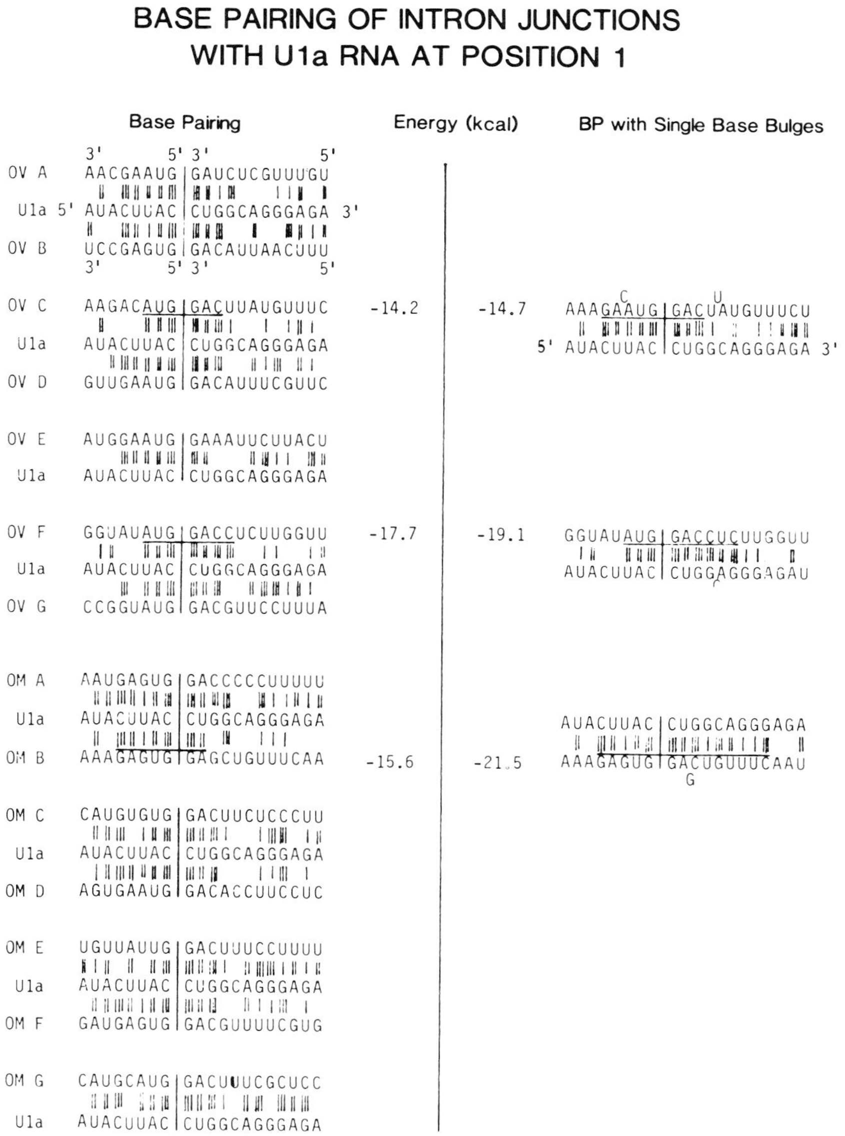

BASE PAIRING OF INTRON JUNCTIONS
WITH U1a RNA AT POSITION 1

FIG. 3. Left panel, Intron search strings of ovalbumin and ovomucoid genes are aligned with the 20 bases at the 5' end of U1a-RNA. Complementary base pairs are indicated by vertical lines; GC pairs are shown with triple lines, AU with double, and GU with single lines. The orientation of the sequences are marked with 5' and 3'. The vertical lines dividing all sequences delineate the putative splice sites. Right panel, Base pairing of certain intron search strings with U1a-RNA sequence allowing for single-base bulges (bubbling out). Center column, Stabilization energy of base pairing was calculated (according to Salser 1977) and given for the underlined region of the indicated introns. Energies for base pairing with and without single-base bulges are placed together for comparison.

position 1 for these introns ranks below that of other positions. Therefore, a search was made to improve the complementarity for these introns. It was found that allowing one-base bulges (bubbling out) will improve the complementarity significantly. These structures are shown in the right panel of Figure 3. In the center column, the energy calculations of base pairing have been done for the underlined regions of original base pairings and the base pairings with single-base bulges.

Frequency Distribution

The frequency distribution of percent complementarity was then plotted to determine whether the intron search strings were preferentially complementary to U1a. The number of searches that resulted in the same number of complementary base pairs (same percent of complementarity) was determined for the control, for each of the introns, and for all the introns combined (cumulative). Since 20-base sequences were being compared, 20 was the maximum number of complementary base pairs. Only the control and the cumulative frequency distribution are shown. The frequencies for ovalbumin, ovomucoid, and the control are graphed in Figure 4.

For an arbitrary search string searching a random sequence, the most frequent score for (−GU) should be 25%, or 5/20, and for (+GU) 37.5%, or 7.5/20. As clearly shown in Figure 4, the (−GU) curves reach maximum between 4.5 and 5, and the (+GU) curves between 7 and 8. The ovalbumin and ovomucoid curves almost coincide with the control curves. These two observations lead to the following conclusions. The search strings and/or the searched sequences do not contain gross anomalies, such as long runs of single nucleotides, and the searched sequences do not contain many repetitive search-string sequences. The base composition of the search strings is not preferentially complementary to that of the searched sequences. Therefore, high complementarity between the search string and the searched sequence must exist at very few of the positions in the searched sequence.

Other snRNAs

Ovalbumin and ovomucoid intron junctions were used to search through U2, U3b, U6, and 4.5S RNAs also. No single position stood out as being more complementary in any of the snRNAs for any of the intron junctions (Table 3). The highest percent of complementarity did not differ significantly from that of the control.

Exon Junctions with snRNAs

Exon search strings were formed by concatenating eight base sequences from both the 5' and 3' ends of the exon junctions. These search strings were used

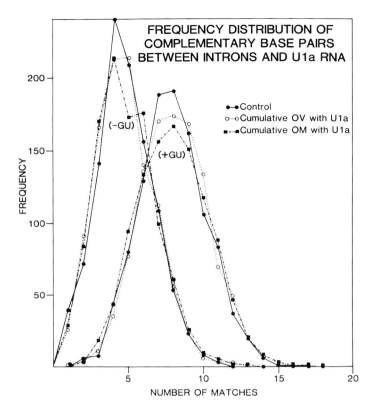

FIG. 4. The frequency distribution of the positions in U1a-RNA is plotted with respect to the number of complementary base pairs (number of matches) formed. Frequency distributions excluding (−) and including (+) GU as complementary base pairs are shown. The frequency distributions cumulative over seven randomly selected sequences from ovalbumin are shown as control (● —— ●). Distributions cumulative over seven intron junctions of ovalbumin gene are shown as cumulative OV with U1a (O----O). Distributions cumulative over the seven intron junctions of ovomucoid gene are shown as cumulative OM with U1a (■ — · — ■).

to search through all the snRNAs for base pairings excluding (−) and including (+) GU base pairs. Results for ovalbumin and ovomucoid exon junctions after a search through U1a-RNA, scoring only for Watson-Crick base pairs, are shown in Table 4. Cumulative percent complementarity was found for each of the snRNAs. Again, no outstanding features were noted. Frequency distribution plots (data not shown) show no significant deviation from control.

Exon Junctions with Introns

The exon junction search strings were used to search through the entire ovalbumin and ovomucoid genes. Although positions were found that had com-

TABLE 3. *Percent complementarity for each position on U2-RNA and 4.5S RNA cumulative over seven introns*

	Rank →						
U2 RNA							
OV (−GU)	42%	38.6%	36.4%	35.7%	35%	34.3%	34.3%
Positions	169	33,73	8	19,75,152	12,14,100,139	7,24,74,92,107	
OV (+GU)	60.7%	58.6%	57.9%	57.1%	56.4%	55.7%	53.6%
Positions	8	73,98	106	100,107,169	14	101	15,33,113
OM (−GU)	43%	42.1%	40%	37.9%	37.1%	36.4%	35.7%
Positions	170	107	173	113,139,152	73	19,74,146	55,56
OM (+GU)	58.6%	57.1%	56.4%	55%	54.3%	52.9%	
Positions	107	100	9,98,106,114	8,15,170	73,113	152	
4.5S RNA							
OV (−GU)	40%	37.1%	36.4%	35.7%	35%	33.6%	33.6%
Positions	40	12,27,74	28	14,58	20,69	13,19,38	20,22,27,69
OV (+GU)	51.4%	50%	49.3%	47.9%	47.1%	46.4%	45.7%
Positions	12	75	9,12,40	28	15,42,58	7,14	37,75
OM (−GU)	42%	40%	36.4%	35.7%	35%	34.3%	33.6%
Positions	12	40	23,58	22	38	14	61
OM (+GU)	51.4%	50.7%	48.6%	47.9%	47.1%	46.4%	65.7%
Positions	12,23	16	22,44,58	7	13,40	61	14,42

Cumulative percent complementarity for U2-RNA and 4.5S RNA calculated and tabulated in the same manner as in Table 2.

TABLE 4. Complementarity of exon junctions with various positions of Ula-RNA

	Complementarity				
	68.7% (11/16)	62.5% (10/16)	56.2% (9/16)	50% (8/16)	43.7% (7/16)
Ovalbumin exons					
I/II	36	52	38,56,68,96,134	16,122	1,33,71,86,127,131,146
II/III				17,68,71,87,113	28,29,32,73,89,114,127
III/IV		2		35,37,130	15,27,41,145,147,149
IV/V				57,124	3,13,23,58,62,70,81,135,138,151
V/VI					15,16,24,31,33,58,72,84,90,100,116,129
VI/VII			65	20,40,80,100	4,5,39,46,50,94,121
VII/VIII			77	44,94,103,105	9,20,62,64,80
Ovomucoid exons					
I/II			22	1,127,139	15,57,71,72,73,87,103,105,113,124,129,136,147
II/III				60,65	17,39,69,75,79,80,85,94,100,119
III/IV				17,33,54,77, 117,127,129	26,38,108,115,118,130
IV/V				17,75	2,6,60,64,79,99,108,124,141
V/VI			54	24,40,56,84 124,125	42,45,68,71,92,94,101,117,120,122,134,152
VI/VII				79	2,17,19,23,37,56,64,153
VII/VIII			42,73,114	2,58,83,86, 89,98,99,131	17,23,37,38,39,102,141

Exon junctional search strings from ovalbumin and ovomucoid genes were scored for percent of sequence complementarity with every position in Ula-RNA. The exon search strings consist of 16 bases as contrasted with the 20-base intron search strings (see Figures 1 and 2). The data are expressed in the same format as in Table 1.

plementarity greater than 75%, the positions were found in exon regions as well as intron regions (data not shown). When 16-base sequences extracted from random positions of ovalbumin were used as control, complementarity above 75% was also found. Consequently, beyond stating that complementarity does exist between intron and exon junctions, little else could be inferred.

DISCUSSION

Several models concerning the role of RNA-RNA interactions in the splicing mechanism for introns may be tested with the present analysis. The model that has received the most interest recently is that of the interaction of snRNAs with intron junctions to allow for stable juxtapositioning of splice sites (Lerner et al. 1980, Murray and Holliday 1979, Rogers and Wall 1980). This model is diagrammatically represented in Figure 5. Although the various intron junctions have limited homology with each other, it is unlikely that such homology is specific enough to allow a single enzyme or a small number of splicing enzymes to recognize the splice sites. However, the class of snRNAs may contain enough different species such that each intron junction could be specifically recognized by an RNA species. Once recognition takes place, the splicing activity could then splice out the intron and ligate the exons together. Several authors support this model with evidence from the Ad2 system (Murray and Holliday 1979) and with evidence for junctional consensus sequences complementary to U1a-RNA (Lerner et al. 1980, Rogers and Wall 1980).

Our analysis shows that primary sequence data may support this model. The 20-base sequence of U1a-RNA beginning immediately following the highly modified cap structure certainly shows high complementarity with most intron junctions of both ovalbumin and ovomucoid genes. On the average, 53% of the U1a sequence is complementary to intron junctions of ovalbumin and 54.3% to that of ovomucoid (see Table 2). This is to be contrasted with 25% complemen-

MODEL I

INTRON JUNCTIONS
BRIDGED WITH snRNA's

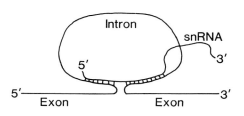

FIG. 5. Model I: Intron junctions bridged with snRNAs.

tarity for arbitrary 20-base strings and random sequences. These percentages increase to 66.4% and 73%, respectively, when GU pairs are included (control is 37.5%). When single-base bulges were created to improve base pairing, no intron junction had less than 55% complementarity to U1a, and the majority had 80% complementarity or better when GU base pairs were included.

Similar analyses using other snRNAs were carried out. U1a, U2, U6, and 4.5S RNAs are all known to be localized in the extranucleolar portion of the nucleus (Ro-Choi and Busch 1974). U1a and U2, among others, have been found within snRNPs precipitated with anti-RNP antibodies from patients with lupus erythematosus (Lerner et al. 1980). Therefore, all these sequences were analyzed for a possible role in splicing. Nevertheless, a high degree of complementarity between the snRNA and the junction regions of ovalbumin and ovomucoid genes was only noted for U1a-RNA.

Another model proposes the interaction of snRNA with exon junctional sequences to create stable juxtapositioning of splice sites (Figure 6). One possible advantage to binding across the exon junctional sequences is in the ligation step of the splicing reaction; in addition to bringing the adjacent junctions together for splicing, the ligation step of the splicing reaction must require that the cleaved ends of the hnRNA be localized in close proximity to each

MODEL II
EXON JUNCTIONS
BRIDGED WITH snRNA's

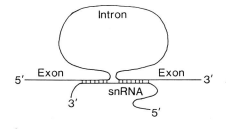

FIG. 6. Model II: Exon junctions bridged with snRNAs.

MODEL I with MODEL II

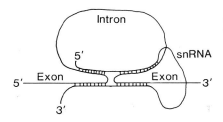

MODEL III

EXON JUNCTIONS
BRIDGED WITH INTRONS

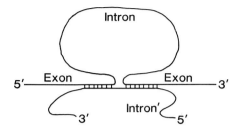

FIG. 7. Model III: Exon junctions bridged with introns.

MODEL I with MODEL III

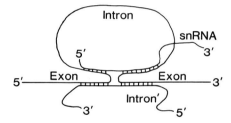

other. Therefore, the snRNA could hybridize to the exon junctions to serve this function. It has been suggested by Murray and Holliday (1979) that the bridging RNA for the intron junction may also fold back and bridge across the exon junction. This combination of Model I and Model II is shown in Figure 6. From our analysis, we could detect no obvious sequence in any of the snRNAs that might serve to bind exon junctions together. However, each snRNA species might have its own target set of exon junctions, in which case our analysis would probably not detect the interaction.

A third model proposes that the exon junction binding sequence be carried within the introns (Figure 7). In studies of the cytochrome genes of yeast mitochondria, mutation in one spot of the intron has been found to interfere with the splicing of an RNA segment located far away from it (Halbreich et al. 1980). The authors suggest that some RNA region away from the splice-junction regions may serve a guide function, either in the precursors or as free introns, to direct a splicing enzyme correctly (Church et al. 1979). This model is shown in the upper portion of Figure 7. An alternative rationale is as follows. Since no real consensus amino acid sequence has been found around junctions, it is difficult to conceive that the many exon junctions of proteins could all hybridize to a few species of snRNAs. It is more conceivable that each protein carries its own exon "bridging" sequences. The most likely place is in its introns. This model does not preclude the possibility of snRNAs holding the intron junctions together, as shown in the lower portion of Figure 7.

This possibility was checked by using exon junction search strings to search through the entire ovalbumin and ovomucoid genes. Since some regions of the ovomucoid introns have not been completely sequenced, an exhaustive search could not be done on ovomucoid genes. However, even with the ovalbumin, little could be determined beyond affirming that complementarity does exist between introns and exon junctions.

In order to extend this analysis further, several considerations will have to be taken into account. Although snRNAs have been shown to be highly conserved between mouse, rat, and HeLa cells (Lerner et al. 1980), more definitive comparisons would be possible if chicken snRNAs could be used in these analyses. The base-pair interactions could also be influenced by the secondary structure of both the hnRNAs and the snRNAs. The role of snRNA-protein particles may also be paramount in determining the secondary and tertiary structure of hnRNAs and snRNAs, in defining the splicing reaction sites, and in controlling the temporal order in which various introns are removed.

ACKNOWLEDGMENTS

This work was supported by NIH grant HD-8188 and the Baylor Center for Population Research and Reproductive Biology.

REFERENCES

Abelson, J. 1979. RNA processing and the intervening sequence problem. Annu. Rev. Biochem. 48:1035–1069.

Berget, S. M., C. Moore, and P. A. Sharp. 1977. Spliced segments at the 5' terminus of adenovirus 2 late mRNA. Proc. Natl. Acad. Sci. USA 74:3171–3175.

Bina, M., R. J. Feldman, and R. G. Deeley. 1980. Could poly(A) align the splicing sites of messenger RNA precursors? Proc. Natl. Acad. Sci. USA 77:1278–1282.

Breathnach, R., C. Benoist, K. O'Hare, F. Gannon, and P. Chambon. 1978. Ovalbumin gene: Evidence for a leader sequence in mRNA and DNA sequence at the exon-intron boundaries. Proc. Natl. Acad. Sci. USA 75:4853–4857.

Breathnach, R., J. L. Mandel, and P. Chambon. 1977. Ovalbumin gene is split in chicken DNA. Nature 270:314–319.

Catterall, J. F., B. W. O'Malley, M. A. Robertson, R. Staden, Y. Tanaka, and G. G. Brownlee. 1978. Nucleotide sequence homology at 12 intron-exon junctions in the chick ovalbumin gene. Nature 257:510–513.

Chow, L. T., R. E. Gelinas, T. R. Broker, and R. J. Roberts. 1977. An amazing sequence arrangement at the 5' ends of adenovirus 2 messenger RNA. Cell 12:1–8.

Church, G. M., P. P. Slonimski, and W. Gilbert. 1979. Pleiotropic mutations within two yeast mitochondrial cytochrome genes block mRNA processing. Cell 18:1209–1215.

Dugaiczyk, A., S. L. C. Woo, E. C. Lai, M. L. Mace, L. McReynolds, and B. W. O'Malley. 1978. The natural ovalbumin gene contains seven intervening sequences. Nature 274:328–333.

Halbreich, A., P. Pajot, M. Foucher, C. Grandchamp, and P. Slonimski. 1980. A pathway of cytochrome b mRNA processing in yeast mitochondria: Specific splicing steps and an intron-derived circular RNA. Cell 19:321–329.

Jeffreys, A. J., and R. A. Flavell. 1977. The rabbit β-globin gene contains a large insert in the coding sequence. Cell 12:1097–1108.

Knapp, G., R. C. Ogden, C. L. Peebles, and J. Abelson. 1979. Splicing of yeast tRNA precursors: Structure of the reaction intermediate. Cell 18:37–45.

Konkel, D. A., S. M. Tilghman, and P. Leder. 1978. The sequence of the chromosomal mouse β-globin major gene: Homologies in capping, splicing, and poly(A) sites. Cell 15:1125–1132.

Lerner, M. R., J. A. Boyle, S. M. Mount, S. L. Wolin, and J. A. Steitz. 1980. Are snRNPs involved in splicing? Nature 283:220–224.

Murray, V., and R. Holliday. 1979. Mechanism for RNA splicing of gene transcripts. FEBS Lett. 106:5–7.

Nordstrom, J. L., D. R. Roop, M.-J. Tsai, and B. W. O'Malley. 1979. Identification of potential ovomucoid mRNA precursors in chick oviduct nuclei. Nature 278:328–331.

Peebles, C. L., R. C. Ogden, G. Knapp, and J. Abelson. 1979. Splicing of yeast tRNA precursors: A two-step reaction. Cell 18:27–35.

Prestayko, A. W., M. Tonato, and H. Busch. 1970. Low molecular weight RNA associated with 28s nucleolar RNA. J. Mol. Biol. 47:505–515.

Reddy, R., D. Henning, and H. Busch. 1979. Nucleotide sequence of nucleolar U3B RNA. J. Biol. Chem. 254:11097–11105.

Ro-Choi, T. S., and H. Busch. 1974. Low molecular weight nuclear RNA's, in The Cell Nucleus, Vol. III, H. Busch, ed., Academic Press, New York, pp. 151–208.

Ro-Choi, T. S., and D. Henning. 1977. Sequence of 5'-oligonucleotide of U1 RNA from Novikoff hepatoma cells. J. Biol. Chem. 252:3814–3820.

Ro-Choi, T. S., R. Reddy, D. Henning, T. Takano, C. W. Taylor, and H. Busch. 1972. Nucleotide sequence of 4.5S ribonucleic acid$_I$ of Novikoff hepatoma cell nuclei. J. Biol. Chem. 247:3205–3222.

Rogers, J., and R. Wall. 1980. A mechanism for RNA splicing. Proc. Natl. Acad. Sci. USA 77:1877–1879.

Roop, D. R., J. L. Nordstrom, S. Y. Tsai, M.-J. Tsai, and B. W. O'Malley. 1978. Transcription of structural and intervening sequences in the ovalbumin gene and identification of potential ovalbumin mRNA precursors. Cell 15:671–685.

Salser, W. 1977. Globin mRNA sequences: Analysis of base pairing and evolutionary implications. Cold Spring Harbor Symp. Quant. Biol. 42:985–1002.

Seif, I., G. Khoury, and R. Dhar. 1979. BKV splice sequences based on analysis of preferred donor and acceptor sites. Nucleic Acids Res. 6:3387–3398.

Shibata, H., T. S. Ro-Choi, R. Reddy, Y. C. Choi, D. Henning, and H. Busch. 1975. The primary nucleotide sequence of nuclear U-2 ribonucleic acid. J. Biol. Chem. 250:3909–3920.

Staden, R. 1977. Sequence data handling by computer. Nucleic Acids Res. 4:4037–4051.

Staden, R., and G. G. Brownlee. 1980. Is the primary sequence at junctions sufficient for correct splicing of pre-mRNA? Nature (in press).

Tilghman, S. M., D. C. Tiemeier, F. Polsky, M. H. Edgel, J. G. Sediman, A. Leder, L. W. Enquist, B. Norman, and P. Leder. 1977. Cloning specific segments of the mammalian genome: Bacteriophage λ containing mouse globin and surrounding gene sequences. Proc. Natl. Acad. Sci. USA 74:4406–4410.

Tinoco, I., Jr., P. N. Borer, B. Dengler, M. D. Levine, D. C. Uhlenbeck, D. M. Crothers, and J. Gralla. 1973. Improved estimation of secondary structure in ribonucleic acids. Nature New Biol. 246:40–41.

Weinberg, R. A., and S. Penman. 1968. Small molecular weight monodisperse nuclear RNA. J. Mol. Biol. 38:289–304.

The *Src* Gene

Genes, Chromosomes, and Neoplasia, edited by
Frances E. Arrighi, Potu N. Rao, and Elton Stubblefield.
Raven Press, New York © 1981.

Viral Oncogenes as Pleiotropic Effectors

J. Michael Bishop, Sara Courtneidge, A. Peter Czernilofsky,
Thomas J. Gonda, Arthur D. Levinson, Leon Levintow, Paul
Luciw, Hermann Oppermann, Diana K. Sheiness, Bjorn
Vennstrom, and Harold E. Varmus

*Department of Microbiology and Immunology, University of California,
San Francisco, California 94143*

Viral oncogenesis takes two general forms. Infection by some tumor viruses is apparently only one of several successive events that combine to produce the neoplastic phenotype (Klein 1979). The precise role of the virus in this setting remains an enigma. Other viruses bear genetic loci ("transforming genes" or "oncogenes") whose expression is wholly responsible for the prompt transformation of infected cells (Tooze 1973). Both forms of viral oncogenesis have repaid study, but we owe much of the recent advance in our understanding of neoplastic transformation to the exploration of viral oncogenes: their action on the cell is swift and direct, they can be mutated with invaluable consequences for the experimentalist, the nucleic acids of which they are composed can be isolated and sequenced, and they act by means of proteins whose properties must underlie the mechanism of transformation.

The pleiotropism of viral oncogenes is remarkable—the changes they induce in susceptible cells appear myriad in number, and in some instances, cells of several different embryological lineages are vulnerable to transformation by a single oncogene. How can we account for these features of viral oncogenesis? Are the products of viral oncogenes themselves pleiotropic effectors, acting on many aspects of cellular structure and physiology, or do they by a single action precipitate a cascade of events that culminates in the transformed phenotype? Can a single viral function transform cells of different embryological lineage, or do pluripotent oncogenes contain separate domains of function? The answers to these questions may have ramifications well beyond the immediate concerns of tumor virologists: we hope to learn whether viral oncogenesis is a legitimate model for tumorigenesis by other influences (such as somatic mutations and chemical carcinogens), we anticipate insight into the normal control of cell division, and we may learn something of the molecular mechanisms that regulate the progress of differentiation in eukaryotic organisms.

ONCOGENESIS BY AVIAN SARCOMA VIRUS

Avian sarcoma virus (ASV) induces sarcomas in birds and mammals and transforms fibroblasts in culture. The genome of ASV carries a genetic locus *(src)* whose action is required for both the initiation and maintenance of neoplastic transformation (Vogt 1977). Can this locus act alone? If so, what are the mechanisms by which it alters the control of cellular growth? Does the locus encode a single protein? Do the properties of the gene product(s) fully account for the complexity of the neoplastic phenotype? The answers to most of these questions elude us for the moment, but we have learned enough to perceive at least an outline of the design by which *src* induces neoplastic transformation, and we have some reason to believe that this design is more than an isolated curiosity.

src Encodes a Single Protein

Previous work has demonstrated that *src* encodes a 60,000-dalton phosphoprotein (pp60^{src}) with the enzymatic activity of a protein kinase (Brugge and Erikson 1977, Collett and Erikson 1978, Levinson et al. 1978). Translation in vitro of the only identified mRNA for *src* generates the same protein exclusively (Bishop

THE ONCOGENE OF AVIAN SARCOMA VIRUS

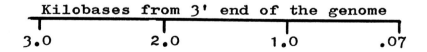

FIG. 1. A DNA fragment containing *src*. The figure illustrates a 3.1-kb *Eco* RI fragment that can be prepared from molecular clones of DNA transcribed from the ASV genome (DeLorbe et al. 1980). The entire nucleotide sequence of the DNA fragment has been determined after subcloning into the single-stranded phage M13mp2 (Czernilovsky, Levinson, Varmus, Bishop, Tischer, and Goodman, manuscript in preparation). The *src* gene has been identified as a single open reading frame of 1,590 nucleotides that initiates with AUG and terminates with UGA; the gene encodes a protein of 530 amino acids with a molecular weight of 58,449. No other portion of the DNA fragment can give rise to a complete protein, although the leftward domain of the fragment apparently harbors the carboxy terminus of the ASV gene *env* in an unexpressible form. The right-ward domain of the fragment is composed of an untranslated region denoted "c" and described in detail elsewhere (Czernilofsky et al. 1980).

et al. 1979). It has therefore generally been assumed that pp60^{src} is the sole product of *src*. Our recent results confirm this assumption. We used molecular cloning to isolate a 3.1-kilobase (kb) fragment of DNA representing the region of the ASV genome that bears the *src* locus (Fig. 1). The nucleotide sequence of this fragment has been determined in full to reveal that *src* should give rise to a single protein with a M_r of 58,450. Chemical and topographical features of the deduced amino acid sequence of this protein conform to established properties of pp60^{src}. No other portion of the DNA fragment could generate a protein of appreciable size without the intervention of complex (and presently undetected) patterns of RNA splicing. We conclude that the *src* locus is a single gene encoding a single protein.

Neoplastic Transformation by *src* Requires No Other Viral Gene

We have used conventional procedures for transfection to introduce the 3.1-kb DNA fragment described above into mammalian cells in culture. Some of the recipient cells become transformed: they undergo characteristic morphological changes, and they acquire the ability to grow in soft agar. Viral DNA derived from the 3.2-kb fragment persists in the transformed cells (probably by virtue of integration into the host genome), and the cells produce pp60^{src}. We conclude that *src* is the only viral gene required for neoplastic transformation of fibroblasts by ASV (although we have yet to demonstrate tumorigenesis by either the isolated *src* DNA or cells transformed by this DNA). The entire mechanism of neoplastic transformation by ASV therefore derives from the biochemical properties of pp60^{src}.

pp60^{src} Phosphorylates Tyrosine in Acceptor Proteins

Immunoprecipitated pp60^{src} phosphorylates the antibody to which it is bound (Collett and Erikson 1978, Levinson et al. 1978), and purified preparations of the viral protein phosphorylate several arbitrarily chosen protein substrates (Erikson et al. 1979, Levinson, Oppermann, Varmus, and Bishop, manuscript submitted). In our experience, the range of proteins phosphorylated by the purified enzyme is not very broad: denatured or native tubulin is an excellent substrate; casein is phosphorylated, but not well; and over 35 other proteins (selected in part because they are rich in tyrosine; see below) failed to serve as substrates. (For a contrasting report, see Erikson et al. 1979.) Otherwise, the properties of the purified protein conform to those attributed originally to pp60^{src} in immunoprecipitates (Collett and Erikson 1978, Levinson et al. 1978, Richert et al. 1979): the phosphotransfer activity is affected by conditional mutations in *src*, but unaffected by either cyclic nucleotides or calcium; the enzymatically active form of pp60^{src} is the monomer protein; and a remarkable variety of nucleoside triphosphates (including deoxynucleoside triphosphates) can be used as phosphate donors.

Both crude (Hunter and Sefton 1980) and purified (Levinson, Oppermann, Varmus, and Bishop, manuscript submitted) preparations of $pp60^{src}$ phosphorylate tyrosine exclusively—a novel form of phosphorylation recently attributed to the transforming proteins of several tumor viruses (Eckhart et al. 1979, Witte et al. 1979a, Hunter and Sefton 1980). Moreover, the amount of phosphotyrosine is increased 5- to 10-fold in cells transformed by the expression of *src* (Hunter and Sefton 1980). It has therefore been proposed that phosphorylation of tyrosine in certain crucial proteins is central to the mechanism of neoplastic transformation by *src*. If correct, this hypothesis could explain the pleiotropic effects of *src:* each cellular protein "attacked" by $pp60^{src}$ might mediate one or more of the separate portions of the transformed phenotype.

$pp60^{src}$ Is an Integral Membrane Protein

Tumor virologists once held that viral oncogenes might transform cells by acting directly on cellular DNA replication (Weinberg 1977). It now appears, however, that this notion may not be correct: the protein most likely responsible for transformation by polyoma virus is affiliated with plasma membranes (Ito 1979); a protein that may mediate oncogenesis by the Abelson murine leukemia virus is an integral protein of the plasma membrane exposed at the cell surface (Witte et al. 1979b); the cytoplasm of enucleated cells can still respond to the transforming influence of *src* (Beug et al. 1978); and studies with both immuno-electron microscopy (Willingham et al. 1979) and biochemical fractionation (Kreuger et al. 1979, Courtneidge et al. 1980) have shown that $pp60^{src}$ is affiliated with the plasma membrane and may not occur at all in the nucleus.

Our present evidence indicates that $pp60^{src}$ can be released from the plasma membrane only by the use of nonionic detergents and must therefore be embedded in the membrane (i.e., as an "integral membrane protein"). Although the entirety of $pp60^{src}$ is decidedly hydrophobic, the membrane sequesters no more than 13,000 daltons at the amino-terminus of the protein (Figure 2); the remainder of the molecule is exposed, probably at the cytoplasmic aspect of the plasma membrane. We have no evidence that $pp60^{src}$ spans the plasma membrane to reach the cell surface. The means by which $pp60^{src}$ makes its way to the plasma membrane pose an amusing puzzle: the protein is not glycosylated, is synthesized on soluble, not membrane-bound, polyribosomes, and moves from the soluble compartment of the cell to the membrane between 5 and 10 minutes following its synthesis.

Structural and Functional Domains in $pp60^{src}$

We have been able to define two distinct functional domains within $pp60^{src}$ (Figure 2). One domain (confined to ca. 13,000 daltons at the amino terminus) anchors the protein to the plasma membrane; the other domain apparently protrudes from the membrane and carries the protein kinase activity described

STRUCTURAL AND FUNCTIONAL TOPOGRAPHY OF pp60^{src}

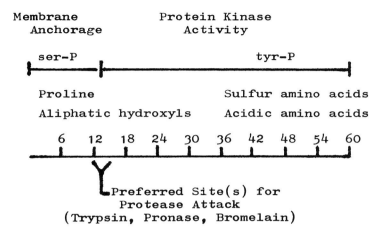

FIG. 2. Structural and functional topography of pp60^{src}. The figure portrays pp60^{src} as a linear array of amino acids demarcated in increments of 6,000 daltons. The division of the molecule into the functional domains for membrane anchorage and kinase activity, and the identification of a site for preferential cleavage by proteases, will be described elsewhere (Levinson, Oppermann, Varmus, and Bishop, manuscript submitted). The concentration of amino acids in one domain or the other was perceived by computer-assisted analysis of the amino acid sequence, deduced in turn from the nucleotide sequence of *src* (Czernilovsky, Levinson, Varmus, Bishop, Tischer, and Goodman, manuscript in preparation).

above (fragments of pp60^{src} that compose most or all of this domain retain full kinase activity even when separated from the amino-terminal domain). These functional domains have correlates in the composition and structure of pp60^{src} (Figure 2): the protein is preferentially susceptible to attack by several proteases in the region that joins the two domains and apparently lies just beyond the confines of the plasma membrane; several amino acids cluster in one domain or the other—proline, serine, and threonine in the amino-terminal domain, and acidic amino acids, cysteine, and methionine in the carboxy-terminal (exposed) domain; the domain that is anchored to the membrane contains one (or perhaps more) phosphoserine, whereas the exposed (kinase) domain contains a phosphotyrosine that has been implicated in phosphotransfer activity (Hunter and Sefton 1980, Levinson, Oppermann, Varmus, and Bishop, manuscript submitted).

We suspect that the one-dimensional topography of pp60^{src} described here has its counterparts in the three-dimensional configuration of the protein (which we already know to be a highly elongated molecule in which the frictional coefficient $f/f_0 = 1.4$; Levinson, Oppermann, Varmus, and Bishop, manuscript submitted). It appears that the protein is designed on the one hand for tethering to the plasma membrane, and on the other, for enzymatic function beyond the confines of the membrane. These inferences lead in turn to the suspicion that we may find "targets" for *src* among at least two classes of proteins—those that are intrinsic to the plasma membrane, and those that merely affiliate with (or "insert" on to) the cytoplasmic surface of membrane.

THE PLURIPOTENT ONCOGENES OF DEFECTIVE LEUKEMIA VIRUSES

The fact that the oncogenic capacity of some avian retroviruses can affect more than one tissue emerged from the study of viruses that transform hematopoietic cells (Graf and Beug 1978). None of these viruses can replicate without complementation by another, "helper" virus—hence, the designation "defective leukemia viruses." The deficiency in replication is attributable in each instance to a deletion that affects one or more of the viral genes required for the production of infectious virions (see Figure 3).

The spectrum of tissues affected by the pathogenicity of defective leukemia viruses is a constant and specific property of individual viral isolates. Prototypes include myelocytomatosis virus (MCV), which transforms fibroblasts, epithelial cells, and macrophages; avian erythroblastosis virus (AEV), which transforms fibroblasts and erythroblasts; and avian myeloblastosis virus (AMV), which transforms myeloblasts exclusively. Although the oncogenic potential of these viruses has only begun to yield to genetic analysis (Royer-Pokora et al. 1979), biochemical studies of the viral genomes have identified loci that probably direct neoplastic transformation of infected cells (see Figure 3). The nucleotide sequence

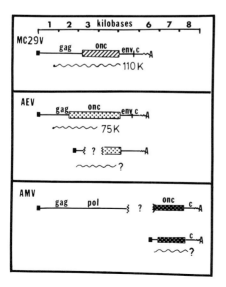

FIG. 3. The oncogenes of defective avian leukemia viruses: modes of expression. The data used to compile the drawing are described in detail elsewhere (Sheiness, Vennstrom, Gonda, Fanshier, and Bishop, manuscript in preparation; Gonda, Sheiness, Fanshier, and Bishop, manuscript submitted). Viral RNAs were extracted from infected cells, fractionated by electrophoresis in agarose gels under denaturing conditions, transferred to chemically activated filter paper, and identified by molecular hybridization with radioactive DNAs prepared with molecular clones derived from specific portions of the viral genomes. Question marks denote regions not yet rigorously characterized. The solid bar at the 5′ ends of the RNAs indicates a leader sequence that is spliced from the 5′ end of the viral genomes to the 5′ ends of the bodies of subgenomic mRNAs. The wavy lines denote possible proteins encoded in viral oncogenes; other viral gene products are not indicated.

of these loci is distinctive for each spectrum of oncogenicity (Roussel et al. 1979), and all viral isolates that share a particular spectrum contain the same locus (Bishop et al. 1979, Duesberg and Vogt 1979). Since each virus contains only one such locus, we can explain pluripotential oncogenicity in only two ways: either single viral gene products are effective in a variety of tissues, or the oncogenic loci are divided into two or more independently expressed domains. By analyzing the mRNAs of defective leukemia viruses, we have learned that both explanations may obtain.

Our experimental strategy was predicated on the assumption that the number of separate mRNAs representing single oncogenic loci would define the minimum number of independently expressed domains in the loci. In order to identify and characterize the mRNAs of defective leukemia viruses, we made use of radioactive cDNAs prepared from specific portions of the viral genomes by molecular cloning. Figure 3 summarizes our results.

MCV

The genome of MCV is itself an mRNA that gives rise to a 110,000-dalton protein (P110), representing both the putative oncogene of the virus and an adjoining portion of the replicative gene *gag* (Figure 3). P110 is the only viral protein that has been found in cells transformed by MCV (Bister et al. 1977), and we find no MCV-specific RNA other than the viral genome in infected cells. We conclude that P110 (or smaller, presently unidentified proteins derived from it) is likely to be the sole mediator of neoplastic transformation by MCV. If correct, this conclusion is remarkable because MCV is one of the more versatile carcinogens in the retrovirus family.

AEV

The genome of AEV also directs the synthesis of a protein (P75) encoded by both *gag* and oncogene (Hayman et al. 1979). In this instance, however, the protein translated from the viral genome appears not to represent the entire pathogenic locus (Lai et al. 1979). It therefore seemed likely that a second genetic domain participates in oncogenesis by AEV. Our description of the mRNAs of AEV has substantiated this presumption (Figure 3). Cells transformed by AEV contain two AEV-specific RNAs: the viral genome, as expected, and a smaller RNA that represents one half to two thirds of the oncogenic locus, including the region of the locus that appears not to be represented in P75. We cannot as yet attribute a specific viral protein to the smaller mRNA, but other investigators have produced a suitable candidate ($M_r \sim 40,000$) by translating fragments of the AEV genome in vitro (Lai et al. 1980, Pawson and Martin 1980).

We conclude that at least two independently expressed domains compose the oncogene of AEV. This conclusion has an interesting correlate in the work of Graf and his colleagues, who have isolated a mutant of AEV that can transform fibroblasts but not erythroblasts (Royer-Pokora et al. 1979). Perhaps the genetic domains distinguished by this mutant are the same as those whose existence we infer from the composition of the viral mRNAs.

AMV

AMV is unlike other defective leukemia viruses in several regards: it transforms only one type of cell, the myeloblast; its genome contains a complete *gag* gene (and most, if not all, of the *pol* gene) (Figure 3); and the putative oncogene is situated between *pol* and a point within a few hundred nucleotides of the 3' end of the genome (Figure 3). As a consequence of these distinguishing features, the expression of AMV genes differs from that of MCV and AEV. First, the *gag* gene is expressed independently, giving rise to mature products normally found in the virions of infectious virus. Second, translation from *gag* and *pol* produces a polyprotein (M_r 160,000–180,000) similar to the precursor of reverse transcriptase in cells infected by other avian retroviruses whose *gag* and *pol* genes are intact. Third, the expression of the putative oncogene appears not to be linked to translation from adjacent genes. Instead, cells transformed by AMV contain a subgenomic mRNA that probably contains the entire oncogene (Figure 3). We suggest that this mRNA represents the sole means by which the oncogene is expressed, and that a single protein may mediate tumorigenesis by AMV.

DISCUSSION

We began this essay with the challenge posed by pleiotropism in the cellular response to viral oncogenes; we conclude with the impression that pleiotropism is achieved by a variety of means. ASV transforms cells through the agency of a single protein whose apparent enzymatic activity could play on numerous aspects of cellular structure and function, but we remain incapable of explaining why the action of this protein is limited to one tissue in the infected animal. MCV is an exceptionally versatile carcinogen, yet its effects may be achieved by the production of a single viral protein whose functional properties are presently unknown. In a third variation on our theme, the dual oncogenicity of AEV may reflect separate genetic domains that give rise to different proteins (although it is also possible that the putative domains are mutually facilitative and are each required for both forms of oncogenesis).

We know very little of the mechanisms by which viral oncogenes transform cells to a neoplastic phenotype. Provisional results suggest elements of a common design: protein phosphorylation as enzymatic mediator, tyrosine as a crucial substrate, the periphery of the cell as site of action. But exceptions exist (for example, efforts to implicate phosphorylation of tyrosine in neoplastic transformation by MCV and polyoma virus have failed), and the game is young.

It now appears that the avian retrovirus oncogenes considered here originated from genetic loci in the avian genome (for a review, see Bishop et al. 1980). These loci are probably essential to the economy of the cell: they have been conserved throughout the course of vertebrate evolution, and they are expressed in all of the vertebrates examined to date. In one instance (*src* of ASV), the proteins encoded by the viral oncogene (pp60^{src}) and its cellular progenitor are remarkably similar in structure and function (Collett et al. 1979, Oppermann et al. 1979); we expect that the same will prove to be true when the cellular homologues of other retrovirus oncogenes have been studied sufficiently. We propose that these findings unite several ostensibly disparate phenomena. First, the biochemical mechanisms that mediate neoplastic transformation by retroviruses may be closely akin to events that participate in the regulation of normal cellular growth and development. Second, retrovirus oncogenes may have been drawn from the cellular genes whose expression or alteration account for oncogenesis by somatic or germ line mutations. Third, the induction of neoplastic transformation by the transfer of DNA from one normal cell to another, as reported recently by Cooper et al. (1980), may represent the activity of genes which, in other circumstances, could give rise to retrovirus oncogenes. It seems likely to us that retrovirus oncogenes will remain at stage center as the exploration of oncogenic mechanisms continues.

ACKNOWLEDGMENTS

We acknowledge with gratitude the collaborative assistance of G. and C. Moscovici, Veteran's Administration Hospital, Gainesville, Florida, and the stenographical assistance of B. Cook. This work was supported by grants from the National Cancer Institute and the American Cancer Society.

REFERENCES

Beug, H., M. Claviez, B. Jockusch, and T. Graf. 1978. Differential expression of Rous sarcoma virus-specific transformation parameters in enucleated cells. Cell 14:843–856.

Bishop, J. M., S. A. Courtneidge, A. D. Levinson, H. Oppermann, N. Quintrell, D. K. Sheiness, S. R. Weiss, and H. E. Varmus. 1979. The origin and function of avian retrovirus transforming genes. Cold Spring Harbor Symp. Quant. Biol. (in press).

Bishop, J. M., T. Gonda, S. H. Hughes, D. K. Sheiness, E. Stubblefield, B. Vennstrom, and H. E. Varmus. 1980. The genesis of avian retrovirus oncogenes, *in* Mobilization and Reassembly of Genetic Information (Twelfth Miami Winter Symposium, January, 1980) (in press).

Bister, K., J. M. Hayman, and P. K. Vogt. 1977. Defectiveness of avian myelocytomatosis virus MC29: Isolation of long-term nonproducer cultures and analysis of virus-specific polypeptide synthesis. Virology 82:431–448.

Brugge, J. S., and R. L. Erikson. 1977. Identification of a transformation-specific antigen induced by an avian sarcoma virus. Nature 269:346–347.

Collett, M. S., and R. L. Erikson. 1978. Protein kinase activity associated with avian sarcoma virus *src* gene product. Proc. Natl. Acad. Sci. USA 75:2021–2024.

Collett, M. S., E. Erikson, A. F. Purchio, J. S. Brugge, and R. L. Erikson. 1979. A normal cell protein similar in structure and function to the avian sarcoma virus transforming gene product. Proc. Natl. Acad. Sci. USA 76:3159–3163.

Cooper, G. M., S. Okenquist, and L. Silverman. 1980. Transforming activity of DNA of chemically transformed cells and normal cells. Nature 284:418–422.

Courtneidge, S. A., A. D. Levinson, and J. M. Bishop. 1980. The protein encoded by the transforming gene of avian sarcoma virus (pp60^{src}) and a homologous protein in normal cells (pp60$^{proto-src}$) are associated with the plasma membrane. Proc. Natl. Acad. Sci. USA (in press).

Czernilofsky, A. P., W. DeLorbe, R. Swanstrom, H. E. Varmus, J. M. Bishop, E. Tischer, and H. M. Goodman. 1980. The nucleotide sequence of "c": An untranslated but conserved domain in the genome of avian sarcoma virus. Nucl. Acid Res. (in press).

DeLorbe, W. J., P. A. Luciw, H. E. Varmus, and J. M. Bishop. 1980. Molecular cloning and characterization of avian sarcoma virus circular DNA molecules. J. Virol. (in press).

Duesberg, P. H., and P. K. Vogt. 1979. Avian acute leukemia viruses MC29 and MH2 share specific RNA sequences: Evidence for a second class of transforming genes. Proc. Natl. Acad. Sci. USA 76:1633–1637.

Eckhart, W., M. A. Hutchinson, and T. Hunter. 1979. An activity phosphorylating tyrosine in polyoma T antigen immunoprecipitates. Cell 18:935–946.

Erikson, R. L., M. S. Collett, E. Erikson, and A. F. Purchio. 1979. Evidence that the avian sarcoma virus transforming gene product is a cyclic AMP-independent protein kinase. Proc. Natl. Acad. Sci. USA 76:6260–6264.

Graf, T., and H. Beug. 1978. Avian leukemia viruses: Interaction with their target cells in vivo and in vitro. Biochim. Biophys. Acta 516:269–299.

Hayman, M. J., B. Royer-Pokora, and T. Graf. 1979. Defectiveness of avian erythroblastosis virus: Synthesis of a 75K *gag*-related protein. Virology 92:31–45.

Hunter, T., and B. W. Sefton. 1980. The transforming gene product of Rous sarcoma virus phosphorylates tyrosine. Proc. Natl. Acad. Sci. USA 77:1311–1315.

Ito, T. 1979. Polyoma virus-specific 55K protein isolated from plasma membrane of productively infected cells is virus-coded and important for cell transformation. Virology 98:261–266.

Klein, G. 1979. Lymphoma development in mice and humans: Diversity of initiation is followed by convergent cytogenetic evolution. Proc. Natl. Acad. Sci. USA 76:2442–2446.

Kreuger, J. G., E. Wang, and A. R. Goldberg. 1979. Evidence that the *src* gene product of Rous sarcoma virus is membrane associated. Virology 101:25–40.

Lai, M. M. C., S. S. F. Hu, and P. K. Vogt. 1979. Avian erythroblastosis virus: Transformation-specific sequences form a contiguous segment of 3.25 kb located in the middle of the genome. Virology 97:366–377.

Lai, M. M. C., J. C. Neil, and P. K. Vogt. 1980. Cell-free translation of avian erythroblastosis virus RNA yields two specific and distinct proteins with molecular weights of 75,000 and 40,000. Virology 100:475–483.

Levinson, A. D., H. Oppermann, L. Levintow, H. E. Varmus, and J. M. Bishop. 1978. Evidence that the transforming gene of avian sarcoma virus encodes a protein kinase associated with a phosphoprotein. Cell 15:561–572.

Oppermann, H., A. D. Levinson, H. E. Varmus, L. Levintow, and J. M. Bishop. 1979. Uninfected vertebrate cells contain a protein that is closely related to the product of the avian sarcoma virus transforming gene *(src)*. Proc. Natl. Acad. Sci. USA 76:1804–1808.

Pawson, T., and G. S. Martin. 1980. Cell-free translation of avian erythroblastosis virus RNA. J. Virol. 34:280–284.

Richert, N. D., P. J. A. Davies, G. Jay, and I. H. Pastan. 1979. Characterization of an immune complex kinase in immunoprecipitates of avian sarcoma virus-transformed fibroblasts. J. Virol. 31:695–706.

Roussel, M., S. Saule, C. Lagron, C. Rommens, H. Beug, T. Graf, and D. Stehelin. 1979. Three new types of viral oncogene of cellular origin specific for hematopoietic cell transformation. Nature 281:452–455.

Royer-Pokora, B., S. Grieser, H. Beug, and T. Graf. 1979. Mutant avian erythroblastosis virus with restricted target cell specificity. Nature 282:750–752.

Tooze, J., ed. 1973. The Molecular Biology of Tumor Viruses. Cold Spring Harbor Laboratory, Cold Spring Harbor, New York.

Vogt, P. K. 1977. The genetics of RNA tumor viruses, in Comprehensive Virology, Vol. 9, H. Fraenkel-Conrat and R. Wagner, eds. Plenum Press, New York, pp.341–455.

Weinberg, R. A. 1977. How does T antigen transform cells? Cell 11:243–246.

Willingham, M. C., G. Jay, and I. Pastan. 1979. Localization of the avian sarcoma virus *src* gene product to the plasma membrane of transformed cells by electron microscopic immunocytochemistry. Cell 18:125–134.

Witte, O. N., A. Dasgupta, and D. Baltimore. 1979a. Abelson murine leukemia virus protein is phosphorylated in vitro to form phosphotyrosine. Nature 283:826–831.

Witte, O., N. Rosenberg, and D. Baltimore. 1979b. Preparation of syngeneic regressor sera reactive with Abelson MuLV encoded P120 molecule at the cell surface. J. Virol. 31:776–784.

Genes, Chromosomes, and Neoplasia, edited by
Frances E. Arrighi, Potu N. Rao, and Elton Stubblefield.
Raven Press, New York © 1981.

Characterization of the Avian Sarcoma Virus Protein Responsible for Malignant Transformation and of its Homologue in Normal Cells

Marc S. Collett,* Eleanor Erikson, Anthony F. Purchio, and Raymond L. Erikson

Department of Pathology, University of Colorado Health Sciences Center School of Medicine, Denver, Colorado 80262

In order to approach a cure for, or a means of controlling, malignant disease, the molecular mechanisms involved in the creation of neoplastic cells must be understood. The genes directing oncogenesis and the products of these genes must be identified and functionally characterized. And, furthermore, the control of their expression and regulation of their function must be understood. This understanding has been one of the major goals of modern viral oncology.

TRANSFORMATION BY AVIAN SARCOMA VIRUSES

The RNA tumor viruses, or retroviruses, offer an exceptional set of agents with which to approach these problems. These viruses appear to be ubiquitous in vertebrates, including primates and possibly man. In susceptible animals, retroviruses cause leukemias, carcinomas, sarcomas, rapid lethal infections, slow degenerative infections, or, often, no disease.

The avian retroviruses have been studied most extensively and have classically served as a prototypical system in tumor virology. One group of avian retroviruses, the avian sarcoma viruses (ASVs), are one of the most efficient carcinogenic agents known. Avian sarcoma viruses are able to induce tumors in birds and in a variety of mammals as well as to transform stably both avian and mammalian cells in culture.

A single ASV gene (termed sarcoma gene, or *src*) has been demonstrated to be responsible for the induction and maintenance of cell transformation in vitro and tumor production in infected animals (Hanafusa 1977, Vogt 1977). This revelation was facilitated by the isolation and subsequent analysis of both noncon-

---

*Present address: Department of Microbiology, University of Minnesota, Minneapolis, Minnesota 55455

ditional deletion mutants and conditional temperature-sensitive (ts) mutants. Virus unable to transform fibroblasts can be obtained from clonal stocks of sarcoma virus. These transformation-defective (td) derivatives, which retain all other properties of the virus, have been shown to be deletion mutants of ASV, lacking approximately 1,500 nucleotides near the 3' end of the 10,000-nucleotide viral genome. Furthermore, these same nucleotide sequences are not found in avian leukosis viruses, which are unable to transform fibroblasts in culture or cause sarcomas in vivo. Avian sarcoma virus mutants temperature sensitive (ts) for the transformed state replicate and produce progeny virus at similar rates at either the permissive temperature (transformed phenotype) or nonpermissive temperature (normal phenotype). The locations of the mutations in these ts transformation mutants of ASV have been mapped within the 1,500-nucleotide sequence deleted in the ASV td mutants. Thus, it has been firmly established that a specific viral gene, the *src* gene, is responsible for neoplastic transformation by ASV.

IDENTIFICATION OF THE PROTEIN PRODUCT OF THE ASV *src* GENE

The fact that ASV mutants temperature sensitive for transformation exist implies that there is a viral protein product involved in ASV-induced oncogenesis. Recently, we have identified the protein product of the ASV *src* gene as a phosphoprotein of apparent molecular weight of 60,000 (Brugge and Erikson, 1977, Purchio et al. 1977, 1978, Brugge et al. 1978). The major tools employed in this work were (1) cell-free translation of that region of the viral RNA that contains the *src* gene and (2) immunoprecipitation carried out with antiserum from rabbits bearing ASV-induced fibrosarcomas (TBR serum). The latter approach is one that has proved useful in the identification of nonstructural proteins encoded by DNA-containing tumor viruses. Determination that the 60,000-dalton protein is actually the product of the *src* gene is based on the following results: (1) It was detected as a nonstructural, transformation-specific antigen in ASV-transformed avian cells, ASV-transformed mammalian cells, and ASV-induced mammalian tumor cells by immunoprecipitation of radiolabeled cell extracts with TBR serum (Figure 1A). (2) In vitro cell-free translation of the 3' third of nondefective ASV viral RNA, the region of the genome that contains the *src* gene, resulted in the synthesis of a 60,000-dalton polypeptide that was immunoprecipitable with TBR serum. Such a protein was not synthesized when td ASV viral RNA was translated (Figure 1B). (3) The polypeptide of 60,000 daltons made in vitro by cell-free translation and the transformation-specific antigen isolated by immunoprecipitation of all types of ASV-infected cell extracts tested are identical as determined by peptide analyses. We believe that it is consistent with these results to conclude that this 60,000-dalton protein is the product of the ASV *src* gene, and have consequently given it the designation pp60[src]. In order to gain additional information concerning this protein, we

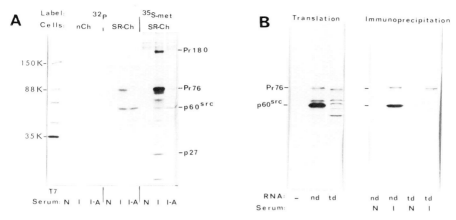

FIG. 1. Identification of the ASV *src* gene product. Panel A, Normal chick embryo fibroblasts (nCh) and chick embryo fibroblasts infected and transformed by the Schmidt-Ruppin strain of ASV (SR-Ch) were labeled in culture with either ^{32}P-orthophosphate (^{32}P) or [^{35}S]methionine (met). Cell extracts were then prepared and aliquots were immunoprecipitated (Brugge and Erikson 1977) with normal rabbit serum (N), immune ASV tumor-bearing rabbit serum (I), and immune serum that had been preadsorbed with ASV structural proteins (I-A). The immunoprecipitated materials were then subjected to electrophoresis in SDS-polyacrylamide gels followed by fluorography as previously described (Purchio et al. 1978). Pr180, Pr76, and p27 represent products of ASV structural genes. Phage T7 virion proteins are included as molecular weight references. p60^{src} represents the product of the ASV *src* gene and appears as a ^{32}P-labeled and [^{35}S]methionine-labeled nonstructural protein specifically precipitated by immune, but not normal, rabbit serum from ASV-transformed cells. Panel B, Polyacrylamide gel electrophoretic analysis of ^{35}S-methionine-labeled polypeptides synthesized in cell-free extracts programmed by virion RNA. Subgenomic (21S) virion RNA generated from nondefective (nd) or transformation-defective (td) ASV was translated in a messenger-dependent reticulocyte lysate system as previously described (Purchio et al. 1977, 1978). –, no RNA added to the system. Immunoprecipitation of the translation products was carried out with normal (N) or immune (I) rabbit serum as previously described (Brugge and Erikson 1977).

have pursued studies involving both the structural and functional characterization of pp60^{src} in the hope of further elucidating the mechanism of ASV-induced cellular transformation.

STRUCTURAL ANALYSIS OF THE *src* PROTEIN

As a means of analyzing various structural features of pp60^{src}, we have used the one-dimensional limited proteolysis procedure of Cleveland et al. (1977). In this procedure the protein of interest prepared by sodium dodecyl sulfate (SDS)-containing polyacrylamide gel electrophoresis is subjected to electrophoresis in a second gel in the presence of a proteolytic enzyme. By use of a variety of proteases in varying amounts, characteristic digestion patterns can be obtained for the pp60^{src} polypeptide (Collett et al. 1979b). The use of *Staphylococcus aureus* V8 protease results in an especially simple partial digestion map. [^{35}S]methionine-labeled pp60^{src}, when subjected to V8 protease partial proteoly-

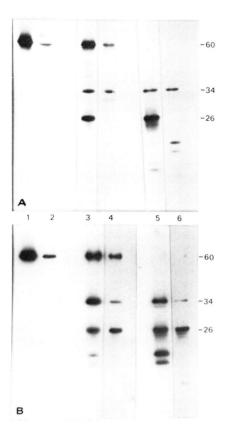

FIG. 2. One-dimensional limited proteolysis mapping of pp60src. The 60,000-dalton ASV *src* protein bands localized by autoradiography of preparative polyacrylamide gels were excised and subjected to partial proteolysis by *S. aureus* V8 protease during re-electrophoresis in a second polyacrylamide gel. Protease concentrations: tracks 1 and 2, no enzyme; tracks 3 and 4, 5 ng of enzyme; tracks 5 and 6, 50 ng of enzyme. The numbers in the margin represent approximate molecular weights in kilodaltons. Panel A, The ASV *src* protein was synthesized in the reticulocyte cell-free system from viral RNA (Purchio et al. 1977, 1978), using either [35S]methionine (tracks 1, 3, and 5) or N-formyl[35S]methionyl-tRNAf (tracks 2, 4, and 6) as radiolabeled precursor. Panel B, [35S]methionine-labeled (tracks 2, 4, and 6) and 32P-labeled (tracks 1, 3, and 5) pp60src were prepared by immunoprecipitation of radiolabeled ASV-transformed cell extracts with TBR serum.

sis, is cleaved into two major fragments having molecular weights of approximately 34,000 and 26,000 (Figure 2, panel B). These results suggested that these digestion products may represent peptides generated by protease cleavage at one site within the pp60src molecule. Consequently, we performed experiments to determine which of these V8 protease fragments represented the amino terminus of the pp60src polypeptide. This was accomplished by the cell-free synthesis of the *src* protein, using N-formyl[35S]methionyl-tRNAf as the radiolabeled precursor. The initiator amino acid N-formyl[35S]methionine can be incorporated only at the NH_2-terminal position of the resultant polypeptide. Thus, by cleaving N-formyl[35S]methionine-labeled *src* protein with V8 protease, we found that only the 34,000-dalton fragment contained the radiolabel (Figure 2, panel A), indicating that this peptide represented the NH_2-terminal 60% of the *src* protein. This result also allowed us to conclude that the 26,000-dalton fragment was derived from the COOH terminus of the protein. (Hereafter, we will refer to these V8 protease digestion fragments as the V8-NH_2 and the V8-COOH peptides, respectively.)

Since pp60^{src} is a phosphorylated protein and this secondary protein modification has been previously shown to be important in the control of a variety of enzymes and regulatory proteins, we undertook studies concerning the characterization of the sites of phosphorylation on the pp60^{src} molecule. By performing similar V8 protease partial digestion analyses on ^{32}P-labeled pp60^{src}, we found two major protease fragments with electrophoretic mobilities identical to those of the [^{35}S]methionine-labeled V8 protease peptides (Figure 2, panel B). These results suggested that the two major ^{32}P-labeled V8 protease fragments were identical to the V8-NH$_2$ and V8-COOH peptides, and further implied that the pp60^{src} protein contained multiple sites of phosphorylation. To further investigate the possibility that pp60^{src} contained more than one site of phosphorylation, we have fractionated exhaustively trypsinized ^{32}P-labeled pp60^{src} in a two-dimensional system. Figure 3, panel A shows the resultant tryptic phosphopeptide map of pp60^{src}. It can be seen that there are two major phosphopeptides. Furthermore, two-dimensional phosphoamino acid analysis of acid-hydrolyzed ^{32}P-labeled pp60^{src} (Figure 4) reveals that this protein contains both phosphoserine and the rare phosphorylated amino acid, phosphotyrosine (Figure 4, panel D). Additional studies have shown that the phosphoserine residue is located in the V8-NH$_2$ peptide and the phosphotyrosine residue is in the V8-COOH peptide (Collett et al. 1979b).

In an effort to understand the phosphorylation of the pp60^{src} protein and, ultimately, the possible functional regulation of the transforming protein by phosphorylation-dephosphorylation modifications, we investigated conditions under which pp60^{src} could be phosphorylated in crude cell-free extracts. After establishing such a system, we found that the basal level of phosphorylation of the *src* protein could be stimulated by cyclic AMP (cAMP). To determine whether cAMP addition resulted in the stimulation of phosphorylation of a

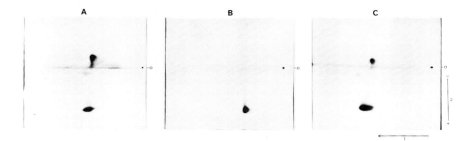

FIG. 3. Two-dimensional tryptic fingerprints of ^{32}P-labeled pp60^{src}. ^{32}P-labeled pp60^{src} proteins were immunoprecipitated from cell extracts and subjected to polyacrylamide gel electrophoresis. After autoradiographic localization, pp60^{src} was eluted and completely digested with TPCK-trypsin (Collett et al. 1979a). Two-dimensional fractionation involved ascending chromatography followed by electrophoresis at pH 6.5. 0, origin. A, ^{32}P-labeled pp60^{src} isolated from ASV-transformed cells radiolabeled in culture; B and C, pp60^{src} phosphorylated in cell-free extracts of ASV-transformed cells in the absence (B) or presence (C) of 10 μM cAMP as previously described (Collett et al. 1979b).

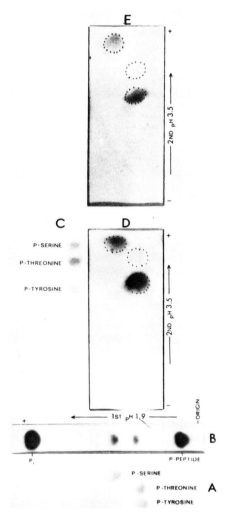

FIG. 4. Phosphoamino acid analyses of pp60^{src} and pp60^{sarc} isolated from ^{32}P-labeled cells in culture. ^{32}P-labeled viral pp60^{src} and normal chick cell pp60^{sarc} were immunoprecipitated from ASV-transformed and normal chick cells, respectively, subjected to SDS-polyacrylamide gel electrophoresis, localized by autoradiography, excised, and then eluted from the gel pieces. After precipitation with 20% trichloroacetic acid, the polypeptides were subjected to acid hydrolysis in 6N HCl at 100° for 3–4 hours. Samples of the hydrolysates were spotted onto Whatman 3MM paper. Electrophoresis at pH 1.9 was as described (Collett et al. 1979b). Electrophoresis at pH 3.5 (pyridine : acetic acid : H$_2$O, 1:10:189) was performed at 4,500 V for 0.5 hour. Samples of authentic phosphoserine, phosphothreonine, and phosphotyrosine were added to all radioactive samples analyzed. Panel A, results of electrophoresis at pH 1.9 of authentic phosphoamino acids (ninhydrin-stained); panel B, electrophoresis at pH 1.9 of an acid hydrolysate of ^{32}P-labeled pp60^{src}; panel C, electrophoresis of authentic phosphoamino acids at pH 3.5 (ninhydrin-stained); panel D, two-dimensional electrophoresis (first at pH 1.9, second at pH 3.5) of an acid hydrolysate of ^{32}P-labeled pp60^{src}; panel E, two-dimensional electrophoresis of an acid hydrolysate of ^{32}P-labeled pp60^{sarc}. The dotted circles show the positions of ninhydrin-stained phosphoamino acid markers.

specific residue on pp60^{src}, we prepared two-dimensional tryptic phosphopeptide maps of pp60^{src} phosphorylated in cell-free extracts with and without the addition of cAMP. Figure 3, panel C, illustrates the fingerprint of the cell-free–phosphorylated pp60^{src} when cAMP was present in the reaction. It can be seen that the same two major tryptic phosphopeptides observed when pp60^{src} was phosphorus-radiolabeled in vivo (Figure 3, panel A) are represented. However, when cAMP was omitted from the cell-free phosphorylation reaction, only one phosphopeptide (the phosphotyrosine-containing one) was phosphorylated (Figure 3, panel B). From these types of experiments we were able to determine that at least two protein kinase activities are involved in the phosphorylation

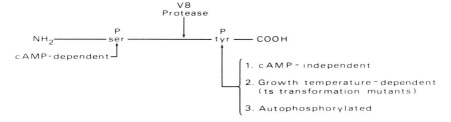

FIG. 5. Characterization of phosphorylation sites in pp60^{src}.

of pp60^{src} in cell-free extracts. One which phosphorylates the NH$_2$-terminal serine residue is a cAMP-dependent protein kinase activity, and the other, acting on the COOH-terminal tyrosine residue, is a cyclic nucleotide-independent phosphotransferase. The results of our structural analyses of the pp60^{src} protein are summarized in Figure 5. This illustration is not meant to specify the particular locations of the two phosphorylated residues within the V8 protease fragments. The points of growth temperature-dependent phosphorylation and autophosphorylation of the tyrosine residue on pp60^{src} are discussed later.

FUNCTION OF THE ASV *src* PROTEIN

In order to understand the mechanism of ASV-induced neoplasia, it will be necessary to determine the function of the transforming protein pp60^{src}. Toward this end, we have determined that pp60^{src} is itself a protein kinase. This function was provisionally ascribed to the ASV *src* protein when we discovered that specific immunoprecipitates from ASV-transformed avian and mammalian cells containing pp60^{src} catalyzed the transfer of phosphate from [γ-^{32}P]-ATP to the heavy chain of rabbit immunoglobulin (Collett and Erikson 1978). Immunoprecipitates formed with immune TBR serum and extracts from normal cells or cells infected with avian leukosis virus or td deletion ASV mutants showed no activity in this assay, nor did complexes formed with normal rabbit serum and normal or transformed cells (Figure 6, panel A). These results demonstrated that the presence of pp60^{src} in immunoprecipitates correlated with the presence of a phosphotransferase activity, suggesting that the ASV-transforming gene product is, or is closely associated with, a protein kinase. Consistent with this conclusion was our observation that the expression of this immune complex protein kinase activity was growth temperature-dependent in cells infected with an ASV mutant (NY68) temperature sensitive for transformation (Collett and Erikson 1978). Furthermore, translation of the viral RNA encompassing the *src* gene, which results in the synthesis of the *src* protein, yields an enzymatically functional protein kinase (as assayed in immunoprecipitates) (Figure 6, panel B; Erikson et al. 1978), strongly suggesting that the product of the ASV *src* gene is a protein kinase. All of the above findings have been confirmed in several

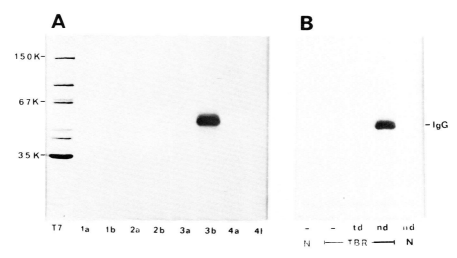

FIG. 6. Detection of protein kinase activity in *src* protein-containing immunoprecipitates. Panel A, Cell extracts were prepared from chick embryo fibroblast cultures that were either uninfected (tracks 1), infected with the avian leukosis virus RAV-2 (tracks 2), infected with SR-ASV (tracks 3), or infected with a td deletion mutant of SR-ASV (tracks 4). Each extract was immunoprecipitated with either normal rabbit serum (a tracks) or TBR serum (b tracks), and a portion of the *S. aureus* (protein A–containing immunoadsorbent bacteria) bound immune complexes was incubated in the protein kinase reaction mixture. After termination of the reaction by heating to 95° C for 1 min in sample buffer and pelleting of the bacteria, the supernatant was subjected to electrophoresis in a discontinuous SDS-polyacrylamide slab gel. The figure represents an autoradiogram of the dried gel. Phage T7 virion proteins are included as molecular weight markers. Panel B, The reticulocyte lysate system without added RNA (−) or with td ASV RNA (td) or nondefective (nd) ASV RNA was employed for the cell-free synthesis of polypeptides as described in the legend to Figure 1B. After immunoprecipitation of the translation products with either normal (N) or TBR serum, a portion of the bacteria-bound immune complexes was assayed for protein kinase activity as described in panel A.

laboratories (Levinson et al. 1978, Rübsamen et al. 1979, Sefton et al. 1979).

Because of the well-recognized role of protein phosphorylation in the functional regulation of a variety of cellular processes, these results had obvious implications concerning the molecular mechanisms of ASV-induced oncogenesis. We have suggested that aberrant phosphorylation of cellular proteins by the transforming gene product pp60[src] may be the mechanism involved in this system of viral transformation (Collett and Erikson 1978). However, the data up to this point, indicating that pp60[src] is capable of protein phosphorylation, have all depended on the immune complex protein kinase assay. The possibility remains that a highly active kinase that cannot be detected by biosynthetic radiolabeling specifically associates with pp60[src] and may be responsible for the phosphotransferase activity described above. Therefore, we have attempted to more directly determine the significance of the IgG-protein kinase activity associated with pp60[src] in immunoprecipitates by partially purifying the pp60[src] protein from cellular extracts. Partially purified preparations of pp60[src] have been obtained, employing either immunoaffinity chromatography (Figure 7) or standard ion exchange chromatography (Erikson et al. 1979a,b). Experiments with these preparations (described below) further support the contention that the ASV transforming gene product itself is a protein kinase.

Using partially purified pp60[src], we have investigated the phosphorylation of commonly used protein kinase substrates. As shown in Figure 8, among those tested, histones (both arginine-rich and lysine-rich), phosvitin, and prot-amine were not phosphorylated by the pp60[src] kinase, whereas casein was phos-phorylated. As expected, normal rabbit IgG was not phosphorylated, whereas immune anti-pp60[src] IgG was. The phosphorylation of casein could be completely inhibited by preincubation of the enzyme fraction with immune, but not normal, IgG, thus demonstrating the pp60[src]-specific nature of the casein phosphory-lation. Our interpretation of this result is that when pp60[src] is sequestered by antibody, it is no longer able to phosphorylate casein. To further demonstrate that the observed phosphorylations were due to pp60[src] itself, we have partially purified by ion exchange chromatography in parallel the pp60[src] proteins from cells infected with a ts *src* gene mutant of ASV (NY68) and the wild-type parental virus. Comparison of the thermolability of the resultant partially purified *src*-specific phosphotransferase activities, employing either immune IgG or casein as substrates, revealed that the enzyme obtained from the ts *src* mutant-infected cells was 7–10 times more thermolabile than the wild-type enzyme (Erikson et al. 1979a,b).

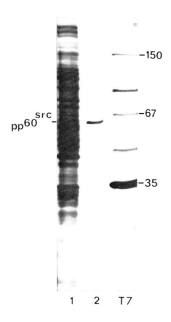

FIG. 7. Purification of pp60[src] by immunoaffinity chroma-tography. Antiserum to pp60[src] was obtained from tumor-bearing rabbits. The immunoglobulin fraction, obtained by ammonium sulfate precipitation, was coupled to Affi-Gel 10 (N-hydroxysuccinimide derivative of Bio-Gel A support, Bio-Rad) and the resultant immunoaffinity matrix used for the purification of pp60[src]. A [35S]methionine-labeled ASV-transformed cell extract was applied to the immunoaffinity column, and the column was extensively washed with a variety of buffers (Erikson et al. 1979a,b), and finally eluted with 1.5 M KSCN. Fractions of the KSCN eluate con-taining radioactivity were pooled and dialyzed against glycerol-containing buffer and then stored at −20°C. Samples of the starting material applied to the column (track 1) and of the pooled KSCN eluate (track 2) were subjected to SDS-polyacrylamide gel electrophoresis along with [35S]methionine-labeled phage T7 virion pro-teins as molecular weight references (displayed as kilodal-tons). The figure is a fluorographic representation of the polyacrylamide gel.

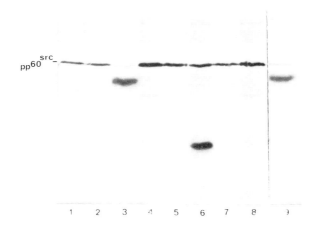

FIG. 8. Phosphorylation of various proteins by partially purified pp60src kinase. Immunoaffinity column-purified pp60src (Figure 7) was incubated in standard protein kinase reaction mixtures with a variety of added exogenous proteins. After termination of the reaction, samples were subjected to SDS-polyacrylamide gel electrophoresis followed by autoradiography. Track 1, partially purified pp60src kinase alone, and with: track 2, normal rabbit serum IgG; track 3, tumor-bearing rabbit serum IgG; track 4, lysine-rich histones; track 5, arginine-rich histones; track 6, casein; track 7, phosvitin; track 8, protamine; and track 9, tumor-bearing rabbit serum IgG and casein.

We have noticed that in all enzymatic reactions with these pp60src preparations, whether or not exogenous proteins are added, pp60src itself becomes phosphorylated (Figure 8). This phosphorylation is inhibited by immune IgG and appears to represent phosphorylation of pp60src by itself (autophosphorylation). The site on the pp60src molecule that is phosphorylated in this apparent autophosphorylation reaction is the same phosphotyrosine residue located on the V8-COOH peptide described earlier (Erikson et al. 1979b). Consistent with this phosphorylation being due to the pp60src kinase itself are our analyses of the pp60src isolated from cells infected with a ts *src* gene mutant (NY68). We found that phosphorylation of the V8-COOH peptide tyrosine residue of pp60src was severely reduced when infected cells were grown at the nonpermissive temperature, whereas a phosphorylation pattern characteristic of wild-type pp60src was observed at the permissive temperature (Collett et al. 1979b). This lack of tyrosine phosphorylation correlated with the greatly reduced protein kinase activity of this protein (Collett and Erikson 1978).

Furthermore, the addition of cAMP to reactions containing the partially purified pp60src kinase does not stimulate autophosphorylation or any other *src*-specific phosphorylation studied to date (Erikson et al. 1979a,b). Finally, when phosphoamino acid analyses were carried out on casein, anti-pp60src IgG, and all subsequent *src*-specific substrates we have identified (Erikson et al. 1979b, Collett et al. 1980), only phosphotyrosine residues were phosphorylated by the pp60src kinase (Figure 9).

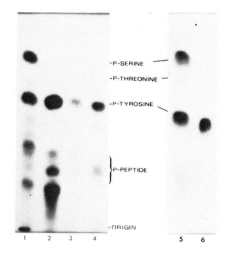

FIG. 9. Phosphoamino acid analyses of proteins phosphorylated by partially purified pp60[src] protein kinase. Various proteins phosphorylated as described in Figure 8 were eluted from polyacrylamide gels, subjected to acid hydrolysis, and analyzed by electrophoresis at pH 3.5. Authentic phosphoamino acid markers were included in all samples. Track 1, pp60[src] isolated by immunoprecipitation from ^{32}P-radiolabeled ASV-transformed cells; track 2, pp60[src] autophosphorylated in vitro; track 3, TBR IgG; track 4, casein; track 5, 34,000-dalton transformation-specific cellular phosphoprotein prepared from radiolabeled SR-ASV-transformed chicken embryo fibroblasts; and track 6, 34,000-dalton cellular protein phosphorylated in vitro as described in Figure 11.

To summarize our initial studies on the enzymatic activity of partially purified pp60[src], we find that the ASV transforming protein is itself a cAMP-independent protein kinase, capable of autophosphorylation, and able to phosphorylate tyrosine residues only. Despite the conclusion that pp60[src] is a cAMP-independent protein kinase, the possibility exists that it has additional functions not revealed by the studies carried out to date. Consequently, additional investigation into other potential functions must be considered. Still, the capacity of protein kinases to influence the function of other proteins via phosphorylation is well recognized, and this activity alone may be responsible for the multiplicity of phenotypic alterations found in the transformed cell.

It is obvious that the phosphorylation of casein and anti-pp60[src] IgG described above do not reflect physiologically significant substrates. Our investigations of the cellular proteins that may serve as substrates for the pp60[src]-kinase activity are at a very preliminary stage. We have shown that various cytoskeletal components (desmin, actin, and tubulin) can act as substrates for phosphorylation by pp60[src]-kinase, and that other cellular proteins cannot (histones, tropomyosin, pyruvate kinase, phosphofructokinase) (Erikson et al. 1979b). However, the mere ability (or inability) of purified pp60[src]-kinase to phosphorylate a protein or protein fraction in vitro provides little information concerning the involvement (or lack of involvement) of that protein in ASV-induced transformation in vivo. Studies must be pursued to determine whether cellular proteins that serve as efficient substrates for pp60[src] phosphorylation in vitro are phosphorylated at the same sites in the transformed cell.

Concerning the analysis of protein phosphorylation in intact untransformed and transformed cells, we (Erikson et al. 1979b) and others (Radke and Martin 1979) have observed the transformation-dependent appearance of a 34,000-dalton

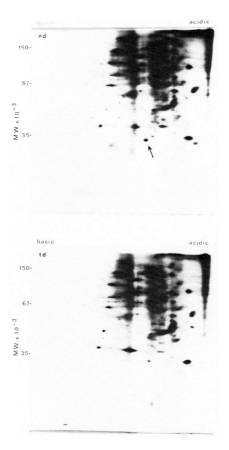

FIG. 10. Two-dimensional gel analysis of phosphoproteins from transformed and untransformed chick embryo fibroblasts. Chick embryo fibroblast cultures transformed by nondefective ASV (nd) or infected with a transformation-defective deletion mutant of ASV (td) were labeled with ^{32}P-orthophosphate. After the labeling period (2 hours) the cells were washed, scraped from the dishes, centrifuged, and then lysed in a urea-containing lysis buffer. A portion of each lysate was analyzed by nonequilibrium pH gradient electrophoresis in the first dimension using pH 3.5–10 Ampholines and SDS-polyacrylamide gel electrophoresis in the second dimension as described by O'Farrell et al. (1977). The arrow points to the transformation-specific 34,000-dalton phosphoprotein.

cellular phosphoprotein by analysis of ^{32}P-labeled total cellular proteins (Figure 10). This phosphoprotein contains a phosphotyrosine residue in addition to a phosphoserine residue (Figure 9). We have subsequently partially purified this 34,000-dalton protein from normal cells and have found that it serves as a phosphate-acceptor for the partially purified pp60^{src} protein kinase activity (Figure 11). The 34,000-dalton protein phosphorylated in vitro by pp60^{src} contains only phosphotyrosine (Figure 9) and, in addition, is phosphorylated on the phosphotyrosine-containing tryptic phosphopeptides corresponding to those found on the 34,000-dalton protein isolated from radiolabeled transformed cells. Thus, this 34,000-dalton cellular protein has many of the characteristics that would be expected of a physiologically significant substrate of the ASV *src* protein kinase. However, future studies must be directed toward determining the functional role of this protein and what effect *src*-specific phosphorylation has on this function. Finally, such a functional modification must be related

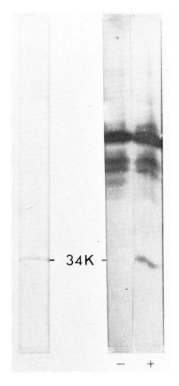

FIG. 11. In vitro phosphorylation of the 34,000-dalton normal cell protein by pp60[src]. A preparation of the 34,000-dalton protein purified from normal chicken embryo fibroblasts was heated at 65°C for 2 minutes and then incubated in the protein kinase reaction mixture as described in the legend to Figure 8. The products of the reaction were analyzed by polyacrylamide gel electrophoresis and autoradiography. Left panel, Coomassie blue–stained SDS-polyacrylamide gel after electrophoresis of the partially purified preparation. Right panel, autoradiogram of SDS-polyacrylamide gel analysis of protein kinase reaction products without (−) and with (+) the addition of the 34,000-dalton protein preparation.

to the transformed phenotype before we may ascertain that the phosphorylation of this particular cellular protein is involved in the mechanism of ASV-induced neoplasia.

NORMAL CELLULAR COUNTERPART OF THE ASV *src* PROTEIN

Molecular hybridization experiments using radioactive DNA specific for the ASV *src* gene have shown that normal uninfected vertebrate cells contain highly conserved nucleotide sequences, denoted *sarc*, which are present in both DNA and RNA and which are related to the viral transforming gene (Stehelin et al. 1976, Wang et al. 1977, Spector et al. 1978a). Furthermore, *sarc*-containing RNA has been found to be associated with polyribosomes (Spector et al. 1978b), suggesting that this RNA may be translated into a *sarc* protein. The close relationship in nucleotide sequence between normal cell *sarc* and viral *src* implies that the putative *sarc* protein may be similar in structure and function to the viral pp60[src]. We have therefore attempted to identify a *src*-related protein in normal, uninfected cells. Recently, we (Collett et al. 1978, Brugge et al. 1979, Collett et al. 1979c) and others (Oppermann et al. 1979, Karess et al. 1979,

Rohrschneider et al. 1979) have found a 60,000-dalton phosphoprotein in normal avian and mammalian cells that is antigenically related to the viral *src* protein (Figure 12). This protein was identified by immunoprecipitation of radiolabeled cell extracts with only certain sera derived from ASV-tumor-bearing rabbits, mice, and marmosets. These rare sera appear to exhibit a higher degree of immunologic cross-reactivity than the majority of the TBR sera used in the identification of the viral pp60^{src} protein. In addition to being antigenically related, all three classes of proteins (viral, avian cell, and mammalian cell) appear to be structurally very similar but not identical (Collett et al. 1978, 1979c, Brugge et al. 1979, Oppermann et al. 1979, Karess et al. 1979, Rohrschneider et al. 1979). Partial proteolysis mapping experiments analogous to those described earlier have been used to demonstrate this structural relatedness (Figure 13). It can be noted in Figure 13 that the V8 protease digestion patterns reveal that the V8-COOH peptide from the normal cell proteins (both avian and mammalian) has a slightly altered electrophoretic mobility when compared to the corresponding viral peptide. Comparative studies of the sites of phosphorylation on these polypeptides have shown that the normal cell *src*-related protein also contains both phosphoserine and phosphotyrosine (Figure 4, panel E), with the former being located on the V8-NH$_2$ peptide and the latter on the V8-COOH fragment (Collett et al. 1979c).

Since the normal cell 60,000-dalton protein is antigenically and structurally so similar to the viral *src* protein, it appears to fit the description expected of

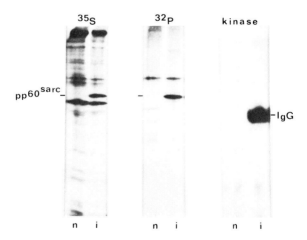

FIG. 12. Detection by immunoprecipitation of a protein related to the ASV *src* protein in normal chick cell extracts. Normal chick embryo fibroblast cultures were labeled with either [^{35}S]methionine or ^{32}P-orthophosphate, cell extracts prepared, and samples immunoprecipitated with either normal (n) or "cross-reacting" tumor-bearing rabbit (i) serum. Immunoprecipitated materials were either subjected directly to SDS-polyacrylamide gel electrophoresis or the bacteria-bound immune complexes were first used in the protein kinase assay, followed by gel electrophoresis.

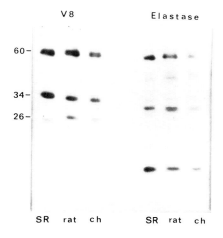

FIG. 13. One-dimensional limited proteolysis mapping of pp60^{src} and pp60^{sarc}. ^{32}P-labeled pp60 proteins were immunoprecipitated from SR-ASV transformed chick cells (SR), normal rat embryo fibroblasts (rat), or normal chick embryo fibroblasts (ch) with standard TBR serum (SR) or tumor-bearing mouse serum (normal cell proteins). These proteins were excised from preparative polyacrylamide gels and subjected to partial proteolysis during re-electrophoresis in a second gel in the presence of either 5 ng of V8 protease or 10 ng of elastase. Numbers represent approximate molecular weight in kilodaltons.

the protein product of the cellular *sarc* sequences, and we have therefore designated it pp60^{sarc} (Collett et al. 1978). This close relationship among the normal endogenous *sarc* polypeptides and the ASV *src* gene product suggests that these apparently highly conserved proteins may have a common role in cellular metabolism. Along these lines, we (Collett et al. 1979c) and others (Oppermann et al. 1979, Karess et al. 1979, Rohrschneider et al. 1979) have identified a protein kinase activity associated with pp60^{sarc} in a manner analogous to that first described for viral pp60^{src} (Collett and Erikson 1978) (Figure 12). The similarity of the results obtained with viral pp60^{src} and normal cell pp60^{sarc} in the immune complex protein kinase assay suggests that the two proteins may be functionally similar. Supporting this possibility are the results obtained by Hanafusa and co-workers involving the recovery of transforming viruses from td virus-infected chickens (Hanafusa et al. 1977). It appears that these viruses are generated by recombination between viral and cellular (presumably *sarc*) sequences to create a functional transforming *(src)* protein (Wang et al. 1978, Karess et al. 1979). The recovered recombinant transforming viruses have obtained at least 75% of the information encoding their pp60^{src} from cellular sequences and produce a transforming gene product which, in the immune complex assay, is associated with a protein kinase activity (Karess et al. 1979). However, to confirm these indirect studies, purified pp60^{src} and purified pp60^{sarc} must be rigorously compared. Studies directed toward this end are currently under way.

ROLE OF pp60^{sarc} IN NORMAL CELL GROWTH AND pp60^{src} IN MALIGNANT TRANSFORMATION

The existence of a normal cellular protein that is structurally and functionally very closely related to the ASV protein responsible for neoplastic transformation raises some intriguing and important questions. These normal cell proteins, present in vertebrates from birds to man, are obviously highly conserved polypeptides, suggesting that they may have a common function in some important facet of cellular metabolism. As it appears that cellular genetic information encoding the avian pp60^{sarc} protein may be the progenitor of the nucleotide sequences specifying the ASV transforming protein pp60^{src}, the role of the normal cell *src*-related protein may be related to some basic mechanism of cell growth control. The finding of an associated protein kinase activity with the normal cell homologues of pp60^{src} implicates phosphorylation-dephosphorylation modification as being critical in this function.

If pp60^{sarc} is functionally similar to pp60^{src}, one might expect the expression of *sarc* to alter the cellular phenotype. To date no evidence suggests that this is the case. Several explanations may be considered for why pp60^{sarc} expression is compatible with normal cellular proliferation. First, as mentioned earlier, there is a chance that the *src* and *src*-related polypeptides have functions other than protein phosphorylation, whereby the roles of *src* and *sarc* differ. Alternatively, both the viral *src* protein and its normal cell homologue do function as protein kinases, but they may have unique substrate specificities. Therefore, different cellular targets of protein phosphorylation may account for the lack of phenotypic transformation by pp60^{sarc}. Finally, cellular transformation may merely be a consequence of a quantitative difference in expression of the two genes. It has been shown that the viral gene product is present in substantially greater amounts (50- to 100-fold) in transformed cells than is the normal cell protein in uninfected cells (Collett et al. 1978, Karess et al. 1979). Thus, it may be the case that the biochemical events in ASV-induced oncogenesis are qualitatively identical to those in normal cells, but occur to a greater degree as the result of pp60^{src} expression, thus producing the transformed phenotype. However, other results suggest that the amount of *src* protein present in a cell is not in itself sufficient to determine whether phenotypic transformation will occur and that host cell factors may play a role (Collett et al. 1979a). There is insufficient information at this time to support rigorously any of the above possible explanations. Additional studies must be pursued in order to obtain further insights into the functional roles of pp60^{src} and pp60^{sarc} in neoplastic transformation and in normal cellular metabolism, respectively.

ACKNOWLEDGMENTS

This investigation was supported by grants CA-21117 and CA-21326 from the National Cancer Institute and grant VC 243 from the American Cancer Society. M.S.C. was a Special Fellow of the Leukemia Society of America and A.F.P. was supported by grant CA-09157 from the National Cancer Institute.

REFERENCES

Brugge, J. S., M. S. Collett, A. Siddiqui, B. Marczynska, F. Deinhardt, and R. L. Erikson. 1979. Detection of the viral sarcoma gene product in cells infected with various strains of avian sarcoma virus and of a related protein in uninfected chicken cells. Virology 29:1196–1203.

Brugge, J. S., E. Erikson, M. S. Collett, and R. L. Erikson. 1978. Peptide analyses of the transformation-specific antigen from avian sarcoma virus-transformed cells. Virology 26:773–782.

Brugge, J. S., and R. L. Erikson. 1977. Identification of a transformation-specific antigen induced by an avian sarcoma virus. Nature 269:346–348.

Cleveland, D. W., S. G. Fischer, M. W. Kirschner, and U. K. Laemmli. 1977. Peptide mapping by limited proteolysis in sodium dodecyl sulfate and analysis by gel electrophoresis. Biol. Chem. 252:1102–1106.

Collett, M. S., J. S. Brugge, and R. L. Erikson. 1978. Characterization of a normal avian cell protein related to the avian sarcoma virus transforming gene product. Cell 15:1363–1369.

Collett, M. S., J. S. Brugge, R. L. Erikson, A. F. Lau, R. A. Krzyzek, and A. J. Faras. 1979a. *src* gene product in both transformed and morphological revertants of ASV-infected mammalian cells. Nature 281:195–198.

Collett, M. S., E. Erikson, and R. L. Erikson. 1979b. Structural analysis of the avian sarcoma virus transforming protein: Sites of phosphorylation. J. Virol. 29:770–781.

Collett, M. S., E. Erikson, A. F. Purchio, J. S. Brugge, and R. L. Erikson. 1979c. A normal cell protein similar in structure and function to the avian sarcoma virus transforming gene product. Proc. Natl. Acad. Sci. USA 76:3159–3163.

Collett, M. S., and R. L. Erikson. 1978. Protein kinase activity associated with the avian sarcoma virus *src* gene product. Proc. Natl. Acad. Sci. USA 75:2021–2024.

Collett, M. S., A. F. Purchio, and R. L. Erikson. 1980. Soluble protein kinase activity of partially purified avian sarcoma virus-transforming protein, pp60^{src}, results in the phosphorylation of tyrosine residues. Nature 285:167–169.

Erikson, E., M. S. Collett, and R. L. Erikson. 1978. In vitro synthesis of a functional avian sarcoma virus transforming gene product. Nature 274:919–921.

Erikson, R. L., M. S. Collett, E. Erikson, and A. F. Purchio. 1979a. Evidence that the avian sarcoma virus transforming gene product is a cyclic AMP-independent protein kinase. Proc. Natl. Acad. Sci. USA 76:6260–6264.

Erikson, R. L., M. S. Collett, E. Erikson, A. F. Purchio, and J. S. Brugge. 1979b. Protein phosphorylation mediated by partially purified avian sarcoma virus transforming gene product. Cold Spring Harbor Symp. Quant. Biol. 44 (in press).

Hanafusa, H. 1977. Cell transformation by RNA tumor viruses, *in* Comprehensive Virology, H. Fraenkel-Conrat and R. P. Wagner, eds. Plenum Publishing Corp., New York, pp. 401–483.

Hanafusa, H., C. C. Halpern, D. L. Buchhagen, and S. Kawai. 1977. Recovery of avian sarcoma virus from tumors induced by transformation defective mutants. J. Exp. Med. 146:1735–1747.

Karess, R. E., W. S. Hayward, and H. Hanafusa. 1979. Cellular information in the genome of recovered avian sarcoma virus directs the synthesis of transforming protein. Proc. Natl. Acad. Sci. USA 76:3154–3158.

Levinson, A. D., H. Oppermann, L. Levintow, H. E. Varmus, and J. M. Bishop. 1978. Evidence that the transforming gene of avian sarcoma virus encodes a protein kinase associated with a phosphoprotein. Cell 15:561–572.

O'Farrell, P. Z., H. M. Goodman, and P. H. O'Farrell. 1977. High resolution two-dimensional electrophoresis of basic as well as acidic proteins. Cell 12:1133–1142.

Oppermann, H., A. D. Levinson, H. E. Varmus, L. Levintow, and J. M. Bishop. 1979. Uninfected vertebrate cells contain a protein that is closely related to the product of the avian sarcoma virus transforming gene *(src)*. Proc. Natl. Acad. Sci. USA 76:1804–1808.

Purchio, A. F., E. Erikson, J. S. Brugge, and R. L. Erikson. 1978. Identification of a polypeptide encoded by the avian sarcoma virus *src* gene. Proc. Natl. Acad. Sci. USA 75:1567–1571.

Purchio, A. F., E. Erikson, and R. L. Erikson. 1977. Translation of 35S and of subgenomic regions of avian sarcoma virus RNA. Proc. Natl. Acad. Sci. USA 74:4661–4665.

Radke, K., and G. S. Martin. 1979. Transformation by Rous sarcoma virus: Effects of *src* gene expression on the synthesis and phosphorylation of cellular polypeptides. Proc. Natl. Acad. Sci. USA 76:5212–5216.

Rohrschneider, L. R., R. N. Eisenman, and C. R. Leitch. 1979. Identification of a Rous sarcoma virus transformation-related protein in normal avian and mammalian cells. Proc. Natl. Acad. Sci. USA 76:4479–4483.

Rübsamen, H., R. R. Friis, and H. Bauer. 1979. *src* gene product from different strains of avian sarcoma virus: Kinetics and possible mechanism of heat inactivation of protein kinase activity from cells infected by transformation-defective, temperature-sensitive mutant and wild type virus. Proc. Natl. Acad. Sci. USA 76:967–971.

Sefton, B. M., T. Hunter, and K. Beemon. 1979. Product of in vitro translation of the Rous sarcoma virus *src* gene has protein kinase activity. J. Virol. 30:311–318.

Spector, D. H., B. Baker, H. E. Varmus, and J. M. Bishop. 1978a. Characteristics of cellular RNA related to the transforming gene of avian sarcoma viruses. Cell 13:381–386.

Spector, D. H., K. Smith, T. Padgett, P. McCombe, D. Roulland-Dussoix, C. Moscovici, H. E. Varmus, and J. M. Bishop. 1978b. Uninfected avian cells contain RNA related to the transforming gene of avian sarcoma virus. Cell 13:371–379.

Stehelin, D., H. E. Varmus, J. M. Bishop, and P. K. Vogt. 1976. DNA related to the transforming gene(s) of avian sarcoma virus is present in normal avian DNA. Nature 260:170–173.

Vogt, P. K. 1977. The genetics of RNA tumor viruses, *in* Comprehensive Virology, H. Fraenkel-Conrat and R. P. Wagner, eds. Plenum Publishing Corp., New York, pp. 341–455.

Wang, L. H., C. C. Halpern, M. Nadel, and H. Hanafusa. 1978. Evidence for recombination between viral and cellular sequences to generate transforming sarcoma virus. Proc. Natl. Acad. Sci. USA 75:5812–5816.

Wang, S. Y., W. S. Hayward, and H. Hanafusa. 1977. Genetic variation in the RNA transcripts of endogenous virus genes in uninfected chicken cells. J. Virol. 24:64–73.

Gene Expression

Genes, Chromosomes, and Neoplasia, edited by
Frances E. Arrighi, Potu N. Rao, and Elton Stubblefield.
Raven Press, New York © 1981.

Genetically Altered Human Mitochondrial DNA and a Cytoplasmic View of Malignant Transformation

Donald L. Robberson, Marion L. Gay, and Cheryl E. Wilkins

*Department of Molecular Biology, The University of Texas System Cancer Center
M. D. Anderson Hospital and Tumor Institute, Houston, Texas 77030*

The transformation of normal cells to malignant genotypes is expected to be accompanied by alteration of cytoplasmically localized components. Changes in cytoplasmic physiologies that accompany transformation may be reflected in either transient or stable alterations of mitochondrial DNA (mtDNA) structure and replication mechanism. Since the molecular biological bases of transformation of human cells to malignancy have not yet been determined, we have initiated studies that use electron microscopy procedures to analyze alterations of the mitochondrial genome structure and replication mechanism in order to further define and then probe the cytoplasmic events that possibly accompany malignant transformation. We describe here certain aspects of our progress using this approach to elucidate the consequences of marrow cell transformation on mitochondrial genome structure and replication mechanism in human acute myelogenous leukemia (AML).

ALTERED MITOCHONDRIAL DNA COMPLEXITIES ASSOCIATED WITH HUMAN MALIGNANCIES

In 1967, David Clayton and the late Professor Jerome Vinograd reported the discovery of one aspect of altered complexity in mtDNA isolated from peripheral blood cells of two patients with chronic myelogenous and lymphocytic leukemia, respectively, before these patients had received chemotherapy (Clayton and Vinograd 1967). The particular type of alteration in complexity of mtDNA that Clayton and Vinograd discovered involved the duplication of the mitochondrial genome to produce circular dimer forms. The electron microscopic appearance of the circular dimer form of human mtDNA, isolated essentially as described (Robberson and Clayton 1972), is illustrated in Figure 1, after nicking the superhelical, covalently closed circular duplex molecules with DNAse I and preparation for microscopy by the aqueous basic protein (Kleinschmidt) spreading procedure (Davis et al. 1971) as modified by Robberson et al. (1971). The circular dimer form is illustrated relative to the sizes of circular monomer

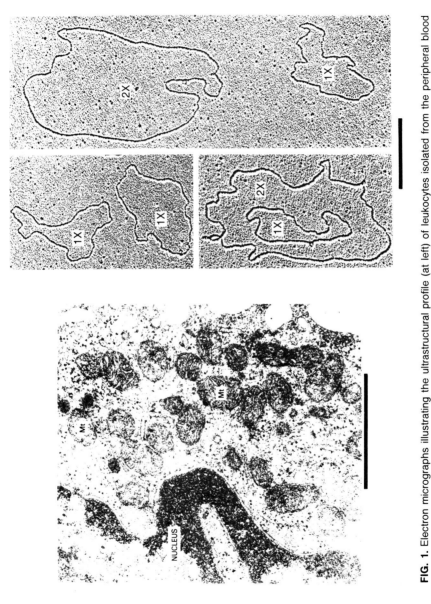

FIG. 1. Electron micrographs illustrating the ultrastructural profile (at left) of leukocytes isolated from the peripheral blood of a patient with AML, and the mitochondrial DNA (mtDNA) molecules (panels at right) from these cells purified as described (Robberson and Clayton 1972). A portion of the nucleus and mitochondria (Mt) are indicated at left. Circular monomer (1X) and circular dimer (2X) mtDNA are indicated at right. In the lower panel at the right, a circular monomer molecule lies completely within the contour of a circular dimer molecule. Magnifications are indicated by lengths of bars, at left (equal to 0.5 microns) and at right (equal to 1.0 microns) below micrographs.

human mtDNA with a length of approximately five microns and containing approximately 16,750 ± 610 base pairs (calculated from the data presented in Robberson et al. 1977).* The sample of mtDNA shown in Figure 1 was isolated from leukocytes in the peripheral blood of a patient with acute myelogenous leukemia. The ultrastructural appearance of mitochondria in these leukocytes is shown in the electron micrograph at the left in Figure 1.

A more extensive study by Clayton and Vinograd (1969) revealed the occurrence of circular dimer mtDNA in peripheral blood cells of patients with either acute or chronic forms of myelogenous leukemia. The circular dimer form was typically found at high frequencies and comprised 48% of all mtDNA molecules in one patient with AML (Clayton and Vinograd 1969). Circular dimer forms were not detected in the mitochondrial isolates from peripheral blood cells of normal individuals (Clayton and Vinograd 1967, 1969) or patients with myeloid metaplasia or leukomoid reaction (Clayton and Vinograd 1969). Furthermore, administration of chemotherapy (primarily Myleran or 6-mercaptopurine) resulted in reduction of the circular dimer frequency in the mitochondrial isolates of peripheral blood cells of patients with acute or chronic myelogenous leukemia (Clayton and Vinograd 1969).

The circular dimer form of mtDNA has also been detected in a variety of human solid tumors (Smith and Vinograd 1973, Kumar and Fox 1974, Clayton and Smith 1975), usually at lower frequencies than observed in most human leukemic leukocyte populations; however, not all human solid tumor tissues or lymphoblastic leukemias (peripheral blood cells or infiltrated tissues) have revealed the presence of circular dimer mtDNA (Clayton and Smith 1975). Likewise, the circular dimer form of mtDNA is not reproducibly detected in viral leukemias in animals (Jordan et al. 1970). Evidence has been presented for exceedingly high frequencies of circular dimer forms of mtDNA in normal beef thyroid and nonmalignant human oncocytomas (Paoletti et al. 1972). The earlier claim of very high frequencies of the circular dimer form of mtDNA in thyroid tissue has not been substantiated in a careful study conducted by Matsumoto et al. (1976). The circular dimer form was detected at a number frequency of 0.1%–0.3% in different animal thyroid tissue, with a slightly elevated value (0.9%) for thyroid tissue of the cow.

More recent studies have revealed that the mtDNA from human peripheral blood leukocytes of healthy young adult males do contain very low frequencies of the circular dimer form (Piko et al. 1978). The leukocytes from three of five individuals contained an average of 0.04% ± 0.03% of the mtDNA molecules as circular dimers. Our own studies have revealed that frequencies of 0.17% ± 0.14% and 0.11% ± 0.10% of the mtDNA mass, respectively, occurred as

---

* Length measurements are reported here as values corrected for the length of the sequence of φx replicative form DNA with 5,386 base pairs (Sanger et al. 1978). The length of φXRF-DNA was measured electron microscopically relative to the length of PM2-DNA (10,260 ± 200 basepairs), which was used as an internal length standard throughout most of these studies.

circular dimer forms in the peripheral blood cells of two healthy young adult males.

Although the mechanism of formation of circular dimer mtDNA is not yet understood, the observations noted above would indicate that finding elevated frequencies of the circular dimer form of mtDNA could possibly serve as an informative marker of altered cytoplasmic physiologies in at least one human malignancy, that of myelogenous leukemia, especially when the acute form of the disease is manifest.

The Bone Marrow Origin of Circular Dimer Mitochondrial DNA in Human AML

Since all previous studies on circular dimer mtDNA in human leukemia have dealt only with peripheral blood cells, we considered it essential to first determine if formation of circular dimer molecules actually occurred in the bone marrow stem cells. When mtDNA components isolated from bone marrow cells are compared to those isolated from peripheral blood cells of the AML patient prior to chemotherapy, it is found that circular dimer forms occur in both samples at similar frequencies, as illustrated by the fields in the micrographs shown in Figure 2. If circular dimer mtDNA is formed in the bone marrow cells and these cells are then released to the peripheral blood, the frequency of the circular dimer form of mtDNA in the marrow cells will always be greater than or equal to the frequency observed in the peripheral blood cells taken from the same patient at the same time. Indeed, this is observed for examination of matched samples of marrow and peripheral blood from ten AML patients, as illustrated in Figure 3. We conclude that circular dimer mtDNA is formed in the bone marrow stem cell population and its frequency in marrow cells can possibly serve to monitor malignant transformation events. Normal human bone marrow cells contain only $0.37\% \pm 0.21\%$ and $0.49\% \pm 0.24\%$ of the mtDNA mass as circular dimer mtDNA for examination of 3,163 molecules in samples from two young healthy individuals, one male and one female. It is interesting to note that analysis of the peripheral blood cells of these same two individuals gave circular dimer frequencies of $0.17\% \pm 0.14\%$ and $0.12\% \pm 0.11\%$ of the mtDNA mass, respectively. Thus, for these two healthy individuals, the frequencies of circular dimer mtDNA are higher in the bone marrow cells than in peripheral blood cells. This is expected if there is a low frequency of such forms in healthy individuals and these forms originate in the marrow stem cell population.

Cytological Correlations of Marrow Cell Populations with Circular Dimer Mitochondrial DNA Frequencies in Human AML

If formation of circular dimer mtDNA does indeed reflect malignant transformation events in selected members of the marrow stem cell population, we

Human AML Mitochondrial DNA

Peripheral Blood   Bone Marrow

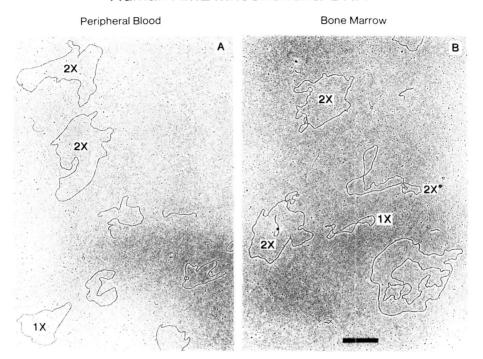

FIG. 2. Electron micrographs illustrating appearance of mtDNA isolated from peripheral blood (A) and bone marrow cells (B) of a patient with AML. Examples of circular monomer (1X) and circular dimer (2X) forms of mtDNA. Magnification is indicated by bar length of one micron (at bottom, right panel).

might anticipate finding a positive correlation between the number of leukemic cells, classified cytologically through clinical diagnostic procedures, and the frequencies of circular dimer mtDNA molecules isolated from the marrow cells of AML patients. When this analysis was performed, we were surprised to find that the results for the 12 AML samples examined were clustered into three distinct groups (Figure 4). Patients with a high frequency of leukemic cells and a correspondingly high frequency of circular dimer forms of mtDNA (Group III in Figure 4) had a very poor prognosis and, even with aggressive chemotherapy, survived only a few months. Marrow smears for one representative member of each group are shown at the right in Figure 4 and illustrate fields containing high frequencies of abnormal myeloblasts for Groups II and III, respectively. The correlations between formation of increased frequencies of circular dimer mtDNA and malignant transformation in these cell populations

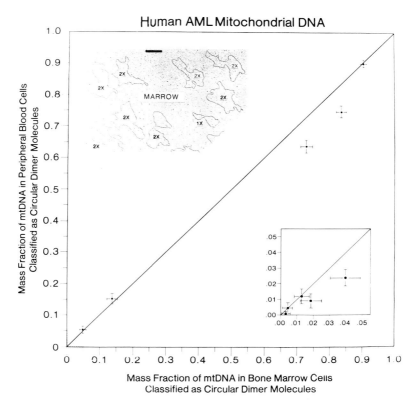

FIG. 3. Distributions of circular dimer mtDNA frequencies in peripheral blood (vertical coordinate) and bone marrow cells (horizontal coordinate) for 10 human AML samples. Plot inset at lower right depicts samples with low frequencies of circular dimer mtDNA (cross-hatched area at lower left on larger plot). Data points that lie on the diagonal indicate samples with equal frequencies of circular dimer mtDNA in peripheral blood and marrow cells. Sampling errors at 95% confidence limits are indicated on data points. Data points that lie to the right of the diagonal indicate samples that contained higher frequencies of circular dimer mtDNA in the marrow cells than in peripheral blood cells. Micrograph inset at top illustrates appearance of mtDNA in one marrow sample containing a very high frequency of circular dimer forms. Magnification of this micrograph is illustrated by bar length of one micron (at top).

appears self-consistent, in so far as a group of patients having a high frequency of circular dimer mtDNA but low frequencies of leukemic cells is not detected (the lower right quadrant of Figure 4).

Elevated Frequencies of Catenated Molecules Associated with Formation of Circular Dimer Mitochondrial DNA in Human AML Marrow Cells

The correlations of discrete groupings of abnormal myeloblasts with circular dimer mtDNA frequencies must reflect substantial alterations in overall cellular

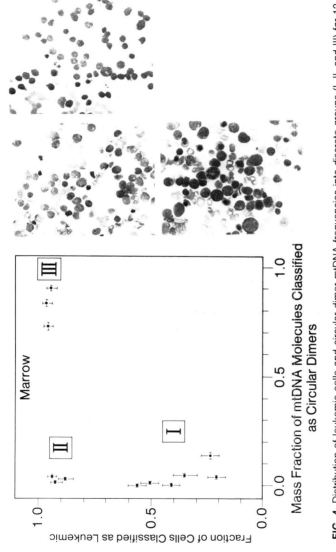

FIG. 4. Distribution of leukemic cells and circular dimer mtDNA frequencies into discrete groups (I, II, and III) for 12 human AML marrow samples. Photomicrographs at right depict Wright-Giemsa stains of marrow smears for one member representative of Groups I (lower left), II (upper left), and III (upper right).

physiology, which could also be manifest as alterations in the mitochondrial physiology of these cells. Physiological as well as morphological changes in the mitochondria of tumor cells have been well documented previously (Paoletti and Riou 1973). In order to extend our observations of altered cellular physiologies to intramitochondrial events that accompany formation of circular dimer mtDNA molecules in malignant cells, we have examined the frequency distribution of catenation of circular mitochondrial genomes in the marrow cells of these AML patients.

Catenanes, in which circular genomes are topologically bonded as for the links in a chain (Hudson and Vinograd 1967), have been found to occur in a variety of both prokaryotic and eukaryotic circular DNAs (Clayton and Smith 1975). Catenanes constitute 6%–9% of the mtDNA molecules in normal human tissues (Clayton and Vinograd 1967) and other mammalian tissues (Clayton et al. 1968). In normal tissues, most of these catenanes involve the topological bonding of two monomer genomes (dimer catenanes), whereas in tumor cells, catenanes of three or more molecules comprise a substantial proportion of all catenated mtDNA molecules (Clayton and Vinograd 1967, Smith and Vinograd 1973). For example, analysis of the mtDNA in two samples of peripheral blood cells from normal individuals has shown that 4% ± 1% and 6% ± 3%, respectively, of the molecules (number frequency) occurred in dimer catenanes, and 5% ± 2% and 1.0% ± 0.9%, respectively, occurred in trimer or higher catenated oligomers (Clayton and Vinograd 1969). In another study, catenated forms of mtDNA were shown to comprise 2%–6% (number frequency) of the molecules in human leukocytes from healthy adults (Piko et al. 1978).

Our own examinations of mtDNA from bone marrow cells of two healthy individuals demonstrated 4.3% and 6.0%, respectively, of the mtDNA mass occurred in dimer catenanes, and 1.4% and 1.6%, respectively, occurred in trimer or higher catenated oligomers for examination of approximately 3,000 molecules in each sample. The frequencies of all forms of catenated mtDNA molecules relative to uncatenated molecules have been reported to be increased in a variety of tumor cells (Smith and Vinograd 1973) and in virally transformed animal cells (Jordan et al. 1970, Van der Eb et al. 1970, Riou and Delain 1971, Riou and Lacour 1971, Nass 1973) as well as in human leukemic leukocytes (Clayton and Vinograd 1967, 1969, Hudson et al. 1968). (For review, see Clayton and Smith 1975.)

Increased frequencies of catenated mtDNA molecules represent an expression of intramitochondrial enzymes involved in replication or recombination events, and the activities of these enzymes may be found to correlate with the degree of metabolic imbalance in the leukemic cell, an imbalance that is reflected by the increased frequencies of circular dimer formation. We therefore examined the frequencies of catenated mtDNA molecules as a function of circular dimer frequencies observed in marrow cells of the same 12 AML patients in this study to find the rather surprising result shown in Figure 5. The results for the same set of patients were also distributed into discrete groups, similar to

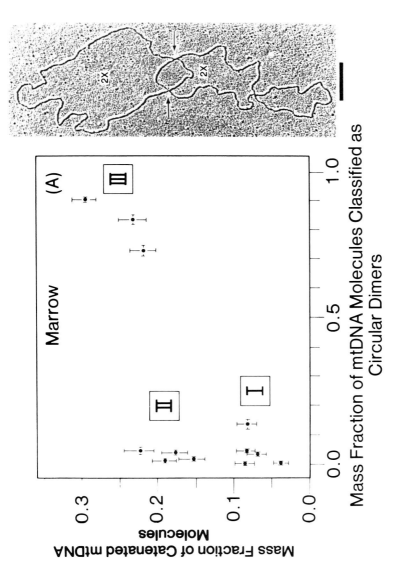

FIG. 5. Distributions of catenated and circular dimer mtDNA frequencies into discrete groups (I, II, and III) for 12 AML marrow samples. Electron micrograph at right illustrates appearance of circular dimer (2X) molecules catenated at positions indicated by arrows.

the pattern we had observed for the relationship between frequencies of leukemic cells and circular dimer mtDNA. It is important to emphasize that although samples in Group III of Figure 4 are likewise in Group III of Figure 5, only one sample in Group II of Figure 4 is assigned to Group II of Figure 5. An example of the appearance of a catenane of two circular dimer molecules is shown at the right in Figure 5. Patients in Group I have normal levels of catenated mtDNA, whereas those in Groups II and III have two to three times the normal level in healthy individuals (Figure 5). Again, we note the significance of not having detected a group of patients having a high frequency of circular dimer mtDNA and only low or normal levels of catenated mtDNA molecules (lower right quadrant of Figure 5). These results suggest a correlation between perturbation of the intramitochondrial enzymology that is active in either formation or removal of catenanes and the concomitant formation of circular dimer mtDNA associated with malignant transformation of the marrow cells.

Higher Oligomeric Forms of Mitochondrial DNA in Human AML Marrow Cells

More complex catenane structures of mtDNA are detected in isolates of human AML cells. An example of a pentamer catenane of circular dimer mtDNA molecules is illustrated in Figure 6. It is possible to detect such complicated structures because we have prepared our samples by the aqueous basic protein (Kleinschmidt) technique (Davis et al. 1971) in which molecules tend to avoid contact in the surface film unless they are topologically bonded (Hudson and Vinograd 1967). For the 12 AML marrow samples analyzed, complex catenated structures consisting of three or more molecules comprised an average of 22.7% ± 10.4% (± one standard deviation) of the *total* catenated mtDNA mass, with the remainder of *catenated* mtDNA consisting of dimer catenanes of monomers with monomers, dimers with dimers, or dimers with monomers. In two samples of normal human marrow, 24.1% and 20.6%, respectively, of the *catenated* mtDNA mass was found in catenanes consisting of three or more molecules. Although our studies of these more complex catenane structures are not yet completed, the results of these preliminary examinations indicate that more complex catenated mtDNA structures may not always be increased in frequency in comparison to mtDNA isolated from normal marrow cells and, accordingly, may not provide a sensitive or useful parameter to monitor malignant transformation.

More complex circular oligomer structures, such as the circular pentamer shown in Figure 7, are detected but occur at very low frequencies. Examination of 5,200 molecules in one AML marrow sample (Group I, Figure 4) revealed the presence of only one circular trimer. Examination of 5,790 molecules in a second AML marrow sample (Group III, Figure 4) did, however, reveal a higher frequency, with detection of 13 circular trimers and three circular tetramers. We have also detected more complex circular oligomers in mtDNA prepa-

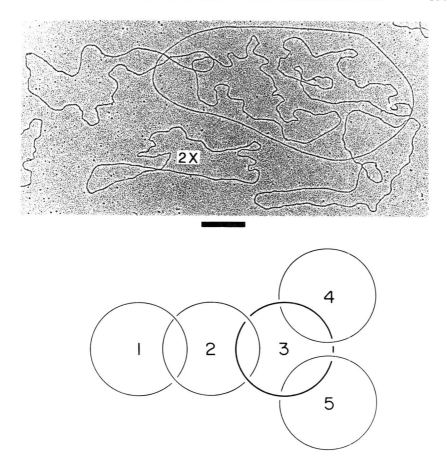

FIG. 6. Electron micrograph illustrating appearance of a catenane of five circular dimer mtDNA molecules. An uncatenated circular dimer (2X) molecule is also present in the field. Drawing below the micrograph schematically illustrates interpretation of the composition of this catenane structure.

rations of normal human marrow cells but at very low frequencies. For example, random sampling of 4,500 molecules in one normal marrow preparation revealed the presence of only one circular tetramer. In combined examinations of both normal and AML samples, we have detected examples of the circular trimer through circular decamer with approximately equally low frequencies. A single circular tetramer of mtDNA was found by Piko et al. (1978) in peripheral blood cells from a healthy individual and until now had been the largest circular oligomer of mtDNA reported. The very low frequencies of such higher circular oligomers make a comparative analysis of these forms in normal and AML conditions, respectively, quite impractical at this time. Since such forms occur

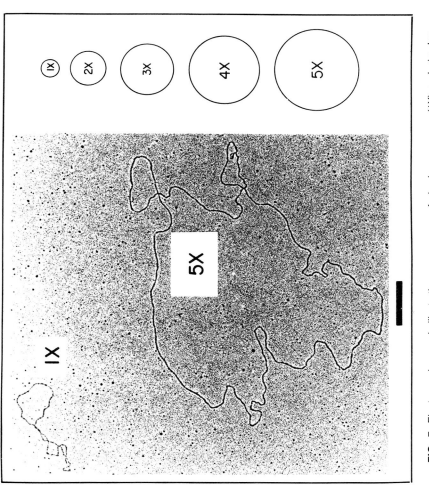

FIG. 7. Electron micrograph illustrating appearance of circular monomer (1X) and circular pentamer (5X) mtDNA. Drawings at right depict, schematically, the relative sizes of monomer through pentamer circular molecules.

in both normal and AML marrow samples and at very low frequencies, it is unlikely that their formation will be found to correlate with malignant transformation. This emphasizes the importance of the circular dimer form of mtDNA as the principal circular oligomeric form that increases in frequency in malignant transformation.

GENOME ALTERATIONS IN CIRCULAR DIMER MITOCHONDRIAL DNA OF HUMAN AML CELLS

From the considerations that have been presented, it is clear that formation of circular dimer mtDNA not only constitutes the major change in topological complexity of mtDNA that accompanies malignant transformation in human AML cells but can also comprise a significant proportion of the total mtDNA mass, for example, up to 90.5% of the mtDNA mass in one of our samples of AML marrow cells (see Figure 3). It is therefore important to determine if the mitochondrial genetic information represented in the circular dimer molecules has been qualitatively altered, in addition to quantitative genetic changes associated with intramitochondrial genome duplication.

We have previously detected microheterogeneity in circular dimer mtDNA molecules isolated from human AML peripheral blood cells (Robberson et al. 1977). Microheterogeneity of DNA sequence content of the circular dimer was first detected in restriction endonuclease cleavage patterns. In addition, microheterogeneity in the sizes of the circular dimer molecules was detected and found to be associated with small deletions or additions of nucleotide sequences, corresponding to approximately 300 ± 140 base pairs (33 molecules measured) out of the 33,800 ± 740 base pairs that compose human circular dimer mtDNA molecules (Robberson et al. 1977). This extent of microheterogeneity would correspond to approximately 50% of the circular dimer molecules in one preparation having at least one such deletion or addition of nucleotide sequence, and approximately 93% of these molecules would contain only one such sequence alteration.

We have extended these studies to determine if the deletion or addition of nucleotide sequences occurs at a unique site on the circular dimer genome. One sample of human AML mtDNA containing 60% circular dimer forms was treated with the restriction enzyme *Eco*RI, which cleaves human mtDNA at three unique sites (Robberson et al. 1974, Brown and Vinograd 1974), and the resulting fragments were then denatured and reassociated. Electron microscopic examination of the heteroduplexes thus formed revealed the presence of small nonhybridized loops of single-stranded DNA on the renatured DNA fragments, approximately 300 ± 140 nucleotides in length, as shown in Figure 8. The three *Eco*RI fragments of human AML mtDNA have lengths of 8,070 ± 170, 7,320 ± 180, and 1,110 ± 140 base pairs, respectively (Robberson et al. 1977), corrected for the length of the sequence of φx replicative form DNA with 5,386 base pairs (Sanger et al. 1978). The two largest *Eco*RI fragments

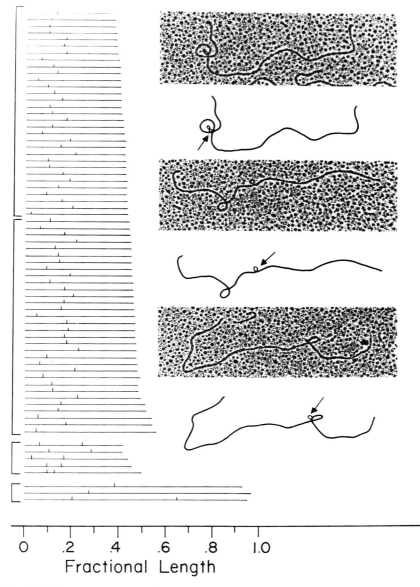

FIG. 8. Deletions or additions of sequences detected in *Eco*RI fragments of circular dimer mtDNA purified from human AML cells. A sample of human AML-mtDNA containing 60% circular dimer forms was treated with *Eco*RI, followed by denaturation and renaturation of strands. Samples were prepared for electron microscopy by the formamide modification of the Kleinschmidt technique (Davis et al. 1971). *Eco*RI cleaves human mtDNA at three unique sites. Deletions or additions of sequences are detected as nonhybridized loops of single-stranded DNA on reassociated strands, as illustrated in micrographs at right at positions indicated by arrows in drawings below the micrographs. The gallery of reassociated strands at

reassociate to a significant extent and permit measurements of the position of the single-strand loop of nonhybridized DNA.

The results of these measurements on 31 reassociated molecules of the second largest *Eco*RI fragment (top bracket in gallery of strands in Figure 8) and 32 reassociated molecules of the largest *Eco*RI fragment (second bracket from top in gallery of strands in Figure 8) indicate that the deletion or addition loop (vertical bars in gallery of strands in Figure 8) can occur at quite variable positions throughout the genome that is represented by the two largest *Eco*RI fragments (Figure 8). Some reassociated strands display two loops of single-strand DNA also at variable positions (third bracket from top in Figure 8). Three partial *Eco*RI digestion fragments also reassociated and are found to display single-strand loops at variable positions (bracket at bottom of gallery of Figure 8). The smallest *Eco*RI fragment of human mtDNA does not reassociate to a sufficient extent, under these conditions of renaturation, to determine if deletions or additions occur in this portion of the genome. It is clear that qualitative changes occur in the mitochondrial genetic information encoded in the circular dimer molecule and that these deletions or additions of nucleotide sequences can occur throughout a large portion of the genome. Of particular interest are deletions or additions of DNA sequences that occur near the center of the largest *Eco*RI fragment (second bracket from top in gallery of Figure 8). These sites are in a region of the genome identified with the origin for displacement synthesis in human mtDNA (Brown and Vinograd 1974, Robberson et al. 1974), and possibly with the initiation sites for duplex synthesis (see next section). Alterations in this region of the genome could possibly provide a selective advantage for replication of the circular dimer molecule.

DISPLACEMENT REPLICATION OF CIRCULAR DIMER MITOCHONDRIAL DNA IN HUMAN AML CELLS

Various aspects of the replication mechanism for circular monomer mtDNA isolated from mouse L cells (Robberson et al. 1972, 1974, Robberson and Clayton 1972, 1973, Kasamatsu et al. 1971, 1974, Berk and Clayton 1974, 1976, 1978, Bogenhagen and Clayton 1978) and mouse tissues (Piko and Matsumoto 1977) have been determined in detail. A similar replication mechanism was developed for mtDNA isolated from Chang rat solid hepatomas and Novikoff rat ascites hepatomas (Wolstenholme et al. 1973). Examination of circular dimer mtDNA

left depicts positions of deletions or additions (indicated by small vertical bars) on second largest (top bracket) and largest (second bracket) *Eco*RI fragments. Reassociated strands with two loops are depicted in third bracket from top. Reassociated strands from *Eco*RI partial digestion fragments that contained one or two loops are shown in fourth bracket from top. Fragments in this gallery have been oriented such that the terminus proximal to the loop is at the left. Mircographs illustrate appearance of reassociated strands of the largest *Eco*RI fragments.

in a culture line of mouse LD cells revealed that a substantial fraction of the circular dimer molecules contained two displacement loops (D-loops) that were diametrically opposed on the 10-micron circular genomes (Kasamatsu et al. 1971). The mode of replication of the circular dimer molecules in mouse LD cells proceeds by unidirectional expansion of only one of the two D-loops (Kasamatsu and Vinograd 1973, Robberson and Clayton 1972). Although the origin of unidirectional displacement replication has been determined for human circular monomer mtDNA (Brown and Vinograd 1974, Robberson et al. 1974) and details of the structure around this initiation sequence have been determined (Gillum and Clayton 1978, 1979, Brown et al. 1978), the details of replication of the complementary strand (duplex synthesis) have not yet been elaborated. It should be noted, however, that the sequence at the single initiation site for duplex synthesis in mouse L cell mtDNA replication has also recently been determined (Martens and Clayton 1979). A comparable analysis of human circular dimer mtDNA has yet to be presented.

A few examples of displacement replicative forms of circular dimer mtDNA have been detected in isolates of human oncocytoma (Paoletti and Riou 1973). We have therefore undertaken studies to define the mechanism of human mtDNA replication, with particular emphasis on possible alterations of this mechanism that may be exhibited by circular dimer forms. Examination of circular dimer mtDNA isolated from human AML cells reveals closed circular molecules without D-loops and molecules with one or two D-loops, respectively (Figure 9), as well as molecules with one expanded D-loop, one D-loop and one expanded D-loop, and with two expanded D-loops (Figure 10). Most interestingly, in contrast to displacement replication of circular dimer mtDNA of mouse LD cells, we have detected examples of human AML circular dimer mtDNA in which both D-loops on the same molecule have been expanded (Figure 11). These particular replicative forms have escaped the normal restrictive control that permits expansion of only one of the two D-loops. Further investigation of these replicative forms could provide insight into the regulatory controls of mtDNA replication that are operative in the human leukemic condition.

Preservation of Unidirectional Displacement Replication in Mitochondrial DNA of Human AML Cells

The finding of displacement replicative forms of the circular dimer molecule in which both D-loops have been expanded has prompted a further examination of this unusual pattern of displacement replication. Length measurements were obtained for the replicated and unreplicated segments (Figure 12) on 28 circular dimer mtDNA molecules isolated from one sample of human AML peripheral blood cells. Circular dimer molecules containing two nonexpanding D-loops are shown to have the D-loops diametrically opposed on the circular dimer genome (Figure 12). The expansion of the D-loop in molecules with one expanded

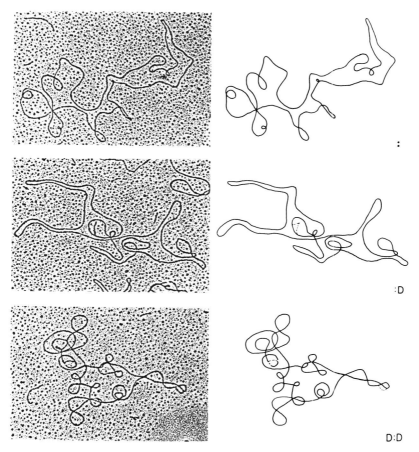

FIG. 9. Displacement replicative forms of circular dimer mtDNA without D-loops (:), with one D-loop (:D), and with two D-loops (D:D). Heavy solid lines depict duplex portions and dashed lines depict single-strand portions of the molecules in drawings at the right. Samples of the covalently closed circular DNA were prepared for electron microscopy by the formamide modification of basic protein (Kleinschmidt) technique (Davis et al. 1971).

D-loop and one nonexpanding D-loop is shown to proceed in a unidirectional fashion (Figure 12). Six molecules with two expanded D-loops were detected in this sampling, but length measurements on segments of these molecules cannot distinguish between unidirectional and bidirectional expansion (Figure 12). For these molecules, bidirectional expansion of both D-loops at the same rate would be indistinguishable from unidirectional expansion of each D-loop. We can, however, utilize this analysis to eliminate the possibility of unidirectional expansion of one D-loop and bidirectional expansion of the other D-loop (Figure 12).

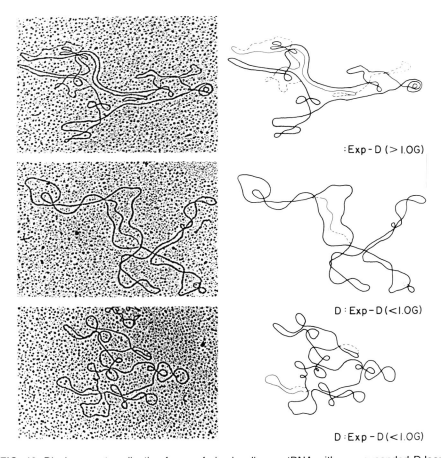

FIG. 10. Displacement replicative forms of circular dimer mtDNA with one expanded D-loop (:Exp-D) for which expansion through more than a monomer genome length (>1.0G) have occurred, and forms with one D-loop and one expanded D-loop (D:Exp-D) for which expansions through less than a monomer genome length (<1.0G) has occurred. Thin solid lines in drawings at the right depict regions of the displaced single strand (dashed lines) where duplex synthesis has occurred.

It is of interest to note that there is a considerable degree of scatter in positioning the origin of replication at an average fractional length of 1.01 ± 0.05. This degree of uncertainty in position is unexpected on the basis of statistical fluctuations for segments of duplex DNA with these lengths, ± 0.02, as measured by the formamide modification of the Kleinschmidt technique (Davis et al. 1971) and also for the three discrete lengths for human 7S DNA initiation sequence that range from 555 to 615 nucleotides (Gillum and Clayton 1978, Brown et al. 1978, Bogenhagen and Clayton 1978). This additional degree of uncertainty in localizing the origin of the expanded D-loop does correspond,

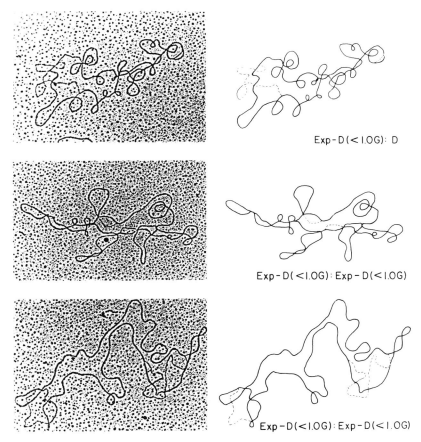

Exp-D(<1.0G): D

Exp-D(<1.0G): Exp-D(<1.0G)

Exp-D(<1.0G): Exp-D(<1.0G)

FIG. 11. Displacement replicative forms of circular dimer mtDNA with one expanded D-loop (<1.0G) and one D-loop (Exp-D:D), and forms with two expanded D-loops (Exp-D:Exp-D) in which expansions through less than a monomer genome length have occured (<1.0G). Thin solid lines in drawings at right depict regions of duplex synthesis.

however, to the magnitude of microheterogeneity in sizes of human circular dimer mtDNA (Robberson et al. 1977).

In order to resolve the question of unidirectional versus bidirectional expansion of both D-loops, this sample of mtDNA was treated with the *Eco*RI restriction enzyme, and the positions of replication loops on the resulting fragments were measured. The results of this analysis along with examples of the electron microscopic appearance of displacement replicative forms after cleavage with *Eco*RI are shown in Figure 13. The collection of *Eco*RI fragments with displacement replication loops can be aligned in the gallery shown in Figure 13, with a result that is consistent with unidirectional expansion of displacement loops. The origin of displacement replication occurs at 0.254 ± 0.011 of the monomer genome

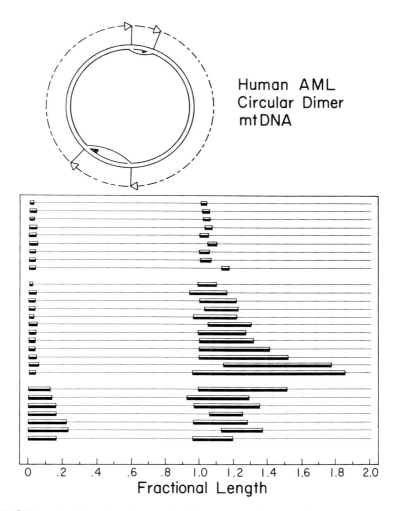

Human AML
Circular Dimer
mtDNA

Fractional Length

FIG. 12. Gallery of replicated and nonreplicated segments of circular dimer mtDNA molecules isolated from human AML cells. Length measurements of replicated and nonreplicated segments, respectively, indicated by open arrows in drawing of circular dimer molecule, begin at the origin of D-loop synthesis (top of drawing) and continue in a clockwise fashion. The length measurements of these segments are then presented in sequence in the linear arrays of strands with D-loop origins at left. Replicated segments are represented as rectangular boxes and their lengths expressed as fractions of the circular dimer genome length.

length on the largest *Eco*RI fragment. The largest *Eco*RI fragment comprises 48.2% ± 1.0% of the monomer genome length, and this origin of replication would occur at a distance representing 52.7% ± 2.3% of the fragment length from the distal terminus. Displacement replication is initiated at this site and proceeds unidirectionally through the second largest *Eco*RI fragment (Figure

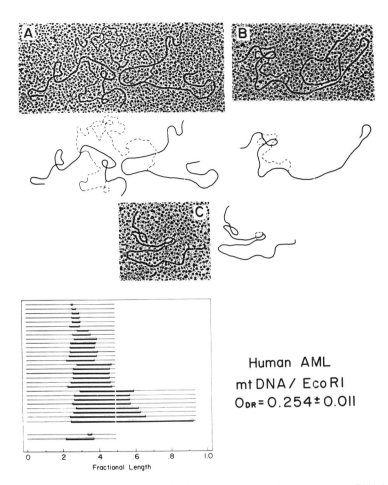

FIG. 13. EcoRI cleavage of replicating circular dimer and circular monomer mtDNA isolated from human AML cells. Electron micrographs at top illustrate appearance of D-loop (C) and expanded D-loop (B) on largest EcoRI fragment. An expanded D-loop with expansion through that portion of the genome represented by the second largest EcoRI fragment is shown in (A). Samples were prepared by the formamide modification of the basic protein (Kleinschmidt) technique (Davis et al. 1971) as modified for examination of replication forms cleaved with restriction enzymes (Robberson et al. 1974). Single-strand segments of D-loop and expanded D-loop are indicated by dashed lines in the drawings below each micrograph. The gallery of strands at bottom represent length measurements of nonreplicated and replicated segments (rectangular boxes) of the EcoRI fragments. The set of strands corresponding to the largest EcoRI fragment containing a replicated segment are shown at the top of this gallery. Displacement replication is initiated near the center of the largest EcoRI fragment at a position that is 0.254 ± 0.011 of the circular monomer genome length and expands unindirectionally through the second largest EcoRI fragment. The two strands at the bottom of this gallery contain origins at positions greater than two standard deviations for the average position determined for the entire population. Strands corresponding to the largest EcoRI fragment have been oriented so that the origin of displacement replication is proximal to the terminus at left.

13). For the combined populations of circular dimer and circular monomer replicative forms in this sample of human AML mtDNA, there is no evidence for bidirectional replication. Despite apparent release from regulatory controls in expansion of both D-loops in a substantial fraction of circular dimer molecules in this particular population, a unidirectional mode of displacement replication is preserved.

Regulation of Duplex Synthesis in Mitochondrial DNA of Human AML Marrow Cells

Mitochondrial DNA replication is distinguished in several important aspects from the replication mechanism of other closed circular duplex genomes (Kasamatsu and Vinograd 1974). One particularly interesting aspect of the replication mechanism pertains to the temporal events in synthesis of complementary strands. In mouse L cells, initiation of synthesis of the light strand is delayed until synthesis of the heavy strand has proceeded through 67% of the genome (Robberson et al. 1972, Martens and Clayton 1979). Also, in mouse L cells, initiation of duplex synthesis occurs predominantly at a single unique site (Robberson et al. 1972, Robberson and Clayton 1972). Additional initiation sites for duplex synthesis have been described for the mtDNA of rat hepatoma cells (Koike and Wolstenholme 1974, Wolstenholme et al. 1973) and mouse tissues (Piko and Matsumoto 1977). Segregation of daughter molecules upon completion of displacement synthesis results, therefore, in one fully duplex daughter molecule and one partially duplex (gapped circular) daughter molecule (Robberson et al. 1972).

Our own studies of mtDNA replicative forms from human AML cells (as well as leukocytes from healthy individuals, healthy human tissue, and established human culture cell lines) have revealed that the basic elements of the displacement replication model (Robberson et al. 1972) apply to human mtDNA replication. Examination of displacement replicative forms (see Figure 10), as well as the gapped circular daughter molecules (Figure 14) indicate that there are multiple initiation sites for duplex synthesis for both circular monomer and circular dimer mtDNA isolated from human AML cells. The combined results of our own studies on replication of human mtDNA from different healthy and malignant cells are summarized in Figure 15. Human mtDNA replication is distinguished from mouse mtDNA replication in displaying multiple sites for initiation of duplex synthesis. We have detected at least five origins for the initiation of duplex synthesis in human mtDNA that are utilized at varying frequencies and that are mapped at different unique sites within 50% of the genome length from the origin of displacement synthesis. It is important to emphasize that even with multiple initiation sites, the replication of the two strands of human mtDNA is still highly asynchronous.

Piko et al. (1978) have recently conducted a careful analysis of duplex synthesis in mtDNA of human leukocytes. In their study, multiple initiation sites were

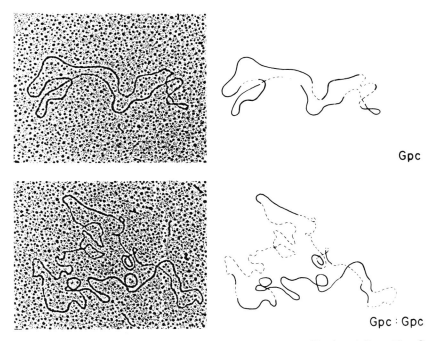

Gpc

Gpc : Gpc

FIG. 14. Electron micrographs of large, gapped circular monomer (Gpc) and dimer (Gpc:Gpc) mtDNA detected in a sample isolated from human AML cells. Single-strand segments are indicated by dashed lines in drawings at right. Samples were prepared by the formamide modification of the Kleinschmidt technique (Davis et al. 1971) as applied by Robberson et al. (1972).

detected in approximately 6% of the expanded loops in expanded D-loop molecules, and approximately 14% of the gapped circular molecules had two or more double-stranded segments. Thus, in healthy cells, most origins for duplex synthesis are not utilized frequently, and as a result, displacement replication of complementary strands is described as being highly asynchronous. However, if initiation at all of the origins for duplex synthesis does happen to occur sequentially in any given molecule, the resulting replicative form will have two apparently fully duplex arms in the replication loop and would be described as a Cairns' form (1963) (see Exp-D[1] in Figure 15) of mtDNA (Wolstenholme et al. 1974). Cairns' forms were only rarely detected (3 Cairns' forms for 98 expanded D-loop molecules examined) in examination of human leukocyte mtDNA of young adults (Piko et al. 1978) but have been detected at high frequencies in sea urchin oocytes (Matsumoto et al. 1974).

We have examined samples of human mtDNA from AML and normal cells for evidence of altered frequencies of initiation of duplex synthesis. To this end we have scored expanded displacement replicative forms that display either highly asynchronous synthesis of strands or, at the other extreme, highly synchro-

Human mtDNA Replication

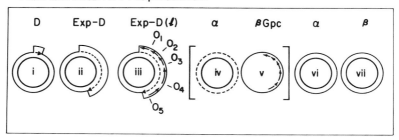

FIG. 15. Model for human mtDNA replication formulated from results of electron microscopy of replicative forms. This model is based on previously established work on the mechanism of replication for mouse mtDNA (see text). Displacement synthesis produces a D-loop (D, structure i) which is then expanded (Exp-D, structure ii). Duplex synthesis can occur on the displaced strand at positions indicated by origins labeled O_1-O_5 to produce Exp-D(1) molecules (structure iii). Completion of displacement synthesis prior to completion of duplex synthesis resulted in segregation of one fully duplex daughter molecule, α (structure iv), and one partially duplex (gapped circular) daughter molecule, β(Gpc structure v). Completion of strand synthesis and closure of phosphodiester bonds results in covalently closed circular α and β daughter molecules (structures vi and vii).

nous synthesis of complementary strands (Cairns' forms). The electron microscopic appearance of Cairns' forms of circular dimer and circular monomer mtDNA is shown in Figure 16. Our examination of two normal human marrow samples indicated that $\sim 17\%$ and $\sim 18\%$, respectively, of the larger replicative forms (with expansions of the D-loop through more than $\sim 20\%$ of the genome) occurred as Cairns' forms. For three samples of human AML mtDNA in Group I of Figure 4, $\sim 1\%$, $\sim 13\%$, and $\sim 22\%$, respectively, of larger displacement replicative forms were detected as circular monomer Cairns' forms. One sample in Group II of Figure 4 had $\sim 3\%$ of the larger replicative forms as circular monomer Cairns' forms. Two samples in Group III of Figure 4 contained $\sim 19\%$ and $\sim 3\%$, respectively, as circular dimer Cairns' forms (Figure 16). Thus, although normal proportions of either circular monomer or circular dimer Cairns' forms can be found among these human AML samples, there is no obvious correlation between the frequencies of these forms and a parameter associated with malignant transformation in AML that we have yet identified. The low frequencies of Cairns' forms in several of the AML samples are comparable to the frequencies of circular monomer Cairns' forms reported by Piko et al. (1978) in normal human leukocytes of the peripheral blood. Our examinations of two samples of normal human leukocytes of the peripheral blood revealed $\sim 1\%$ and $\sim 4\%$ of the larger replicative forms occurred as Cairns' forms. The rather high frequencies of Cairns' forms of mtDNA in normal human marrow cells, as well as the variable frequencies of Cairns' forms of mtDNA in AML marrow cells were unexpected, and further investigation could provide insight into the molecular basis for regulation of initiation of duplex synthesis.

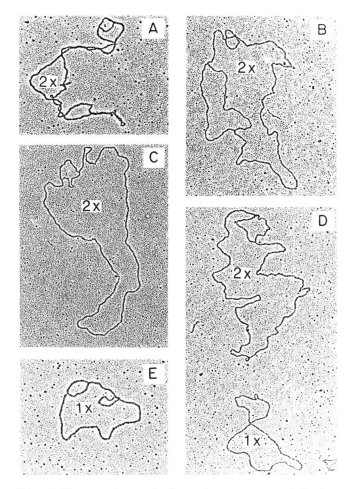

FIG. 16. Electron micrographs illustrating appearance of circular dimer (A-D) and circular monomer (E) mtDNA isolated from human AML marrow cells, which apparently contain fully duplex replication loops (Cairns' forms). The molecule in A is supercoiled and contains a duplex replication loop at top. The molecules in B-E are nicked and display the duplex replication loop near the top. In D, at bottom, a circular monomer (1X) molecule without a replication loop is presented for size comparison with circular dimer (2X) molecules.

DISCUSSION

We have described here some of our progress in further defining the alterations in the human mtDNA complexity, structure, and replication mechanism manifest in acute myelogenous leukemia (AML). Based on earlier studies of mtDNA isolated from the peripheral blood cells of both acute and chronic forms of this disease (Clayton and Vinograd 1969), it was clear that samples would have to be obtained from patients prior to initiation of clinical chemotherapy. Only under these circumstances could one be confident that the alterations detected would be derived through genetic alterations associated directly with the malignancy and not the result of structure-modifying, DNA-reactive agents that are typically used in chemotherapy. Along this same line of reasoning, it was also considered important to first examine alterations in mitochondrial DNA rather than mitochondrial RNA and transcription mechanisms or proteins and translation mechanisms, where alterations might be found but would not necessarily reflect genetic changes associated with malignant transformation.

We note that there is no evidence that the patients we have sampled had previously been exposed to environmental agents that would have substantially modified mtDNA. Under such circumstances, the alterations we have detected would simply be coincidental with, but would not have necessarily accompanied, the actual transformation event(s) associated with AML. For studies of mtDNA, where the genome ploidy is high (approximately 8,800 mtDNA molecules in cultured human cells; Bogenhagen and Clayton 1974), this would seem to be a good assumption. Furthermore, for studies of human AML, we have examined populations of transformed cells that are intermediates in the pathway of hematopoiesis, rather than pluripotent stem cells, which could, in principle, accumulate sufficient lesions in mtDNA over long periods of time to substantially alter mtDNA structural profiles in the daughter cell populations.

We have proceeded first to examine alterations of mtDNA complexity in the marrow cells from AML patients and to extend the previously documented findings of Clayton and Vinograd (1967, 1969) on the peripheral blood cells of leukemic patients. We considered it essential to first demonstrate that increased frequencies of the circular dimer form of human mtDNA truly reflected malignant transformation events in the marrow stem cell population rather than simply being an abnormal physiology associated with extended resident times and therefore elevated levels of proliferating but nondividing precursor cells in the peripheral blood. We observed that the circular dimer form of mtDNA isolated from marrow cells is always found at frequencies greater than or equal to the frequencies observed for mtDNA isolated from the peripheral blood cells of the same AML patient obtained at the same time. This result establishes the bone marrow origin or circular dimer mtDNA in human AML cells. Since marrow cells from healthy individuals also contain very low frequencies of the circular dimer form of mtDNA, it would appear that the transformation event is associated more with an increase in frequency of the circular dimer form than with the de novo formation of this particular oligomeric form of mtDNA.

It has been possible to demonstrate that the increased frequencies of circular dimer mtDNA can be used to probe the alterations of cellular physiology that are reflected in more classical cytological changes identified with abnormal myeloblast populations. When the frequencies of leukemia marrow cells, classified cytologically as abnormal myeloblasts through clinical diagnostic procedures, were examined as a function of the circular dimer mtDNA content of these cells, we were surprised to find that the data for 12 samples segregated into three discrete groups. Furthermore, altered intramitochondrial physiologies, reflected by catenation of mtDNA, increased as a function of increased circular dimer frequencies in some of these same patients and led to the unexpected finding that the data for these samples also segregated into three discrete groups. Although the number of samples in each group is small, this finding is still significant for two reasons. First, there is the observation that groups of data are indeed very well separated at 95% confidence limits of sampling error. Second, it is particularly relevant that we do not find a fourth group containing either low frequencies of leukemic cells or catenated mtDNA and, at the same time, correspondingly high frequencies of circular dimer mtDNA. This argues for a valid association between formation of increased frequencies of circular dimer mtDNA and other physiological events that accompany malignant transformation. We wish to emphasize that only the members of Group III and one member of Group II leukemic cells (see Figure 4) are mapped onto Groups III- and II-catenated mtDNA (see Figure 5), respectively. Since Group I leukemic cells are the largest group and appear less clustered than the other groups (see Figure 4), it is possible that additional discrete groups will subsequently be identified when a larger data base is available for analysis. The highest frequencies of circular dimer mtDNA yet reported are found in our Group III, and for one member of this group, the circular dimer form comprised 90.5% of the total mtDNA mass isolated from the marrow cells. The prognosis for patients in Group III was also very poor, with survivals of less than three months, despite chemotherapy administered subsequent to the time of diagnosis.

The finding of segregation of data into discrete groups has raised the possibility that either distinct types of human AML, perhaps distinctly different diseases, have been detected or that there is only a single disease which develops through discontinuous stages. In studies now in progress, we hope to distinguish between these alternative hypotheses by examinations of Group I patients who progress through the course of the disease without having chemotherapy.

It also appears from these studies that formation of increased frequencies of the circular dimer form is the dominant and perhaps the only significant alteration of circular oligomeric complexity. Only very low frequencies of higher circular oligomers (trimer through decamer) were detected in combined samplings of both normal cells and AML marrow cells. Although the frequencies of catenanes of mtDNA are significantly elevated in half of the human AML samples we examined (Groups II and III in Figure 5), there does not appear to be sufficient alteration in the topological complexity of these catenanes to distinguish AML cell-mtDNA from that of normal cells. In addition, no particular complexity class of higher catenated forms was found that distinguishes human AML-mtDNA.

Since the circular dimer form of mtDNA is found to be the dominant alteration of circular oligomeric complexity, we considered it important to determine if the genetic information in circular dimer mtDNA had been qualitatively altered. Such changes would be expected to alter possibly expression of mitochondrial genetic information and therefore mitochondrial function. The finding of deletions or additions of nucleotide sequences in a significant fraction of the circular dimer mtDNA population (Robberson et al. 1977) prompted us to initiate studies directed toward mapping the genome sites of these lesions. Our initial studies demonstrated that deletions or additions occur at genome positions corresponding to both of the largest *Eco*RI fragments of human mtDNA, which together represent 94% of the genome. The positions at which deletions or additions of sequences can occur are quite variable, although we note that some of these sites lie close to, but not precisely at, the origin of displacement synthesis in the replication of human mtDNA and possibly near sites for initiation of duplex synthesis. Deletions or additions that occur near these sites might provide selective advantage for displacement replication of the circular dimer molecule. Quite apart from genome alterations that may lead to selective replication advantage, it is important to note that the circular dimer molecule contains two copies of the origin for displacement synthesis and, in principle, might be expected to compete more effectively for replicases than would circular monomer mtDNA. Although we do not currently understand the mechanism by which deletions or additions of nucleotide sequences arise in circular dimer mtDNA, a process of abortive replication proposed by Robberson and Fried (1974), in which origins are joined to termini of growing chains on the same replication loop, could generate a deletion near an origin of, for example, duplex synthesis (see Figure 15). This could bring two or more origins in closer proximity, perhaps providing some selective advantage in replicase binding.

Alterations, such as deletions or additions of sequences, could also be expressed in altered patterns of regulation of mtDNA replication. We therefore included in our survey of mtDNA from human AML cells an examination of certain aspects of the mechanism of replication of circular dimer forms.

The replication of the circular dimer form of human mtDNA can be described in most respects by the model for circular monomer as well as circular dimer mtDNA of mouse cells (Robberson et al. 1972, Kasamatsu et al. 1971, Robberson and Clayton 1972). Certain aspects of the circular dimer mtDNA replication mechanism do differ, however, from those of mouse cell mtDNA and have provided an opportunity to further investigate regulatory controls of displacement replication in human AML cells. In circular dimer mtDNA of mouse LD cells, only one of the two D-loops on such molecules is ever found to be expanded (Kasamatsu and Vinograd 1973, Robberson and Clayton 1972).

In contrast to this, a few examples of circular dimer mtDNA with two expanding D-loops were detected in one sample of human AML peripheral leukocytes. If both D-loops on a circular dimer molecule expand, there is the possibility that altered regulatory controls have permitted expansion of these D-loops in a bidirectional fashion, in contrast to unidirectional growth of replication loops in circular monomer and circular dimer mtDNA of mouse cells (Kasamatsu and Vinograd 1974, Robberson and Clayton 1972, Robberson et al. 1974), as well as circular monomer mtDNA of cultured HeLa cells (Brown and Vinograd 1974, Robberson et al. 1974). An extensive analysis presented here revealed, however, no evidence for bidirectional growth of either D-loop in circular dimer mtDNA of human AML cells. Furthermore, it is inferred from these analyses of the total population of displacement replicative forms that expansions of both D-loops, when they do occur, proceed unidirectionally in head-to-tail fashion, with the growing fork of one expanding D-loop moving toward the origin of the other expanding D-loop on the same molecule. It has previously been established that the two monomer genomes that comprise human circular dimer mtDNA are arranged in a head-to-tail fashion (Clayton et al. 1970).

Another interesting feature of the human mtDNA replication mechanism that we have detected pertains to the number and frequencies of initiation of duplex synthesis. In contrast to displacement replication of mouse cell mtDNA replication, where there is predominantly a single unique site for initiation of duplex synthesis (Robberson et al. 1972), human mtDNA replication is characterized as having multiple initiation sites for duplex synthesis, which occur at unique positions on the genome. We have thus far mapped five of these origins in development of a model for human mtDNA replication. Although there are multiple initiation sites for duplex synthesis, the displacement replication of human mtDNA is still highly asynchronous (Piko et al. 1978), as it is for mouse L cells (Robberson et al. 1972). This derives from the fact that on most molecules each of these initiation sites is not utilized with equal frequency in turn as displacement synthesis proceeds. If this were to occur, one would detect displacement replication loops in which both arms of the loop were largely double stranded. Such replicative forms would be described as Cairns' forms, even though replication is proceeding unidirectionally.

We have examined the relative frequencies of Cairns' forms and expanded D-loop molecules that have greater than ~ 20% of the genome replicated as a possible indicator of altered patterns of initiation of duplex synthesis in mtDNA from human AML cells. Unfortunately, our initial studies have not revealed a consistent pattern in the relative frequencies of these two different types of replicative forms that could be correlated with the groups of patients we have described. The variations in relative frequencies of these two classes of replicative forms probably reflects different levels of mtDNA replication activity that are altered by elements of the malignant cell physiology which we do not yet understand. It is of interest to note here that large replicative forms indicative of mtDNA replication activity are absent from mitochondria of the unfertilized mouse egg (Piko and Matsumoto 1976) or sea urchin egg (Matsumoto et al. 1974). A thorough investigation of altered patterns of duplex synthesis could therefore provide insight into the regulation of mtDNA replication in human AML cells.

Altogether, it is quite evident from these studies that alterations in the mtDNA complexity, structure, and replication mechanism do occur in human AML cells. The studies we have described that relate circular dimer mtDNA frequencies to leukemic cell frequencies or catenation frequencies potentially will have diagnostic and perhaps prognostic applications in the clinical setting for human AML. The finding of alterations in the genetic information and replication mechanism associated with the circular dimer form of human mtDNA has also provided a unique opportunity to investigate the mechanisms that regulate mtDNA replication.

ACKNOWLEDGMENTS

We wish to express our gratitude to Dr. Jeanne Hester, Department of Developmental Therapeutics, for her continued support in providing us with access to human leukemic marrow and blood samples. We are also most grateful for assistance provided by Tommie Daniels, Jan Tekell, and Margaret Weber in the Leukapheresis and Bone Marrow Service. We thank Ehsan Youness, Department of Laboratory Medicine, and her staff for providing clinical cytology diagnoses and for making available to us the slides of marrow smears. We thank Mr. John Kuykendall for excellent photomicrography. Special thanks are extended to Ms. Janie Finch for her help in preparation of this manuscript and to Susan Berkley for recent technical assistance. The interest in and discussion of these studies with Drs. Emil J Freireich, Michael Keating, and Anne Marie Maddox, Department of Developmental Therapeutics, are also very much appreciated.

This investigation was supported by Grant Number CA-16527, awarded by the National Cancer Institute, Department of Health, Education, and Welfare and in part by institutional research grants ACS IN-43N and various donors for cancer research.

REFERENCES

Berk, A. J., and D. A. Clayton. 1974. Mechanism of mitochondrial DNA replication in mouse L-cells: Asynchronous replication of strands, segregation of circular daughter molecules, aspects of topology and turnover of an initiation sequence. J. Mol. Biol. 86:801–824.

Berk, A. J., and D. A. Clayton. 1976. Mechanism of mitochondrial DNA replication in mouse L-cells: Topology of circular daughter molecules and dynamics of catenated oligomer formation. J. Mol. Biol. 100:85–102.

Berk, A. J., and D. A. Clayton. 1978. Mechanism of mitochondrial DNA replication in mouse L-cells: Introduction of superhelical turns into newly replicated molecules. J. Mol. Biol. 119:69–81.

Bogenhagen, D., and D. A. Clayton. 1974. The number of mitochondrial deoxyribonucleic acid genomes in mouse L and human HeLa cells. J. Biol. Chem. 249:7991–7995.

Bogenhagen, D., and D. A. Clayton. 1978. Mechanism of mitochondrial DNA replication: Kinetics of synthesis and turnover of the initiation sequence. J. Mol. Biol. 119:49–68.

Brown, W. M., J. Shine, and H. M. Goodman. 1978. Human mitochondrial DNA: Analysis of 7S DNA from the origin of replication. Proc. Natl. Acad. Sci. USA 75:735–739.

Brown, W. M., and J. Vinograd. 1974. Restriction endonuclease cleavage maps of animal mitochondrial DNAs. Proc. Natl. Acad. Sci. USA 71:4617–4621.

Cairns, J. 1963. The chromosome of *Escherichia coli.* Cold Spring Harbor Symp. Quant. Biol. 28:43–46.

Clayton, D. A., R. W. Davis, and J. Vinograd. 1970. Homology and structural relationships between the dimeric and monomeric circular forms of mitochondrial DNA from human leukemic leucocytes. J. Mol. Biol. 47:137–153.

Clayton, D. A., and C. A. Smith. 1975. Complex mitochondrial DNA. Int. Rev. Exp. Pathol. 14:1–67.

Clayton, D. A., C. A. Smith, J. M. Jordan, M. Teplitz, and J. Vinograd. 1968. Occurrence of complex mitochondrial DNA in normal tissues. Nature 220:976–979.

Clayton, D. A., and J. Vinograd. 1967. Circular dimer and catenate forms of mitochondrial DNA in human leukemic leucocytes. Nature 216:652–657.

Clayton, D. A., and J. Vinograd. 1969. Complex mitochondrial DNA in leukemic and normal human myeloid cells. Proc. Natl. Acad. Sci. USA 62:1077–1084.

Davis, R. W., M. Simon, and N. Davidson. 1971. Electron microscope heteroduplex methods for mapping regions of base sequence homology in nucleic acids, *in* Methods in Enzymology XXI, L. Grossman and K. Moldave, eds. Academic Press, New York, pp. 413–428.

Gillum, A. M., and D. A. Clayton. 1978. Displacement-loop replication initiation sequence in animal mitochondrial DNA exists as a family of discrete lengths. Proc. Natl. Acad. Sci. USA 75:677–681.

Gillum, A. M., and D. A. Clayton. 1979. Mechanism of mitochondrial DNA replication in mouse L-cells: RNA priming during the initiation of heavy-strand synthesis. J. Mol. Biol. 135:353–368.

Hudson, B., D. A. Clayton, and J. Vinograd. 1968. Complex mitochondrial DNA. Cold Spring Harbor Symp. Quant. Biol. 33:435–442.

Hudson, B., and J. Vinograd. 1967. Catenated circular DNA molecules in HeLa cell mitochondria. Nature 216:647–652.

Jordan, J. M., A. Van der Eb, and J. Vinograd. 1970. Complex mitochondrial DNA in cells in culture. Fed. Proc. 29:725.

Kasamatsu, H., L. I. Grossman, D. L. Robberson, R. Watson, and J. Vinograd. 1974. The replication and structure of mitochondrial DNA in animal cells. Cold Spring Harbor Symp. Quant. Biol. 38:281–288.

Kasamatsu, H., D. L. Robberson, and J. Vinograd. 1971. A novel closed-circular mitochondrial DNA with properties of a replicating intermediate. Proc. Natl. Acad. Sci. USA 68:2252–2257.

Kasamatsu, H., and J. Vinograd. 1973. Unidirectionality of replication in mouse mitochondrial DNA. Nature New Biol. 241:103–105.

Kasamatsu, H., and J. Vinograd. 1974. Replication of circular DNA in eukaryotic cells. Annu. Rev. Biochem. 43:695–719.

Koike, K., and D. R. Wolstenholme. 1974. Evidence for discontinuous replication of circular mitochondrial DNA molecules from Novikoff rat ascites hepatoma cells. J. Cell Biol. 61:14–25.

Kumar, P. M., and B. W. Fox. 1974. An electron microscope study of mitochondrial DNA in spontaneous human tumors and chemically induced animal tumors. Br. J. Cancer 29:447–461.

Matsumoto, L., H. Kasamatsu, L. Piko, and J. Vinograd. 1974. Mitochondrial DNA replication in sea urchin oocytes. J. Cell Biol. 63:146–159.

Matsumoto, L., L. Piko, and J. Vinograd. 1976. Complex mitochondrial DNA in animal thyroids. Biochim. Biophys. Acta 432:257–266.

Martens, P. A., and D. A. Clayton. 1979. Mechanism of mitochondrial DNA replication in mouse L-cells: Localization and sequence of the light-strand origin of replication. J. Mol. Biol. 135:327–351.

Nass, M. M. K. 1973. Temperature-dependent formation of dimers and oligomers of mitochondrial DNA in cells transformed by a thermosensitive mutant of Rous sarcoma virus. Proc. Natl. Acad. Sci. USA 70:3739–3743.

Paoletti, C., and G. Riou. 1973. The mitochondrial DNA of malignant cells, *in* Progress in Molecular and Subcellular Biology, F. E. Hahn, ed., Vol. 3. Springer-Verlag, Berlin and New York, pp. 203–248.

Paoletti, C., G. Riou, and J. Pairault. 1972. Circular oligomers in mitochondrial DNA of human and beef non-malignant thyroid glands. Proc. Natl. Acad. Sci. USA 69:847–850.

Piko, L., and L. Matsumoto. 1976. Number of mitochondria and some properties of mitochondrial DNA in the mouse egg. Dev. Biol. 49:1–10.

Piko, L., and L. Matsumoto. 1977. Complex forms and replicative intermediates of mitochondrial DNA in tissues from adult and senescent mice. Nucleic Acid Res. 4:1301–1314.

Piko, L., R. Meyer, J. Eipe, and N. Costea. 1978. Structural and replicative forms of mitochondrial DNA from human leucocytes in relation to age. Mech. Ageing Dev. 3:351–365.

Riou, G., and E. Delain. 1971. Mitochondrial DNA from cells transformed by adenoviruses and SV40. Biochimie 53:831–836.

Riou, G., and F. Lacour. 1971. Mitochondrial DNA from cells transformed by myeloblastosis virus. Biochimie 53:47–49.

Robberson, D., Y. Aloni, and G. Attardi. 1971. Electron microscopic visualization of mitochondrial RNA-DNA hybrids. J. Mol. Biol. 55:267–270.

Robberson, D. L., and D. A. Clayton. 1972. Replication of mitochondrial DNA in mouse L cells and their thymidine kinase minus derivatives: Displacement replication on a covalently closed circular template. Proc. Natl. Acad. Sci. USA 69:3810–3814.

Robberson, D. L., and D. A. Clayton. 1973. Pulse-labeled components in the replication of mitochondrial DNA. J. Biol. Chem. 248:4512–4514.

Robberson, D. L., D. A. Clayton, and J. F. Morrow. 1974. Cleavage of replicating forms of mitochondrial DNA by EcoRI endonuclease. Proc. Natl. Acad. Sci. USA 71:4447–4451.

Robberson, D. L., and M. Fried. 1974. Sequence arrangements in clonal isolates of polyoma defective DNA. Proc. Natl. Acad. Sci. USA 71:3497–3501.

Robberson, D. L., H. Kasamatsu, and J. Vinograd. 1972. Replication of mitochondrial DNA. Circular replicative intermediates. Proc. Natl. Acad. Sci. USA 69:737–741.

Robberson, D. L., C. E. Wilkins, D. A. Clayton, and J. N. Doda. 1977. Microheterogeneity detected in circular dimer mitochondrial DNA. Nucleic Acid Res. 4:1315–1338.

Sanger, F., A. R. Coulson, T. Friedmann, G. M. Air, B. G. Barrell, N. L. Brown, J. C. Fiddes, C. A. Hutchinson III, P. M. Slocombe, and M. Smith. 1978. The nucleotide sequence of bacteriophage ϕx174. J. Mol. Biol. 125:225–246.

Smith, C. A., and J. Vinograd. 1973. Complex mitochondrial DNA in human tumors. Cancer Res. 33:1065–1070.

Van der Eb, A. J., J. M. Jordan, and J. Vinograd. 1970. Complex mitochondrial DNA in cultured cells and in experimentally induced leukemias and tumors, in 20th Yearbook for Cancer Research and Fight Against Cancer in the Netherlands, J. H. DeBussy, Amsterdam, pp. 53–67.

Wolstenholme, D. R., K. Koike, and P. Cochran-Fouts. 1973. Single strand-containing replicating molecules of circular mitochondrial DNA. J. Cell Biol. 56:230–245.

Wolstenholme, D. R., K. Koike, and P. Cochran-Fouts. 1974. Replication of mitochondrial DNA: Replicative forms of molecules from rat tissues and evidence for discontinuous replication. Cold Spring Harbor Symp. Quant. Biol. 38:267–280.

Genes, Chromosomes, and Neoplasia, edited by
Frances E. Arrighi, Potu N. Rao, and Elton Stubblefield.
Raven Press, New York © 1981.

The Nature of Gene Variation and Gene Transfer in Somatic Cells

Louis Siminovitch

Hospital for Sick Children, Department of Medical Genetics, University of Toronto, Toronto, Ontario, Canada

Over the last decade, our laboratory has been interested in developing genetic systems to facilitate the study of structure, function, and regulation in somatic cells. For the most part, we have chosen to use a line of Chinese hamster ovary (CHO) cells for these investigations. This choice was predicated on two major properties of these cells. First, the cells have extremely favorable culture properties, and although the karyotype is not normal, it is relatively stable (Deaven and Peterson 1973, Worton 1978, Worton et al. 1977). Secondly, the ability to isolate mutants is a major prerequisite for genetic studies, and Puck and his colleagues had shown that auxotrophic mutants could be obtained relatively easily from CHO cells (Puck and Kao 1967, Kao and Puck 1967, 1968, 1969, Kao et al. 1969).

A major thrust of our work has been to develop a variety of genetic systems with CHO cells and to use these systems both to examine the functions involved in the lesion and to obtain information on genetic variations in somatic cells. Since much of our work has been published, I do not propose to describe the properties of the mutants we have studied in any detail. Instead, in order to present an overview, and to provide a basis for discussion of the problems of genetic variation, I shall first present a listing (Tables 1 and 2) of most of the genetic systems that have been developed in CHO cells in Toronto and elsewhere and make some general comments about the information contained in the tables.

A SURVEY OF GENETIC SYSTEMS AVAILABLE IN CHO CELLS

Recessive

I have divided the mutants studied in CHO cells arbitrarily on the basis of whether they behave recessively or dominantly in somatic cell hybrids. Several points should be made about the phenotypically recessive mutations listed in Table 1. First, there is considerable evidence that many of the mutants represent

TABLE 1. Phenotypically recessive mutations in CHO cells*

Class	Phenotype	Lesion	Number of Complementation Groups	References
Auxotroph	Requirement for:			
	glycine		4	Kao and Puck 1968, 1969, Kao et al. 1969.
	proline		1	Kao and Puck 1967.
	adenine		9	Patterson et al. 1975, Patterson 1975, 1976a, 1976b, Tu and Patterson 1977, Oates and Patterson 1977, Chu et al. 1972, Feldman and Taylor 1975a, 1975b, Holmes et al. 1976, Irwin et al. 1979.
	glycine, adenine, thymidine		1	McBurney and Whitmore, 1974a, 1974b, Patterson et al. 1975, Patterson 1975.
	adenine, thymine		1	Patterson et al. 1975, Patterson 1975.
	thymidine (resistance to ara C)		1	Meuth et al. 1979.
	uridine		1	Patterson and Carnwright 1977, Davidson et al. 1979.
	serine		1	Jones and Puck 1973.
	glutamate		1	Hankinson 1976.
	alanine		1	Hankinson 1976.
	inositol		1	Kao and Puck 1968.
	asparagine		1	Waye and Stanners 1979.
Conditional lethal temperature-sensitive	aminoacyl-tRNA synthetase		8	Thompson et al. 1973, Stanners and Thompson 1974, Thompson et al. 1975, 1977, 1978, Wasmuth and Caskey 1976, Stanners et al. 1978, Adair et al. 1978, 1979, Ashman 1978, Haars et al. 1976, Farber and Deutscher 1976, Andrulis et al. 1978.
	mitotic apparatus		1	Thompson and Lindl 1976.
	undefined		~10	Crane and Thomas 1976, Siminovitch and Thompson 1978, Patterson et al. 1976.

TABLE 1 (Cont'd). *Phenotypically recessive mutations in CHO cells*

Class	Phenotype	Lesion	Number of Complementation Groups	References
Resistant	Resistant to: lectins	membrane glycoproteins & glycolipids	8–9	Wright 1972, Gottlieb et al. 1974, 1975, Stanley et al. 1975a,b,c, Juliano and Stanley 1975, Briles et al. 1977, 1980, Stanley and Siminovitch 1977, Krag et al. 1977.
	adenosine, toyocamycin, or tubericidin	adenosine kinase	1	McBurney and Whitmore 1975, Gupta and Siminovitch 1978d, Rabin and Gottesman 1979.
	8-azaguanine or 6-thioguanine	hypoxanthine phosphoribosyl transferase	1	Szybalski and Smith 1959, Littlefield 1963, Chu et al. 1969, Gillin et al. 1972, Sharp et al. 1973.
	8-azaadenine or 6-mercaptopurine	adenosine phosphoribosyl transferase	1	Chasin 1974, Jones and Sargent 1974.
	emetine, cryptopleurine or tylocrebrine	40S ribosome subunit	1	Gupta and Siminovitch 1976, 1977a, 1977b, 1978a, 1978b.
	trichodermin	60S ribosome subunit	1	Gupta and Siminovitch 1978c.
	chromate	sulfate transport	1	Campbell and Worton (in preparation).
	methotrexate	transport	1	Flintoff et al. 1976a, 1976b.
	5-bromodeoxyuridine	thymidine kinase	1	Littlefield 1965, Kit et al. 1963, Clive et al. 1972.
	2-deoxygalactose	galactokinase	1	Thirion et al. 1976.
	methylglyoxal bis guanyl hydrazone	polyamine transport	1	Mandel and Flintoff 1978.
	oligomycin, rutamycin, venturicidin and antimycin	glycolysis	1	Lagarde and Siminovitch 1979
	pactomycin	morphology	1	Gupta and Siminovitch 1980c.

* Some of the references refer to mutants selected in non-CHO cell lines. However, such mutants have been observed in CHO cells as well.

TABLE 2. *Phenotypically dominant or codominant mutations in CHO cells*

Marker*	Phenotype (resistant to:)	Lesion	References
Oua^R	ouabain	Na$^+$/K$^+$ ATPase	Baker et al. 1974.
Ama^R	α-amanitin	RNA polymerase II	Chan et al. 1972, Lobban and Siminovitch 1975, 1976, Ingles et al. 1976a, 1976b, Ingles 1978, Gupta et al. 1978b.
Mtx^{RI}	methotrexate	dihydrofolate reductase (structure)	Flintoff et al. 1976a, 1976b, Gupta et al. 1977.
Mtx^{RIII}	methotrexate (derived from Mtx^{RI})	increased activity of altered dihydrofolate reductase	Flintoff et al. 1976a, 1976b.
CH^R	colchicine	membrane	Ling and Thompson 1974, Ling 1975, Bech-Hansen et al. 1976, Juliano et al. 1976, Ling and Baker 1978.
CM^R	Colcemid	microtubule protein	Ling et al. 1979.
Hyd^R	hydroxyurea	ribonucleotide reductase	Wright and Lewis 1974, Lewis and Wright 1974, 1978, 1979.
Ric^{RII}	ricin	unknown	Stanley et al. 1975b, Stanley and Siminovitch 1977.
AHA I	aspartyl-hydroxamate	asparaginyl tRNA modification	Andrulis et al. 1978, 1979.
AHA II	aspartyl-hydroxamate	increased levels of asparagine synthetase	Andrulis et al. 1978, 1979.

* In the paper, R refers to a dominant or codominant mutation whereas r represents a recessive marker.

structural gene lesions. Secondly, using a variety of selection methods, a large number of mutants have been isolated in CHO cells and characterized, and these involve a wide spectrum of phenotypes. Auxotrophs, conditional lethal temperature-sensitive mutants (ts), cells resistant to a variety of drugs, and cells sensitive to radiomimetic agents constitute the major classes. At least 20 complementation groups have been ascertained among the auxotrophs, and mutants exist for at least nine of the steps leading to the synthesis of AMP (adenine requiring, Table 1). In fact, for this class of isolates, because of the absolute nutritional requirement for cells in culture, the spectrum of mutants that have been recovered represents a large fraction of the potentially selectable lesions. In addition, about 20 complementation groups of the conditional lethal ts class have been delineated, and Thompson and his group have succeeded in isolating almost half of the possible conditional lethal ts mutants showing lesions in tRNA synthetase (Thompson et al. 1973, 1975, 1977, 1978, Adair et al. 1978, 1979).

As can be seen from Table 1, the number of mutant types that have been isolated in the class resistant to drugs is also very large. For lectin resistance alone, Stanley and her colleagues, and others, have been able to identify eight different complementation groups (cf. Table 1), and more recently she has expanded this list and shown that there are five separate complementation groups for wheat germ agglutinin resistance alone. Finally, Rosenstraus and Chasin have been able to obtain CHO mutants with altered glucose-6-phosphate dehydrogenase activity (Rosenstraus and Chasin 1977), and Thompson and co-workers have recently described a series of mutants of CHO cells with increased sensitivity to agents such as ultraviolet (UV) light and mitomycin (Thompson et al. 1980).

The list shown in Table 1 is not complete and further isolates that behave recessively are being obtained all the time. For example in our laboratory, Mento has obtained several different mutants that are resistant to Sindbis virus (Mento and Siminovitch, in preparation), and Debenham has selected mutants whose growth is independent of the presence of glucose in the culture medium (Debenham and Siminovitch, in preparation). In addition, phenotypically recessive CHO mutants are found that show defects in either protein synthesis or their membranes among diphtheria toxin–resistant isolates (Moehring and Moehring 1977, 1979, Gupta and Siminovitch 1978e, 1980a), and Arpaia has found two complementation classes of hydroxy 6-(p-hydroxyphenylazo)uracil-resistant mutants that show pyrimidine auxotrophy and which may carry lesions in ribonucleotide reductase (Arpaia and Siminovitch, in preparation). In other recent studies, Gottesman and his colleagues (Gottesman et al. 1980) have been able to select a variety of cyclic AMP–resistant mutants in CHO cells that apparently includes isolates of both the recessive and dominant kind.

The general conclusion to be drawn from the above discussion is that a large number and variety of recessive mutants can be obtained in quasi-diploid CHO cells. It is probable that these loci are distributed widely across the genome. I have commented elsewhere on the seeming paradox represented by the success in obtaining recessive mutations in quasi-diploid cells (Siminovitch 1976, 1979a,b), and I shall discuss it further later in this paper.

A third important point should be made about the mutant systems shown in Table 1. Considerable progress has been made recently in the ability to transfer genetic markers between somatic cells with DNA (Graham and Van Der Eb 1973, Bacchetti and Graham 1977, Wigler et al. 1977, 1979a, 1979b, Minson et al. 1978, Maitland and McDougall 1977) and chromosomes (McBride and Ozer 1973, Burch and McBride 1975, Willecke and Ruddle 1975, Willecke et al. 1976) or micro cells (Fournier and Ruddle 1977a, 1977b). There is of course great interest in these technologies because of their potential importance in developing understanding of gene structure and function in mammalian cells. However, the number of markers that have been available for such studies have been limited so far to a few recessive lesions and one or two dominant markers. The availability of the auxotrophs and the conditional lethal mutants provides a much increased set of recipients for gene transfer. Clearly, the wild type allele derived from any sort of cell will act dominantly to the mutant allele in such experiments, and therefore should be selectable.

Dominant

In addition to the extensive series of recessive mutants that have been selected in CHO cells, a number of others behave dominantly. The properties of most of these systems, shown in Table 2, have been described before, and I shall again only make some generalizations about this whole set of isolates. In nearly all cases, there are good indications that the loci involve structural gene mutations. In some cases, such as resistance to methotrexate (MTx^{RIII}) and hydroxyurea (Hyd^R), the increased resistance seems to be due to gene duplication, a phenomenon that may occur rather commonly in somatic cells (Alt et al. 1976, Schimke et al. 1978, Kemp et al. 1976, Lewis and Wright 1978). Since the mutations act dominantly, it should be possible to develop gene transfer systems with these markers readily, and this has already been accomplished with MTx^{RIII} cells (Lewis et al. 1980).

STUDIES OF MUTAGENESIS IN CHO CELLS

There has long been a need for somatic cell systems in which to study quantitative mutagenesis. Many of the mutational loci described in Tables 1 and 2 are ideal for this purpose, since the mutants are well characterized and are observed with reasonable frequency, and the systems can be manipulated and described quantitatively. Dr. Gupta and I have examined this question using some of the markers shown in Table 1. These include mutants resistant to emetine (Emt^r), diphtheria toxin (Dip^r), thioguanine (Thg^r), toyocamycin (Toy^r), ouabain (Oua^R), and methylglyoxal bis-guanylhydrazone (Mbg^r). These results have been published elsewhere (Siminovitch 1979b, Gupta and Siminovitch 1980b) and I shall only summarize them here in order to indicate the potential use to which these systems can be put for studies in mutagenesis.

We first showed that a characteristic expression time of 2 to 6 days could be identified for each of the above systems. Using such expression times, we then found that there was a linear dose relationship between mutagen concentration (ethyl methanesulfonate; EMS) and mutation frequency for all the markers except Toy^r over the range of 10-700 $\mu g/ml$ of the drug. Since Toy^r mutants are observed at very high frequencies, contain little or none of the enzyme affected by the lesion, and are difficult if not impossible to make revert, it seems probable that they do not represent standard types of structural gene mutations. The failure to observe increased yields of Toy^r cells after EMS mutagenesis is therefore not surprising.

It was then of interest to determine whether the same CHO systems could be used to examine the specificity of mutagenesis. Decay of ^3H can act as a mutagen by causing chromosome breaks (Person and Bockrath 1964, Cleaver 1977). It would therefore not be expected to be effective for markers in which essential cellular functions are involved, such as Oua^R, Emt^r, and Dip^r. The Thg^r and Toy^r mutations, however, affect salvage enzymes, and therefore their frequency should be sensitive to ^3H decay mutagenesis. The exact lesion involved in Mbg^r cells is not known, so it was difficult to predict how it would behave. The results were as expected: Emt^r, Dip^r, and Oua^r were unaffected by the ^3H decay, whereas Thg^r and Toy^r frequencies were increased (Siminovitch 1979b, Gupta and Siminovitch 1980b). Mbg^r turned out to behave similarly to the former class.

These results indicate that the CHO genetic cell systems should provide an excellent model for studies of mutagenesis. As indicated above, the cells have several advantageous culture characteristics, the markers are selected readily, all resistant clones show similar genetic and biochemical phenotypes, and the frequencies of mutation are highly responsive to mutagens in a dose-dependent manner. It seems important, therefore, that this system be considered as one of general applicability in tests for mutagens and carcinogens.

FUNCTIONAL HEMIZYGOSITY IN CHO CELLS

As pointed out earlier, one remarkable feature of the work with CHO cells is the large number and variety of recessive mutants that have been obtained in cells that are supposedly diploid. In earlier papers I have postulated that these findings could be explained if, in fact, CHO cells were *not* diploid over their whole genome and if the cells were extensively functionally hemizygous (Siminovitch 1976, 1979a). In addition to the fact that recessive mutations are rather frequent in CHO cells, I have presented other considerable support for this view (Siminovitch 1979a). This evidence, in brief, is as follows. First, dominant mutants of CHO cells resistant to α amanitin (cf. Table 2) only contain resistant RNA polymerase II, and no sensitive enzyme (Lobban and Siminovitch

1975, Ingles et al. 1976a,b, Gupta et al. 1978b), whereas it would be expected that both types of enzyme would be found in equal quantities if cells contained two copies of the gene. All other types of α amanitin–resistant cells contain both types of enzyme (Siminovitch 1979a). Secondly, the frequency with which recessive Emt^r mutants can be obtained in CHO cells is about 100-fold higher than in five other Chinese hamster lines, whereas the frequencies of a dominant marker, Oua^R, and an X-linked marker (Thg^r) are the same for all six lines (Campbell and Worton 1979). This would be expected if the Emt^r locus is autosomal, and if CHO only contains one copy of the gene whereas the other lines contain two copies. Although one would expect frequencies of 10^{-x} and 10^{-2x} for the single versus the double events, this does not seem to be the case in comparisons of haploid and diploid slime molds (Williams 1976), and in yeasts, where the frequencies are of the order of 10^{-x} and $10^{-1.2x}$. The similar frequencies found among all six lines for Oua^R and Thg^r are also to be expected since only one locus need be altered in each case to give rise to the mutant cell.

Parenthetically, it is of interest that another CHO cell line, the glyB mutant of CHO K1 (Kao et al. 1969), behaves similarly to the five Chinese hamster cell lines that show low frequencies of mutations for the Emt^r marker. The frequencies of Toy^r mutants in the Toronto line of CHO cells is also much higher than that found in most other Chinese hamster lines (Gupta and Siminovitch 1978d, Rabin and Gottesman 1979). Finally, Thompson and his colleagues have used mutation frequency observations, obtained by using a tRNA synthetase locus, to provide evidence for functional hemizygosity (Adair et al. 1979). They observed originally that conditionally lethal *ts* mutations affecting the glutamyl tRNA synthetase were much less frequent than those affecting leucyl or asparagine tRNA synthetase. They argued that this result could be due to the need to mutate two copies of the allele in the former case and one copy in the latter. To obtain evidence for this view, they first isolated a revertant of the *ts* glutamyl tRNA synthetase. According to their hypothesis, this revertant should then have one mutated allele and one normal allele. They then examined the ease with which new *ts* glutamyl tRNA synthetase mutants could be selected from the revertant. As predicted, assuming that only one functional allele was now present, the frequencies of such mutants were much higher than in the wild type population (Adair et al. 1979).

Another indication of functional hemizygosity at the Emt^r locus has been obtained by segregation analysis. Since this locus behaves recessively, hybrids between CHO wild type and CHO Emt^r cells are sensitive to emetine. Cells resistant to emetine can be reselected from such hybrids at high frequency, and there is considerable evidence that the new Emt^r cells arise not by mutation but by either segregation or inactivation of the wild type allele (Campbell and Worton, in preparation, Farrell and Worton 1977, Chasin and Urlaub 1976). The frequency with which resistant cells are obtained from the hybrid should therefore depend on whether the wild type cell contains two copies or one copy of the relevant allele, since two events would be required if two copies are present and one event if only one copy is present. When such segregation frequencies were measured in hybrids between CHO Emt^r cells and a series of Chinese hamster lines, it was found that the CHO × CHO hybrids yielded much higher frequencies of Emt^r segregants than the other hybrids, a result consistent with the hypothesis that the emetine locus is present in one copy in CHO cells and two copies in the other lines (Gupta et al. 1978a).

Campbell and Worton have obtained cytological evidence for this view in that they have located the Emt^r locus on chromosome 2 of the CHO lines and have found that the section of the chromosome that probably contains the locus is deleted in one of the number 2 chromosomes (Campbell and Worton, in preparation).

We have recently obtained information on the evolution of functional hemizygosity. In the original investigation of Campbell and Worton (Campbell and Worton 1979), one of the lines that yielded about 1/100th of the CHO frequency of Emt^r was the V79/V6 line. One possible explanation for this result was that 1% of the V79/V6 cell population was hemizygous for the Emt^r locus and that the resistant cells that were observed arose from this minority population. In order to examine this possibility we isolated about 300 independent clones of V79 cells, subjected each one to EMS mutagenesis, and then tested for emetine resistance. As expected, most of the clones gave very low mutation frequencies. However, about one in 100 clones yielded frequencies of the same order as in CHO cells (Siminovitch 1979a). The property of yielding high frequencies of Emt^r cells after mutagenesis was maintained in these specific clones even after long periods of culture. Thus, there seems to be a spontaneous process in V79/V6 cells that gives rise to cells of greater susceptibility for mutation to Emt^r. One plausible explanation for this phenomenon is that an alteration to hemizygosity has occurred in the original V79 wild type population.

It is of interest that during these studies we found that some of the Emt^r mutants obtained from the high frequency clones complemented the Emt^r isolates obtained earlier by Gupta and Siminovitch (1976). This result does not invalidate the general conclusions of the experiment described above, since the new Emt^r mutants also behave recessively. However, since the new Emt^r also seems to involve a lesion in protein synthesis, there appear to be at least two loci that can give rise to emetine resistance in V79 and CHO cells.

The evolution of sublines that can give higher frequencies of mutation for a particular locus is not unusual in cultured cells. We have found a subline of CHO cells that yields 100- to 1000-fold higher frequency of tk$^-$ cells (Siminovitch 1979a), and sublines exist that yield an increased number of aprt$^-$ cells (Taylor et al. 1977). It seems reasonable to believe that all of these cases can be attributed to the evolution of lines hemizygous for the locus in question.

GENE TRANSFER IN SOMATIC CELLS

As I indicated earlier, the development of techniques for gene transfer in mammalian cells represents one of the most important and promising advances in the field of somatic cell genetics in recent years. A priori, the CHO cell line represents excellent experimental material for such studies because of the wide spectrum of dominant and conditional lethal mutations available for gene transfer. We therefore initiated studies in this area a short while ago, with the major aim at the outset of developing the optimum conditions for transfer of DNA and chromosomes and determining whether major differences in frequency existed between these two vectors. These results have been published recently and will only be summarized here (Lewis et al. 1980). We worked with CHO cells as the donor and L cells as the recipient and examined the efficiency of transfer of the recessive thymidine kinase locus (Tk) and the dominant marker for methotrexate resistance, Mtx^{RIII}. The latter is a particularly suitable system for gene transfer since it involves a double mutation (cf. Table 2), and the chance of observing a new mutation in the recipient cells (rather than the transferred gene) is very small (about 10^{-12}–10^{-14}). Using these systems, we were able to develop maximum conditions of adsorption, gene expression, and gene dosage, and optimum conditions for the use of dimethylsulfoxide (DMSO) as a facilitator, for both DNA and chromosome transfer. We found that no significant differences could be observed in the frequencies of transfer for the two markers, but that the frequencies were about 10-fold higher when chromosomes were used than when DNA was used (Lewis et al. 1980).

Although these results are rather encouraging, one discouraging aspect has arisen from our work, as well as the studies of others. All of the above experiments were carried out with L cells as the recipients. Because most of the mutant systems have been developed in CHO cells, it would of course be preferable to work in this latter system. However, despite extensive efforts we have had very limited success in achieving reasonable frequencies of transfer of chromosomes or DNA to CHO cells. Another hamster line, V79, and some CHO sublines give somewhat higher frequencies, but none of these are comparable to those found with L cells. Thus, there seems to be a great deal of specificity in cells' abilities to act as recipients for gene transfer. Because most of the mutants exist in the CHO line and seem to be easier to obtain in these cells, unless methods are developed to overcome the problem, there will be severe limitations on the general immediate application of gene transfer technology. It may in fact be necessary to develop the selection systems in L cells. Although the latter line may not be extensively hemizygous, it may still be feasible to obtain recessive mutations using large numbers of cells if the frequency of change to hemizygosity in the population is similar to that found in V79 cells (Siminovitch 1979a). Some preliminary work in our laboratory indicates that it is indeed possible to obtain conditional lethal mutations of the auxotrophic and *ts* kind in L cells.

The fact that the ability to act as a recipient for gene transfer is relatively specific has other negative ramifications. One obvious exploitation of this technology would be to transfer and study genes involving differentiated functions. Depending on the problem to be examined, this would require the availability of a wide spectrum of recipients. However, until one can solve the problem of recipient specificity, the general application of gene transfer technology may be seriously limited.

DISCUSSION

Several aspects of recent advances in somatic cell genetics have been discussed in this paper. Primarily I have described the large number and wide spectrum of both recessive and dominant mutants that have been obtained in CHO cells. In general, most of these probably represent structural gene changes, and reports on the isolation of regulatory genes have been extremely rare. The paucity of the latter isolates may reflect an inadequacy in selection methods, or perhaps that such regulatory mutants are rare in somatic cells. This question represents a future challenge in the field.

The relative success in mutant isolation using CHO cells may reflect some unique property of this line or the fact that it has enjoyed so much attention by those working in the field. Certainly, several different types of recessive and dominant mutants have been obtained using V79 Chinese hamster cells, and isolated reports have appeared describing mutants of all classes shown in Tables 1 and 2 in many other cell lines. But in our own limited experience, recessive mutants seem to be obtained with greater ease and frequency in CHO cells.

As described in the text and elsewhere (Siminovitch 1976, 1979a), the relative facility with which recessive mutants are obtained in CHO cells may be due to the existence of extensive functional hemizygosity in this line. Assuming the validity of this hypothesis, there is very little information on how such hemizygosity arises. Because of the extensive chromosomal rearrangement that has occurred in the CHO cell, one obvious possibility is the existence of deletions in chromosomal regions of one of the two autosomes. This seems to be the fact for the emetine and chromate loci in CHO cells. Functional hemizygosity could also occur by gene inactivation on one of the autosomes. Such inactivation could occur in several other ways, although very few, if any, have been well documented in somatic cells.

Our observations on the events leading to emetine resistance in V79/V6 cells are of interest in this regard since they indicate that the processes leading to presumed functional hemizygosity occur at a relatively high frequency in that cell line. The fact the "high frequency" lines for other markers are also observed rather frequently provides support for this conclusion. Whatever the mechanism, our studies indicate that it should be feasible to obtain recessive mutants in some cell lines even if two copies of the gene are present, providing sufficient numbers of cells are placed at risk. This conclusion is supported by the results from Thompson's laboratory in which they were able to obtain *ts* glutamyl-tRNA synthetase mutants even when the cells were presumably not hemizygous at this locus (Adair et al. 1979). This result also underlines the fact that although some loci in CHO cells seem to be functionally hemizygous, CHO does not contain a complete haploid genome. There is considerable evidence to the contrary, based on the above studies by Thompson, on work by Siciliano and his colleagues (Siciliano et al. 1978), and on the behavior of some other markers. In fact, at this stage, it is difficult to ascertain how many of the isolates described in Table 1 were obtained as a result of one event at a functional hemizygous locus, or because of two distinct events.

Although CHO cells seem to be the "ideal" cell line for mutant selection and characterization, the line suffers from at least two drawbacks. Firstly, it obviously cannot be used effectively to study many important cell biology problems involving, for example, carcinogenesis and differentiation. The possible application of the selection technologies developed with CHO cells may depend on whether the primary event leading to presumed functional hemizygosity occurs with reasonable frequency in other cell lines.

As indicated earlier, a second major limitation of the CHO system involves the low frequency of gene transfer found with that line. There is little information on why CHO is an ineffective recipient for such experiments. In preliminary experiments, Debenham has shown that labeled DNA is taken up by CHO cells as well as with L cells, both into cytoplasm and nucleus. Also CHO cells serve as good recipients in micronuclei transfer where whole chromosomes are involved. Whatever the difficulty, unless it is resolved, the plethora of mutants available may have limited application for gene transfer work.

The future challenge therefore is either to develop methods to improve the frequencies observed with CHO cells and to attempt to generalize the methodology, or to determine whether mutants similar to those shown in Tables 1 and 2 can be obtained with facility in L cells. Because of the rapid progress that has occurred in the field of somatic cell genetics over the last 10 or so years, the latter objective may not represent an unattainable objective.

ACKNOWLEDGMENTS

I am indebted to my colleagues in the Toronto area and to my postdoctoral fellows and technicians who performed much of the work described in this paper. I am particularly thankful to Dr. P. Ip for his key contributions to several aspects of the work in my own laboratory. The studies were supported by grants from the Medical Research Council and National Cancer Institute of Canada and by a contract from the National Cancer Institute of the USA.

REFERENCES

Adair, G. M., L. H. Thompson, and P. A. Lindl. 1978. Six complementation classes of conditionally lethal protein synthesis mutants of CHO cells selected by ^{3}H-amino acids. Somatic Cell Genetics 4:27–44.

Adair, G. M., L. H. Thompson and S. Fong. 1979. (^{3}H) Amino acid selection of aminoacyl-tRNA synthetase mutants of CHO cells: Evidence of homo- vs. hemizygosity at specific loci. Somatic Cell Genetics 5:329–344.

Alt, F. W., R. E. Kellems, and R. T. Schimke. 1976. Synthesis and degradation of folate reductase in sensitive and methotrexate-resistant lines of S-180 cells. J. Biol. Chem. 251:3063–3074.

Andrulis, I. L., G. S. Chiang, S. M. Arfin, T. A. Miner and G. W. Hatfield. 1978. Biochemical characterization of a mutant asparaginyl-tRNA synthetase from Chinese hamster ovary cells. J. Biol. Chem. 253:58–62.

Andrulis, I. L., G. W. Hatfield, and S. M. Arfin. 1979. Asparaginyl-tRNA aminoacylation levels and asparagine synthetase expression in cultured Chinese hamster ovary cells. J. Biol. Chem. 254:10629–10633.

Ashman, C. R. 1978. Mutations in the structural genes of CHO cell histidyl-, valyl-, and leucyl-tRNA synthetases. Somatic Cell Genetics 4:299–312.

Bacchetti, S., and F. L. Graham. 1977. Transfer of the gene for thymidine kinase to thymidine kinase-deficient human cells by purified herpes simplex viral DNA. Proc. Natl. Acad. Sci. USA 74:1590–1594.

Baker, R. M., D. M. Brunette, R. Mankovitz, L. H. Thompson, G. F. Whitmore, L. Siminovitch, and J. E. Till. 1974. Ouabain resistant mutants of mouse and hamster cells in culture. Cell 1:9–21.

Bech-Hansen, N. T., J. E. Till and V. Ling. 1976. Pleiotropic phenotype of colchicine-resistant CHO cells: Cross-resistance and collateral sensitivity. J. Cell. Physiol. 88:23–39.

Briles, E. G., E. Li, and S. Kornfeld. 1977. Isolation of wheat-germ agglutinin-resistant clones of CHO cells deficient in membrane sialic acid and galactose. J. Biol. Chem. 252:1107–1116.

Briles, E. G., E. Li, and S. Kornfeld. 1980. J. Cell. Biol. (In press).

Burch, J. W., and O. W. McBride. 1975. Human gene expression in rodent cells after uptake of isolated metaphase chromosomes. Proc. Natl. Acad. Sci. USA 72:1977–1981.

Campbell, C. E., and R. G. Worton. 1979. Evidence obtained by induced mutation frequency analysis for functional hemizygosity at the *emt* locus in CHO cells. Somatic Cell Genetics 5:51–65.

Chan, V. L., G. F. Whitmore, and L. Siminovitch. 1972. Mammalian cells with altered forms of RNA polymerase II. Proc. Natl. Acad. Sci. USA 69:3119–3123.

Chasin, L. A. 1974. Mutations affecting adenine phosphoribosyl transferase activity in Chinese hamster cells. Cell 2:43–54.

Chasin, L. A. and G. Urlaub. 1976. Mutant alleles for hypoxanthine phosphoribosyl transferase: Codominant expression, complementation and segregation in hybrid Chinese hamster cells. Somatic Cell Genetics 2:453–467.

Chu, E. H. Y., P. Brimer, K. B. Jacobson and E. Merriam. 1969. Mammalian cell genetics. I. Selection and characterization of mutations auxotrophic for L-glutamine or resistant to 8-azaguanine in Chinese hamster cells in vitro. Genetics 62:359–377.

Chu, E. H. Y., N. C. Sun, and C. C. Chang 1972. Induction of auxotrophic mutations by treatment of Chinese hamster cells with 5-bromodeoxyuridine and black light. Proc. Natl. Acad. Sci. USA 69:3459–3463.

Cleaver, J. E. 1977. Induction of thioguanine- and ouabain-resistant mutants and single-strand breaks in the DNA of Chinese hamster ovary cells by ³H-thymidine. Genetics 87:129–138.

Clive, D., W. G. Flamm, M. R. Machesko, and N. J. Bernheim. 1972. A mutational assay system using the thymidine kinase locus in mouse lymphoma cells. Mutat. Res. 16:77–87.

Crane, M. S. J., and D. B. Thomas. 1976. Cell-cycle, cell-shape mutant with features of the Go state. Nature 261:205–208.

Davidson, J. N., D. V. Carnright, and D. Patterson. 1979. Biochemical genetic analysis of pyrimidine biosynthesis in mammalian cells: III. Association of carbamyl phosphate synthetase, aspartate transcarbamylase, and dihydroorotase in mutants of cultured Chinese hamster cells. Somatic Cell Genetics 5:175–191.

Deaven, L. L., and D. F. Peterson. 1973. The chromosomes of CHO, an aneuploid Chinese hamster cell line: G-band, C-band and autoradiographic analysis. Chromosoma 41:129–144.

Farber, R. A., and M. P. Deutscher. 1976. Physiological and biochemical properties of a temperature-sensitive leucyl-tRNA synthetase mutant (tsH1) and revertant from Chinese hamster cells. Somatic Cell Genetics 2:509–520.

Farrell, S. A., and R. G. Worton 1977. Chromosome loss is responsible for segregation at the HPRT locus in Chinese hamster cell hybrids. Somatic Cell Genetics 3:539–551.

Feldman, R. K., and M. W. Taylor. 1975a. Purine mutants of mammalian cell lines. II. Identification of a phosphoribosylpyrophosphate amidotransferase-deficient mutant of Chinese hamster lung cells. Biochem. Genet. 13:227–234.

Feldman, R. K., and M. W. Taylor. 1975b. Purine mutants of mammalian cell lines. I. Accumulation of formylglycinamide ribotide by purine mutants of Chinese hamster ovary cells. Biochem. Genet. 12:393–405.

Flintoff, W. F., S. V. Davidson, and L. Siminovitch. 1976a. Isolation and partial characterization of three methotrexate-resistant phenotypes from Chinese hamster ovary cells. Somatic Cell Genetics 2:245–261.

Flintoff, W. F., S. M. Spindler, and L. Siminovitch. 1976b. Genetic characterization of methotrexate-resistant Chinese hamster ovary cells. In Vitro 12:749–757.

Fournier, R. E. K., and F. H. Ruddle. 1977a. Microcell-mediated transfer of murine chromosomes into mouse, Chinese hamster, and human somatic cells. Proc. Natl. Acad. Sci. USA 74:319–323.

Fournier, R. E. K., and F. H. Ruddle. 1977b. Stable association of the human transgenome and host murine chromosomes demonstrated with trispecific microcell hybrids. Proc. Natl. Acad. Sci. USA 74:3937–3941.

Gillin, F. D., D. J. Roufa, A. L. Beaudet, and C. T. Caskey. 1972. 8-Azaguanine resistance in mammalian cells. I. Hypoxanthine-guanine phosphoribosyltransferase. Genetics 72:239–252.

Gottesman, M. M., A. LeCam, M. Bukowski, and I. Pastan. 1980. Isolation of multiple classes of mutants of CHO cells resistant to cyclic AMP. Somatic Cell Genetics 6:45–61.

Gottlieb, C., A. M. Skinner, and S. Kornfeld. 1974. Isolation of a clone of Chinese hamster ovary cells deficient in plant lectin-binding sites. Proc. Natl. Acad. Sci. USA 71:1078–1082.

Gottlieb, C., J. Baenziger, and S. Kornfeld. 1975. Deficient uridine diphosphate-N-acetylglucosamine: Glycoprotein N-acetyl-glycosaminyl-transferase activity in a clone of Chinese hamster ovary cells with altered surface glycoproteins. J. Biol. Chem. 250:3303–3309.

Graham, F. L., and A. J. Van der Eb. 1973. A new technique for the assay of infectivity of human adenovirus 5 DNA. Virology 52:456–467.

Gupta, R. S., D. Y. H. Chan, and L. Siminovitch. 1978a. Evidence for functional hemizygosity at the Emt locus in CHO cells through segregation analysis. Cell 14:1007–1013.

Gupta, R. S., D. H. Y. Chan, and L. Siminovitch. 1978b. Evidence for variation in the number of functional gene copies at the Ama^R locus in Chinese hamster cell lines. J. Cell. Physiol. 97:461–468.

Gupta, R. S., W. F. Flintoff, and L. Siminovitch. 1977. Purification and properties of dihydrofolate reductase from methotrexate-sensitive and resistant Chinese hamster ovary cells. Canad. J. Biochem. 55:445–452.

Gupta, R. S., and L. Siminovitch. 1976. Isolation and preliminary characterization of mutants of CHO cells resistant to the protein synthesis inhibitor emetine. Cell 9:213–219.

Gupta, R. S., and L. Siminovitch. 1977a. The molecular basis of emetine resistance in Chinese hamster ovary cells. Alteration in the 40S subunit. Cell 10:61–66.

Gupta, R. S., and L. Siminovitch. 1977b. Mutants of CHO cells resistant to the protein synthesis inhibitors cryptopleurine and tylocrebrine: Genetic and biochemical evidence for common site of action of emetine, cryptopleurine, tylocrebrine and tubulosine. Biochemistry 16:3209–3214.

Gupta, R. S., and L. Siminovitch. 1978a. Mutants of CHO cell resistant to the protein synthesis inhibitor emetine: Genetic and biochemical characterization of second step mutants. Somatic Cell Genetics 4:77–94.

Gupta, R. S., and L. Siminovitch. 1978b. An in vitro analysis of the dominance of emetine sensitivity in CHO cell hybrids. J. Biol. Chem. 253:3978–3982.

Gupta, R. S., and L. Siminovitch. 1978c. Genetic and biochemical characterization of mutants of CHO cells resistant to the protein synthesis inhibitor trichodermin. Somatic Cell Genetics 4:355–375.

Gupta, R. S., and L. Siminovitch. 1978d. Genetic and biochemical studies with the adenosine analogs toyocamycin and tubercidin: Mutation at the adenosine kinase locus in Chinese hamster cells. Somatic Cell Genetics 4:715–736.

Gupta, R. S., and L. Siminovitch. 1978e. Diphtheria toxin resistant mutants of CHO cells defective in protein synthesis: A novel phenotype. Somatic Cell Genetics 4:553–571.

Gupta, R. S., and L. Siminovitch. 1980a. Diphtheria toxin resistance in Chinese hamster cells: Genetic and biochemical characteristics of the mutants affected in protein synthesis. Somatic Cell Genetics 6:361–379.

Gupta, R. S., and L. Siminovitch. 1980b. Genetic markers for quantitative mutagenesis studies in Chinese hamster ovary cells. Mutat. Res. 69:113–126.

Gupta, R. S., and L. Siminovitch. 1980c. Pactamycin resistance in CHO cells: Morphological changes induced by the drug in wild type and mutant cells. J. Cell. Physiol. 102:305–316.

Haars, L., A. Hampel, and L. H. Thompson. 1976. Altered leucyl-transfer RNA synthetase from a mammalian cell culture mutant. Biochim. Biophys. Acta 454:493–503.

Hanggi, U. J., and J. W. Littlefield. 1976. Altered regulation of the rate of synthesis of dihydrofolate reductase in methotrexate resistant cells. J. Biol. Chem. 251:3075–3080.

Hankinson, O. 1976. Mutants of the Chinese hamster ovary cell line requiring alanine and glutamate. Somatic Cell Genetics 2:497–507.

Holmes, E. W., G. L. King, A. Layva, and S. C. Singer. 1976. A purine auxotroph deficient in phosphoribosylpyrophosphate amidotransferase and phosphoribosylpryophosphate aminotransferase activities with normal activity of ribose-5-phosphate aminotransferase. Proc. Natl. Acad. Sci. USA 73:2458–2461.

Ingles, C. J. 1978. Temperature sensitive RNA polymerase II mutations in Chinese hamster ovary cells. Proc. Natl. Acad. Sci. USA 75:405–409.

Ingles, C. J., A. Guialis, J. Lam, and L. Siminovitch. 1976a. α-Amanitin-resistance of RNA polymerase II in mutant Chinese hamster ovary cell lines. J. Biol. Chem. 251:2729–2734.

Ingles, C. J., M. L. Person, M. Buchwald, B. C. Beatty, M. M. Crerar, A. Guialis, P. E. Lobban, L. Siminovitch, and D. C. Somers. 1967b. α-Amanitin-resistant mutants of mammalian cells and the regulation of RNA polymerase II activity, in RNA Polymerase, M. Chamberlin and R. Losick, eds. Cold Spring Harbor Laboratory, Cold Spring Harbor, N.Y., pp. 835–853.

Irwin, M., D. C. Oates, and D. Patterson. 1979. Biochemical genetics of Chinese hamster cell mutants with deviant purine metabolism: Isolation and characterization of a mutant deficient in the activity of phosphoribosylaminoimidazole synthetase. Somatic Cell Genetics 5:203–216.

Jones, C., and T. T. Puck. 1973. Genetics of somatic mammalian cells. XVII. Induction and isolation of Chinese hamster cell mutants requiring serine. J. Cell. Physiol. 81:299–304.

Jones, G. E., and P. A. Sargent. 1974. Mutants of cultured Chinese hamster cells deficient in adenine phosphoribosyl transferase. Cell 2:37–41.

Juliano, R., V. Ling, and J. Graves. 1976. Drug-resistant mutants of Chinese hamster ovary cells possess an altered cell surface carbohydrate component. J. Supramol. Struc. 2:728–736.

Juliano, R. L., and P. Stanley. 1975. Altered cell surface glycoproteins in phytohemagglutinin-resistant mutants of Chinese hamster ovary cells. Biochim. Biophys. Acta 389:401–406.

Kao, F. T., L. Chasin, and T. T. Puck. 1969. Genetics of somatic mamalian cells. X. Complementation analysis of glycine-requiring mutants. Proc. Natl. Acad. Sci. USA 64:1284–1291.

Kao, F. T., and T. T. Puck. 1967. Genetics of Chinese hamster cell mutants with respect to the requirement for proline. Genetics 55:513–529.

Kao, F. T., and T. T. Puck. 1968. Genetics of somatic mammalian cells. VII. Induction and isolation of nutritional mutants in Chinese hamster cells. Proc. Natl. Acad. Sci. USA 60:1275–1281.

Kao, F. T., and T. T. Puck. 1969. Genetics of somatic mammalian cells. IX. Quantitation of mutagenesis by physical and chemical agents. J. Cell. Physiol. 74:245–258.

Kemp, T. D., E. A. Swyryd, M. Bruist, and G. R. Stark. 1976. Stable mutants of mammalian cells that overproduce the first three enzymes of pyrimidine nucleotide biosynthesis. Cell 9:541–550.

Kit, S., D. R. Dubbs, L. J. Piekarski, and T. C. Hsu. 1963. Deletion of thymidine kinase activity from L cells resistant to bromodeoxyuridine. Exp. Cell Res. 31:297–312.

Krag, S. S., M. Cifone, P. W. Robbins, and R. M. Baker. 1977. Reduced synthesis of (14C)mannosyl oligosaccharide-lipid by membranes prepared from concanavalin A-resistant Chinese hamster ovary cells. J. Biol. Chem. 252:3561–3564.

Lagarde, A. E., and L. Siminovitch. 1979. Studies on Chinese hamster ovary mutants showing multiple cross-resistance to oxidative phosphorylation inhibitors. Somatic Cell Genetics 5:847–871.

Lewis, W. H., and J. A. Wright. 1974. Altered ribonucleotide reductase activity in mammalian tissue culture cells resistant to hydroxyurea. Biochem. Biophys. Res. Commun. 60:926–933.

Lewis, W. H., and J. A. Wright. 1978. Genetic characterization of hydroxyurea-resistance in Chinese hamster ovary cells. J. Cell. Physiol. 97:73–86.

Lewis, W. H., and J. A. Wright. 1979. Isolation of hydroxyurea-resistant cells with altered levels of ribonucleotide reductase. Somatic Cell Genetics 5:83–96.

Lewis, W. H., P. R. Srinivasan, N. Stokoe, and L. Siminovitch. 1980. Parameters governing the transfer of the genes for thymidine kinase and dihydrofolate reductase into mouse cells using metaphase chromosomes or DNA. Somatic Cell Genetics 6:333–348.

Ling, V. 1975. Drug resistance and membrane alteration in mutant mammalian cells. Canad. J. Genet. Cytol. 17:503–515.

Ling, V., J. E. Aubin, A. Chan, and F. Sarangi. 1979. Mutants of Chinese hamster ovary (CHO) cells with altered colcemid-binding affinity. Cell 18:423–430.

Ling, V., and R. M. Baker. 1978. Dominance of colchicine resistance in hybrid CHO cells. Somatic Cell Genetics 6:193–200.

Ling, V., and L. H. Thompson. 1974. Reduced permeability in CHO cells as a mechanism of resistance to colchicine. J. Cell. Physiol. 83:103–116.

Littlefield, J. W. 1963. The inosinic acid pyrophosphorylase activity of mouse fibroblasts partially resistant to 8-azaguanine. Proc. Natl. Acad. Sci. USA 50:568–576.

Littlefield, J. W. 1965. Studies on thymidine kinase in cultured mouse fibroblasts. Biochim. Biophys. Acta 95:14–22.

Lobban, P. E., and L. Siminovitch. 1975. α-Amanitin resistance: A dominant mutation in CHO cells. Cell 4:167–172.

Lobban, P. E., and L. Siminovitch. 1976. The RNA polymerase II of an α-amanitin-resistant Chinese hamster ovary cell line. Cell 8:65–70.

Maitland, N. J., and J. K. McDougall. 1977. Biochemical transformation of mouse cells by fragments of herpes simplex virus DNA. Cell 11:233–241.

Mandel, J., and W. I. Flintoff. 1978. Isolation of mutant mammalian cells altered in polyamine transport. J. Cell. Physiol. 97:335–344.

McBride, O. W., and H. L. Ozer. 1973. Transfer of genetic information by purified metaphase chromosomes. Proc. Natl. Acad. Sci. USA 70:1258–1262.

McBurney, M., and G. F. Whitmore. 1974a. Isolation and biochemical characterization of folate deficient mutants of Chinese hamster cells. Cell 2:173–182.

McBurney, M., and G. F. Whitmore. 1974b. Characterization of a Chinese hamster cell with a temperature-sensitive mutation in folate metabolism. Cell 2:183.

McBurney, M. W., and G. F. Whitmore. 1975. Mutants of Chinese hamster cells resistant to adenosine. J. Cell. Physiol. 85:87–100.

Meuth, M., M. Trudel, and L. Siminovitch. 1979. Selection of Chinese hamster cells auxotrophic for thymidine by 1-B-D-arabinofuranosyl cytosine. Somatic Cell Genetics 5:303–318.

Minson, A. C., P. Wildy, A. Buchan, and G. Darby. 1978. Introduction of the herpes simplex virus thymidine kinase gene into mouse cells using virus DNA or transformed cell DNA. Cell 13:581–587.

Moehring, J. M., and T. J. Moehring. 1979. Characterization of the diphtheria toxin-resistance system in Chinese hamster ovary cells. Somatic Cell Genetics 5:453–468.

Moehring T. J., and J. M. Moehring. 1977. Selection and characterization of cells resistant to diphtheria toxin and pseudomonas exotoxin A: Presumptive translational mutants. Cell 11:447–454.

Oates, D. C., and D. Patterson. 1977. Biochemical genetics of Chinese hamster cell mutants with deviant purine metabolism: Characterization of Chinese hamster cell mutants defective in phosphoribosylpyrophosphate amidotransferase and phosphoribosylglycinamide synthetase and an examination of alternatives to the first step of purine biosynthesis. Somatic Cell Genetics 3:561–577.

Patterson, D. 1975. Biochemical genetics of Chinese hamster cell mutants with deviant purine metabolism: Biochemical analysis of eight mutants. Somatic Cell Genetics 1:91–110.

Patterson, D. 1976a. Biochemical genetics of Chinese hamster cell mutants with deviant purine metabolism. III. Isolation and characterization of a mutant unable to convert IMP to AMP. Somatic Cell Genetics 2:41–53.

Patterson, D. 1976b. Biochemical genetics of Chinese hamster cell mutants with deviant purine metabolism. IV. Isolation of a mutant which accumulates adenylosuccinic acid and succinylaminoimidazole carboxamide ribotide. Somatic Cell Genetics 2:189–203.

Patterson, D., and D. V. Carnright. 1977. Biochemical genetic analysis of pyrimidine biosynthesis in mammalian cells. I. Isolation of a mutant defective in the early steps of de novo pyrimidine synthesis. Somatic Cell Genetics 3:483–495.

Patterson, D., F. T. Kao, and T. T. Puck. 1975. Genetics of somatic mammalian cells: Biochemical genetics of Chinese hamster cell mutants with deviant purine metabolism. Proc. Natl. Acad. Sci. USA 71:2057–2061.

Patterson, D., C. Waldren, and C. Walker. 1976. Isolation and characterization of temperature-sensitive mutants of Chinese hamster ovary cells after treatment with UV and X-irradiation. Somatic Cell Genetics 2:113–123.

Person, S., and R. S. Bockrath. 1964. Differential mutation production by the decay of incorporated tritium compounds in E. coli. Biophys. J. 4:355–365.

Puck, T. T., and F. T. Kao. 1967. Genetics of somatic mammalian cells. V. Treatment with 5-bromodeoxyuridine and visible light for isolation of nutritionally deficient mutants. Proc. Natl. Acad. Sci. USA 58:1227–1234.

Rabin, M. S., and M. M. Gottesman. 1979. High frequency of mutation to tubercidin resistance in CHO cells. Somatic Cell Genetics. 5:571–583.

Rosenstraus, M. J., and L. A. Chasin. 1977. Mutants of Chinese hamster ovary cells with altered glucose-6-phosphate dehydrogenase activity. Somatic Cell Genetics 3:323–333.

Schimke, R. T., R. J. Kaufman, F. W. Alt, and R. F. Kellams. 1978. Gene amplification and drug resistance in cultured murine cells. Science 202:1051–1055.

Sharp, J. D., N. E. Capecchi, and M. R. Capecchi. 1973. Altered enzymes in drug-resistant variants of mammalian tissue culture cells. Proc. Natl. Acad. Sci. USA 70:3145–3149.

Siciliano, M. J., J. Siciliano, and R. M. Humphrey. 1978. Electrophoretic shift mutants in Chinese hamster ovary cells: Evidence for genetic diploidy. Proc. Natl. Acad. Sci. USA 4:1919–1923.

Siminovitch, L. 1976. On the nature of hereditable variation in cultured somatic cells. Cell 7:1–11.

Siminovitch, L. 1979a. On the origin of mutants of somatic cells, *in* Eucaryotic Gene Regulation, ICN-UCLA Symposia on Molecular and Cellular Biology, R. Axel, T. Maniatis, and C. F. Fox, eds. Academic Press, New York, pp. 433–443.

Siminovitch, L. 1979b. Studies of mutation in CHO cells, *in* Mammalian Cell Mutagenesis. The Maturation of Test Systems, A. W. Hsu, J. P. O'Neill, and V. K. McElheny, eds. Cold Spring Harbor Laboratories, Cold Spring Harbor, pp. 15–21.

Siminovitch, L., and L. H. Thompson. 1978. The nature of conditionally lethal temperature-sensitive mutations in somatic cells. J. Cell Physiol. 95:361–366.

Stanley, P., V. Caillibot, and L. Siminovitch. 1975a. Stable alterations at the cell membrane of Chinese hamster ovary cells resistant to the cytotoxicity of phytohemagglutinin. Somatic Cell Genetics 1:3–26.

Stanley, P., V. Callibot, and L. Siminovitch. 1975b. Selection and characterization of eight phenotypically-distinct lines of lectin-resistant Chinese hamster ovary cells. Cell 6:121–128.

Stanley, P., S. Narasimhan, L. Siminovitch, and H. Schachter. 1975c. Chinese hamster ovary cells selected for resistance to the cytotoxicity of phytohemagglutinin are deficient in a UDP-N-acetyl-glucosamine-glycoprotein N-acetyl-glucosaminyl-transferase activity. Proc. Natl. Acad. Sci. USA 72:3323–3327.

Stanley, P., and L. Siminovitch. 1977. Complementation between mutants of CHO cells resistant to a variety of plant lectins. Somatic Cell Genetics 3:391–405.

Stanners, C. P., and L. H. Thompson. 1974. Studies on a mammalian cell mutant with a temperature-sensitive leucyl-tRNA synthetase, *in* Control of Proliferation in Animal Cells, Cold Spring Harbor Conferences on Cell Proliferation, Vol. 1, B. Clarkson, and R. Baserga, eds. Cold Spring Harbor Laboratories, Cold Spring Harbor, pp. 191–203.

Stanners, C. P., T. M. Wightman, and J. L. Harkins. 1978. Effect of extreme amino acid starvation on the protein synthetic machinery of CHO cells. J. Cell Physiol. 95:125–138.

Szbalski, W., and M. J. Smith. 1959. Genetics of human cell lines. I. 8-azaguanine resistance, a selective "single-step" marker. Proc. Soc. Exp. Biol. Med. 101:662–666.

Taylor, M. W., J. H. Pipkorn, M. K. Tokito, and R. O. Pozzatti, Jr. 1977. Purine mutants of mammalian cells. III. Control of purine biosynthesis in adenine phosphoribosyl transferase mutants of CHO cells. Somatic Cell Genetics 3:105–206.

Thirion, J. P., D. Banville, and H. Noel. 1976. Galactokinase mutants of Chinese hamster somatic cells resistant to 2-deoxyglucose. Genetics 83:137–147.

Thompson, L. H., J. L. Harkins, and C. P. Stanners. 1973. A mammalian cell mutant with a temperature-sensitive leucyl-transfer RNA synthetase. Proc. Natl. Acad. Sci. USA 70:3094–3098.

Thompson, L. H., and P. A. Lindl. 1976. A CHO-cell mutant with a defect in cytokinesis. Somatic Cell Genetics 2:387–400.

Thompson, L. H., D. J. Lofgren, and G. M. Adair. 1977. CHO cell mutants for arginyl-, asparagyl-, glutaminyl-, histidyl-, and methionyl-transfer RNA synthetases: Identification and initial characterization. Cell 11:157–168.

Thompson, L. H., D. J. Lofgren, and G. M. Adair. 1978. Evidence for structural gene alterations affecting aminoacyl-tRNA synthetases in CHO cell mutants and revertants. Somatic Cell Genetics 4:423–435.

Thompson, L. H., J. S. Rubin, J. E. Cleaver, G. F. Whitmore, and K. Brookman. 1980. A screening method for isolating DNA repair-deficient mutants of CHO cells. Somatic Cell Genetics 6:391–405.

Thompson, L. H., C. P. Stanners, and L. Siminovitch. 1975. Selection by (^{3}H) amino acids of CHO-cell mutants with altered leucyl- and asparagyl-transfer RNA synthetases. Somatic Cell Genetics 1:187–208.

Tu, A. S., and D. Patterson. 1977. Biochemical genetics of Chinese hamster cell mutants with deviant purine metabolism. VI. Enzymatic studies of two mutants unable to convert inosinic acid to adenylic acid. Biochem. Genet. 15:195–210.

Wasmuth, J. J., and C. T. Caskey. 1976. Selection of temperature-sensitive CHL asparagyl t-RNA synthetase mutants using the toxic lysine analog, S-2-aminoethyl-L-cysteine. Cell 9:655–662.

Waye, M. M. Y., and C. P. Stanners. 1979. Isolation and characterization of CHO cell mutants with altered asparagine synthetase. Somatic Cell Genetics 5:625–639.

Wigler, M., A. Pellicer, S. Silverstein, R. Axel, G. Urlaub, and L. Chasin. 1979a. DNA-mediated transfer of the adenine phosphoribosyl-transferase locus into mammalian cells. Proc. Natl. Acad. Sci. USA 76:1373–1376.

Wigler, M., S. Silverstein, L-S. Lee, A. Pellicer, Y-C. Cheng, and R. Axel. 1977. Transfer of purified herpes virus thymidine kinase gene to cultured mouse cells. Cell 11:223–232.

Wigler, M., R. Sweet, S. Gek Kee, B. Wold, A. Pellicer, E. Lacy, T. Maniatis, S. Silverstein, and R. Axel. 1979b. Transformation of mammalian cells with genes from procaryotes and eucaryotes. Cell 16:777–785.

Willecke, K., R. Lange, A. Kruger, and T. Reber. 1976. Co-transfer of two linked human genes into cultured mouse cells. Proc. Natl. Acad. Sci. USA 73:1274–1278.

Willecke, K., and F. H. Ruddle. 1975. Transfer of the human gene for hypoxanthine-phosphoribosyl-transferase via isolated human metaphase chromosomes into mouse L-cells. Proc. Natl. Acad. Sci. USA 72:1792–1794.

Williams, K. L. 1976. Mutation frequency of a recessive locus in haploid and diploid strains of a slime mould. Nature 260:785–786.

Worton, R. G. 1978. Karyotypic heterogeneity in CHO cell lines. Cytogenet. Cell Genet. 21:105–110.

Worton, R. G., C. C. Ho, and C. Duff. 1977. Chromosome stability in CHO cells. Somatic Cell Genetics 3:27–45.

Wright, J. A. 1972. Pleiotrophic changes in lines of Chinese hamster ovary cells resistant to concanavalin and phytohaemagglutinin. J. Cell Biol. 56:666–675.

Wright, J. A., and W. H. Lewis. 1974. Evidence for a common site of action for the antitumor agents hydroxyurea and guanazole. J. Cell. Physiol. 83:437–440.

Genes, Chromosomes, and Neoplasia, edited by
Frances E. Arrighi, Potu N. Rao, and Elton Stubblefield.
Raven Press, New York © 1981.

Genetic and Biochemical Analysis of Dedifferentiated Variants of a Rat Hepatoma

Mary C. Weiss, Linda Sperling, Emma E. Moore,*
and Jean Deschatrette

Centre de Genetique Moleculaire du C.N.R.S., 91190 Gif-sur-Yvette, France

Just as physicians know that an understanding of how physiological processes can malfunction in diseased states helps clarify normal physiology, geneticists study defective mutants to help them understand normal processes. We have used this approach to study the problem of cell differentiation, reasoning that the analysis of stable dedifferentiated variants derived from differentiated cells in culture will lead to the elucidation of some of the control mechanisms involved in the regulation of cell differentiation.

The cell type that we study is a model for differentiation of the hepatocyte and represents a favorable system from several points of view. First, the hepatocyte is very rich in the spectrum of differentiated functions it expresses. Second, the functions are initially expressed at different times during ontogenesis; some are expressed very early, others during mid-gestation, and some only just prior to birth or after. Finally, the hepatocyte is an anabolic cell, able to synthesize, via liver-specific metabolic pathways, some of the nutrients such as amino acids and sugars required by other cells of the body. It is thus possible to devise special culture media that support the growth only of cells in which the metabolic pathways in question are active.

Since hepatocytes do not grow actively in tissue culture, it is convenient to make use of the wide variety of transplantable hepatomas that continue to express many functions of the normal hepatocyte. From one such tumor of the rat, the Reuber H35 hepatoma (Reuber 1961), Pitot et al. (1964) derived a clonal cell line, H4IIEC3, whose cells express many functions of the hepatocyte of the adult rat. A number of years ago we began to study this line and to derive independent subclones into which we introduced different selective markers useful for somatic cell genetics. The cells of all of these cloned lines, like the original H4IIEC3 cells, produce a wide spectrum of proteins peculiar to the hepatocyte, among which we study seven.

---

* Department of Pathology, University of Colorado Health Sciences Center School of Medicine, Denver, Colorado 80262

From time to time we found in the cultures cells that differed in morphology from the original cells. Isolated and subcloned, these unusual-looking cells were revealed to be stable, dedifferentiated variants that failed to express the whole group of functions characteristic of the cells of origin (Deschatrette and Weiss 1974). In this paper, we will discuss three independent clones of these variant cells.

Our analysis of the dedifferentiated variants has involved several steps. First, the cells were analyzed for their expression of seven different liver-specific proteins. Second, their karyotypes were compared to those of the differentiated cells of origin. Third, in order to determine whether one state is dominant over the other, analysis was carried out on somatic hybrids between the dedifferentiated variant cells and the differentiated cells from which they were derived. Fourth, the variant cells were examined for their potential to give rise to revertants that again express the hepatic functions. Fifth, one aspect of the chromatin structure of the various lines was compared.

RESULTS AND DISCUSSION

Phenotypic Properties of Dedifferentiated Hepatoma Cell Variants

Figure 1 shows the filiation of the cloned cell lines that we will discuss: each of the three dedifferentiated variant lines derives from a separate differentiated line, and we can therefore be certain that we are dealing with variants of independent origin. Table 1 shows the expression of the various hepatic functions by each of these lines; cells of the well-differentiated clones of origin express each of the functions, with some quantitative variations being observed from one clone to another. In contrast, the dedifferentiated clones are deficient in expression of each of them, with the single exception of the clone FaoflC2, whose cells continue to produce both basal and hormone-induced tyrosine amino-transferase (TAT).

The fact that cells of each of these clones no longer express an entire set of tissue-specific functions indicates that we are dealing with regulatory changes having pleiotropic effects and not with mutations in the structural genes coding for the various enzymatic or secretory proteins.

The karyotype of each of the variant lines has been compared with that of its differentiated line of origin. In no case have we been able to correlate a specific, karyotypic modification with the drastic change in phenotype (Deschatrette and Weiss 1974, and unpublished results).

Since a change in the activity of some regulatory element is most likely to be at the origin of the dedifferentiated phenotype of each of the variant lines, one question that must be posed is whether the dedifferentiated state is dominant or recessive compared to the differentiated state. If it is recessive, we would be led to hypothesize that dedifferentiation results from the loss of some element necessary to maintain the expression of differentiated functions.

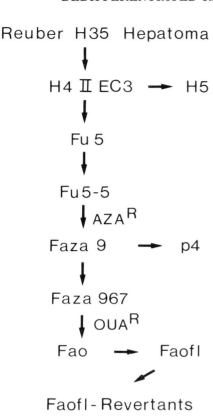

Reuber H35 Hepatoma

FIG. 1. Filiation of clones derived from line H4IIEC3. Each arrow signifies one clonal generation. The clones in the center are well differentiated; those to the right are dedifferentiated. AZA[R] and OUA[R] signify, respectively, resistance to 12 μg/ml of 8-azaguanine and 3 mM ouabain.

In order to answer this question, we isolated somatic hybrids between cells of each of the variant lines and the differentiated cells from which they were derived.

"Dominance" of the Dedifferentiated State

p4 × Fu5–5

Hybrids between cells of these two lines showed the morphological features of the dedifferentiated p4 parent. The majority of clones presented a karyotype near to that expected from the fusion of one cell of each parent. As shown in Table 2, the p4 genome extinguished the expression by its partner of the liver-specific functions. It is clear that the phenotype of the p4 parent is dominant (Deschatrette and Weiss 1975) and that dedifferentiation of p4 cells is associated with the production of diffusible regulatory factors whose final effect is negative and which act to prevent the expression of the very functions that characterized the precursor differentiated cells.

TABLE 1. *Properties of differentiated clones and of dedifferentiated variant cells*

Cell lines	Number of chromosomes	Albumin (μg/10^6/24 hr)	Aldolase B	L-ADH	AAT		TAT		FDPase	PEPCK
					B	I	B	I		
Fu5-5	53	0.5	+	+	100	450	35	250	3	40
Faza 967	52	2.5	+	+	100	320	60	330	16	85
Fao	52	2.5	+	+	90	280	50	400	13	80
H5	52	<0.001	−	−	<5	<5	1	1	<0.5	19
p4	48	<0.001	−	−	<6	12	1	1	NT	NT
FaoflC2	52	≤0.006	−	−	≤10	70	20	240	<0.1	19

Data taken from Deschatrette and Weiss (1974), Deschatrette et al. (1979), and unpublished results from this laboratory. Enzyme activities are specific activities (milliunits/mg protein in the extracts), as detailed in the original publications. Abbreviations used: AAT, alanine aminotransferase; TAT, tyrosine aminotransferase; FDPase, fructose diphosphatase; PEPCK, phosphoenolpyruvate carboxykinase; B and I, basal and hormone-induced specific activities; NT, not tested.

TABLE 2. *Expression of liver-specific functions by hybrids between dedifferentiated and differentiated hepatoma cells*

Cell line	Albumin	Aldolase B	L-ADH	AAT		TAT		Growth in G⁻ medium
				B	I	B	I	
p4 × Fu5-5	0/5	0/10	2/10	2/10	8/10	1/10	0/10	
H5 × Fao	4/6	0/6	5/6	5/6	4/6	0/6	0/6	0/6
FaoflC2 × Fu5-5								
"Flat"	0/2	0/2	2/2	0/2	0/2	—	—	0/2
"Hepatoma-like"	0/1	1/1	1/1	1/1	1/1	—	—	0/1
"Round"	7/8	8/8	8/8	8/8	8/8	—	—	8/8

The number of clones expressing greater than 12% of the parental activity over the number of independent hybrid clones tested is indicated. For TAT activity, since both FaoflC2 and Fu5–5 parental cells express this function, the activities of the hybrid cells are not included. Abbreviations: as in Table 1.

H5 × Fao

Like the hybrids described above, those resulting from this cross presented the morphological features of the dedifferentiated parent and a chromosome number near to the sum of the parental numbers. When examined for the expression of hepatic functions, it was found that the hybrids showed extinction of expression of many of the functions of the Fao parent. However, for three of the functions, albumin secretion, activity of the liver isozyme of alcohol dehydrogenase (L-ADH), and activity and inducibility of alanine aminotransferase (AAT), some of the hybrids showed an intermediate phenotype. Since the majority of functions were not expressed by these hybrid cells at or near to the levels of Fao cells, or not at all, it was concluded that, as in the case of p4, the dedifferentiated state is dominant (Deschatrette et al. 1979). For neither p4 nor H5 does the hypothesis of simple loss of some factor necessary to maintain the differentiated state seem to be valid.

FaoflC2 × Fu5–5

Hybrids resulting from this cross were found to show an evolution in their morphology. During the first 10 days after fusion, the hybrid cells resembled the FaoflC2 parent; thereafter, they began to resemble more closely the differentiated Fu5–5 parent, and still later to take on an entirely new form different from that of both parents. These morphological transitions are illustrated in Figure 2.

That the morphological state reflects the expression of liver-specific functions has been demonstrated in two ways. First, hybrids of each of the three morphological forms were assayed for the liver-specific functions characteristic of the differentiated Fu5–5 parent (see Table 2). "Flat" hybrids that resemble the FaoflC2 parent show extinction of the majority of functions; the hybrid that is similar to Fu5–5 in its morphology shows an intermediate state, and the "round" hybrid cells show full expression of all the functions (Deschatrette et al. 1979).

A second test of the state of expression of the different types of hybrids was obtained by determining their ability to grow in a special selective medium. As mentioned at the outset, one of the advantages of studying hepatic differentiation is that liver cells are able to synthesize certain nutrients that other cells cannot. For example, there is a special pathway in the liver, the gluconeogenic pathway, that permits the synthesis of glucose by a reversal of the glycogenic pathway, the irreversible steps of which are compensated for by the activity of the specific enzymes fructose diphosphatase (FDPase) and phosphoenolpyruvate carboxykinase (PEPCK). As shown by Bertolotti (1977a,1977b), any cell that produces these enzymes is able to proliferate in medium lacking glucose (G⁻ medium). The well-differentiated hepatoma cells, such as Fu5–5, produce these enzymes and proliferate actively in G⁻ medium; the dedifferentiated variant

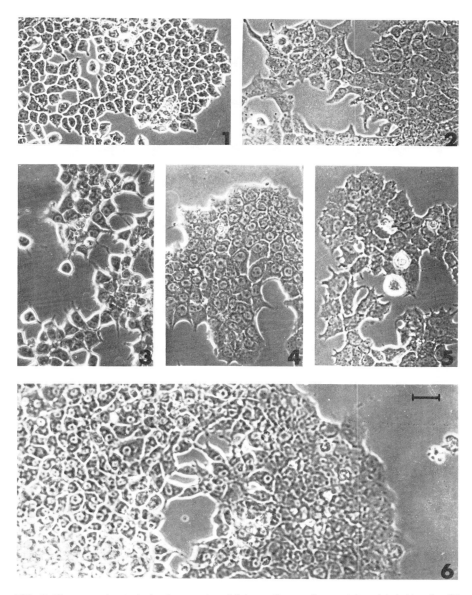

FIG. 2. Phase contrast photomicrographs of living cultures of parental and hybrid cells. (1) Fu5–5 well-differentiated hepatoma cells: very regular epithelial form, phase-dense cytoplasm, and a large centrally located nucleolus are characteristic of these cells. (2) FaoflC2 dedifferentiated variant cells are characterized by pale cytoplasm, small nucleoli, and a flat appearance. (3–6) Hybrids of Fu5–5 × FaoflC2. The newly formed hybrids (panel 5) resemble the FaoflC2 parent. With time, the cells evolve to Fu5–5 like morphology (panel 4), and, finally, to a refractile round form (panel 3). (6) The edge of a hybrid colony showing all three morphological forms.
Reprinted from Deschatrette et al. 1979, with the permission of Plenum Press.

cells do not, and they die in G⁻ medium. Whether or not hybrid cells proliferate in G⁻ medium is an indirect measure of whether FDPase and PEPCK are produced.

At various times after fusion, hybrid colonies were challenged with G⁻ medium; as Table 3 shows, the appearance of hybrids that survive in this medium corresponds to the time when there is a transition in morphology, from "flat" to "hepatoma-like" and "round" morphology (Deschatrette et al. 1979).

These results show that soon after fusion, FaoflC2 × Fu5–5 hybrids show extinction of the liver functions, but this extinction is only *transitory* and is followed by full expression of the functions characteristic of the differentiated Fu5–5 parent. It thus appears that FaoflC2 cells, like p4 and H5 cells, produce factor(s) that cause extinction. However, while the extinction imposed by p4 and H5 is stable and heritable, that caused by FaoflC2 is only transitory. We have proposed that the period of transitory extinction reflects the time necessary for inactivation of the extinguishing factor(s) and that these factors are no longer produced by the FaoflC2 genome after fusion with Fu5–5 cells. By contrast, production of these factors appears to be maintained by p4 and H5 cells, even after fusion with the well-differentiated cells from which they were derived.

TABLE 3. *Correlation between ability of FaoflC2 × Fu5–5 hybrid cells to proliferate in G⁻ medium and their morphology*

Day after fusion of medium change	Number of colonies per plate after 50 days		% Colonies on day of medium change containing:	
	G⁺	G⁻	Hepatoma-like cells	Round cells
6	57	0	0	0
8	52	0	0	0
10	57	0	7	0
12	54	1.5	10	0
14	73	8.5	25	0
16	145	24	31	0
20	185	47	45	0
22	175	58	55	3
24	153	44	62	3
26	NT	52	66	7
34	NT	NT	69	21
40	NT	NT	72	38

Cells were fused and inoculated into 50 plates (10⁵ cells/plate) in standard selective medium containing fetal calf serum, HAT medium, and ouabain. At various times after fusion, groups of plates were changed to G⁻ or G⁺ medium (both supplemented with dialyzed serum plus HAT and ouabain). Fifty days after fusion, all plates were fixed, and the number of colonies was counted (144 colonies per plate were present in standard selective medium). At the beginning of the experiment 6 days after fusion), 31 colonies were marked and examined every 2 days for morphological evolution; these cells were maintained in standard selective medium throughout the course of the experiment. Note that only about one-fourth of the presumed hybrids grow in G⁺ medium for the first 2 weeks after fusion.

The fact that dedifferentiated hepatoma cells produce some elements that inhibit the expression of liver-specific functions is perhaps surprising and suggests that this capability is one aspect of cell differentiation. In other words, a differentiated cell may have the potential for producing regulatory molecules that act not only to insure the expression of the spectrum of functions characteristic of its differentiation but also to prevent their expression under particular physiological, pathological, or developmental conditions. In fact, this hypothesis is entirely consistent with what is known about development. During ontogeny, a cell becomes determined to assume a particular differentiation later on, and, in addition when the tissue-specific proteins are first produced, they may not all be expressed at the same time but in a discontinuous and sequential fashion, as is the case for the hepatocyte. The idea that the extinguishing capacity of the dedifferentiated variants reflects one aspect of hepatocyte differentiation implies that these cells have not lost their hepatocyte determination. If this is true, they should be able to revert back to the differentiated state.

Revertants of FaoflC2

The potentiality of FaoflC2 cells to revert to the differentiated state was tested by challenging cultures with G^- medium; under these conditions, FaoflC2 cells die, but any revertant cell that again produces FDPase and PEPCK should survive. Since a selective medium was available, it was possible to quantitate the reversion event, and to do this it was necessary to test several independent clones of Faofl cells. The results of these tests are given in Table 4; among five independent clones of Faofl tested, three gave rise to colonies able to grow progressively in G^- medium, at frequencies ranging from 6×10^{-9} to 6×10^{-8}.

Faofl cells have indeed retained their "determination," judging by the criterion of their capacity to give rise to cells that produce FDPase and PEPCK. However, since not only hepatocytes but also cells of the kidney cortex and the intestinal mucosa produce these enzymes, it was necessary to examine the revertant cells for hepatocyte functions whose expression would not confer a selective advantage

TABLE 4. *Reversion of Faofl cells*

Cell line	Frequency of revertants	Number of revertant colonies characterized	Number of functions expressed/ analyzed
Faofl 1	$<1.3 \times 10^{-8}$	—	—
Faofl C2	6.9×10^{-9}	6	7/7
Faofl 3	$<1.3 \times 10^{-8}$	—	—
Faofl C4	6.7×10^{-8}	5	7/7
Faofl 5	5.5×10^{-8}	2	7/7

in G⁻ medium. Tests carried out on Faofl revertant cells (Deschatrette et al. 1980) revealed that they produce not only the gluconeogenic enzymes FDPase and PEPCK but also each of the other hepatic proteins whose expression characterizes the original differentiated line (see Table 1). Thus, whatever is the nature of the regulatory modification undergone by Faofl cells, it has a pleiotropic effect; when Faofl cells revert to the differentiated state, a mechanism having a pleiotropic effect is again implicated.

Table 5 shows the results of karyotype analyses carried out on each of the Faofl clones and the revertants derived from them in order to determine whether a loss or a gain in chromosomes is correlated with the reversion event. Fifteen or more metaphases of each clone were analyzed, and the chromosomes of each assigned to one of three categories that can be recognized in the rat. We then looked for a systematic change in the number of chromosomes of the various types. As can be seen in Table 5, some of the revertant clones have fewer chromosomes than the Faofl clone of origin, but this difference does not hold for all clones. Most of the revertants show either a slight loss of small metacentric chromosomes, or a slight increase in the number of telocentric chromosomes, but the differences are so small that they could reflect random clonal fluctuations. In conclusion, no convincing evidence was obtained in favor of the idea of a chromosomal basis for reversion.

Cells of clone FaoflC2 were treated with mutagenic agents, and the surviving cells challenged with G⁻ medium; no enhancement in the frequency of revertants was observed (see Deschatrette et al. 1980, for detailed data). These experiments provided no argument in favor of the hypothesis of mutation as the event responsible for reversion.

The reversion of Faofl cells confronts us with a stable pleiotropic change that is rare, is not provoked by mutagens, and does not seem to result from gross chromosomal changes. Since many developmental changes are pleiotropic, clearly not due to mutations or to chromosomal changes, we are led to propose that the events observed may be a reflection of regulatory changes similar to or identical to those that operate during development. The possibility that this may be the case has led us to continue study of the dedifferentiated variant cells, with the ultimate aim of identifying the molecular basis of the differences between the dedifferentiated variant cells and the differentiated cells from which they were derived.

It may be estimated that simple comparisons of cells of different phenotypes is a risky undertaking, for it is difficult to evaluate which differences are critical and which are simply coincidental. However, for such an undertaking, the material we use presents two advantages. First, all of the cell lines are clonal descendants of a single cell; thus, trivial differences that may exist between cell lines of independent origin will not intervene. Second, the techniques of genetic analysis permit a functional test of the significance of any differences identified. The experiments to be described below illustrate this point.

TABLE 5. Karyotypes of Faofl and revertant clones*

Cell line	Total number of chromosomes	Corrected total	Corrected telocentric	Small metacentric	Ratio of telocentric: small metacentric
Fao	51.9 (50–55)	51.9 (50–55)	27.1	21.8	1.24
FaoflC2	51.7 (50–54)	51.9 (50–54)	25.6	23.3	1.10
FaoflC2–41°	50.9 (49–54)	51.3 (50–54)	25.0	23.4	1.07
C2-Rev 5	55.6 (54–58)	55.6 (54–58)	25.9	26.7	0.97
C2-Rev 7	50.4 (48–53)	51.1 (49–54)	27.1	21.2	1.28
C2-Rev 4	49.5 (48–52)	50.5 (49–53)	25.9	21.7	1.19
C2-Rev 3	49.7 (48–50)	49.7 (48–50)	25.9	20.9	1.24
C2-Rev 1	46.5 (45–48)	47.4 (46–49)	22.3	21.6	1.06
C2-Rev b	44.1 (43–45)	46.3 (45–48)	25.0	18.3	1.37
FaoflC4	50.9 (49–53)	51.1 (47–53)	22.3	25.5	0.88
C4-Rev 1	51.6 (49–54)	52.1 (50–54)	24.3	24.9	0.98
C4-Rev 4b	50.1 (47–52)	51.0 (48–53)	24.7	23.4	1.06
C4-Rev 2	49.7 (48–51)	50.7 (49–52)	24.9	22.8	1.09
C4-Rev 4a	49.7 (48–51)	50.7 (49–52)	23.9	23.8	1.01
C4-Rev 3	48.9 (47–51)	49.9 (48–52)	24.1	22.9	1.05
Faofl 5	52.7 (50–55)	53.1 (50–55)	26.1	23.5	1.11
5-Rev 2	50.3 (49–52)	52.8 (50–56)	27.8	22.9	1.21
5-Rev 1	49.4 (46–51)	50.8 (48–53)	26.0	21.8	1.19
Faofl 1	51.2 (49–54)	53.6 (52–56)	29.5	21.3	1.39
Faofl 3	53.5 (51–56)	53.5 (51–56)	25.1	25.3	0.99

* Based on the analyses of 15 or more metaphases of each clone. Means are given, with ranges in parentheses.

Chromatic Repeat Length of Hepatoma Cells, Their Dedifferentiated Variants and Somatic Hybrids

It has been known for several years now that chromatin consists of DNA wrapped around histone cores to form a chain of repeating subunits known as nucleosomes (for a review of chromatin structure, see Kornberg 1977). It is possible, using certain nucleases, to cut the DNA in chromatin preferentially in a certain region (the "linker") of the nucleosome and thus obtain DNA fragments corresponding to 1, 2, 3, etc. nucleosomes. Measurements of the sizes of these fragments may be used to calculate the *repeat length* of the chromatin of a given population of cells, that is, the *average* amount of DNA in a nucleosome.

Measurements of repeat length that may be found in the literature show that this parameter may vary widely depending not only upon the species but also upon the tissue of origin of the chromatin. That there may be a relationship between repeat length and the mechanisms implicated in the establishment and maintenance of tissue-specific patterns of gene expression is suggested by the fact that different tissues of the same organism have different repeat lengths: 198 base pairs for chicken liver but 210 base pairs for chicken erythrocytes (Morris 1976, Compton et al. 1976); about 200 base pairs for rabbit liver, glia, and cerebellar neurons, but 160 base pairs for rabbit cortical neurons (Thomas and Thompson 1977); 218 base pairs for sea urchin gastrula but 241 base pairs for sea urchin sperm (Spadafora et al. 1976). In order to test the hypothesis that chromatin repeat length is correlated with the state of differentiation of cells, we chose to examine the chromatin repeat length of the differentiated and dedifferentiated hepatoma clones (Sperling and Weiss 1980).

In order to carry out the proposed study it was necessary to improve the methods for determining repeat length so that we could detect very small differences. The technique adopted has been described in detail (Sperling et al. 1980): as judged by repeated measurements on the same cell line, the accuracy is ± 1 base pair. It is important to note that the values we measure are operationally defined as *average* repeat lengths and that they are a property of the entire genome (e.g., the differences in repeat length to be described cannot arise from changes in the repeat length of only a part of the genome).

Figure 3 shows the results of these analyses. Each of the three differentiated clones examined shows the same repeat length (189 base pairs ± 1); the extensive overlap of the individual probability distributions of average repeat length for the three lines provides a visual demonstration of this fact. When the repeat length of the various dedifferentiated lines was determined, we found that the value of 189 base pairs no longer applies. Cells of line p4 show a repeat length of 192, those of H5 a repeat length of 184, and FaoflC2 cells, a value of 188. Thus, not only do the variants show a repeat length either shorter or longer than that of the well-differentiated cells, but they differ from one another as well.

It was surprising to find that a property that reflects the packaging of the entire genome can differ in derivatives of a single cell. The change in repeat length might be intimately correlated with the state of phenotypic expression of the cells, or it could be unrelated and purely coincidental. If the former

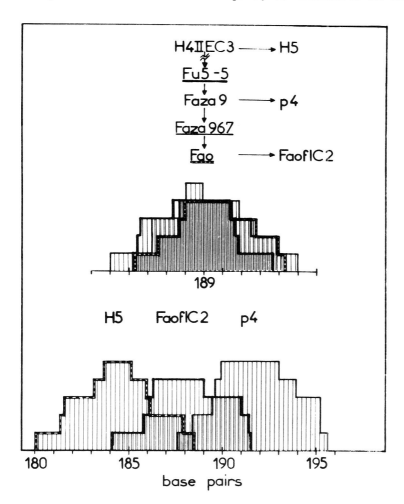

FIG. 3. Chromatin repeat lengths of hepatoma clones. The repeat length data are represented as probability distributions for the average repeat length of a given clone. The degree of overlap between two distributions indicates the significance of differences in peak position. (For details of the measurements and calculations, see Sperling et al. 1980).

The filiations of the clones are given, at the top of the figure with the dedifferentiated variants on the right. The probability distributions for the well-differentiated clones (in the middle) are seen to overlap almost entirely, all three showing a repeat length of 189 base pairs. In the lower part of the figure are given the probability distributions for the three dedifferentiated clones; there is only a tiny region in the middle where all three distributions overlap.

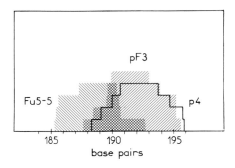

FIG. 4. Repeat length probability distributions for p4 and Fu5–5 (striped areas) as well as for a hybrid (pF3, outlined) between them.

hypothesis is correct, a similar correlation should exist between repeat length and phenotypic expression in somatic hybrids between the two types of cells.

It has already been noted that hybrids between p4 and Fu5–5 cells show extinction of expression of the liver specific functions. Figure 4 shows that the cells of hybrid clone pF3 have a repeat length similar to or identical to that of the parent p4. Thus, the dominance of the p4 parent holds not only for the state of phenotypic expression but for repeat length as well.

H5 cells also extinguish the expression of differentiated functions when hybridized with well-differentiated Fao cells. Figure 5 shows that HF1 hybrid cells have a repeat length similar to that of the H5 parent. Having found that hybrids showing extinction of the liver functions present a repeat length like that of the nonexpressing parent, it became important to determine whether the repeat length of hybrid cells changes when the functions are re-expressed. In order to isolate hybrid cells showing re-expression of the functions of the Fao parent, HF1 cells were challenged in G$^-$ medium. Under these conditions, the vast majority of these hybrid cells died, and a few surviving colonies were isolated.

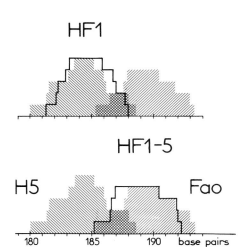

FIG. 5. Probability distributions for H5 × Fao hybrids. HF1 and HF1–5 (outlined) are compared to the parental clones H5 and Fao (striped areas).

When characterized for the group of liver functions we study, such cells able to grow in G$^-$ medium were found to re-express six of the seven functions analyzed (Deschatrette et al. 1979). In addition, they had undergone a 25% chromosome loss. The repeat length of one such segregated hybrid clone, HF1–5 is shown in Figure 5: HF1–5 cells re-express not only the Fao phenotype, but the Fao repeat length as well.

The similarity of the repeat length of FaoflC2 and the differentiated cells has prevented us from carrying out equivalent analyses of FaoflC2 × Fu5–5 hybrid cells.

The nature of the relationship between repeat length and phenotypic expression has not yet been elucidated. Two kinds of mechanisms can be proposed to account for the observations. First, it is possible that the regulatory factors whose expression is responsible for maintenance or loss of the differentiated state also affect repeat length. Second, it may be that the way in which DNA is packaged in the chromatin influences the way in which it interacts with and hence reponds to specific regulatory factors.

CONCLUSIONS

The analyses we have carried out on dedifferentiated variants of rat hepatoma cells have provided new information not only on the consequences of the change from the differentiated to the dedifferentiated state but have also provided us with a material that has permitted the demonstration that chromatin repeat length is well correlated with phenotypic expression, suggesting that chromatin structure is intimately involved in cell differentiation.

Four major points have emerged from these studies. First, when the expression of cell differentiation is modified, an entire set of functions may be affected. Second, the loss of expression of differentiation may be correlated with the acquisition of production of new regulatory factors whose final effect is negative and which thus act to prevent the expression of differentiation. Third, when reversion back to the differentiated state occurs, it is due to a mechanism that has a pleiotropic effect. Finally, maintenance of the differentiated state appears to require the conservation of a constant chromatin repeat length.

The pleiotropy of dedifferentiation and its reversion indicates that we are seeing a reflection of regulatory mechanisms that act to govern the expression of an entire set of tissue-specific functions. That such "key" control mechanisms might exist has long been predicted, but their nature has escaped analysis. The present studies show that at least one of them appears to be mutagen insensitive. It is surprising that a regulatory phenomenon that must originate from the genome can be neither reversed nor mimicked by mutation. While it would be premature to generalize before similar studies are carried out on other cell types, we suggest that it is worthwhile to consider the possibility that reversion from the dedifferentiated to the differentiated state involves a series of changes that can be triggered only by environmental conditions. In accordance with this hypothesis, experiments that will be detailed elsewhere (Deschatrette, in preparation) have shown that a partial reversion of Faofl cells can be provoked by cell aggregation, that such partial revertants express only some hepatic functions and only those that are expressed early during liver ontogeny.

The fact that the dedifferentiated state is "dominant" over the differentiated state was unexpected and requires two comments. First, it suggests that *all* phenotypic states are actively regulated and do not result simply from the absence, for example, of activating factors. Second, the fact that derivatives of hepatocytes may produce diffusible regulatory substances that act to prevent the expression of liver-specific proteins suggests that these same mechanisms may be involved during certain phases of normal development. Further experimental evidence in support of this hypothesis has emerged from the analysis of mouse hepatoma × rat hepatoma hybrids (Cassio and Weiss 1979). The parental cells used for this cross were chosen as models of fetal and adult hepatocytes, respectively. The mouse hepatoma cells produce several liver-specific functions but only those that are first expressed before mid-gestation; the rat hepatoma cells express functions that first appear not only early during ontogenesis but also several that are expressed only at birth. The hybrid cells resulting from this cross express all of the "early" functions, while the expression of late ones is extinguished. It thus appears that the mouse hepatoma cells, which produce only fetal liver proteins, fail to express "late" functions because they produce diffusible, regulatory molecules that act to prevent their expression. It is entirely plausible to postulate that the same mechanisms intervene during the course of normal liver ontogeny.

Turning to the last aspect of the results presented, it has been found that the maintenance of the differentiated state is associated with a constant repeat length of the chromatin and that loss of the differentiated state is associated with the acquisition of a new repeat length which may be shorter or longer than that of the cell of origin. Moreover, this correlation was found to hold for somatic hybrids between the two types of cells: extinction of the liver functions is correlated with a repeat length similar to that of the dedifferentiated parent; when the functions are re-expressed, the repeat length of the differentiated parent is re-expressed as well. The latter observations show that the inheritance of the *potential* to express differentiated functions is independent of the repeat length: in HF1 hybrid cells, the Fao parental genome is packaged like that of H5, yet the potential to express differentiation is not lost, for segregated subclones re-express nearly the entire group of functions. We can thus conclude that *determination* is inherited not only independently of the actual state of phenotypic expression but also of the chromatin repeat length.

REFERENCES

Bertolotti, R. 1977a. A selective system for hepatoma cells producing gluconeogenic enzymes. Somatic Cell Genetics 3:365–380.

Bertolotti, R. 1977b. Expression of differentiated functions in hepatoma cell hybrids: Selection in glucose-free media of segregated hybrid cells which reexpress gluconeogenic enzymes. Somatic Cell Genetics 3:579–602.

Cassio, D., and M. C. Weiss. 1979. Expression of fetal and neonatal hepatic functions by mouse hepatoma-rat hepatoma hybrids. Somatic Cell Genetics 5:719–738.

Compton, J. L., M. Bellard, and P. Chambon. 1976. Biochemical evidence of variability in the DNA repeat length in the chromatin in higher eukaryotes. Proc. Natl. Acad. Sci. USA 73:4382–4386.

Deschatrette, J., E. E. Moore, M. Dubois, and M. C. Weiss. 1979. Dedifferentiated variants of a rat hepatoma: Analysis by cell hybridization. Somatic Cell Genetics 5:697–718.

Deschatrette, J., E. E. Moore, M. Dubois, and M. C. Weiss. 1980. Dedifferentiated variants of a rat hepatoma: Reversion analysis. Cell 19:1043–1051.

Deschatrette, J., and M. C. Weiss. 1974. Characterization of differentiated and dedifferentiated clones of a rat hepatoma. Biochimie 56:1603–1611.

Deschatrette, J., and M. C. Weiss. 1975. Extinction of liver specific functions in hybrids between differentiated and dedifferentiated rat hepatoma cells. Somatic Cell Genetics 1:279–292.

Kornberg, R. D. 1977. Structure of chromatin. Annu. Rev. Biochem. 46:931–954.

Morris, N. R. 1976. A comparison of the structure of chicken erythrocyte and chicken liver chromatin. Cell 9:627–632.

Pitot, H. C., C. Periaino, P. A. Morse, and V. R. Potter. 1964. Hepatoma in tissue culture compared with adapting liver in vivo. Natl. Cancer Inst. Monogr. 13:229–242.

Reuber, M. D. 1961. A transplantable bile-secreting hepatocellular carcinoma of the rat. J. Natl. Cancer Inst. 26:891–899.

Spadafora, C., M. Bellard, J. L. Compton, and P. Chambon. 1976. The DNA repeat lengths in chromatins from sea urchin sperm and gastrula cells are markedly different. FEBS Lett. 69:281–285.

Sperling, L., A. Tardieu, and M. C. Weiss. 1980. Chromatin repeat length in somatic hybrids. Proc. Natl. Acad. Sci. USA 77:2716–2720.

Sperling, L., and M. C. Weiss. 1980. Chromatin repeat length correlates with phenotypic expression in hepatoma cells, their dedifferentiated variants and somatic hybrids. Proc. Natl. Acad. Sci. USA 77:3412–3416.

Thomas, J. O., and R. J. Thompson. 1977. Variations in chromatin structure in two cell types from the same tissue: A short repeat length in cerebral cortex neurons. Cell 10:633–640.

Genes, Chromosomes, and Neoplasia, edited by
Frances E. Arrighi, Potu N. Rao, and Elton Stubblefield.
Raven Press, New York © 1981.

Enzyme Gene Expression in Human Tumor Cell Lines

Michael J. Siciliano

Department of Medical Genetics, The University of Texas System Cancer Center M. D. Anderson Hospital and Tumor Institute, Houston, Texas 77030

Human tumor cells in culture represent an excellent source material for studying the basic characteristics of neoplastic cells as well as being superb models for experimental manipulation. Biochemical genetic analysis of enzyme gene expression in such cells has been conducted in our laboratory in order to: (1) validate their specific identity and integrity, and (2) study the patterns and mechanisms of gene expression in them.

Two levels of genetic variation of enzyme genes are exploited in approaching these problems. The first is electrophoretic variation in the products of enzyme gene loci that tend to be ubiquitously expressed in most all mammalian tissues and species. Between species, genetic drift has produced electrophoretic differences between these gene products so that they are excellent chromosomal markers in experiments in which cells of different species are fused to produce interspecific hybrids. However, even within human beings there has been mutation and fixation of alleles that specify electrophoretically variant products. Loci at which this has happened and at which variant alleles have been retained at reasonable frequency (>0.01) are referred to as polymorphic loci. These can be used to assign to an individual (or cells derived from that individual) a genetic signature. An example of such a locus is shown in Figure 1.

The second type of enzyme gene variation is also electrophoretically detectable. These are tissue-to-tissue variations of gene expression for enzymes encoded by multiple genetic loci. By utilizing them, one can study the pattern of gene expression in human tumor cells in culture and relate that pattern to the suspected tissue of origin of the neoplasm. Genetic dissection of the control of such genetic loci is then approachable by somatic cell fusion experiments and biochemical genetic analysis of hybrids so produced. Such data enable us to determine the dominant-recessive relationships of such control loci, as well as their chromosomal location relative to the genetic loci they control.

MATERIALS AND METHODS

As indicated above, the types of genetic variation of enzyme genes studied are electrophoretically detectable. Our biochemical genetic techniques therefore

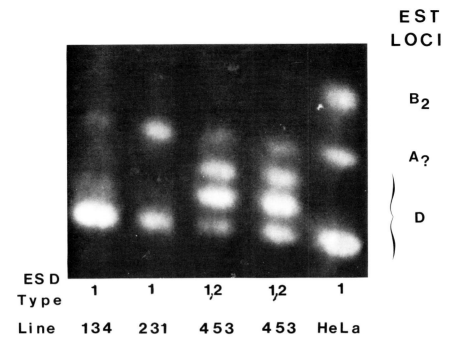

EST
LOCI

B₂

A?

D

ES D
Type   **1   1   1,2   1,2   1**

Line   **134   231   453   453   HeLa**

FIG. 1. Slice of a starch gel following electrophoresis and histochemical staining for esterases. Products of three esterase loci are resolved on this system —Est_{B_2}, an unknown Est_A, and Es_D (positions of the isozymes are indicated in the right margin). Homogenates from three human breast tumor lines—134, 231, and 453 (repeated)—and HeLa cells were run. The *EsD* locus is polymorphic. Lines 134, 231, and HeLa are homozygous for the most common allele, *EsD*[1]. Line 453 can be seen to be heterozygous, *EsD*[1,2], for the variant allele. The three banded patterns for the gene product in heterozygotes is consistent with the dimeric structure of the enzyme.

involve starch gel electrophoresis of cleared cell homogenates and histochemical staining for enzyme gene products. These procedures as followed in our laboratory were generally described by Siciliano and Shaw (1976) and Harris and Hopkinson (1976). Our specific procedures for identifying variant alleles at human polymorphic enzyme loci were described by Siciliano et al. (1979).

Cell culture, fusion, and hybrid selection are standard procedures and were generally described by Siciliano et al. (1978).

RESULTS AND DISCUSSION

Genetic Signature Analysis

As more and more human cell lines become established and disseminated to various laboratories, the question of maintaining the integrity and specific identity of cell lines becomes critical. Cross-contaminations of cell lines were first recognized as a problem when several lines supposedly derived from Caucasians were found to contain the *a* electrophoretic form of glucose-6-phosphate dehydrogenase (G_6PD; Gartler 1968). G_6PD^a is present exclusively in blacks and is the form present in HeLa cells (derived from a black woman; Gey et al. 1952). G_6PD typing was therefore shown to be an excellent monitor for determining HeLa contamination. However, the need for monitoring cell lines for many polymorphic enzyme loci and ascribing to them genetic signatures based on multiple phenotypes (Povey et al. 1976, O'Brien et al. 1977) became obvious as additional good-growing cell lines were derived from blacks and as many more Caucasian cell lines were developed that have the capability of cross-contaminating each other.

For instance, Pathak et al. (1979) reported a breast adenocarcinoma cell line (MDA-415), derived from a black woman, that had several chromosome markers similar to those of HeLa cells as well as the G_6PD^a allele. It might easily have been classified as a HeLa contaminant except that it was also studied for an additional 14 polymorphic enzyme loci. Two others of these had phenotypes different from HeLa, indicating that MDA-415 differs from HeLa.

Studies of this kind become essential as many cell lines derived from the same tumor type from different individuals are developed. Attempts to demonstrate characteristics in common between the cell lines and to suggest that those common characteristics might be attributes of cells derived from that tumor type depend upon the ability to show that the cell lines have not been cross-contaminated and are in fact of independent origin from different individuals. Such a situation has recently arisen with respect to 19 human breast adenocarcinoma cell lines established from pleural effusions from 19 women (Cailleau et al. 1978). These cell lines tended to show the same chromosomal marker—a translocation involving chromosome 1 (Cruciger et al. 1976). Genetic signature analysis of these cell lines for as many as 17 polymorphic enzyme loci (Siciliano et al. 1979) indicated all but two lines had different genetic signatures and therefore were of independent origin.

The two lines (MDA-134 and MDA-309) that could not be distinguished by their genetic signatures were studied for 15 enzyme loci. The phenotypes for each of these loci and the frequency with which each of these phenotypes might be expected to be encountered in Caucasian populations (individuals from whom these two cell lines were established were Caucasian) are listed in Table 1. Except at *GlyI,* the phenotype at each locus is the most common. The frequency in the Caucasian population of the genetic signature (phenotypes at all 15 loci) is calculated as the product of the frequencies at each locus. That frequency is 0.024. If these were the only cell lines studied, the probability of their being independently derived and not contaminants of each other would be the square of the frequency of that genetic signature, or 5.8×10^{-4}. Such a result would imply that these were not independently derived, but contaminants. However, the probability of encountering two such signatures by chance in a sample of

TABLE 1. *Phenotypes and frequency of those phenotypes in Caucasian populations of MDA-134 and -309 for 15 enzyme loci*

Enzyme locus	Phenotype	Frequency of phenotype in caucasian populations*
G_6PD	b	1.00†
GR	1	1.00†
GOT_m	1	.96
6PGD	a	.90
Pep_A	1	1.00
Pep_C	1	.98
Pep_D	1	.98
EsD	1	.81
Glyl	1	.16
PGM_1	1	.56
PGM_3	1	.56
AK_1	1	.92
ADA	1	.81
$ACON_S$	1	1.00
$\alpha GLUC$	1	.94

* From Harris and Hopkinson (1976).

† Not polymorphic among Caucasians but included in the study since they are polymorphic in blacks from whom several other cell lines in the study were derived.

G6PD, glucose 6 phosphate dehydrogenase; GR, glutathione reductase; GOT, glutamate oxaloacetate dehydrogenase; 6PGD, 6 phosphogluconate dehydrogenase; Pep, peptidase; Es, esterase; Gly, glyoxylase; PGM, phosphoglucomutase; AK, adenylate kinase; ADA, adenosine deaminase; ACON, aconitase; αGLUC, αglucosidase.

19 independently derived cell lines is best described by a binomial distribution and may be calculated by:

$$\frac{n!}{i!(n-i)!} (P)^i (1-P)^{n-i}$$

where n is the number of cell lines studied, i is the number of cell lines with identical signatures, and P is the frequency of that signature in the human population. That probability is 0.07, which indicates that the identity of the genetic signatures of MDA-134 and MDA-309 might be due to chance and that they can be considered independent.

A different conclusion was arrived at by Leibovitz et al. (1979). Eight of approximately 100 lines had the same genetic signature, the frequency of which was so low (6×10^{-8}) that they more likely represented contaminants of each other than independently derived lines.

Rutzky et al. (1980) used the technique to demonstrate that two adenocarcinoma cell lines believed derived from the same individual were in fact from the same individual even though there were chromosomal differences between the lines. The lines had identical genetic signatures, the frequency of which was low (2.6×10^{-4}), leading to a probability of them having independent origin of $<10^{-8}$. Clearly, they must have been derived from the same individual so that the chromosomal differences observed between them were not accountable by contamination with another cell line.

Gene Expression in Tumor Cells

Isozymic forms in tumor material can aid in the early diagnosis of malignancy and can provide clues to the tissue of origin of malignant disease (Siciliano 1980). Since lactate dehydrogenase (LDH) is an enzyme with multiple, tissue-specific genetic loci that are differentially expressed in mammalian tissues and the products of which are electrophoretically separable, it has often been regarded as an enzyme that could prove useful in such analysis.

Five common forms of LDH are identifiable in cleared homogenates from most mammalian tissues and in cells subjected to electrophoresis and histochemical staining. These five isozymes have been demonstrated to be the result of the random combination into tetramers of electrophoretically variant subunits specified by two different genetic loci (*LDH-A* and *LDH-B*). The subunit structures of the five isozymes from most anodal to most cathodal are B_4, B_3A, B_2A_2, B_1A_3, and A_4, known as LDH-1, -2, -3, -4, and -5, respectively. The relative activities of the isozymes are skewed in the direction of LDH-1 or -5 depending on the relative expression of the loci in the cells or tissues sampled (Markert 1963, 1968, Shaw and Barto 1963). A third genetic locus for LDH, designated *LDH-C,* is expressed only in primary spermatocytes, producing a protein called LDH-X, which migrates anodal to LDH-4 (Blanco and Zinkham 1963, Blanco et al. 1964).

We have investigated LDH isozyme expression in over 60 human tumor cell lines derived from a wide variety of human neoplasms (Siciliano et al. 1980). The most interesting results were obtained from the human choriocarcinoma cell lines JEG-3 derived by Kohler and Bridson (1971).

Following electrophoresis and histochemical staining of their homogenates for LDH, HeLa cells and mammary carcinoma cell lines displayed patterns typical of tissues with greater expression of *LDH-A* than *LDH-B* (Figure 2). Homogenates from the choriocarcinoma cell line JEG-3 had greater expression of *LDH-B* as indicated by the great intensity of LDH-1 and the inability to detect the A_4 homotetramer (LDH-5). Its most interesting feature, however, was the extra band of LDH activity cathodal to LDH-2 (Figure 2). We will refer to this band as LDH-Z. It had the same lactate dependence and general mobility observed in the JEG-3 cells studied by Edlow et al. (1975).

LDH patterns exhibited by six subclones of JEG-3 are also shown in Figure 2. A total of 25 subclones were produced, and all had LDH-Z at approximately the same intensity. The subclones fell into two classes with respect to the expres-

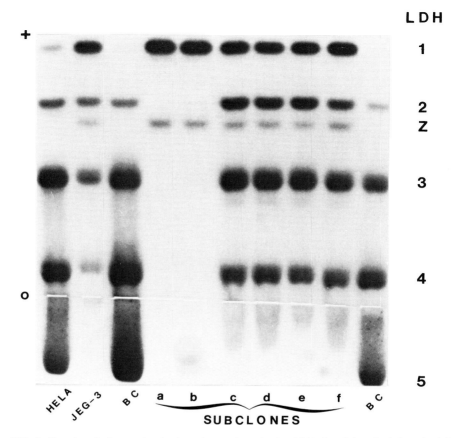

FIG. 2. Starch gel slice stained, after electrophoresis, for LDH. Anodal ends (+) and origin (0) are marked to the left. Positions of the five usual LDH isozymes (1 to 5) and LDH-Z are marked on the right. The samples run (indicated along the bottom) are: HeLa cells, the choriocarcinoma cell line JEG-3, two different breast carcinoma cell lines (BC), and six subclones of JEG-3 (a to f). (Reprinted from Siciliano et al. 1980 with permission of *Cancer Research*.)

sion of *LDH-A* as visualized by the activity of the heterotetramers responsible for LDH-2, -3, and -4. Class 1 is typified by the subclone samples a and b (Figure 2). In these, *LDH-A* expression was essentially nil, as indicated by the lack of heterotetramer activity (none even in LDH-2). A total of 11 class 1 subclones were produced. The remaining 14 subclones fell into class 2 and are typified by subclone samples c to f (Figure 2). In these, *LDH-A* was expressed as indicated by the activity of the LDH-2, -3, and -4 isozymes. Therefore, LDH-Z was clearly expressed in all JEG-3 subclones, independent of the expression of the products of *LDH-A*. This result, combined with those of Ferrell et al. (1980), which indicated that LDH-Z was made up of subunits with greater antigenic similarity to the products of *LDH-A* than of *LDH-B*, suggest LDH-Z may be the product of a new *LDH* locus.

LDH-Z was present in three additional, independently derived (by R. Pattillo, Medical College of Wisconsin, Milwaukee, Wisconsin) choriocarcinoma cell lines. Genetic signature analysis verified their independent origin. Autopsy samples of choriocarcinoma metastasized to the liver, lung, and spleen revealed the presence of LDH-Z, while it was not present in surrounding uninvolved tissue. Association between the malignant trophoblastic disease, choriocarcinoma, and LDH-Z was therefore established (Siciliano et al. 1980).

We and others have also showed the isozyme in placental stages (Edlow et al. 1971, Prasad et al. 1977, Siciliano et al. 1980) as well as in hydatidiform mole (Van Bogaert et al. 1967, Siciliano et al. 1980), an intrauterine hyperplastic growth of trophoblastic origin that may become invasive and is believed to be precursor for 50% of the cases of choriocarcinoma of gestational origin (Hammond and Parker 1970).

The evidence from these experiments leads us to suggest that LDH-Z is the product of a genetic locus independent of the previously reported *LDH-A, -B,* and *-C* loci, and that the expression of this locus is restricted to trophoblast-derived tissues, both normal and malignant. We are continuing to investigate both the biochemical properties and genetic expression of LDH-Z with the hope of understanding its role in normal and neoplastic tissues, and to develop this isozyme not only as a sensitive aid to early diagnosis but also as a possible target for immunotherapy for malignant disease. The presence of LDH-Z in choriocarcinoma illustrates that unexpected isozymic forms in tumor material may be present for reasons other than altered genetic regulation of tumor cells, but may be markers for the cells of origin of tumors that develop in tissues of heterogeneous cell types (such as trophoblast). In such cases, presence of "fetal" or earlier stage proteins may be a consequence of the expansion of relatively undifferentiated cells that were a minority in the tissue of origin.

Tumor cell lines represent a highly enriched population of such cells so that such markers can be better resolved. More importantly, the presence of such markers in living cells in culture opens up studies on the genetic basis of their regulation, since the cells can be manipulated by somatic cell genetic techniques.

Somatic Cell Genetic Analysis of Gene Expression

By fusing cells having certain gene expression characteristics with cells having different expression for the loci under study, one can examine the resultant hybrid clones to determine the dominance-recessiveness relationship as well as the formal genetic mapping of genes responsible for regulation of expression. We have conducted such studies with JEG-3 cells because they express several interesting characteristics—LDH-Z and others. JEG-3 cells were fused with mouse Clld cells using inactivated Sendai virus or polyethelene glycol. Since the human cells are ouabain sensitive and the mouse cells HAT (hypoxanthineaminoptyrin-thymidine)-sensitive (since they lack thymidine kinase), hybrids were selected in HAT-ouabain medium. Parental cells were killed, but in hybrids, each parental genome complements the other's deficiencies, allowing survival.

Surviving clones were verified as true hybrids by chromosome and isozyme analysis. The isozyme analysis was particularly critical to this verification. Many enzyme loci products differ in electrophoretic mobility between man and mouse. In hybrids retaining both the mouse and human chromosomes bearing the locus for such an enzyme, both mouse and human forms will be observed, as will intermediately migrating heteropolymeric forms (the number and spacing of which depend upon the number of subunits making up the enzyme). Since under conditions in which these experiments were run heteropolymers will be produced only in cells in which both human and mouse gene products are being actively synthesized, the presence of heteropolymers is taken as the best evidence of hybrid formation. (See Figure 3 for an example of such data.)

Forty-five independent hybrids were isolated. While human chromosomes segragated from these hybrids, in the panel of 45 hybrids every human chromosome was represented at least once. However, LDH-Z (the trophoblastic isozyme present in the human choriocarcinoma cells used in the fusion) could be identified in none of them. Since LDH-Z can be considered a differentiated phenotype (trophoblastic), the extinguishment of its expression in hybrids formed with undifferentiated cells (mouse fibroblasts) is a phenomenon that has been observed in many other systems (see review in Ringertz and Savage 1976).

In the JEG-3 × Clld system, one has the opportunity to test the corollary to that phenomenon. If an enzyme gene that is *not* normally expressed in the cells of the human differentiated parent of the hybrid cells is expressed in the undifferentiated fusion partner, one might expect the activation of the human locus

G P I

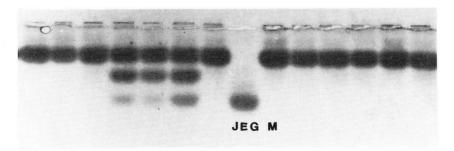

JEG   M

FIG. 3. Starch gel slice stained for glucose phosphate isomerase (GPI) following electrophoresis from origin (0) to cathode(−) of homogenates from putative hybrid clones isolated after fusion of JEG cells (channel marked "JEG") and Clld cells (channel marked "M"). This enzyme verified the 4th, 5th, and 6th putative hybrid clones from the left as true hybrids since they have not only the mouse and human forms of the enzyme but also the heteropolymeric molecule of intermediate migration expected for this dimeric enzyme.

in the hybrid clones that retain the human chromosome bearing the structural locus for the enzyme.

We have had an opportunity to test this hypothesis in the panel of hybrids under discussion (Siciliano et al. 1978). JEG-3 cells have very much reduced activity for the enzyme adenosine deaminase (ADA). The activity, about 2% of that observed in control cells, was insufficient to be detected on starch gels. This observation is consistent with the low ADA levels measured in the fetal components of placenta (Brady and O'Donovan 1965, Hayashi 1965, Sim and Maguire 1970).

Of the 45 JEG-3 × CIId hybrids, 30 produced human ADA and all expressed mouse ADA (Figure 4). Hybrids missing human chromosome 20, the site of the structural gene for ADA (Tischfield et al. 1974), did not express human

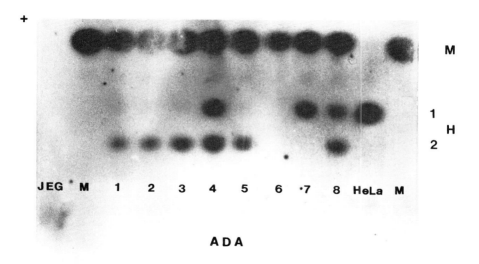

ADA

FIG. 4. Starch gel slice histochemically stained for ADA after electrophoresis of cleared cell homogenates. Left-band margin indicates the origin (0) of the gels and their anodal ends (+). Right-band margin indicates the migration positions of the mouse form of ADA (M) as well as two additional (nonmouse) forms seen in hybrids and human material. The positions of these latter forms are indicated by 1 and 2. ADA 1 and 2 are the products of two alleles for this enzyme present in human populations (Spencer et al. 1968). Channels containing samples from the parental cells used in the fusions—human JEG and mouse M—are appropriately indicated. A channel containing human HeLa cells, serving as a marker for the position of human ADA 1, is also indicated. The channels marked 1–8 contain samples from eight independently derived JEG-3 × CIId hybrid clones. Extra ADA forms (1 and/or 2) are seen in hybrids but not in the parental cells used for fusion. Some hybrids have human ADA-1 and -2 (Nos. 4 and 8), one only ADA-1 (No. 7), some only ADA-2 (Nos. 1,2,3, and 5), and one no human ADA (No. 6).

ADA, which was present only in hybrids containing at least one chromosome 20. These results indicate that: (1) JEG-3 cells are heterozygous for the two electrophoretically distinguishable alleles of *ADA* present in human populations, (2) both alleles of the *ADA* locus are unexpressed in JEG-3 cells, (3) fusion with mouse fibroblasts expressing ADA results in the expression of both human alleles in hybrids retaining the structural genes, and (4) the lack of ADA expression behaves as a recessive trait in hybridization schemes.

These results show that the *ADA* locus in human tumor cells is genetically manipulatable and opens the way for a variety of somatic cell and biochemical approaches to understanding the genetic basis for the control of such enzymes. These studies are in progress.

In this report we have attempted to demonstrate biochemical genetic and somatic cell genetic approaches to the understanding of gene expression in human tumor material. We have indicated the value and usefulness of human tumor cell lines in such studies and indicated the need for and means of maintaining quality control in such systems.

ACKNOWLEDGMENTS

Collaborators for various portions of this work include Dr. Relda Cailleau, Department of Medicine, The University of Texas System Cancer Center M. D. Anderson Hospital and Tumor Institute, Dr. Robert E. Ferrell, Center for Demographic and Population Genetics, Graduate School of Biomedical Sciences, The University of Texas Health Science Center at Houston, Lynn P. Rutzky, Department of Surgery, The University of Texas Medical School at Houston, and Dr. Mary E. Riser, Department of Cell Biology, Baylor College of Medicine, Houston, Texas.

This investigation was supported by grant number ES01287, awarded by the Institute for Environmental Health Sciences, Department of Health, Education, and Welfare, and a gift from the Kleberg Foundation.

REFERENCES

Blanco, A., and W. H. Zinkham. 1963. Lactate dehydrogenases in human testes. Science 139:601–602.

Blanco, A., W. H. Zinkham and L. Kupchyk. 1964. Genetic control and ontogeny of lactate dehydrogenase in pigeon testes. J. Exp. Zool. 156:137–152.

Brady, T. G., and C. I. O'Donovan. 1965. A study of the tissue distribution of adenosine deaminase in six mammalian species. Comp. Biochem. Physiol. 14:101–120.

Cailleau, R., M. Olive, and Q. V. J. Cruciger. 1978. Long-term human breast carcinoma cell lines of metastatic origin: Preliminary characterization. In Vitro 14:911–915.

Cruciger, Q. V. J., S. Pathak, and R. Cailleau. 1976. Human breast carcinomas: Marker chromosomes involving 1q in seven cases. Cytogenet. Cell Genet. 17:231–235.

Edlow, J. B., J. F. Huddleston, G. Lee, W. F. Peterson, and J. C. Robinson. 1971. Placental enzymes: Specific activities and isoenzyme patterns during early and late gestation. Am. J. Obstet. Gynecol. 111:360–364.

Edlow, J. B., T. Ota, J. R. Relacion, P. O. Kohler, and J. C. Robinson. 1975. Enzymes of normal and malignant trophoblast: Phosphoglucose isomerase, phosphoglucomutase, hexokinase, lactate dehydrogenase, and alkaline phosphatase. Am. J. Obstet. Gynecol. 121:674–681.

Ferrell, R. E., M. Goode, and M. J. Siciliano. 1980. Trophoblastic lactate dehydrogenase (LDH-Z): Evidence for a separate genetic locus. Cancer Bull. 32:58–60.

Gartler, S. M. 1968. Apparent HeLa cell contamination of human heteroploid cell lines. Nature 217:750–751.

Gey, G. O., W. D. Coffman, and M. T. Kubicek. 1952. Tissue culture studies of the proliferative capacity of cervical carcinoma and normal epithelium. Cancer Res. 12:264–265.

Hammond, C. B., and R. T. Parker. 1970. Diagnosis and treatment of trophoblastic disease—a report from the Southeastern Regional Center. Obstet. Gynecol. 35:132–143.

Harris, H., and D. A. Hopkinson. 1976. Handbook of Enzyme Electrophoresis in Human Genetics. North-Holland Publishing Co., Oxford.

Hayashi, T. T. 1965. Studies on placental metabolism. Am. J. Obstet. Gynecol. 93:266–268.

Kohler, P. O., and W. E. Bridson. 1971. Isolation of hormone producing clonal lines of choriocarcinoma. J. Clin. Endocrinol. 32:683–687.

Leibovitz, A., J. Fogh, S. Pathak, M. J. Siciliano, W. C. Wright, W. P. Daniels, and H. Fogh. 1979. Detection and analysis of a G-6-PD phenotype B cell line contamination. J. Natl. Cancer Inst. 63:635–645.

Markert, C. L. 1963. Lactate dehydrogenase isozymes: Dissociation and recombination of subunits. Science 140:1329–1330.

Markert, C. 1968. The molecular basis for isozymes. Ann. N. Y. Acad. Sci. 151:14–40.

O'Brien, S. J., G. Kleiner, R. Olsen, and J. E. Shannon. 1977. Enzyme polymorphisms as genetic signatures in human cell cultures. Science 195:1345–1348.

Pathak, S., M. J. Siciliano, R. Cailleau, C. L. Wiseman, and T. C. Hsu. 1979. A human breast adenocarcinoma with chromosome and isoenzyme markers similar to those of the HeLa line. J. Natl. Cancer Inst. 62:263–271.

Povey, S., D. A. Hopkinson, H. Harris, and L. M. Franks. 1976. Characterizations of human cell lines and differentiation from HeLa by enzyme typing. Nature 264:60–63.

Prasad, R., A. Werch, and R. H. Kaufman. 1977. A new isozyme of lactate dehydrogenase in the human placenta (?). Am. J. Obstet. Gynecol. 129:104.

Ringertz, N. R., and R. E. Savage. 1976. Phenotypic expression in hybrid cell, in Cell Hybrids, vol. 11. Academic Press, New York, pp. 180–212.

Rutzky, L. P., C. I. Kaye, M. J. Siciliano, M. Chao, and B. D. Kahan. 1980. Longitudinal karyotype and genetic signature analysis of cultured human colon adenocarcinoma cell lines LS180 and LS174T. Cancer Res. 40:1443–1448.

Shaw, C., and E. Barto. 1963. Genetic evidence for the subunit structure of lactate dehydrogenase isozymes. Proc. Natl. Acad. Sci. USA 50:211–214.

Siciliano, M. J. 1980. Isozymes and cancer: an overview. Cancer Bull. 32:43–44.

Siciliano, M. J., P. E. Barker, and R. Cailleau. 1979. Mutually exclusive genetic signatures of human breast tumor cell lines carrying a common cytogenetic marker. Cancer Res. 39:919–922.

Siciliano, M. J., M. E. Bordelon, and P. O. Kohler. 1978. Expression of human adenosine deaminase after fusion of adenosine deaminase-deficient cells with mouse fibroblasts. Proc. Natl. Acad. Sci. USA 75:936–940.

Siciliano, M. J., M. E. Bordelon-Riser, R. Freedman, and P. O. Kohler. 1980. LDH-Z: a human trophoblastic isozyme associated with choriocarcinoma. Cancer Res. 40:283–287.

Siciliano, M. J., and C. R. Shaw. 1976. Separation and visualization of enzymes on gels, in Chromatographic and Electrophoretic Techniques, vol. 2, Ed. 4, I. Smith, ed. Wm. Heinemann Medical Books Ltd., London, pp. 185–209.

Sim, M. K., and M. H. Maguire. 1970. Variation in placental adenosine deaminase activity during gestation. Biol. Reprod. 2:291–298.

Spencer, N., D. A. Hopkinson, and H. Harris. 1968. Adenosine deaminase polymorphism in man. Ann. Hum. Genet., Lond. 32:9–14.

Tischfield, J. A., R. P. Creagan, E. A. Nichols, and F. H. Ruddle. 1974. Assignment of a gene for adenosine deaminase to human chromosome 20. Hum. Hered. 24:1–11.

Van Bogaert, E. C., E. DePeretti, and C. A. Villee. 1967. Electrophoretic studies of human placental dehydrogenases. Am. J. Obst. Gynecol. 98:919–923.

Genes, Chromosomes, and Neoplasia, edited by
Frances E. Arrighi, Potu N. Rao, and Elton Stubblefield.
Raven Press, New York © 1981.

The Expression, Arrangement, and Rearrangement of Genes in DNA-Transformed Cells

R. Sweet, J. Jackson, I. Lowy, M. Ostrander, A. Pellicer, J. Roberts, D. Robins, G.-K. Sim, B. Wold, R. Axel, and S. Silverstein

Cancer Center, College of Physicians and Surgeons, Columbia University, New York, New York 10032

Transformation permits the introduction of new genetic information into a cell. The expression of newly acquired genes frequently results in phenotypic changes in the recipient. Thus, DNA obtained from either viruses or eukaryotic cells can be used to transfer thymidine kinase (tk) (Wigler et al. 1977, Maitland and McDougall 1977, Wigler et al. 1978), adenine phosphoribosyl transferase (aprt) (Wigler et al. 1979a), and hypoxanthine guanine phosphoribosyl transferase (Willecke et al. 1979, Graf et al. 1979) to mutant mouse L cells deficient in these enzyme functions.

We have previously demonstrated that cells transformed with selectable biochemical markers are often cotransformed with other physically unlinked DNA sequences at high frequency (Wigler et al. 1979b). In this way, we have selected mammalian cells cotransformed with a viral *tk* gene and defined prokaryotic and eukaryotic genes. This system utilizes a recessive gene as a vector and allows the introduction of virtually any cloned gene into mutant recipients. More recently, we have transformed cells with a dominant acting marker, the gene coding for an altered dihydrofolate reductase (Flintoff et al. 1976). Transformation with this gene renders cells resistant to high levels of methotrexate (Wigler et al. 1980). The use of this gene as a vector in cotransformation experiments may permit the transfer of virtually any genetic element into a variety of suitable cellular environments.

We have cotransformed four distinct cell types using the viral *tk* gene as vector: murine fibroblasts, erythroleukemia cells, teratocarcinoma cells, and rat hepatoma cells. In each of these cell types, the viral *tk* is expressed. Analysis of the transcripts of the heterologous cotransformed genes suggests that intervening sequences are processed properly; however, these transcripts frequently display aberrant 5' termini (Wold et al. 1979).

In this report, we describe a class of unstable tk mutant cell sets that switch phenotypes from tk^+ to tk^- to tk^+ . . . at frequencies that exceed the observed rate of mutation of a hemizygous gene (DeMars 1974). Since the *tk* gene has been cloned in bacterial plasmids and the tk$^+$ and tk$^-$ phenotypes may be selected under appropriate restrictive growth conditions, it is possible to analyze phenotypic switching at the molecular level. Transformation studies in concert with restriction endonuclease mapping suggest that the determinants responsible for phenotypic switching result from frequent but reversible alterations in the DNA.

RESULTS AND DISCUSSION

Cotransformation of Mouse tk$^-$ Fibroblasts with Defined Genetic Elements

We have utilized the *tk* gene as a vector in cotransformation experiments to construct mouse cell lines containing prokaryotic and eukaryotic genes for which no selective criteria exist (Wigler et al. 1979b). A representative analysis of cotransformants obtained with a variant human growth hormone gene (Fiddes et al. 1979) is shown in Figure 1. DNA from five cotransformants was analyzed for the presence of growth hormone sequences by blot hybridization. The results derived from these and additional experiments with several other genes permit the following conclusions: (1) The frequency of cotransformation is high; 80% of the tk$^+$ transformants contain sequences homologous to the nonselectable gene. (2) Most cotransformed sequences are localized in the high molecular weight DNA fraction within the cell. (3) The number of integrated copies varies from one to over 100 in independent clones. (4) The extent of the cotransformed sequences maintained within a given transformant is variable; some clones integrate only 20% of the donor DNA sequences while others contain over 90% of the donor sequences. (5) The genotype of cotransformed sequences, as reflected by blot hybridization profiles, is stable for many generations when maintained under selective pressure. (6) All of the cloned eukaryotic genes we have studied can be introduced into mammalian cells with roughly equal efficiency. The properties of cotransformants, noted above, appear to be independent of the cloned DNA sequence.

Expression of Cotransformed Genes

The introduction of cloned eukaryotic genes into animal cells provides an in vivo system to study the functional significance of various features of DNA sequence organization. Our earlier studies demonstrated that DNA from various species of mammals and birds transfer either tk or aprt activity to mutant murine cells with equivalent efficiencies. These studies suggest that the transcriptional and translational apparatus of the mouse is capable of recognizing, transcribing, and translating these DNA sequences into a functional protein without regard to vertebrate species of origin.

Cotransformants containing cloned eukaryotic genes permit us to study the transcription and subsequent processing of these sequences in a heterologous

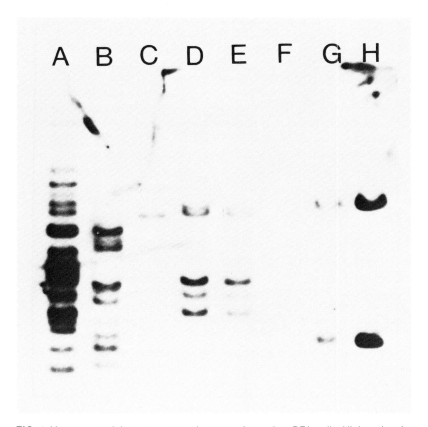

FIG. 1. Human growth hormone genes in cotransformed rat BRL cells. High molecular weight DNA from five BRL clones cotransformed with ptk and pHGH was digested with Eco RI and separated electrophoretically on a 0.8% agarose gel. The DNA was denatured in situ, transferred to nitrocellulose, and hybridized with a ^{32}P-labeled 2.6-kb Eco RI fragment containing the variant human growth hormone gene (Fiddes et al. 1979). Lanes A-E: DNA from five tk+ BRL cell transformants digested with Eco RI. Lane F: BRL tk⁻ cell DNA. Lanes G and H: 25 pg and 250 pg, respectively, of pHGH digested with Eco RI (to visualize the 2.6-kb growth hormone gene insert) mixed with an equal amount of undigested pHGH (6.8 kb).

host. To date, these studies have largely been restricted to transformed mouse fibroblasts. In one study, we have examined the expression of the rabbit β-globin gene in six independent transformants that contain 1–20 copies of the cloned globin gene (Wold et al. 1979). Rabbit globin transcripts were detected in two of these transformants at steady state concentrations of five and two copies per cell. Solution hybridization experiments in concert with RNA blotting techniques indicate that, in one transformed line, the rabbit globin sequences are present in the cytoplasm as a polyadenylated, 9S species. Further analysis indicates that the two intervening sequences present in the original globin transcript are correctly processed. Surprisingly, 45 nucleotides present at the 5' terminus of mature rabbit mRNA are absent from the β-globin RNA sequence detected in the cytoplasm of this transformed fibroblast. Appropriate processing of the rabbit β-globin gene has also been observed in tk$^+$ mouse cell transformants in which the globin and tk plasmids have been ligated prior to transformation (Mantei et al. 1979). Similar results have also been obtained by using viral vectors to introduce the rabbit globin gene into monkey cells (Mulligan et al. 1979, Hamer and Leder 1979). Taken together, these results suggest that nonerythroid cells from heterologous species contain the enzymes to correctly process intervening sequences of a rabbit gene whose expression is usually restricted to erythroid cells.

The appearance of an aberrant yet unique 5' terminus suggests the possibility of incorrect initiation of transcription in this mouse cell line. A similar analysis of cotransformants constructed by ligation of the *tk* and globin gene reveals transcripts with at least two discrete 5' termini: RNA with 5' termini indistinguishable from that of mature rabbit mRNA and RNA with 5' termini also lacking about 45 nucleotides (Mantei et al. 1979, C. Weissman, personal communication).

Preliminary analyses of rat cells cotransformed with insulin and human growth hormone genes reveal the presence of transcripts several hundred nucleotides longer than mature insulin or growth hormone mRNA (unpublished studies from our laboratory in collaboration with A. Efstratiadis and P. Seeberg). The introduction of the chick ovalbumin gene to mouse L cells results in the synthesis of immunoprecipitable ovalbumin protein (Lai et al. 1980). In a separate study, analysis of the RNA indicates that while all seven intervening sequences may be precisely removed from this avian gene by murine enzymes, the transcript is 650 nucleotides longer at the 5' terminus than mature ovalbumin (Breathnach et al. 1980).

These studies suggest that if the 5' terminus of mature mRNA also reflects the site of initation of a primary transcript, murine fibroblasts do not consistently support correct initiation of heterologous genes whose expression is usually restricted to other differentiated cell types. The mere presence of a promoter sequence may not insure accurate transcription initiation in this in vivo system, since alternate promoters are apparently chosen. Correct initiation may be dictated by promoters acting in concert with transcription control factors that are only provided by an appropriate cellular environment. Alternatively, it is possible that correct initiation is sensitive to the relative position of the specific gene in the eukaryotic genome.

Cotransformation of Diverse tk$^-$ Cell Lines

We have recently developed cotransformation systems that permit the introduction of defined genes into murine erythroleukemia (MEL) cells and teratocar-

TABLE 1. *Transformation and cotransformation frequency of diverse tk⁻ cell lines*

Cell line*	Transformation frequency†	Cotransformation frequency‡
Mouse L	50,000	80% (>100)
BRL	5	60% (22)
TCC	0.5	27% (11)
MEL	0.5	6% (50)

* See text for explanation of abbreviations.

† Units are in colonies/μg/10^6 cells.

‡ Expressed as the frequency with which tk⁺ transformants also contain at least one cotransformed element. The number in parentheses reflects the number of tk⁺ transformants analyzed.

cinoma (TCC) cells (Pellicer et al. 1980), as well as Buffalo rat liver (BRL) cells. The properties of these three recipients are unique and quite different from those displayed by the mouse L cell (Table 1). All three cell lines show a dramatic reduction in tk transformation efficiency when compared to mouse L cells. Transformation of L cells generates one surviving colony per 20 pg of purified gene per 5×10^5 cells; in MEL cells, one transformant is obtained per 3 μg of gene. Intermediate values are obtained for TCC and BRL cell lines. While inordinately high concentrations of gene are required to effect transformation in these cells, the availability of a cloned *tk* gene nonetheless permits the construction and selection of numerous tk⁺ transformants.

We have cotransformed these cell lines with cloned eukaryotic genes. The frequency of cotransformation in rat liver cells is high, but is reduced significantly in the murine erythroleukemia and teratocarcinoma cell lines (Table 1). The number of cotransformed sequences introduced into rat liver cells ranges from one to several hundred copies per cell. However, only one to ten cotransformed sequences have been introduced into TCC or MEL cells.

We can therefore discern three classes of recipients: cells that transform and cotransform with high efficiency (L cells); cells that transform at low efficiency but cotransform at high efficiency (BRL cells); and cells that both transform and cotransform at low efficiency (TCC and MEL cells). At present, we do not understand the wide variation in transformation frequency among different cell lines, since the biological basis for competence in eukaryotic cells is unknown. In tk⁻ L cells, at least 2% of the culture can be transformed to the tk⁺ phenotype. Competence in this line is not stably heritable and may therefore be a transient property whose expression varies greatly among different cell types.

Phenotypic Switching of tk Activity in Transformants

The addition of the herpes virus thymidine kinase gene to mouse L tk⁻ cells results in the appearance of numerous surviving colonies expressing the viral enzyme. Transformants obtained in this way contain at least one copy of the viral *tk* gene integrated into high molecular weight DNA in the transformed cell. These cells have maintained the tk^+ phenotype for several hundred generations under selective conditions. The newly introduced *tk* gene is particularly amenable to mutational analysis, since the tk^+ phenotype can be selected in growth medium containing HAT (hypoxanthine, aminopterin, and thymidine), and the tk^- phenotype can be back selected in medium containing bromodeoxyuridine (BUdR).

We have constructed two transformants, K_1 and K_2, and studied the stability of both the tk phenotype and *tk* genotype in several independent tk^- revertants by transformation in concert with blot hybridizations (Table 2). tk^+ transformants derived from individual surviving colonies are phenotypically stable under selective pressure and have been maintained for over 300 generations. When plated into counterselective medium (BUdR), however, surviving tk^- colonies are obtained with a frequency of 10% for K_1 and 1% for K_2 (Table 2), values far greater than the expected mutation rates. Several tk^- revertant clones were chosen from both K_1 and K_2 and are designated K_1B_n and K_2B_n. These revertant cell lines are stable in BUdR, but when plated into HAT, surviving colonies are again observed. The frequency with which tk^- revertants give rise to tk^+ re-revertants (designated $K_1B_nH_n$ and $K_2B_nH_n$) varies among individual revertants. Thus, K_2B_6 re-reverts to the tk^+ phenotype at a frequency of 1.0% while K_1B_6 switches phenotype with a frequency close to 50% (Table 2). tk^+ re-revertants can again be plated in BUdR, continuing the switching cycle from tk^+ to tk^- to tk^+. . . . The frequency with which transformant, revertants, and re-revertants switch phenotypes varies among different clones, but remains constant for each individual clone.

TABLE 2. *Stability of the tk phenotype*

Cell line	Relative cloning efficiency in selective medium*	
	HAT	BUdR
K_1	0.98	0.13
K_1B_6	0.5	0.52
$K_1B_6^N$	<0.0001	0.87
$K_1B_6H_2$	0.93	0.47
K_2	1.0	0.016
K_2B_6	0.015	0.97
K_2B_7	<0.0001	0.87
$K_2B_6H_4$	0.79	0.08

* 100 to 1,000 cells were plated in triplicate into either neutral medium (HT) or selective medium (HAT or BUdR). Cell lines are described in text.

Phenotypic Switching Is Associated with Changes in DNA

The unusual pattern of phenotypic switching may be caused by self-reinforcing phenotypic changes transmissable to progeny in the absence of concomitant changes in DNA. Alternatively, this phenotypic switching may result from frequent but reversible changes in DNA. To distinguish between these alternatives, we have performed a series of transformation experiments with purified DNA from transformant, revertant, and re-revertant cell lines. If reversion to the tk⁻ phenotype occurs without genetic changes, DNA from these tk⁻ cells should transfer the *tk* gene to L tk⁻ recipients as efficiently as DNA from tk⁺ cells. On the other hand, if alterations in DNA are associated with the phenotypic switch, transfer would be far less efficient with DNA from the tk⁻ revertants. As expected, DNA from the tk⁺ transformant or re-revertant cells is an efficient donor of the *tk* gene, generating from 10–30 tk⁺ colonies/80 μg DNA/5 × 10^5 cells. In contrast, DNA from two independent revertants generated 50- to 100-fold fewer colonies than observed with DNA from tk⁺ lines (Table 3). The potential inhibition of DNA transfer by BUdR substitution in revertant DNA was avoided by growing the revertant lines in neutral Dulbecco's modified Eagle's medium for five generations prior to isolation.

Thus, the relative efficiency of gene transfer with tk⁻ DNA roughly correlates with the frequency at which the tk⁻ revertants plate in HAT medium (Tables 2 and 3). This observation suggests that a specific but reversible genetic alteration of the *tk* gene accompanies phenotypic switching.

TABLE 3. *Transformation efficiency of transformant, revertant, and re-revertant DNA*

Cell line	tk⁺ Colonies*	Relative tk transformation efficiency†
K_1	75	1.0
K_1B_5	4	0.05
K_1B_6	15	0.20
$K_1B_6{}^N$	0	—
$K_1B_6H_1$	48	0.65
K_2	26	1.0
K_2B_6	1	0.03
$K_2B_6H_4$	42	1.6

* Number of surviving tk⁺ colonies observed in three dishes after transformation of L tk⁻ cells with 80 μg of DNA from transformant, revertant, and re-revertant cell lines.

† Relative transformation activity reflects the number of tk⁺ colonies obtained with tk⁻ revertant or tk⁺ re-revertant DNA compared with the number of colonies obtained with the original tk⁺ transformants.

Organization of the *tk* Gene in tk$^+$ and tk$^-$ Transformants

We have analyzed the organization of the viral *tk* gene sequences in cell sets composed of the original transformant and its revertants and re-revertants by restriction endonuclease mapping and blot hybridization. A restriction map of the 5-kilobase (kb) Kpn fragment of HSV DNA containing the *tk* gene is shown in Figure 2. This map derives from single and double digestions of cloned fragments with eight enzymes and is in accord with previously published data (Pellicer et al. 1978, Colbere-Garapin et al. 1979). The 2.0-kb Pvu II fragment maintains efficient transformation activity and therefore provides maximal boundaries for the functional *tk* gene (Colbere-Garapin et al. 1979).

A restriction map of the *tk* gene within the DNA of the transformants, K$_1$ and K$_2$, and their families of revertants and re-revertants was obtained by blot hybridization (Figure 2). These results reveal three discrete classes of tk$^-$ rever-

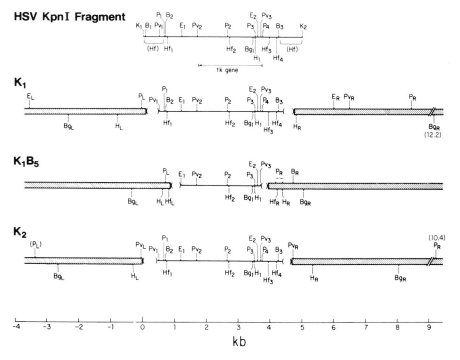

FIG. 2. Restriction endonuclease maps of the HSV *tk* gene region in the 5.0 kb Kpn I fragment from HSV, the transformant K$_1$, the revertant K$_1$B$_5$, and the transformant K$_2$. Viral sequences are represented by (——) and nonviral flanking DNA sequences by (▨). The region of the breakpoint between these sequences is bounded by parentheses. The functional *tk* gene (arrow) resides between Pvu II sites 2 and 3 (see text). Restriction sites whose position is uncertain are shown in parentheses. Numbers in parentheses beside restriction sites are map distances in kb. B-Bam HI, Bg-Bgl II, E-Eco RI, H-Hinc II, Hf-Hinf I, K-Kpn I, P-Pst I, Pv-Pvu II.

tants: clones with restriction patterns identical to the original tk$^+$ transformants, a clone in which the *tk* gene has undergone a transposition, and a clone which has deleted the *tk* gene. All re-revertants, however, maintained restriction patterns identical to those of the tk$^-$ clones from which they were derived.

tk$^-$ Revertants with Restriction Patterns Identical to Transformants

Illustrative restriction mapping data obtained with the enzyme Bgl II, which cleaves once within the *tk* gene, are shown in Figure 3. K_1, two of its tk$^-$ revertants, and numerous tk$^+$ re-revertants reveal identical annealing fragments, 5.8 and 8.7 kb (Figure 3A); K_2, two of its revertants, and numerous re-revertants also share identical annealing fragments (Figure 3B). Analysis of the fragment profiles obtained after digestion with seven additional enzymes mapping 30 observable sites generate profiles that permit the following conclusions: (1) Within the limitations of this approach, four of six revertants and their corresponding re-revertants maintain a restriction pattern identical to that of the original transformant; (2) the donor tk fragment is integrated into DNA within the transformed cell; and (3) the *tk* gene is integrated at a single site in transformed cell DNA and therefore exists in a hemizygous state.

Transposition of the tk Gene

In one revertant, K_1B_5, changes in sequence organization have occurred at both sides of the integrated viral fragment, suggesting that K_1B_5 results from

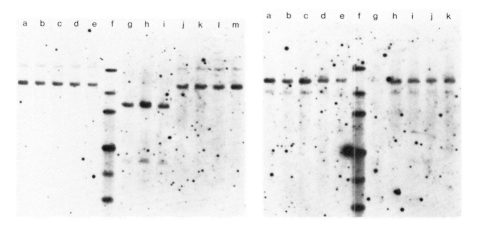

FIG. 3. Bgl II profiles of the tk fragment in sets of transformant, revertant, and re-revertant cell lines. A, (left) K_1 family: Lanes a–e: K_1, K_1B_4, $K_1B_4H_1$, $K_1B_4H_2$, $K_1B_4H_3$; lanes g–i: K_1B_5, $K_1B_5H_2$, $K_1B_5H_3$; lanes j–m: $K_1B_6{}^N$, K_1B_6, $K_1B_6H_1$, $K_1B_6H_2$; lane f, markers (8.0, 5.0, 3.6, 2.0, 1.3, and 0.88 kb). B, (right) K_2 family: Lanes a–e: K_2, K_2B_6, $K_2B_6H_4$, $K_2B_6H_5$, $K_2B_6H_6$; lane g: K_2B_7; lanes h–k: K_2B_8, $K_2B_8H_4$, $K_2B_8H_5$, $K_2B_8H_6$; lane f: markers as in A. The lower band in lane e is readily visible in the original autoradiograph.

a transposition of the *tk* gene from one integrated location to another. The restriction maps of K_1 and K_1B_5 are shown in Figure 2, with illustrative blot hybridization data in Figure 3A. Cleavage of DNA from K_1B_5 with Bgl II generates two annealing fragments, 4.1 kb and 1.6 kb in length. Both these fragments differ from the fragments observed after restriction endonuclease digestion of K_1 DNA. Since Bgl II cleaves only once within the *tk* gene, the molecular weight of the fragments generated reflects the distribution of Bgl II sites in DNA flanking the integrated *tk* gene. These data suggest that sequences flanking the *tk* gene on both ends differ in K_1 and K_1B_5. Significant additional mapping data (Figure 2) support the conclusion that a transposition had occurred and further indicate that the viral sequences in K_1B_5 have undergone deletions at both ends of the *tk*-containing fragment during this rearrangement process. tk^+ re-revertants can be selected from K_1B_5 (Table 2), and these re-revertants maintain restriction profiles identical to K_1B_5. Therefore, expression of the *tk* gene can occur subsequent to transposition and the initial rearrangement may not be related to the loss of tk activity.

Loss of the tk Gene

Revertant K_2B_7 is unique among the tk^- lines derived in this study since it never re-reverts to the tk^+ phenotype ($<10^{-6}$). Cleavage of K_2B_7 DNA with either Bgl II or Hin II and subsequent annealing with tk probes shows no evidence of hybridizing fragments (Figure 3B). This mutant therefore results from a deletion of most, if not all, of the integrated *tk* gene from the tk^+ parent, K_2.

Methylation Profiles of the *tk* Gene

Analysis of four independent revertants and eleven tk^- re-revertants at 30 restriction endonuclease cleavage sites reveals no sequence alterations associated with the reversible switching between tk^+ and tk^- phenotypes. However, short sequence alterations such as inversions or single base mutations and large rearrangements that may influence gene activity at a distance could escape detection. Alternatively, the genotypic switch may reflect DNA modification.

At present, methylation of cytosine within the DNA sequence pXpCpGpY is the predominant postsynthetic modification in mammalian DNA (Grippo et al. 1968, Burdon and Adams 1969). Hpa II and Msp I recognize and cleave the sequence pCpCpGpG; methylation in the pCpG sequence inhibits Hpa II but not Mpa I (Bird and Southern 1978, Mann and Smith 1977). A comparison of the fragment profiles generated by this enzyme pair permits conclusions about the pattern of methylation around specific genes in cellular DNA. We therefore compared the Hpa II/Msp I restriction profiles for the *tk* gene in the HSV genome and in the DNA of tk^+ and tk^- transformants. Hpa II and Msp I recognize at least 26 sites within the 3.6 kb Bam segment of the *tk* gene (Figure

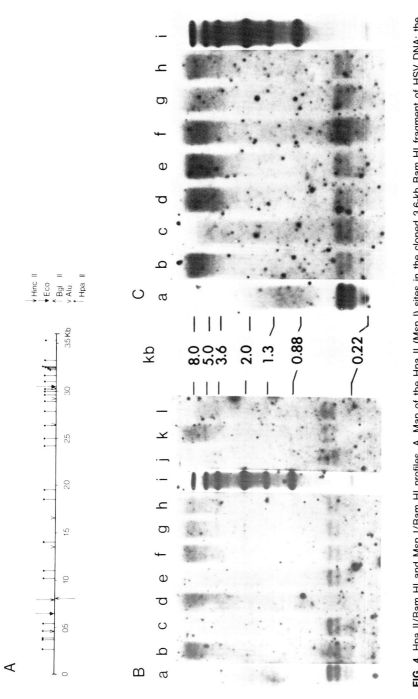

FIG. 4. Hpa II/Bam HI and Msp I/Bam HI profiles. A, Map of the Hpa II (Msp I) sites in the cloned 3.6-kb Bam HI fragment of HSV DNA; the orientation is opposite of that in Figure 2. B, K_1 family: Hpa II/Bam HI digests of HSV (a), transformant K_1 (b), revertants K_1B^N (d), K_1B_6 (f), K_1B_4 (k), and K_1B_5 (l), and re-revertants $K_1B_6H_{1-2}$ (g, h); markers as in Figure 3 (j). C, K_2 family: Hpa II/Bam HI digests of K_1 (c) and K_1B_5 (j); Msp I/Bam HI digests of K_2 (b), revertants K_2B_6 (d) and K_2B_8 (h) and re-revertants $K_2B_6H_{4-6}$ (e–g); Msp/Bam HI digests of K_2; markers Bam digests of HSV (a), transformant K_2 (b), revertants K_2B_6 (d) and K_2B_8 (h) and re-revertants $K_2B_6H_{4-6}$ (e–g); Msp/Bam HI digests of K_2; markers as above (i). The diffuse blackening at the top of the Hpa II lanes results from nonspecific binding of the labeled probe.

4A). In our blots, we resolve only the largest four or five bands, 250–500 bp. With one exception, discussed below, viral DNA, tk$^+$ transformants, and tk$^-$ mutants reveal identical fragments with Hpa II and Msp I (Figure 4). Thus, the *tk* gene in the virus and in tk$^+$ and tk$^-$ transformants is unmodified. Hypermethylation is therefore not associated with genetic inactivity of the *tk* gene.

Methylation in a Nonreverting Line

tk$^-$ revertants maintained in culture for long periods of time can lose the ability to re-revert to tk$^+$. These mutants probably emerge as variants with a growth advantage in the selective medium. K_1B_6 reverts to the tk$^+$ phenotype with a frequency of 40%; upon long-term culture in BUdR, this line lost its ability to switch to the tk$^+$ phenotype ($<10^{-6}$). This variant, $K_1B_6^N$, is "locked" into the tk$^-$ phenotype. The restriction maps of the *tk* gene in K_1B_6 and its variant, $K_1B_6^N$, are identical. Analysis of DNA modification, however, reveals that the *tk* gene in $K_1B_6^N$ is extensively methylated. Digestion of $K_1B_6^N$ DNA with Msp II reveals the characteristic set of small molecular weight fragments consistent with a limit digest of the *tk* gene. With Hpa II, however, this set of fragments is replaced by a series of larger DNA fragments, which must result from modification at many Hpa II cleavage sites within the *tk* gene (Figure 4B, lanes d, e). The number, size, and intensity of these fragments suggest a cellular heterogeneity in the sites of methylation. Corresponding results were obtained with two other enzymes sensitive to methylation, Hinc II and Sma I; partial digestion patterns were obtained for $K_1B_6^N$ under conditions that give complete digestion with all other cell lines (not shown).

For comparison, we have examined the Hpa II and Msp I patterns of the β-globin genes in these same cell lines. In both K_1 and $K_1B_6^N$, Hpa II yields an identical set of fragments that differs from those obtained with Msp I. Thus, while certain Hpa II sites are modified in the β-globin sequences, the extent and position of this methylation is the same in K_1 and $K_1B_6^N$, as well as in all other tk$^+$ and tk$^-$ transformants. These results suggest that specific methylation about the viral *tk* gene has occurred in $K_1B_6^N$, a cell line that has lost its ability to switch phenotypes.

CONCLUSIONS

Transformation permits the introduction of new genetic information into a cell and frequently results in a change in phenotype. We have developed cotransformation systems to study several aspects of eukaryotic gene expression: (1) to study in vivo the functional significance of various features of DNA sequence organization; (2) to purify genes where now classical routes dependent on mRNA enrichment are not feasible; (3) to analyze the molecular basis of mutations in somatic cells; and (4) to examine the fluidity of the eukaryotic chromosome.

We have studied the tk phenotype and genotype in two tk$^+$ transformants and several independently derived mutants. The tk$^-$ mutants fall into two categories: a stable class that does not re-revert to tk$^+$ at measurable frequencies and an unstable class that undergoes phenotypic switching from tk$^+$ to tk$^-$ to tk$^+$. One stable tk$^-$ revertant examined in this study is typical of three additional tk$^-$ clones previously examined in our laboratory. It results from a loss of most, if not all, of the integrated viral gene. In the unstable class, our data suggest that phenotypic switching results from heritable but unstable alterations in the *tk* gene. In one mutant, a transposition has occurred, resulting in a change in position of the *tk* gene from one integrated location to another. In four cell sets, the restriction map is invariable from transformant to revertant to re-revertant. These studies are far from exhaustive and significant rearrangements such as inversions about the gene or large translocations may have escaped detection. Thus, in at least two of six tk$^-$ mutants, loss or rearrangement of large blocks of DNA, including the *tk* gene sequences, has occurred.

We have also examined the pattern of methylation about the *tk* gene. Methylation in our clones is not associated, in an obligatory manner, with genetic inactivity of the *tk* gene. In one clone, however, extensive methylation is associated with a nonreverting tk$^-$ phenotype. Therefore, methylation of the *tk* gene may prevent genetic rearrangements that result in phenotypic switching. The genome of somatic cells may be quite fluid, and subtle alterations in DNA sequence organization may provide a mechanism for gene regulation. Perhaps specific genes must be free of modification to undergo genetic rearrangements that lead to gene activation.

We have described the development of cotransformation systems that may now permit the transfer of virtually any cloned gene into cultured cells from diverse tissues. The introduction of wild type genes along with native and in vitro constructed mutants into cultured cells may provide an assay for the functional role of specific DNA sequences. Studies on the expression of cotransformed genes, restricted thus far to murine fibroblasts, reveal transcripts that are correctly processed but frequently display aberrant 5' termini. If the 5' terminus reflects the site of transcription initiation, these results suggest that multiple potential promoter sequences may exist and their choice may be dictated by the differentiated state of the host cell.

These studies require that we extend the generality of cotransformation to more appropriate recipient cell lines. To this end, in collaboration with V. Parker, P. Mellon, and T. Maniatis, we have constructed a murine erythroleukemia cell containing several copies of the human β-globin gene and are currently analyzing the expression of these sequences in an erythroid cell. In addition, with Dr. B. Mintz we have cotransformed murine teratocarcinoma cells, which may provide a unique vector for the introduction of heterologous genes into the developing mouse embryo.

ACKNOWLEDGMENTS

This work was supported by grants from the National Institutes of Health to R.A. (CA-16346, CA-23767) and to S.S. (CA-17477 and RCDA CA-00492).

REFERENCES

Bird, A. P., and E. M. Southern. 1978. Use of restriction enzymes to study eukaryotic DNA methylation: I. The methylation pattern in ribosomal DNA from *Xenopus laevis*. J. Mol. Biol. 118:27–47.

Breathnach, R., N. Mantei, and P. Chambon. 1980. Correct splicing of a chicken ovalbumin gene transcript in mouse L cells. Proc. Natl. Acad. Sci. USA 77:740–744.

Burdon, R. H., and R. L. P. Adams. 1969. The in vivo methylation of DNA in mouse fibroblasts. Biochim. Biophys. Acta 174:322–329.

Colbere-Garapin, F., S. Chousterman, F. Horodniceanu, P. Kourilsky, and A. -C. Garapin. 1979. Cloning of the active thymidine kinase gene of herpes simplex virus type 1 in *Escherichia coli* K-12. Proc. Natl. Acad. Sci. USA 76:3755–3759.

DeMars, R. 1974. Resistance of human fibroblasts and other cells to purine and pyrimidine analogues in relation to mutagenesis detection. Mutat. Res. 24:335–364.

Fiddes, J. C., P. H. Seeberg, F. M. DeNoto, R. A. Hallewell, J. D. Baxter, and H. M. Goodman. 1979. Structure of genes for human growth hormone and chorionic somatomammotropin. Proc. Natl. Acad. Sci. USA 76:4294–4298.

Flintoff, W. F., S. V. Davidson, and L. Siminovitch. 1976. Isolation and partial characterization of three methotrexate resistant phenotypes from Chinese hamster ovary cells. Somatic Cell Genetics 2:245–261.

Graf, L. H., G. Urlaub, and L. Chasin. 1979. Transformation of the gene for hypoxanthine phosphoribosyl transferase. Somatic Cell Genetics 5:1031–1044.

Grippo, P., M. Iaccarino, E. Parisi, and E. Scarano. 1968. Methylation of DNA in developing sea urchin embryos. J. Mol. Biol. 36:195–208.

Hamer, D., and P. Leder. 1979. Expression of the chromosomal mouse $\beta^{maк}$-globin gene cloned in SV-40. Nature 281:35–39.

Lai, E. C., S. L. Woo, M. E. Bordelon-Riser, T. H. Fraser, and B. W. O'Malley. 1980. Ovalbumin is synthesized in mouse cells transformed with the natural chicken ovalbumin gene. Proc. Natl. Acad. Sci. USA 77:244–248.

Maitland, N., and J. McDougall. 1977. Biochemical transformation of mouse cells by fragments of herpes simplex virus DNA. Cell 11:233–241.

Mann, M. B., and H. O. Smith. 1977. Specificity of Hpa II and Hae III DNA methylases. Nucleic Acids Res. 4:4211–4221.

Mantei, N., W. Boll, and C. Weissman. 1979. Rabbit β-globin mRNA production in mouse L cells transformed with cloned β-globin chromosomal DNA. Nature 281:40–46.

Mulligan, R. C., B. H. Howard, and P. Berg. 1979. Synthesis of rabbit β-globin in cultured monkey kidney cells following infection with a SV-40 β-globin recombinant genome. Nature 277:108–114.

Pellicer, A., E. F. Wagner, A. El-Kareh, M. J. Dewey, A. J. Reuser, S. Silverstein, R. Axel, and B. Mintz. 1980. Introduction of a viral thymidine kinase gene and the human β-globin gene into developmentally multipotential mouse teratocarcinoma cells. Proc. Natl. Acad. Sci. USA (in press).

Pellicer, A., M. Wigler, R. Axel, and S. Silverstein. 1978. The transfer and stable integration of the HSV thymidine kinase gene into mouse cells. Cell 14:133–141.

Wigler, M., A. Pellicer, S. Silverstein, and R. Axel. 1978. Transfer of single copy eukaryotic genes using total cellular DNA as donor. Cell 14:725–731.

Wigler, M., A. Pellicer, S. Silverstein, R. Axel, G. Urlaub, and L. Chasin. 1979a. DNA-mediated transfer of the adenine phosphoribosyl transferase locus into mammalian cells. Proc. Natl. Acad. Sci. USA 76:1373–1376.

Wigler, M., M. Perucho, D. Kurtz, S. Dana, A. Pellicer, R. Axel, and S. Silverstein. 1980. Transformation of mammalian cells with an amplifiable dominant acting gene. Proc. Natl. Acad. Sci. USA 77:3567–3570.

Wigler, M., S. Silverstein, L.-S. Lee, A. Pellicer, Y.-C. Cheng, and R. Axel. 1977. Transfer of purified herpes virus thymidine kinase gene to cultured mouse cells. Cell 11:223–232.

Wigler, M., R. Sweet, G.-K. Sim, B. Wold, A. Pellicer, E. Lacy, T. Maniatis, S. Silverstein, and R. Axel. 1979b. Transformation of mammalian cells with genes from prokaryotes and eukaryotes. Cell 16:777–785.

Willecke, K., M. Klomfass, R. Mirau, and J. Dohmer. 1979. Intraspecies transfer via total cellular DNA of the gene for hypoxanthine phosphoribosyl transferase into cultured mouse cells. Mol. Gen. Genet. 170:179–185.

Wold, B., M. Wigler, E. Lacy, T. Maniatis, S. Silverstein, and R. Axel. 1979. Introduction and expression of a rabbit β-globin gene in mouse fibroblasts. Proc. Natl. Acad. Sci. USA 76:5684–5688.

Gene Amplification

Genes, Chromosomes, and Neoplasia, edited by
Frances E. Arrighi, Potu N. Rao, and Elton Stubblefield.
Raven Press, New York © 1981.

Double Minutes and C-Bandless Chromosomes in a Mouse Tumor

Albert Levan, Göran Levan,* and Nils Mandahl

*Institute of Genetics, University of Lund, Sweden, *Institute of Genetics, University of Gothenburg, Sweden*

The SEWA mouse tumor was induced on September 25, 1959, by subcutaneous inoculation of polyoma virus in a newborn A.SW mouse (Sjögren et al. 1961). It started as an osteosarcoma but lost its differentiation during the first 10 passages and is now an anaplastic sarcoma carried serially as an ascites tumor. Its chromosomes were first studied by Hellström et al. (1962) in passage 7, while it still maintained its capacity for bone formation. Its stem line had 46 chromosomes, including a large marker with a secondary constriction. In 1973, Fenyö et al. described the SEWA karyotype and presented a photomicrograph of a metaphase (Fenyö et al. 1973, Figure 2, p. 1874). The stem line number was now s = 43, and in addition to the marker with the secondary constriction they found one "abnormally" long chromosome and "several minute chromosomes or fragments." It can be seen from their photograph that these minute chromosomes were genuine double minutes (DM). Hellström et al., in their paper just quoted, had reported minute markers in another of the polyoma-induced tumors they studied, but did not mention minutes in relation to SEWA. Jonasson et al. (1977), in connection with their comprehensive studies on cell fusion, worked out the karyotype of the SEWA stem line, for the first time on the basis of G-banded preparations. They gave the stem line number s = 45. Of the normal chromosomes one was nullisomic, seven were monosomic, one had a deletion, two were trisomic, and in addition 14 marker chromosomes were found. The authors did not mention whether DM were present.

Our interest in the SEWA chromosomes was aroused when in September 1975 we happened to look at some slides of the SEWA tumor that Dr. Urszula Bregula had brought from Dr. George Klein's laboratory in Stockholm. At that time, almost every metaphase contained DM, often in high numbers. In November 1975, Dr. Klein sent us a sample of passage 46 of the SEWA tumor. This line is still carried in serial in vivo transfer at our laboratory under the name of SEWA stock 1. Since July 1976 it has also been carried in mice of strain A, which is related to A.SW but carries different alleles at the H-2 locus. In August 1979, we received two more samples of SEWA from the cell bank of Dr. Klein. They represented earlier passages of the tumor, passages 31 and 38, counted from the primary tumor. They were named SEWA stock 4 and stock 3, respectively.

The four stocks are represented in the upper branches of the genealogical tree of Figure 1. They have been different in their content of DM. In stock 1, cells with DM gradually became less frequent and after a year it was uncertain whether they had become very small or had disappeared. During 1979 we were able to see them only in single cells. Probably this was related to the recent finding that in G-banding late passages of stock 1 have one or two homogeneously staining regions (HSR). In stock 1 in strain A nearly 100% of the cells with many small DM have been maintained all the time. In stock 3, only occasional cells had DM, but these were large DM, while stock 4 had high numbers of DM in most cells. As mentioned, DM were not seen in passage 7 (according to Hellström et al. 1962) but were abundant in passage 31 after supposedly half a year of transfer. We have found that even in those SEWA lines in which DM are usually not seen, one can always find cells with DM. We therefore hesitate to conclude that DM originated in SEWA between passages 7 and 31. We rather tend to assume that the potential for DM formation is inherent in the SEWA genotype from the beginning.

During our early acquaintance with the SEWA tumor we performed various experiments to characterize the DM. We found that they were C-band, G-band, and N-band negative. They were Feulgen positive and took up thymidine and bromodeoxyuridine. They behaved as ordinary chromosomes under harlequin staining. We conducted a number of experiments to test whether the replication cycle of the DM was synchronized with that of the chromosomes, or whether the DM underwent extra, unscheduled replication cycles. The latter would explain why the DM are maintained or may even increase in number in certain populations in spite of their elimination during mitosis. We found no signs of any extra cycles of the DM. The DM replicate their DNA early during the S period. During prolonged in vitro culture, cells with DM tended to disappear, but they reappeared after a period in vivo.

In March 1977 we observed a dramatic change in one specific SEWA subline (Levan et al. 1978). In passage 32 of the line SEWA1R (Figure 1), the average chromosome number per cell suddenly increased from 44 to 51, which was caused by the addition of 1 to 12 copies of a new medium-sized metacentric chromosome to almost every cell. At the same time we noticed that the DM that would normally have been present in most of the cells of this line had disappeared. Even before C-banding, the new chromosomes were a very striking feature of the karyotype. After C-banding it was revealed that the new chromosomes were totally devoid of the centromeric C-bands that are so characteristic of ordinary mouse chromosomes. The new chromosomes had faint G-bands; therefore, we hesitated to designate them chromosomes formed by HSR. We decided provisionally to refer to them as C-bandless or CM (C minus) chromosomes.

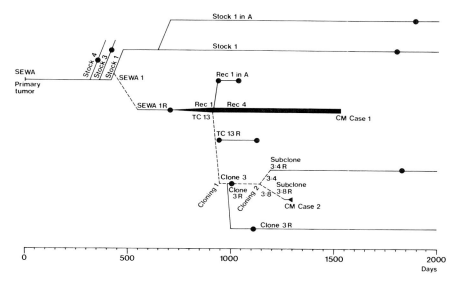

FIG. 1. Genealogical tree of the SEWA lines studied. The upper branches represent stock tumors, the lower branches represent the SEWA1 pedigree and include the accumulations of CM chromosomes of Cases 1 and 2. The time scale at the bottom indicates days in serial transfer in vivo (unbroken lines) and in vitro (broken lines) with periods in cell bank omitted. Dots = findings of single cells with CM.

Besides their lack of C-bands, the CM had a number of other characteristics that emphasized their kinship to the DM rather than to the ordinary chromosomes of the mouse. They varied in number and were absent in certain cells. In G-banding they took a dull shade with vague but unquestionable bands. In harlequin staining they behaved as ordinary chromosomes; their incidence of sister chromatid exchanges was at the same average level as in ordinary chromosomes. They were early replicating, and cells with CM gradually disappeared when kept in culture. All these facts, together with the observation that, in certain cases at least, DM and CM were mutually exclusive, led to the conclusion that they represented similar genetic materials and had a common origin.

Ever since we found the first case of heavy accumulation of CM at a certain stage of development of SEWA1R, we have been on the lookout for other similar cases. What had originally led to the discovery of the first case was the increase in chromosome number, so we decided to count the chromosomes routinely in all materials examined with the hope that this would lead us to new cases of CM. We soon learned that pronounced cases of CM must be rare indeed. In April 1978 we struck paydirt, however. In a set of 16 tumors initiated after the inoculation in vivo of the same clonal in vitro population, a clear increase in chromosome number was exhibited by two tumors, one of which could be studied further. Even before C-banding it was evident that in this new case the smallest chromosomes were in excess, and C-banding revealed that 88% of the cells had additional small telocentric chromosomes without C-bands, thus being in fact CM. Unfortunately, this tumor was discarded before the CM had been detected.

The interrelations between the only two cases of CM we have found so far should be clear from Figure 1. Both of them belonged to the SEWA1R pedigree. The tumor of the second case was a descendant of tissue culture TC13, which was started from SEWA1R at the very moment the CM of the first case had reached maximum level. In the lineage of SEWA1R, Rec1, Rec4 (Figure 1), this level was maintained for a long period of time, perhaps permanently. The second case was separated from the first by one year of cell culture. It is known from several experiments that both DM and CM will be practically eliminated by as few as 100 days in culture (Levan et al. 1977, 1978). Since during the year in culture, the stem cells of case 2 were submitted to two consecutive clonings (Figure 1), it was reasonable to assume that the line no longer had any CM chromosomes left from case 1. These observations actually touch upon one of the main problems of the present investigation: Do cells with DM or CM persist after clonings of populations that had very few cells with DM or CM? As we mentioned above, it has always been possible to find single cells with DM or CM in any line of SEWA if a sufficient number of cells are scanned. One alternative to explain this apparent inconsistency would be that cells with DM and CM are carried over by the stem line of the population, the alternative being iterated de novo formation of DM and CM. We shall return to this question later on.

After this survey of the general background of our material, we shall now discuss in more detail some cytogenetic features of the two cases of CM so far available. After that, the incidence of CM in other lines of the SEWA1R pedigree will be examined, and eventually the same question will be taken up in relation to the four SEWA stocks, which, except for the fact that the culture that gave rise to the SEWA1R tumor emanated from stock 1, have had a completely separate existence from the SEWA1R pedigree.

THE ORIGINAL CASE OF C-BANDLESS CHROMOSOMES

The Antagonism between CM and DM

Since we considered the origin of the new CM chromosomes an event worthy of careful analysis, we have grown a fair number of CM-carrying sublines and examined them from various viewpoints. One such study concerns the interrelations of DM and CM. In Figure 2, percentages of cells with DM and CM are recorded for five sublines of SEWA, including the original SEWA1R. As seen from Figure 2a, the transition from a population with mainly DM-carrying cells to one with mainly CM-carrying cells took place quickly; on day 86 7% of the cells had CM and 80% had DM, and on day 169 96% of the cells had CM and 7% had DM. As soon as this transition was recognized, a big sample

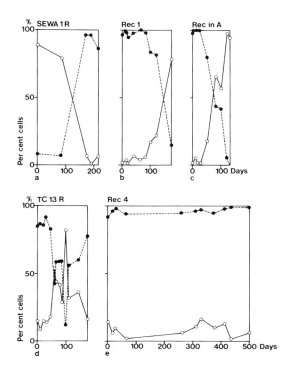

FIG. 2. Percentage of cells with DM (open circles, unbroken lines) and CM (filled circles, broken lines) in five SEWA lines of Case 1.

of the ascites was frozen, and the four other sublines of Figure 2 were all derived from this sample, b, c, and e directly, and d after 1 month of in vitro growth (see Figure 1). Accordingly, these samples all started out with high levels of CM and low levels of DM. In their continued in vivo growth we could discern three patterns of behavior.

1. The first pattern is represented by Figures 2b and c, in which the level of CM underwent a steep decrease and that of DM an equally steep increase. These changes, after some 100 to 200 days, resulted in the restitution of the original characteristics of the SEWA1R. A difference was noticed between Figures 2b and c: In b, syngeneic hosts, there was a lag period of about 100 days, whereas in c, allogeneic hosts, the change started directly. Two separate attempts to repeat the latter experiment with the same allogeneic host have failed, because the tumor in both cases regressed after one or two passages. This was somewhat surprising, since the stock 1 tumor had grown excellently in strain A mice. Another experiment, not shown in Figure 2, was performed with Rec1, inoculated into Swiss mice in May 1977. Of three mice inoculated, one died immediately and the other two developed ascites that were gradually resorbed during the regression of the tumor. Fixations from days 2, 4, 12, and 40 could be analyzed. The percentage of cells with CM was high during days 2 and 4 (79% and 93%) but went down to 50% on day 12 and to 2% on day 40. No correspond-

ing increase in DM was seen, the percentage of cells with DM being below 10% all the time.

2. The second pattern of behavior is seen in Figure 2d, in which the situation is more confused. The number of cells with CM first decrease, as in the preceding pattern, but the behavior oscillates a few times and eventually the trend is a rising number of cells with CM. The DM behave characteristically throughout these fluctuations: the DM curve is a mirror image of the CM curve.

3. The third pattern (Figure 2e) is represented so far only by one subline, Rec4, started in July 1978 and still in animal transfer. Here, the initial pattern has persisted for more than 500 days. The cells with CM have never gone below 92%, and those with DM never above 18%. Rec4 is peculiar in another respect: an unusually high fraction of its DM are large and often appear as rings or rods (Levan and Levan 1980b).

The three types of behavior illustrated have one thing in common: there exists a certain antagonism between cells with CM and those with DM. If one is high the other is low, and so far no population has stabilized with equal percentages of both.

This observation leads to the next question: Do CM and DM exclude each other from the same cell? The answer is that they very definitely prefer to inhabit separate cells, but they do occur together in a minority of the cells. For the five materials of Figure 2, data from 5,702 metaphases have been surveyed from this viewpoint. Of them 92% had either CM or DM, 4% had neither, and 4% had both. In this material, cells with CM were almost four times more frequent than those with DM, indicating that in these sublines CM stimulated viability more than DM, at least during certain periods.

Figure 3 represents the covariation of DM and CM in the same five sublines as shown in Figure 2, except the order of the sublines has been arranged according to falling percentage of cells with DM in the zero class of CM. Also, in SEWA1R

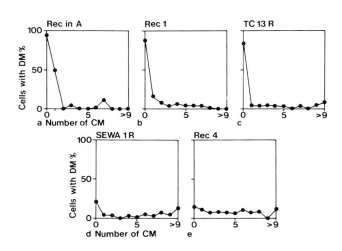

FIG. 3. Covariation between number of CM per cell and percentage of cells with DM in five SEWA lines of Case 1.

(Figure 3d) only the three last fixations could be included, i.e., those after the finding of the CM. The high percentages of DM for 0 CM of Figures 3a-c are due to the final samples of Rec in A and Rec1 and to certain middle samples of TC13R (see Figure 2b-d).

The Appearance of the CM Chromosomes, Their Variability, and Some Evolutionary Trends

The previous section was based on the total material available, unbanded and banded. Since it was recognized early that C-banding would be important for the identification and evaluation of the new CM chromosomes, a fair number of C-bandings have been performed in the CM-containing lines. In these, the CM chromosomes were first recognized as medium-sized metacentric to submetacentric chromosomes ("msm" according to the nomenclature of Levan et al. (1964), which will be adhered to in this paper). The CM always occurred as an addition to the stem line set of ordinary chromosomes. In most sublines belonging to the first case of CM, this type is absolutely predominant. It will be referred to as the standard CM, and its appearance is shown in the two cells of Figure 4. They were from a sample of TC13R (see Figure 1) fixed 1 month after the CM were detected, actually from the second sample of Figure 2d. In the 50 metaphases analyzed on this occasion, the chromosome numbers varied from 41 to 56 (average 49.1) and the number of CM varied from 0 to 11 (average 5.4). The two cells of Figure 4 had 45 (a, b) and 52 chromosomes (c, d). In Figure 4a, b, there were also two medium-sized DM.

Among the ordinary C-band positive chromosomes, several markers were identified. They are common to most SEWA lines and will be listed here under the names we provisionally use for them. Recently, we have identified by G-banding the normal chromosomes, from which some of the markers have been derived. Even though we shall reserve to a later occasion the G-band analysis of the SEWA karyotype, some data will be given here concerning those markers that are easily distinguished in C-banding (Figures 4 and 5).

The largest chromosome, mar11, has been identified as consisting of the main part of chromosome 3, to which an unidentified terminal segment has been added. The distal half of this segment may be an HSR. This is probably the long marker reported by Fenyö et al. (1973). The smallest chromosome is called mar3 when it has no visible arm; in C-banding it appears as a spherical, strongly stained body. When there is a short arm, as in Figures 4 and 5, it is called mar4. Its identity is unknown. The big marker with a heavy secondary constriction, ubiquitous in the SEWA tumor and first reported by Hellström et al. (1962), is called mar2 and has a heavy C-band at the middle of its arm at the site of the constriction. This marker may also appear as mar2-l, in which the segment distal to the constriction is twice as long (Figures 5a, f, i). Mar2 has been identified as a tandem translocation of chromosomes 19 and 18. In mar2-l, the end segment of mar11, including the HSR, has been added distally. One

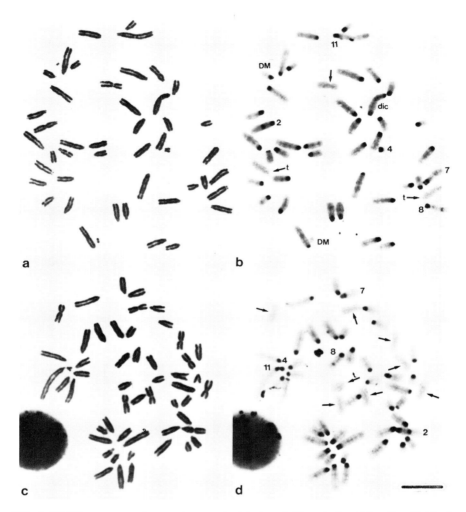

FIG. 4. TC13R, two metaphases photographed before (left) and after C-banding (right). C-band positive markers are indicated by Arabic numerals. Plain arrows = standard CM, arrow and t = telocentric CM, dic = dicentric. Bar at bottom right = 10 μm.

medium-sized chromosome, mar7, is present in most SEWA materials. It has an extra C-band close to the centromere and corresponds to one homologue of chromosome pair 13. Another medium-sized marker, mar8, has a tiny C-band at the distal third of its arm. As in mar2, this C-band corresponds to a constriction, but this constriction is much less pronounced. The proximal part of the X chromosome forms the segment between the centromere and the interstitial C-band of mar8.

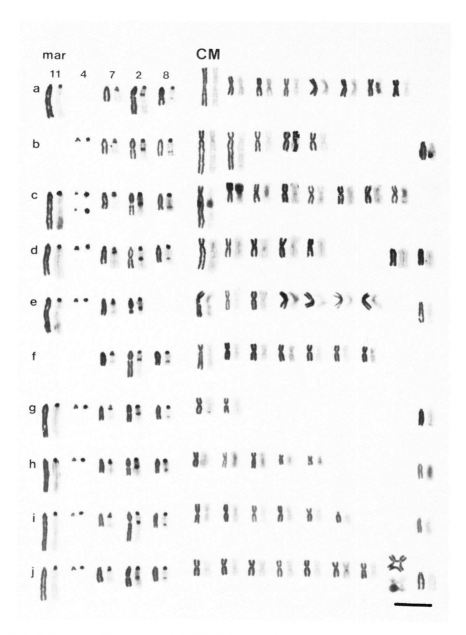

FIG. 5. Condensed karyotypes of SEWA1R (a, d-j), TC13R (b, c); C-band positive markers (left), CM chromosomes (right). Standard CM are in the center, small smst CM are to their right, and t CM are to the extreme right. Each chromosome was photographed before and after C-banding. Note the variation in size and centromeric location among the CM. Most cells contain several standard CM and one t CM. In j, a quadriradial was formed by two CM. Bar at bottom right = 10 μm.

In the upper cell of Figure 4, there were three CM, from which it immediately was clear that the CM may vary in type: only one of them was a standard CM, the other two were of about the same length but had terminal centromeres (t CM). In the latter case, the small segments with the centromere were set off by a constriction, as is often seen in the ordinary mouse chromosomes. In the lower cell of Figure 4 there were eight CM: seven were standard and one was a medium-sized t CM. This cell was rather typical of the populations with a high frequency of CM-carrying cells during the first months after the CM were detected. In the sample of 50 cells, from which the two cells of Figure 4 were taken, the average karyotype included six CM, five of which were standard and one a t CM.

The morphologic variation of the CM is represented more fully in Figure 5. Considerable size differences among the CM are shown in this figure, with very large metacentric (m) and subtelocentric (st) CM types in Figure 5 a-f and small sm and st ones in Figure 5 h, i. The standard CM type, represented in the figure by some 50 copies, appeared very uniform, although minor variations in size and in centromeric location were frequent. That CM do participate in structural rearrangements, both among themselves and with C-band positive chromosomes, is obvious. Direct observation of interchange figures, as the quadriradial of Figure 5j, is rare, but the variation in size and shape continually observed is good evidence. Interchange between CM and C-band positive chromosomes has been made likely by the presence of biarmed chromosomes, in which just the one arm has a C-band. Some biarmed chromosomes from six cells of the same sample of TC13R have been collected in Figure 6, in which most of the biarmed chromosomes have double centric C-bands, one on each side of the centromere. A few, however, had C-bands only on one side (left ones in Figure 6a, c, e, and the one in 6b). These may represent either fusions of CM with ordinary chromosomes or translocations of C-bandless segments from CM or ordinary chromosomes to the centromeric regions of ordinary chromosomes.

An instance of the latter kind was found in a new big st chromosome that originated in the Rec4 subline some 450 days later, thus in a direct descendant of SEWA1R. This new marker had a C-band only on the short-arm side of the centromere. Since at the same time the medium-sized t CM had disappeared, it seemed likely that a fusion had taken place between the t CM and a small ordinary chromosome. It was found by G-banding, however, that the new st marker consisted of the distal half of mar11 joined to normal chromosome 17. This tallied well with the observation that in 66 cells analyzed, mar11 and the new st marker were found in the following combinations and frequencies: both mar11 and the st marker, 1 cell; only mar11, 7 cells; only the st marker, 55 cells; neither, 3 cells. The translocation (17;mar11) evidently involved an evolutionary advantage.

Evolutionary changes involving CM chromosomes were also found. Thus, in TC13R a gradual substitution of a medium-sized st CM for some of the ordinary

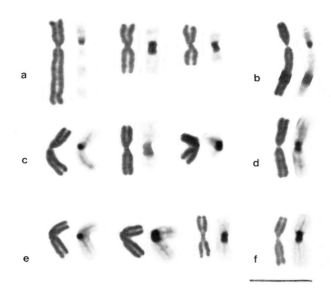

FIG. 6. SEWA1R, instances of biarmed chromosomes photographed before and after C-banding. C-bands may occur on one or on both sides of the centromere. Bar at bottom right = 10 μm.

standard CM could be followed through 23 passages. The new st CM was a type that was seen occasionally in several of the sublines (see Figure 5h, i). In TC13R, however, the proportion of cells with st CM steadily increased during the serial transfer of this tumor. Two metaphases with this type of CM are pictured in Figure 7, the upper cell with five copies (only four are visible, since one copy of the CM and mar2 fell outside the viewing field) and the lower cell with four copies. Figure 8 clearly suggests that during the development of this subline a selection was acting in favor of cells with a good share of the variant CM.

After this mainly qualitative analysis of the CM of case 1, it should be expedient to record some quantitative data from the six occasions on which C-banded materials of the SEWA1R pedigree were analyzed. Table 1 reports the incidence of cells with CM and the percentages of the different CM morphologic types during the 649 days such data were collected. SEWA1R, passage 15, corresponds to the first sample plotted in Figure 2a and thus precedes the finding of the CM by 169 days. After the CM had been found, attempts were made to explore the ancestry of SEWA1R in a search for CM. Close scrutiny of C-banded material from passage 15 showed that a low but regular incidence of CM was indeed a characteristic of SEWA1R at that time. Since the incidence seemed to be higher in cells with the double stem line number (2s) than in s cells (38.5% versus 9.2%), the two stem line levels were recorded separately in the table. These observations are evidence that CM had begun to accumulate in the population even at that early time. It was also found that the proportions of different

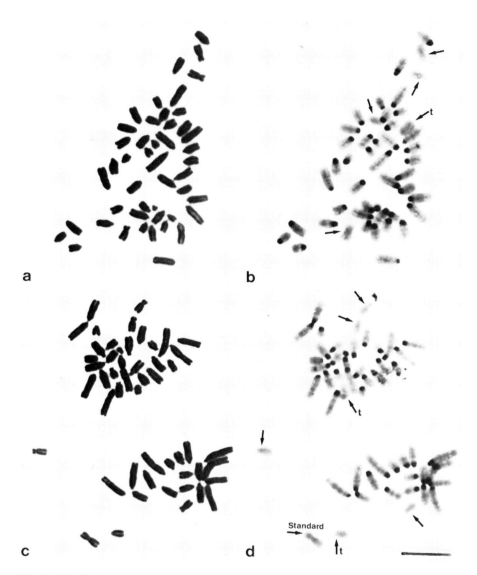

FIG. 7. TC13R. The two cells with variant CM were photographed before (left) and after C-banding (right). Plain arrows = variant CM, arrow and t = telocentric CM. One standard CM is in bottom cell. Bar at bottom right = 10 μm.

morphologic types of CM differed from those present later on, after CM had spread to practically the entire cell population. Thus, the standard type was found at a much lower frequency at the earlier time than later on, and the medium-sized t CM was found at a higher frequency.

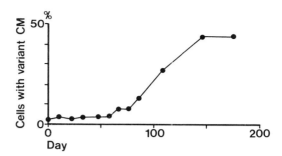

The pattern of the different CM types became fairly uniform (samples 2 to 5, Table 1) and largely agreed with the situation represented by Figure 4 c, d, that is, some five CM per cell of standard type and one of t type. For sample 6, which was taken more than a year after sample 5, the percentage of cells with CM was still high (97%), but the total number of CM had dropped from five–six per cell to four per cell; the standard CM that earlier composed some 70%–80% of all CM now composed 94%, whereas the t CM had dropped from around 15% to 1.4%.

Keeping the SEWA1R pedigree under observation through almost 2 years has led to the revelation of specific trends in the presence and behavior of the CM that can only mean that they are significant factors in the viability of the tumor. From an uncertain and vague appearance in single cells, CM accumulated in certain lineages of the tumor, and cells with them were capable of outgrowing cells without CM. Again and again the populations divided into two cell types, those with mainly DM and those with mainly CM. Cells with both were seen but never became important. The equilibrium between the two types was always labile, and the balance eventually settled in favor of one or the other. If, as seems reasonable, DM and CM consist of essentially similar genic material, it is difficult to see why they should not be able to cooperate efficiently in the same cell. Evidently, their harmonious cooperation is disturbed by the different ways in which their genic material is organized.

Selective forces were revealed that responded directly to challenges from the CM chromosomes. Among these were the accumulation of a new morphologic type of CM in a specific subline and the disappearance in another subline of a t CM that had been a regular member of the stem line for a long time. This kind of evolutionary trend suggests that structural remodeling of CM chromosomes is significant in maintaining the general viability of the cell. This may also be concluded from the fact that a specific structural pattern of the CM chromosomes—a number of standard CM and one t CM—persisted for such a long time. If the genic content of the CM types had been qualitatively identical and interchangeable, one would expect that the cells would have become specialized, carrying one type of CM chromosome structure and not continuing to carry two. All this leads to the conclusion that the CM chromosomes, in spite

TABLE 1. Percentage incidence of the different CM types in the SEWA1R pedigree

| No. | SEWA subline | Passage | Day* of fix. | Number of cells | | Total CM analyzed | Percentage of different CM types | | | | | | | | | CM per CM-carrying cell |
| | | | | Total | With CM | | msm | | | st | | | t | | | |
							Big	Med.	Small	Big	Med.	Small	Big	Med.	Small	
1	SEWA1R	15 s	279	109	10	18	27.8	5.6	—	11.1	—	—	11.1	38.9	5.6	1.80
		2s		13	5	17	47.1	—	—	—	—	—	11.8	41.2	—	3.40
		s + 2s		122	15	35	37.1	2.9	—	5.7	—	—	11.4	40.0	2.9	2.33
2	Rec1	3	479	50	44	276	1.5	80.1	1.1	2.2	1.1	0.4	0.7	12.0	1.1	6.27
3	TC13R	2	513	50	43	267	2.6	70.4	0.8	1.1	6.7	—	—	16.9	1.5	6.21
4	TC15R	10	568	23	11	64	—	79.7	—	—	3.1	—	—	17.2	—	5.82
5	TC13R	13	590	27	22	107	0.9	79.4	—	3.7	2.8	—	—	12.2	0.9	4.86
6	Rec4	64	928	100	97	354	0.3	94.4	0.9	1.7	0.3	—	—	1.4	1.1	3.65

* Day 0 = December 18, 1975 (start of SEWA1 = day 433 in Figure 1); periods in the cell bank have been deducted.

of their apparent origin as a multiple copy of a single chromosome that in turn supposedly contains the same DNA as the DM, are composed of genetically differentiated materials. This conclusion is compatible with the presence of vague G-bands in the CM. Incidentally, traces of such G-bands often penetrated through the C-banding, as in the long CM of Figure 5, especially the two long st CM in Figure 5b.

CM in Clones within the SEWA1R Pedigree

It was mentioned in the introduction that DM and CM in SEWA are eliminated in vitro. During the present work, however, they have been seen in single cells even after long in vitro periods and after clonings. This persistence has been observed too many times to be ignored. Five such cases are recorded in Table 2, whose materials were all derivatives of SEWA1R. (All are included in the pedigree of Figure 1, except clones 4 and 4R, which are sister lines to clones 3 and 3R.) All of them had been started in tissue culture on day 469 and #1, #2, #3, and #4 had been cloned on day 512, whereas #5 had been cloned on days 512 and 714. For the reason explained in the footnote of Table 1, these day nos., as well as those of Tables 1 and 2, are convertible into the day nos. of Figure 1 by the addition of 433. Numbers 1 and 2 of Table 2 were examined after 100 days in vitro; #3, #4, and #5 were reimplanted into animals on days 543, 539, and 762, respectively, and the samples in Table 2 were taken after 9, 10, and 63 passages in vivo.

Only one of these sublines, #3, had recovered a regular and high frequency of DM-carrying cells, but all of them had single cells with CM. The frequencies of cells with CM varied from 2% to 6% in the samples, and it was obvious that in no case had CM-carrying cells attained selective superiority. Whether they represented remainders of the old CM population of SEWA1R that had been successful in passing one or two clonings or were newly arisen CM chromosomes is an open question. Altogether, 15 CM (29.9%) in #1, #3, and #5 were apparently of the standard type, and 13 CM (22.4%) in #1, #2, #3, and #5 were of the medium-sized t type, which perhaps speaks in favor of the former alternative.

In Figure 9, four cells from subclone 3:4R (Table 2, #5) are presented. Subclone 3:4R is characterized by several centric fusions by which the stem line number has been brought down from the usual 43–44 to 39. Most of the biarmed chromosomes have C-bands on both sides of the centromere. One st marker, however, has a narrow C-band only on the long arm side of the centromere and none on the other side. This is mar1 (Figure 9), which occurs widely among the SEWA1R clones. This marker was present in most cells of the old SEWA1R fixations of 1976. It was still found in a few cells of the fixation in March 1977, when the CM were first detected, but has not been seen in SEWA1R since then. It did not occur in any of the SEWA1R derivatives, Rec1, TC13R, Rec in A, or Rec4, but suddenly reappeared in all clones and subclones isolated

TABLE 2. Percentage incidence of CM types in clonal derivatives of SEWA1R

No.	SEWA subline	Passage	Day* of fix.	Number of cells		Total CM analyzed	Percentage of different CM types									CM per CM-carrying cell
				Total	With CM		msm			st			t			
							Big	Med.	Small	Big	Med.	Small	Big	Med.	Small	
1	Clone 3	1	569	157	8	22	36.4	27.3	—	—	—	—	9.1	9.1	18.2	2.75
2	Clone 4	1	569	91	2	10	—	—	—	—	—	—	10.0	10.0	80.0	5.00
3	Clone 3R	9	684	90	5	12	—	33.3	—	—	25.0	—	—	25.0	16.7	2.40
4	Clone 4R	10	684	18	1	1	—	—	—	—	—	—	—	—	100	1.00
5	Subclone 3:4R	63	1,397	202	10	22	9.1	22.7	—	—	9.1	—	—	31.8	27.3	2.20

* Same chronology as in Table 1.

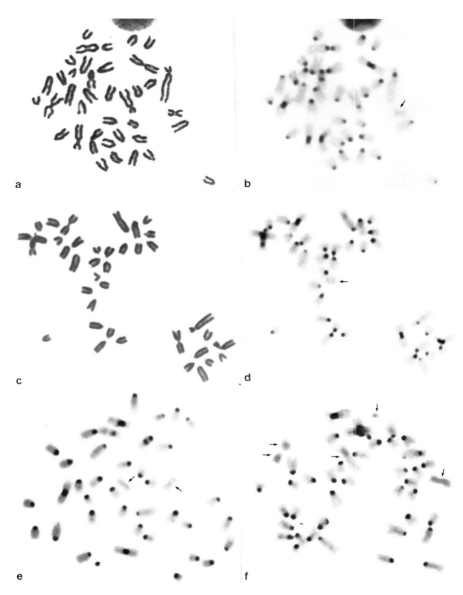

FIG. 9. Subclone 3:4R, passage 63 in vivo. The four cells have CM (arrows). Typical standard CM occur in b, e, f; one large m CM in f; one medium-sized t CM in c, f; and one small t CM in f. (This fixation corresponds to Table 2, No. 5.) Magnification, 1,150×.

in TC13. It has not been observed in SEWA stock 1, from which the SEWA1R was derived, nor in any of the other stocks unrelated to SEWA1R. It has been ascertained by G-banding that mar1 consists of chromosome 1 provided with a short arm of unknown origin.

In Figure 9, four cells are pictured in which one or more CM were present. The unbanded pictures of a and c leave no doubt that the CM have regular centromeres. If the CM are new products, centromeres must also have been provided somehow. As pointed out before, the CM always occur as an addition to the regular stem line set. Accordingly, seven of the 10 cells with CM in Table 2 had chromosome numbers above the stem line number. The four cells reproduced in Figure 9 had $2n = 39$, 40 (each with one CM), 41 (two CM), and 46 (five CM).

THE SECOND CASE OF C-BANDLESS CHROMOSOMES

Discovery and Incidence of the CM Chromosomes

The second case of an exceptionally great accumulation of CM in a SEWA tumor was accidentally stumbled upon during the analysis of a set of 16 tumors, grown for an unrelated purpose. The tumors were part of an experiment to elucidate the appearance of normal host mitoses among the tumor mitoses in the ascites. Normal mitoses are regularly found in tumors induced by SEWA cells that have had their malignant capacity attenuated by long-term exposure to in vitro conditions before implantation into animal hosts (Levan and Levan 1980a).

This experiment was undertaken with a sister clone to the 3:4 subclone of the preceding section. This clone, 3:8, had the same history as the 3:4 (cf. Figure 1). The subline 3:8R2, in which the second CM case was found, had been in tissue culture 357 days before being implanted in animals, which was 64 days more than 3:4R. Sixteen tumors resulted, 13 of which yielded good fixations. Total chromosome number per cell and the number of the smallest chromosomes were determined in 1,050 metaphases. It was known that the stem line number of subclone 3:8 is $s = 42$ and that the stem line includes one small marker of the mar3 type, i.e., without arms and forming a small strongly stained sphere in C-banding.

Of the 13 tumors examined, 11 were perfectly normal both in total chromosome number per cell and in number of mar3. Two tumors deviated. One had $s = 44$ with one mar3, the other had $s = 44$ but an elevated number of mar3 (average 2.1 per cell). These deviations were found on day 33 after implantation in vivo; on day 15 the values of both tumors had been in the normal range. Figure 10 clearly shows that the two deviating tumors fell outside of the cluster of normal tumors and thus were suspects for having regular CM chromosomes. Only one of them could be analyzed by C-banding, and both were unfortunately lost before their exceptional properties were recognized.

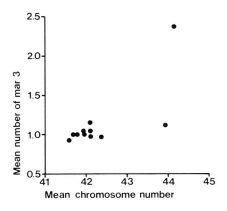

FIG. 10. Subclone 3:8R. Relation between chromosome number and number of mar3 in 11 tumors in which at least 25 metaphases were analyzed. Two tumors deviated from the cluster formed by the others. One of them (the uppermost) represented CM case 2.

The Appearance of the CM

One metaphase from case 2 is pictured in figure 11, before (a) and after (b) C-banding. Among the large chromosomes, mar1 and mar2 were found. The former agreed well with mar1 of subclone 3:4R; the latter was somewhat different from the usual mar2. Its proportions were the same as in the short form of mar2 (Figure 5), but in subclone 3:8 it had a small second arm formed as a couple of spheres, somewhat smaller in diameter than the chromatid. In Figure 11, the short arm is C-band negative distal of the strongly positive centromeric region.

Even in unbanded slides, differences were found among the smallest chromo-

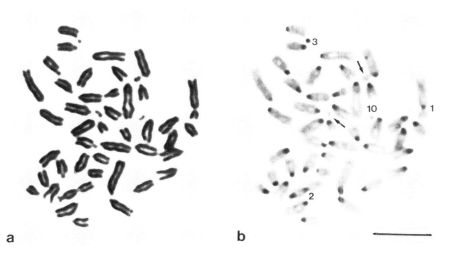

FIG. 11. Subclone 3:8R, metaphase before and after C-banding. Arrows = two small telocentric CM, Arabic numerals = C-band positive markers. Bar at bottom right = 10 μm.

TABLE 3. *Combinations of CM types in case 2*

	st	t		
No.	Medium-sized	Medium-sized	Small	No. cells
1	1	1	—	1
2	—	2	3	1
3	—	2	2	2
4	—	1	2	2
5	—	1	—	3
6	—	—	2	26
7	—	—	1	9

somes that were present in abnormally high numbers. One of them was smaller than the rest and strongly C-band positive. It was mar3 proper. One to three of them were slightly larger and completely C-band negative. They were genuine CM and were found in most cells. Another small marker, in size between mar3 and the small CM, had a tiny speck of C-band positive matter in its centromeric region and was therefore considered C-band positive, although no such chromosome had been seen before among the ordinary SEWA chromosomes. It was given the name mar10 and was present in most cells.

Fifty C-banded cells were analyzed in detail; 44 of them contained one to five CM. Only three morphologic types of CM were distinguished, medium-sized st and t and small t CM. The standard CM of case 1 was never seen here, and the st CM was seen only once. The different combinations of CM are recorded in Table 3, the predominant combination being that of two small t CM, found in 26 of the 50 cells. Altogether 85 CM were observed at the following frequencies: medium-sized st, 1 (1.2%); medium-sized t, 12 (14.1%); and small t, 72 (84.7%).

Plotting the number of CM per cell against the total chromosome number resulted in an apparently rectilinear correlation (Figure 12), indicating that the number of C-band positive chromosomes remained fairly constant and that

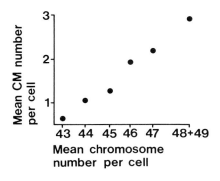

FIG. 12. CM case 2, relation between chromosome number and number of CM per cell.

the CM were mainly responsible for the variation in number. A similar calculation for case 1, comprising SEWA1R and Rec1, was presented previously (Levan et al. 1978, Figure 1, p. 15).

CM IN SEWA LINES NOT INVOLVED WITH THE SEWA1R PEDIGREE

The Material

All the materials so far dealt with were direct descendants of the SEWA1R line. We became concerned about whether the capacity to form CM was limited to the SEWA1R pedigree or whether it was a general capacity of all SEWA lines, inherited from the primary tumor. In order to approach this question, we performed C-bandings in the summer of 1979 in the four SEWA stock tumors that were briefly mentioned in the introduction (cf. Figure 1). Some further data on their history will be given here.

The four stocks represent different levels of in vivo transfer of the tumor. Stock 1, from which SEWA1R emanated, has been carried in our laboratory since November 29, 1975. When the material presented below and recorded in Table 4 was examined, it had been carried in 152–155 passages, corresponding to a period of 1,411–1,574 days. Before that, it had been maintained for 46 passages in Stockholm, which makes its total age approximately 1,800 days. After the SEWA1 cell culture had been set up on December 18, 1975, stock 1 has been carried without any involvement with the SEWA1R pedigree.

Stock 1 was implanted into mice of the A strain on July 19, 1976, and has since then been carried in strain A hosts as an independent line. The samples of this stock in Table 4 represent passages 159–163 in Lund, and the total age of the line can be estimated to be 1,900 days.

Stocks 3 and 4 were received in Lund in August 1979 after having been grown in Stockholm for 38 and 31 passages, respectively. Their ages at the time of the fixations of Table 4 were between 321 and 467 days. The four stocks available for the study of the CM were thus maintained in vivo for from less than 1 year to more than 5 years after the induction of the primary tumor.

Incidence and Appearance of the CM

In Table 4 the SEWA stocks have been listed in order of increasing age in serial in vivo transfer. (The chronology of Table 4 follows Figure 1 and thus differs from the chronology of Tables 1 and 2.) Table 4 summarizes results from 19 fixations. Slides were scanned for the presence of CM, and all cells with one or more CM were analyzed. Altogether 8,541 cells were examined, 178 of which (2.8%) had at least one CM. The incidence of cells with CM varied in the individual stock tumors from 0.6% in stock 1 in A to 4.2% in stock 4. These frequencies are of the same order as in the clonal derivatives

TABLE 4. Percentage incidence of CM types in four SEWA stocks

No.	SEWA stock no.	Passages	Days in animal transfer*	Number of cells Total	Number of cells With CM	Total CM analyzed	msm Big	msm Med.	msm Small	st Big	st Med.	st Small	t Big	t Med.	t Small	Acentric Big	Acentric Med.	Acentric Small	CM per CM-carrying cell
1	Stock 4	33–35	350	647	27	99	20.2	25.3	4.0	—	—	—	3.0	9.1	16.2	9.1	5.1	8.1	3.67
2	Stock 3	39–50	420	2,544	59	195	23.1	30.3	18.0	4.1	2.1	3.1	1.5	8.7	5.6	2.1	1.0	0.5	3.31
3	Stock 1	198–201	1,800	3,764	82	274	21.9	31.8	14.6	—	—	—	1.1	13.9	13.5	2.9	0.4	—	3.35
4	Stock 1 in A	223–227	1,900	1,586	10	30	16.7	30.0	—	—	—	—	—	26.7	26.7	—	—	—	3.00

* These values include the exact number of days of serial passage in Lund and a tentative estimate of the number of days previous to the period in Lund, allowing 100 days for passages 0 and 1 and 7 days for each subsequent passage.

of SEWA1R (Table 2). In the 178 CM-containing cells, 598 CM were found (3.37 per cell). As expected, the number of CM per cell was higher in the cells with chromosome numbers above the s region. The ratio of cells with 28 or more chromosomes to those with s chromosomes varied from 1.38 in stock 1 in A to 2.16 in stock 3, with an average of 1.67 for all four stocks. It was often difficult to decide in the material treated with colchicine whether or not a chromosome was centric, but in several cases the CM were obviously acentric. Therefore, in Table 4 unquestionable acentrics have been collected under a separate heading.

The finding of acentric chromosomes directs attention to a possible source of error: Acentric CM may easily be, and probably sometimes are, ordinary mouse chromosomes that have lost their centromere and centromeric C-band. That this type of "false" CM does occur is likely, especially since dicentrics and other signs of structural changes were not too rare. That the majority of CM were really centric is perfectly certain, however. Often the centromeres were directly seen, and in some cases, in which the CM were revealed in plain Giemsa preparations, the centromeres were indisputable. One such case is shown

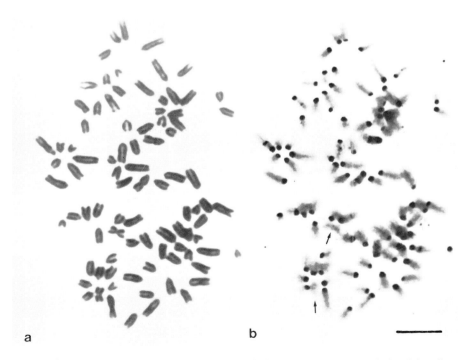

a b

FIG. 13. Stock 1, passage 201, double stem line cell photographed before and after C-banding. Note two CM similar to the standard CM of case 1. (This fixation corresponds to Table 4, No. 3.) Bar at bottom right = 10 μm.

TABLE 5. *Mean chromosome numbers in cells without and with CM*

Material	No CM		CM		t value	P
	Counts	Mean 2n	Counts	Mean 2n		
Stock 4	85	43.5 ± 0.12	11	46.5 ± 0.80	3.6	<0.001
Stock 3	442	43.9 ± 0.07	21	46.6 ± 0.39	6.6	<0.001
Stock 1	101	42.4 ± 0.10	21	45.9 ± 0.63	5.4	<0.001
Stock 1 in A	75	44.7 ± 0.14	8	47.8 ± 0.55	5.4	<0.001

in Figure 13, in which a metaphase with two CM of standard type was photographed before and after C-banding. A strong indication that most CM were of the same nature as those occurring with high incidence in cases 1 and 2 is the fact that cells with CM persistently had higher average chromosome numbers than cells without CM. Instances from the four stock tumors are given in Table 5, from which it is evident that a highly significant difference existed in all materials. The CM-carrying cells had an average of 2.6 to 3.4 chromosomes more than the cells without CM. Precisely as in all the materials described earlier, the CM were an addition to the stem line set of chromosomes.

The incidence of CM was characterized by another striking feature, viz. the preponderance of even numbers of CM per cell. If the four stocks were combined, the percentages of cells with one, two, three, four, five, or six CM were 22, 37, 7, 16, 1, and 6, respectively (11% having more than six CM), and all cells with odd numbers of CM per cell constituted 36% and those with even numbers 64% of the 178 cells with CM. This together with the observation that the CM of a certain cell were usually of the same morphologic type and were

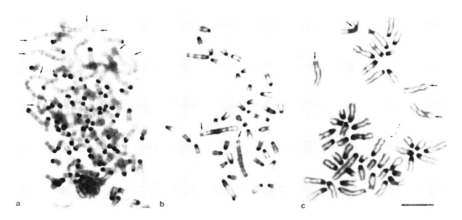

FIG. 14. a: Stock 4, passage 34 (Table 4, No. 1), b, c: Stock 3, passage 41 (Table 4, No. 2). Arrows = big, probably acentric CM. Bar at bottom right = 10 μm.

often located beside each other leads to the conclusion that the CM often undergo nondisjunction.

Some instances of cells with CM are given in Figures 14 and 15. Figure 14a was from a prometaphase with 91 chromosomes. The eight long CM were clearly arranged into four pairs, and they appeared to be acentric. In a cell with 48 chromosomes (Figure 14b), seven of which were outside the viewing field, the two largest chromosomes were CM and might have had a terminal centromere but were more likely acentric. The third cell (Figure 14c) had 49 chromosomes, including four, big, supposedly acentric CM (one of them was outside the viewing field and has been inserted to the left in the metaphase).

In Figure 15, four cells are represented, each of which contained one or more centric CM and all of which were from stock 3. The remarkable cell of Figure 15a had 50 chromosomes, including seven big metacentric CM, six of which were scattered outside the metaphase plate. In spite of the colchicine pretreatment, the centric chromosomes maintained their characteristic orientation, with the centromeres pointing towards the center and the arms arranged radially. The position of the CM indicated that they had been gathered in one corner of the plate and that the colchicine action had caused them to spread more extremely than the ordinary chromosomes. The other three cells of Figure 15 were from one fixation and were rather similar in appearance. They had 47, 44, and 52 chromosomes. The former two cells had each one CM, the latter cell had eight CM loosely arranged into four pairs. All the CM of these three cells were medium-sized msm chromosomes and resembled closely the standard CM of case 1.

CONCLUSIONS

The aim of this paper has been to survey work in our laboratory during the last 4 years dealing with double minutes and C-bandless chromosomes in the SEWA tumor of the mouse. Although the C-bandless chromosomes, the CM, show certain differences from what is considered the typical appearance of HSR, there is no doubt that they belong to the same category of cytogenetic phenomena. Because of the method we have used, searching for chromosomes without C-bands, the prevailing type of HSR in our study has been chromosomes formed by stretches of HSR and provided with a centromere. Only during the last year, after we started systematic G-band analyses, have we found another manifestation of HSR in several SEWA lines, namely ordinary C-band positive chromosomes with stretches of HSR forming an entire arm or a segment of an arm, in the latter case usually a distal segment. Both in cells with CM and in cells with HSR segments in ordinary chromosomes, there is a clear antagonism against DM: either HSR or DM, rarely both, may occur in the same cell.

After the first finding of CM chromosomes, it seemed natural to assume that they were unique and characteristic of one specific lineage of the tumor. Our work since then has made us modify this idea. We found a second case

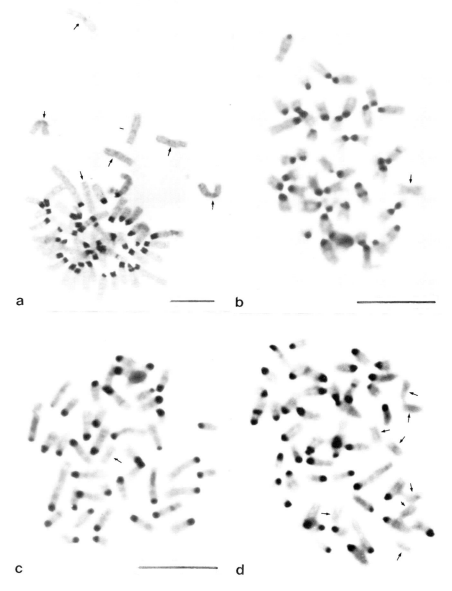

FIG. 15. Stock 3 (Table 4, No. 2). a: passage 43; seven big CM with submedian centromere; b-d: passage 38, medium-sized CM with submedian centromere, similar to the standard CM of case 1. In d are eight CM, vaguely arranged into four pairs. Bars = 10 μm.

of accumulation of CM chromosomes in a SEWA line. This line was separated from that of the first case by a long period in vitro, during which two clonings had been performed. Therefore, it seemed improbable that the CM of the second case had been directly carried over from the first case, and it appeared necessary to assume de novo formation of CM.

Next it was discovered that CM could be found even in SEWA materials that had never had any part in the ancestry of the earlier two cases. CM were found in SEWA lines that differed in age from less than 1 to more than 5 years from the induction of the tumor. In these circumstances, CM were always rare; one had to scan several hundreds of C-banded mitoses to detect a few cells with CM. These cells had interesting characteristics: Most of them contained more than one CM, and two, four, and six CM were more frequent than one, three, and five, respectively. Also, it was the rule that all CM of one cell belonged to the same morphologic type. These data have been interpreted as indicating that in any SEWA population a minority of cells belonging to one or a few clonal lines, carry CM, which often undergo nondisjunction at mitosis. The fact that this minority of CM-carrying cells are maintained in all SEWA lines studied shows in our opinion that the CM carry some viability advantage, and this is also borne out by the fact that twice during our 4 years of experimentation we have found sublines in which all or most cells have been invaded by CM. This also indicates that under specific, unknown conditions the CM constitute a more favorable form of amplification than the DM of the genic information embodied in the extra DNA of CM and DM. Even though the functional difference between DM and CM must be looked for at the molecular level, there is one mechanical factor that may favor CM. Their mitotic stability should be much greater than that of the DM. This should be even more the case with HSR attached to ordinary chromosomes, as has actually been shown recently by Kaufman et al. (1979). In their case, the resistance to methotrexate of amplified Chinese hamster cells remained stable in cells with HSR, but in cells with DM the resistance was lost at the same time as the DM.

The most significant question raised by the present experiments, whether DM and CM may be formed de novo in cells devoid of them or whether they are permanently present in a part of the population, is still moot. The strongest evidence for de novo formation is their presence after clonings of populations that to begin with had a very low incidence of them. The alternative view extended to its extremity would require that the CM (or DM) only originate once and that all CM occurring later stem from this early predecessor. This would also mean that the scarce cells with CM in the various SEWA stock populations constitute the real "germ line" of the population. They may spend most of their life out of the limelight; when the population is reduced to one single cell at clonings, the only cell that fills the requirements is the one with DM and CM. Put into such extreme terms, this alternative seems rather unrealistic. Probably the actual situation is somewhere in between: DM, CM, and HSR may disappear and again reappear. This implies that the SEWA cells carry in their genotype—intra- or extra-chromosomally—the machinery necessary for the production of DM and HSR.

The functional significance for the SEWA tumor of the DNA carried by DM and CM is still completely unknown. In Chinese hamster cells, the work of Biedler and Spengler (1976) and of Nunberg et al. (1978) has related the HSR to the induction of resistance to an antimetabolite, and Kaufman et al. (1979) showed that DM filled the same function. Since the absolute majority of cases with DM and HSR have been found in malignant cells, it seems reasonable to assume that their function is related to some challenge that malignant cells are exposed to. One fact in support of this idea is that DM in some cases have been found to disappear when their carrier cells are transferred from in vivo to in vitro conditions (Donner and Bubeník 1968, Levan et al. 1977). In the latter of these two investigations, the DM-carrying cells were again recovered when the tissue culture was reimplanted in vivo.

ACKNOWLEDGMENTS

Financial support of this investigation from the Swedish Cancer Society, the John and Augusta Persson Foundation for Medical Research, and the Cancer International Research Co-Operative (CANCIRCO) is gratefully acknowledged. We also wish to thank Professor George Klein for inbred animals and cell samples of the SEWA tumor from different stages of in vivo growth, and Professor Hans Olov Sjögren for data on the early history of the SEWA tumor.

REFERENCES

Biedler, J. L., and B. A. Spengler. 1976. Metaphase chromosome anomaly: Association with drug resistance and cell-specific products. Science 191:185–187.

Donner, L., and J. Bubeník. 1968. Minute chromatin bodies in two mouse tumours induced in vivo by Rous sarcoma virus. Folia Biol. 14:86–88.

Fenyö, E. M., F. Wiener, G. Klein, and H. Harris. 1973. Selection of tumor-host cell hybrids from polyoma virus- and methylcholanthrene-induced sarcomas. J. Natl. Cancer Inst. 51:1865–1875.

Hellström, I., K. E. Hellström, and H. O. Sjögren. 1962. Further studies on superinfection of polyoma-induced mouse tumors with polyoma virus in vitro. Virology 16:282–300.

Jonasson, J., S. Povey, and H. Harris. 1977. The analysis of malignancy by cell fusion. VII. Cytogenetic analysis of hybrids between malignant and diploid cells and of tumours derived from them. J. Cell Sci. 24:217–254.

Kaufman, R. J., P. C. Brown, and R. T. Schimke. 1979. Amplified dihydrofolate reductase genes in unstable methotrexate-resistant cells are associated with double minute chromosomes. Proc. Natl. Acad. Sci. USA 76:5669–5673.

Levan, A., K. Fredga, and A. A. Sandberg. 1964. Nomenclature for centromeric position on chromosomes. Hereditas 52:201–220.

Levan, A., and G. Levan. 1980a. The induction of host cell mitoses in a transplantable ascites tumor. Cytogenet. Cell Genet. 26:76–84.

Levan, A., and G. Levan. 1980b. Large double minutes with ring-shape and rod-shape. Hereditas 92:259–265.

Levan, A., G. Levan, and N. Mandahl. 1978. A new chromosome type replacing the double minutes in a mouse tumor. Cytogenet. Cell Genet. 20:12–23.

Levan, G., N. Mandahl, B. O. Bengtsson, and A. Levan. 1977. Experimental elimination and recovery of double minute chromosomes in malignant cell populations. Hereditas 86:75–90.

Nunberg, J. H., R. J. Kaufman, R. T. Schimke, G. Urlaub, and L. A. Chasin. 1978. Amplified dihydrofolate reductase genes are localized to a homogeneously staining region of a single chromosome in a methotrexate-resistant Chinese hamster ovary cell line. Proc. Natl. Acad. Sci. USA 75:5553–5556.

Sjögren, H. O., I. Hellström, and G. Klein. 1961. Transplantation of polyoma virus-induced tumors in mice. Cancer Res. 21:329–337.

Genes, Chromosomes, and Neoplasia, edited by
Frances E. Arrighi, Potu N. Rao, and Elton Stubblefield.
Raven Press, New York © 1981.

Amplification of Mammalian Ribosomal RNA Genes and Their Regulation by Methylation

Orlando J. Miller,*·†·‡ Umadevi Tantravahi,* Richard Katz,§
Bernard F. Erlanger,‡·§ and Ramareddy V. Guntaka‡·§

Department of Human Genetics and Development, Department of Obstetrics and
Gynecology,† Cancer Center/Institute for Cancer Research,‡ Department of Microbiology,§
College of Physicians and Surgeons, Columbia University, New York, New York 10032*

In most organisms, the 18S and 28S ribosomal RNA (rRNA) genes are present in multiple copies arranged in tandem arrays, with the transcribed portions separated by non-transcribed spacer regions (Beyer et al. 1979). When these genes are transcriptionally active, nucleoli are formed. The location of rRNA gene clusters (rDNA) in the chromosome set can be detected most directly by in situ hybridization, using radiolabeled rRNA. Such studies show that rDNA can be concentrated at a single chromosomal site in the genome or can occur at multiple sites (Hsu et al. 1975). Sometimes this site is marked by a secondary constriction, called the nucleolus organizer region (NOR), or simply the nucleolus organizer, because of its close physical association with the nucleolus (McClintock 1934). The NOR fails to stain by the Feulgen reaction or standard chromatin stains, indicating the presence of a very low concentration of DNA. The chromosome has a different ultrastructure in the NOR, more or less intermediate between the active lampbrush configuration of the kinetochore and the inactive configuration of the rest of the chromosome (Hsu et al. 1967, Goessens and Lepoint 1974).

A relationship between the presence of a secondary constriction and rRNA gene activity has long been clear. Nawaschin (1934) demonstrated the phenomenon of nucleolar dominance in interspecific hybrids of the plant genus *Crepis.* He found that secondary constrictions and their associated nucleoli occurred on chromosomes of only one of the two species in the hybrid. Nucleolar dominance has been observed in interspecific *Drosophila* hybrids (Durica and Krider 1977) and in interspecific *Xenopus* hybrids, in which repression of nucleolus organizer activity, i.e., disappearance of secondary constrictions, is associated with suppression of transcriptional activity of the rRNA genes of one species (Honjo and Reeder 1973, Cassidy and Blackler 1974). Unfortunately, in some species active rRNA gene clusters are not usually visible as secondary constrictions (Goodpasture and Bloom 1975).

The state of activity of rRNA genes can be determined by another cytological technique with more accuracy than that provided by simply looking at the size of the secondary constrictions. Howell et al. (1975) discovered that NORs can be stained with ammoniacal silver. Goodpasture and Bloom (1975) and Hsu and his associates (1975) showed the close correspondence between sites of Ag staining and sites of hybridization to radiolabeled rRNA probe. Ag staining is not affected by removing DNA and RNA from the chromosomes but is abolished by proteolytic enzymes, indicating that a protein or proteins are responsible for reducing Ag^+ to Ag. The proteins involved are related to rRNA transcriptional activity, as shown by suppression of Ag staining of metaphase chromosomes of one species in interspecific mammalian hybrids (Miller et al. 1976a,b) and associated absence of cytoplasmic 28S rRNA of that species (Croce et al. 1977). Both of these are due to suppression of transcription of the 45S precursor of rRNA (Perry et al. 1979), not to loss of all the rDNA of the species whose rRNA genes are inactive (Arnheim and Southern 1977).

Although Ag staining of NORs on metaphase chromosomes reflects the functional activity of the rRNA genes during the preceding interphase (rRNA transcription ceases during cell division), the amount of Ag staining generally parallels the amount of rDNA in each site (Warburton et al. 1979). Occasionally there is an exception to this rule and a chromosome has much less Ag staining than expected from the large amount of rDNA present. As a result, the frequency of NOR associations, which also depends on prior rRNA gene activity, is more closely associated with the amount of Ag staining than with the amount of rDNA (Miller et al. 1977).

The amount of rDNA is highly variable from chromosome to chromosome, presumably as a result of unequal crossing-over. This variation is readily apparent even in species with multiple NORs, such as the mouse (Elsevier and Ruddle 1975) and the human (Warburton et al. 1976). In one family, a human chromosome 14p+ has been observed that has eight times as much rDNA as the average amount in the other NORs (Figure 1) (Miller et al. 1978). If we assume there are about 200 copies of the rRNA gene per cell, distributed over 10 chromosomes with an average of 20 per chromosome, this 14p+ chromosome would have about 160 copies and the individuals with this variant might have somewhat less than twice the average gene multiplicity. Interestingly, this variant chromosome (Figure 2) is present unchanged in seven individuals, all normal, in three generations of this family. A similar example of a human chromosome 14 with amplified rDNA has been reported (Lau et al. 1979), and numerous cases are known in which the short arm of one or another of the acrocentric NOR chromosomes has the same appearance as the 14p+ in these two families.

In both families, the amplified rRNA genes are present in a G+C-rich, largely homogeneously staining region (HSR) that differs from the HSRs containing amplified dihydrofolate reductase genes by having one or more unstained secondary constrictions, most commonly seen at each end. These constrictions can be stained with Ag, indicating that they are sites of active rRNA genes (Figure 2). However, the amount of Ag staining is less than expected, considering the

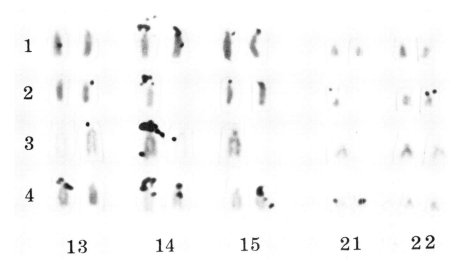

FIG. 1. The acrocentric chromosomes from three cells of an individual with a 14p+ variant. Autoradiographic silver grains after in situ hybridization to radiolabeled probe demonstrate the increased number of rRNA coding sequences in the short arm of the 14p+ variant. (Reproduced from Miller et al. 1978, with permission of Springer-Verlag.)

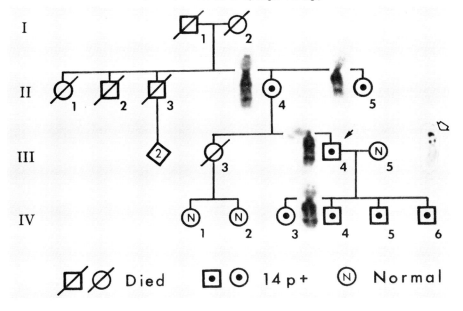

FIG. 2. Pedigree of the 14p+ variant human chromosome. The G-banded chromosomes show one clear secondary constriction and sometimes a smaller one. Silver staining (open arrow) is restricted to the area of these constrictions. (Modified from Miller et al. 1978, with permission of Springer-Verlag.)

amount of rDNA. Most of the amplified rRNA genes appear to have been inactivated and to be present in an altered conformation, producing condensed chromatin of the HSR. Thus, the number of active gene copies is probably little greater than the number present in individuals without such an amplified rRNA gene cluster. It is unclear whether the HSR contains genes other than rDNA; its large size, relative to the modest increase in rRNA gene multiplicity, suggests that it does. If so, then either unequal sister chromatid exchange can lead to an increase in more than one family of DNA sequences, or the HSRs have arisen by a different mechanism, e.g., integration of extrachromosomal DNA.

It is possible that rRNA gene dosage is kept within fairly strict limits. In species in which this has been studied, the presence of too few gene copies has a deleterious or even lethal effect, e.g., the "bobbed" mutant in *Drosophila melanogaster* (Ritossa 1976) and the "anucleolate" mutant in *Xenopus laevis* (Wallace and Birnstiel 1966). The presence of too many gene copies does not appear to be deleterious, perhaps in some cases because there is a mechanism for inactivating all copies above a certain theshold level. The evidence to be reviewed here indicates that, if such a mechanism is operating, it involves methylation of cytosine residues in the DNA.

5-Methylcytosine (5MeC) is the only modified base in the DNA of most higher eukaryotes. It is found mainly (up to 90% of the total) in the CpG dinucleotide (Grippo et al. 1968, Harbers et al. 1975), and reaches concentrations of about one mole percent in the DNA of various mammals (Vanyushin et al. 1970). It is considerably more concentrated in the non-transcribed, highly repetitive satellite DNA sequences of heterochromatin. This has been demonstrated by chemical methods (Grippo et al. 1968) and by immunocytochemical methods using antibodies to 5MeC and either an indirect immunofluorescence (Miller et al. 1974, Schreck et al. 1977) or immunoperoxidase technique (Lubit et al. 1974). Insects have extremely low levels of 5MeC in the DNA of adults and early embryos (Adams et al. 1979, Argyrakis and Bessman 1963); however, the transcriptionally inactive bands of polytene chromosomes in *D. melanogaster* and various other dipteran species contain abundant 5MeC residues as shown immunocytochemically (Eastman et al. 1980). In both mammals and birds, evidence is accumulating that indicates that specific host or viral genes are enriched in 5MeC in cells or tissues in which the genes are inactive but not in cells in which the genes can be expressed. This is true for avian β-globin genes and egg protein genes (McGhee and Ginder 1979, Mandel and Chambon 1979) and for various viral DNA sequences that are methylated only when they are in nonproductive mammalian cells: herpesvirus (Desrosiers et al. 1979), avian sarcoma virus (Guntaka et al. 1980), and adenovirus (Sutter and Doerfler 1980). Furthermore, the transcriptionally active, amplified rRNA genes in amphibian *(X. laevis)* oocytes are unmethylated, while the somatic rRNA genes are methylated to the extent that 13% of cytosine residues are replaced by 5MeC (Dawid et al. 1970, Bird and Southern 1978). In view of these findings, it would not be surprising if there were enhanced methylation of the inactive, amplified rRNA genes in the condensed chromatin of the HSR between the Ag-stained secondary constrictions (AgNORs) marking the sites of active genes on the 14p+ chromosome. Using an indirect immunoperoxidase technique and specifically purified antibodies, we have shown that DNA of the HSR as a whole is highly enriched in 5MeC (Figure 3 and Tantravahi et al. 1980). Similar results have been obtained in a 22p+ chromosome with an even larger HSR and small AgNORs (R. Bernstein, B. Dawson, J. Griffiths, and U. Tantravahi, unpublished data).

More extreme examples of rDNA amplification than those described thus far have been observed in cancer cells. Cells of the Rous sarcoma virus–induced rat sarcoma cell line XC have a threefold to fourfold increase in the amount of rDNA. The cells contain two different HSRs; one of these has an Ag-stained NOR near each end, while the other does not. The DNA in one HSR is rich in 5MeC (Tantravahi et al. 1980). Cells derived from the H4-IIE-C3 rat hepatoma line (H4) show a still higher degree of rRNA gene amplification (Figure 4 and Miller et al. 1979b). Both filter hybridization and quantitative autoradiography after in situ hybridization to metaphase chromosomes indicate that there is at least a 10-fold increase in DNA coding for 18S and 28S rRNA. The

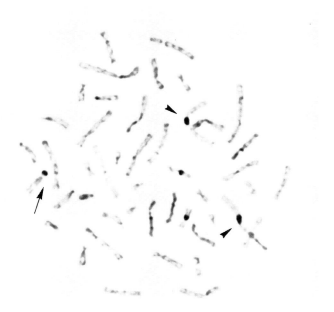

FIG. 3. Metaphase cell stained by an indirect immunoperoxidase technique to show major sites of anti-5MeC binding. The arrow points to the 5MeC-rich short arm of the 14p+ variant. Arrowheads point to the 5MeC-rich satellite DNA on chromosome 1.

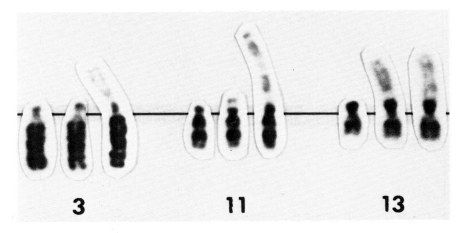

FIG. 4. G-banded chromosomes from an H4 cell showing DSR on the short arms of chromosomes 3, 11, and 13. (Reproduced from Miller et al. 1979b, with permission of Springer-Verlag.)

increase is so great that grain density after the usual exposure times used for in situ hybridization are far too high to permit counting (Figure 5); exposure of only a few hours is required for quantitative study. The amplified rDNA is present in multiple HSR-like regions, which we have called differentially staining regions (DSRs) because they contain unstained secondary constrictions (sometimes five or more) and are thus not homogeneously stained (Figure 4). The secondary constrictions, which occur at both regular and irregular intervals along the length of the DSRs, stain with Ag (Figure 6), indicating the presence of active rRNA genes at restricted sites along the DSRs. DSRs are present on the short arms of chromosomes 3 and 11, i.e., usual sites of rDNA in the rat (Tantravahi et al. 1979), and in two unusual sites: the short arm of chromosome 13 and the long arm of an unidentified small chromosome. On the average, the DSRs make up about 5% of the total chromosome complement, but they vary a great deal in number and length from cell to cell. If we assume there are about 200 rRNA gene copies in the ACl rat, from which H4 cells were derived, with an average of 33 per cluster on the three pairs of NOR chromosomes, there are at least 2,000 copies per H4 cell, with perhaps 100 in three normal NORs and more than 600 in the average DSR. The DSRs are so variable in length, they probably contain anywhere from about 50 to 1,500 rRNA gene copies.

The DSRs in H4 cells appear to be quite unstable, with acentric fragments, rings, chromatid breaks, and unequal sister chromatid interchanges common (Figure 7). Most of these lead to loss of DSR material, but the last type can lead to an increase. As a result, the amplification process can be progressive. When we first studied the H4 line (Miller et al. 1972) there was a single DSR. In a more recent report on another subline of H4, three different DSRs were

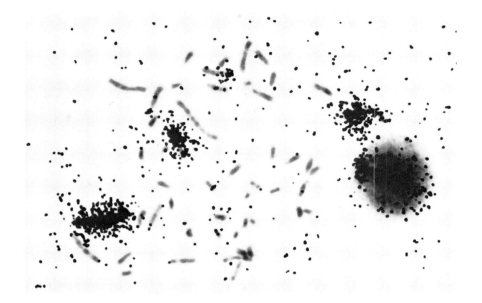

FIG. 5. H4 metaphase spread after hybridization in situ to ^{125}I-labeled 18S and 28S rRNA. Three sites show extremely heavy labeling. (Reproduced from Miller et al. 1979b, with permission of Springer-Verlag.)

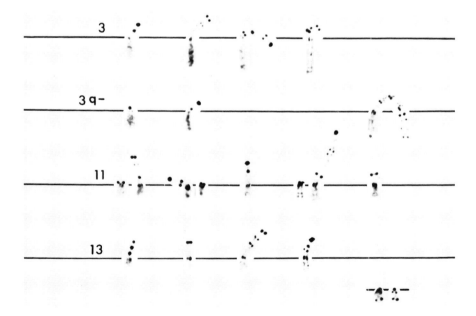

FIG. 6. DSR chromosomes from five H4 cells stained with Ag to show sites of active rRNA genes. (Reproduced from Miller et al. 1979b, with permission of Springer-Verlag.)

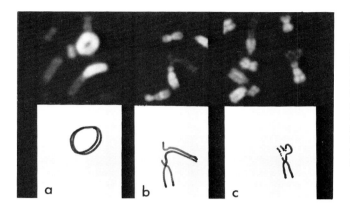

FIG. 7. Unstable configuration of DSRs in H4 cells. The DSR ring chromosome is brightly fluorescent by acridine orange R-banding. The other DSRs show faint quinacrine fluorescence. Interpretive drawings are given for each configuration.

shown (Mullen and Barnet 1976), and our still more recent report described an even larger array of DSRs (Miller et al. 1979b). The increase of DSRs in the H4 cell line despite their instability may be the result of in vitro selection for cells with an increase in active rRNA genes.

Before discussing this possibility, it may be worthwhile to compare the mechanisms of rRNA gene amplification that appear to be operating in meiosis and in mitosis. Unequal crossing-over in meiosis may amplify only the rRNA gene, although the widespread distribution of intermediately repetitive sequences provides a basis for the more general amplification apparently present in the 14p+ cases described earlier. Unequal sister chromatid exchange can also amplify the redundant gene (Tartof 1974) or perhaps a longer sequence. What we have seen in H4 cells, breakage and intrachange between chromatids, can amplify non-rDNA as well as rDNA.

Although the amplification of some non-rRNA gene might be responsible for the selective advantage that H4 and similar cancer cells have in cell culture, we think it more likely that selection is based on the presence of an increased number of active rRNA genes. The increase does not have to be very great. Thus, the H4 cells are larger than control hepatocytes, with larger nuclei and nucleoli, but the increase is slight when contrasted to the 10-fold increase in rRNA genes. The rate of transcription of 45S precursor rRNA is also not much increased (J. Lipszyc, D. Engelhardt, and O. J. Miller, unpublished observations), and the total number of Ag-stained NORs is only slightly increased over that in control hepatocytes (Miller et al. 1979b). The DNA in the HSRs lying between the Ag-stained NORs is more highly condensed than that in the NORs (secondary constrictions) and is presumed to contain inactive rRNA genes. This DNA is highly methylated, as shown by binding of anti-5MeC after denaturing the DNA by either photo-oxidation or UV irradiation (Figure 8 and Tantravahi et al. 1980).

In contrast to the high degree of methylation of the DNA in HSRs containing inactive rRNA genes, the content of 5MeC is very low in the DNA of HSRs containing active dihydrofolate reductase genes (Tantravahi et al. 1980). Since the DNA in the latter class of HSRs is also G+C rich, and 5MeC content is generally proportional to G+C content, it seems likely that the striking difference in the extent of methylation in the two types of HSRs is causally related to the difference in gene activity, with a high level of methylation of the inactive rRNA genes but not of the active folate reductase genes or, by an extension of this reasoning, of the active rRNA genes. However, the concentration of DNA is so much lower in secondary constrictions than in the adjacent chromosome arms that the absence of detectable anti-5MeC binding in the constrictions does not rule out the presence of abundant 5MeC in the active rDNA found there. Furthermore, rDNA probably makes up only a small percentage, perhaps no more than 20%, of the DNA in the DSRs of H4 cells (Miller et al. 1979b). The extent of methylation of any of the rDNA in the DSRs is therefore uncertain despite cytological evidence of extensive anti-5MeC binding to the DNA.

In order to resolve this problem, we have analyzed the DNA of various rat

FIG. 8. Part of an H4 metaphase spread stained (a) by G-banding to show the DSRs (arrows), and (b) by an indirect immunoperoxidase technique to demonstrate that the DSRs (arrows) are major sites of anti-5MeC binding.

cell lines, including H4, after digesting the DNA with various restriction endonucleases. The DNA fragments produced by digestion with one or two restriction enzymes have been separated according to molecular weight by agarose gel electrophoresis. The fragments are analyzed directly in the gel by ethidium bromide staining and autoradiography, or are transferred to nitrocellulose filters for hybridization to radiolabeled 18S and 28S rRNA or cDNA probes. After EcoRI digestion and agarose gel electrophoresis of DNA from H4 cells (but not of DNA from control rat hepatocytes), ethidium bromide fluorescence reveals fragments of amplified DNA in five molecular weight classes. The DNA in three of these classes, 11.0, 6.6, and 4.6 kilobase pairs (kb) in size, contains 18S or 28S rRNA coding sequences (Figure 9). HindIII digestion of H4 DNA produces fragments of amplified DNA of three sizes, and two of these, 10.2 and 5.9 kb in size, contain 18S or 28S coding sequences. Thus, most of the amplified DNA detectable by restriction enzyme analysis of H4 DNA is rRNA coding sequences and its flanking nontranscribed spacer sequences.

Digestion of H4 DNA with both EcoRI and HindIII endonucleases leads to a reduction in size of some of the amplified fragments (Figure 9). Because of limited heterogeneity in the restriction fragments produced by digestion of H4 or control DNA by EcoRI, HindIII, BamHI, or SacI enzymes, it has been possible, using single and double digests, to construct a restriction map of the most common repeating units in amplified and unamplified rDNA. A partial

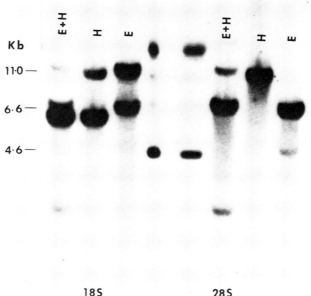

FIG. 9. Autoradiograph of EcoRI, HindIII, and EcoRI plus HindIII restriction endonuclease fragments after hybridization to radiolabeled 18S or 28S probe. The numbers refer to the sizes (in kb) of the fragments.

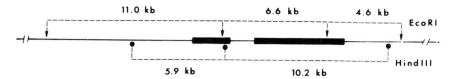

FIG. 10. Partial restriction map of the major rDNA repeat unit found in both H4 and K22 DNA.

map, based only on the results with EcoRI and HindIII, is shown in Figure 10.

Residues containing 5MeC are abundant in the various restriction fragments that contain the amplified 18S and 28S rRNA coding sequences in the DNA of H4 cells but are not in those of control DNA. This has been demonstrated autoradiographically, using DNA from cells grown in the presence of [*methyl*-^{14}C]methionine to label the 5MeC (Figure 11), and by the resistance of the H4 DNA to cleavage by the methylation-sensitive endonucleases HpaII and

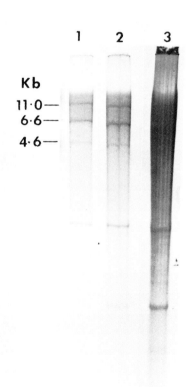

FIG. 11. Autoradiographs of EcoRI restriction fragments after agarose gel electrophoresis. Lane 1: 1 µg of H4 DNA. Lane 2: 2 µg of H4 DNA. Lane 3: 20 µg of K22 (control) DNA. Note the absence of methylation of specific 11.0-, 6.6-, and 4.6-kb fragments in the control DNA. The highly methylated bands of smaller size are satellite DNA.

HhaI. HpaII cleaves both strands of DNA within the sequence 5 ' C-C-G-G 3 ', but only if the internal C is not methylated (Mann and Smith 1977). Its isoschizomer, MspI, on the other hand, will cleave DNA in the C-C-G-G sequence whether or not the internal C is methylated (Waalwijk and Flavell 1978). We have compared the effect of HpaII and MspI on the EcoRI restriction fragments that contain 18S and 28S rRNA coding sequences in H4 and control DNA (Figure 12). Both HpaII and MspI cleave all the EcoRI fragments of control DNA containing ribosomal sequences, indicating that the C-C-G-G sequences are not methylated. MspI also cleaves these fragments in H4 DNA, but HpaII fails to cleave the bulk of the DNA fragments of any of the three sizes, indicating that all the C-C-G-G sequences in these fragments are methylated in H4 DNA. A heterogeneous 2.1-kb fragment containing ribosomal coding sequences, present in H4 DNA but not in control DNA, is totally cleaved by both HpaII and MspI, indicating that it contains C-C-G-G sites that are not

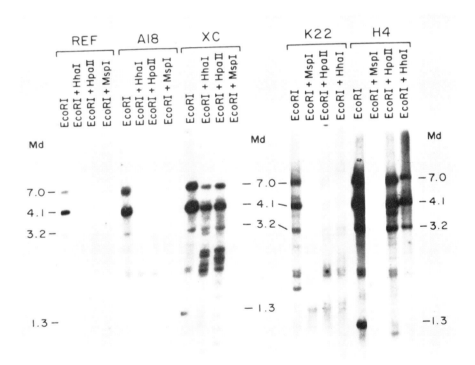

FIG. 12. Distribution of restriction enzyme fragments containing 18S or 28S coding sequences in DNA from rat embryo fibroblasts (REF), adenovirus-transformed rat fibroblasts (A18), XC cells, K22 hepatocytes, and H4 cells. The molecular sizes in megadaltons (Md) correspond to 11.0, 6.6, 4.6, and 2.1 kb.

methylated. The relative amount of rDNA in this size class suggests that these may be the active rRNA genes.

A different methylation-sensitive enzyme, HhaI, cleaves DNA in the recognition sequence 5 ' G-C-G-C 3 ', but only if the internal C is not methylated. Digestion of EcoRI fragments of H4 DNA with HhaI fails to cleave a large proportion of the three major fragments containing ribosomal sequences but does cleave the 2.1-kb fragments (Figure 12). After digestion of H4 DNA with HhaI alone, more than 90% of the ribosomal sequences are in high molecular weight fragments, resistant to HhaI digestion (Figure 13). A small proportion of the sequences are in DNA fragments only 4.5 kb in size. In DNA from control hepatocytes, none of the ribosomal coding sequences are in the uncleaved fraction; instead they are found in fragments 4.5 kb in size, and these fragments are less abundant in control DNA than in H4 DNA. If these represent the active rRNA genes, such genes are somewhat more abundant in the H4 cancer cells than in the normal hepatocytes. Autoradiography of the agarose gels demonstrates that 5MeC residues are concentrated in the high molecular weight DNA that is uncleaved by HhaI. These fragments are detectable after autoradiographic exposures of only a few days. The 4.5-kb fragments, on the other hand, are not detectable autoradiographically even after a 6-month exposure, indicating the extremely low level of methylation of what are presumed to be the active ribosomal genes.

The presence of ribosomal coding sequences in a 2.1-kb EcoRI restriction

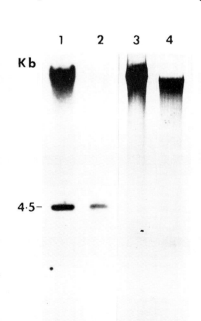

FIG. 13. HhaI restriction fragments of H4 (lanes 1 and 3) and K22 control DNA (lanes 2 and 4). Lanes 1 and 2: 18S or 28S rRNA coding sequences revealed by hybridization of Southern transfers to radiolabeled probe. Lanes 3 and 4: sites of highly methylated DNA revealed by autoradiography, using DNA from cells grown in [*methyl*-14C] methionine.

fragment in H4 DNA but not in K22 rat hepatocyte DNA is probably an indication of heterogeneity in the DNA. This view is supported by studies of the DNA from other cells. The 2.1-kb fragment is present in DNA from cells of the XC Rous sarcoma virus–induced rat tumor cell line, but absent from a rat embryo fibroblast line and an adenovirus-transformed rat cell line. This fragment has been traced to a specific chromosome site with active rRNA gene clusters, as shown by Ag staining. When metaphase chromosomes are isolated from XC cells and separated by density gradient centrifugation, the dot-like chromosomes are concentrated in the lightest fractions at the top of the gradient, and so is the DNA containing the 2.1-kb fragment with its ribosomal sequences (R. Katz et al., unpublished data). It should be possible to clone this fragment in bacterial plasmids and use the resultant probes to determine the chromosomal site, or sites, of this fragment in H4 DNA and in various rat strains.

Another interesting feature of rDNA methylation has been revealed by studies of XC DNA. When EcoRI restriction fragments are digested with HpaII or HhaI, there is partial cleavage of the two major fragments containing ribosomal sequences. However, not all are cleaved into very small fragments; some are cleaved into fragments ranging from about 2 kb to 3 kb in size (Figure 12). Such large fragments are not seen after digestion with MspI, that is, the extent of methylation of C-C-G-G, and presumably G-C-G-C, sequences is less in XC DNA than in H4 DNA. The significance of this is unclear, other than the implication that perhaps very few of these sites need be methylated to inactivate the rRNA genes. Mapping the sites of methylation in cell lines like XC may disclose the position of important regulatory sites controlling the expression of these genes.

A model illustrating the proposed relationship between rRNA gene activity and DNA methylation in amplified rDNA is shown in Figure 14. Active genes are unmethylated while inactive genes are highly methylated, both in the coding sequences and in the spacer sequences in the nontranscribed as well as the transcribed. It is unclear whether gene inactivation precedes or follows DNA methylation, but the association itself is striking. It is also uncertain whether gene amplification precedes or follows DNA methylation.

Several characteristics of the methylation of rRNA genes warrant consideration. Methylation is limited to the amplified rRNA gene cluster and does not inactivate all copies of the gene in that cluster. This indicates rather precise regulation of the process of methylation. The methylated state is stable through both mitotic and meiotic divisions, supporting the extensive biochemical evidence for semiconservative copying of a methylation pattern, once it is established (Bird 1978). If amplification precedes DNA methylation, there may be a threshold rRNA gene dosage for the induction of a DNA methylase that produces a new (inactivating) pattern of DNA methylation. This might account for the maintenance of multiple rDNA sites in some species, i.e., the threshold for DNA methylation may be too low for a sufficient number of active gene copies to be assembled on a single chromosome pair in a species with multiple sites.

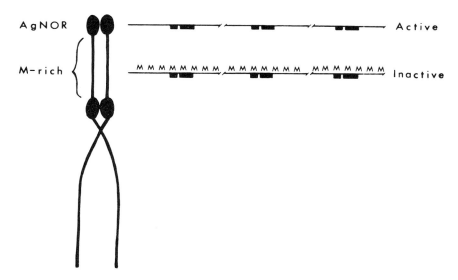

FIG. 14. Model illustrating the relationship between rRNA gene activity and the extent of DNA methylation. M signifies 5MeC residues in the DNA.

On the other hand, DNA methylation may precede rRNA gene amplification, leading to a functional deficiency of active ribosomal genes. In *Drosophila melanogaster,* the presence of too few copies of these genes can lead to a compensatory increase in somatic rDNA (Procunier and Tartof 1978). A similar mechanism may operate in mammals, although it has never been demonstrated.

Amplification of rRNA genes may thus be either a cause or an effect of DNA methylation. In either case, we think that amplification of rRNA genes may be fairly common in cancer cells, and may play a role in carcinogenesis. Many tumors and tumor cell lines have long been known to have large nucleoli, increased numbers of cytoplasmic ribosomes, and a high growth rate (Busch et al. 1977). It is unknown whether this is frequently or only rarely due to amplification of rRNA genes. A subline of the Yoshida rat sarcoma arose in 1960–62 (Matsushima and Yosida 1971) that had markers whose appearance (HSR-like) and behavior (involvement in satellite associations) suggest they contain amplified rDNA. Transmissible venereal sarcomas in dogs have an HSR whose staining reactions (Oshimura et al. 1973, Sasaki et al. 1974) are compatible with its containing amplified rDNA. Mice infected with murine erythroleukemia virus (Friend and Rauscher) show enhancement in number and size of secondary constrictions (Tsuchida and Rich 1964, Sofuni et al. 1967). Mouse erythroleukemia cell lines sometimes show enlargement of a specific NOR (Miller et al. 1979a). Further studies are necessary to clarify the importance of rRNA gene amplification in the origin or progression of malignant tumors.

ACKNOWLEDGMENTS

We wish to thank Drs. Lawrence Chasin and I. Bernard Weinstein for providing cell lines with amplified dihydrofolate reductase genes, Dr. B. Lubit for providing specifically purified anti-5-methylcytidine, and Dr. J. Beard for providing reverse transcriptase. This investigation was supported in part by Grant Number CD-66A from the American Cancer Society, Grant Number I-366 from the March of Dimes-Birth Defects Foundation, Grant Number AI06860 from the National Institute for Arthritis and Infectious Diseases, Grant Number GM25193 from the National Institute for General Medical Sciences, Grant Number CA27655 and Contract Number 1CP71055, awarded by the National Cancer Institute, Department of Health, Education and Welfare.

REFERENCES

Adams, R. L. P., E. L. McKay, L. M. Craig, and R. H. Burdon. 1979. Methylation of mosquito DNA. Biochim. Biophys. Acta 563:72–81.

Argyrakis, M. P., and M. J. Bessman. 1963. Analysis of the base composition of the DNA from *Drosophila*. Biochim. Biophys. Acta 72:122–124.

Arnheim, N., and E. M. Southern. 1977. Heterogeneity of the ribosomal genes in mice and man. Cell 11:363–370.

Beyer, A. L., S. L. McKnight, and O. L. Miller, Jr. 1979. Transcriptional units in eukaryotic chromosomes, *in* Molecular Genetics Part III. Chromosome Structure. J. H. Taylor, ed. Academic Press, New York, pp. 117–175.

Bird, A. P. 1978. Use of restriction enzymes to study eukaryotic DNA methylation: II. The symmetry of methylated sites supports semi-conservative copying of the methylation patterns. J. Mol. Biol. 118:49–60.

Bird, A. P., and E. M. Southern. 1978. Use of restriction enzymes to study eukaryotic DNA methylation: I. The methylation pattern in ribosomal DNA from *Xenopus laevis*. J. Mol. Biol. 118:27–47.

Busch, H., N. R. Ballal, R. K. Busch, Y. C. Choi, F. Davis, I. L. Goldknopf, S. I. Matsui, M. S. Rao, and L. I. Rothblum. 1977. Controls of nucleolar function in cancer cells. Adv. Exp. Med. Biol. 91:125–137.

Cassidy, D. M., and A. W. Blackler. 1974. Repression of nucleolar organizer activity in an interspecific hybrid of the genus *Xenopus*. Dev. Biol. 41:84–86.

Croce, C. M., A. Talavera, C. Basilico, and O. J. Miller. 1977. Suppression of production of mouse 28S ribosomal RNA in mouse-human hybrids segregating mouse chromosomes. Proc. Natl. Acad. Sci. USA 74:694–697.

Dawid, I. B., D. D. Brown, and R. H. Reeder. 1970. Composition and structure of chromosomal and amplified ribosomal DNA's of *Xenopus laevis*. J. Mol. Biol. 51:341–360.

Desrosiers, R. C., C. Mulder, and B. Fleckenstein. 1979. Methylation of herpesvirus saimiri DNA in lymphoid tumor cell lines. Proc. Natl. Acad. Sci. USA 76:3839–3843.

Durica, D. S., and H. M. Krider. 1977. Studies on the ribosomal RNA cistrons in interspecific *Drosophila* hybrids. I. Nucleolar dominance. Dev. Biol. 59:62–74.

Eastman, E. M., R. M. Goodman, B. F. Erlanger, and O. J. Miller. 1980. 5-Methylcytosine in the DNA of the polytene chromosomes of the diptera *Sciara coprophila, Drosophila melanogaster* and *D. persimilis*. Chromosoma 79:225–239.

Elsevier, S. M., and F. H. Ruddle. 1975. Location of genes coding for 18S and 28S ribosomal RNA within the genome of *Mus musculus*. Chromosoma 52:219–228.

Goessens, G., and A. Lepoint. 1974. The fine structure of the nucleolus during interphase and mitosis in Ehrlich tumour cells cultivated in vitro. Exp. Cell Res. 87:63–72.

Goodpasture, C., and S. E. Bloom. 1975. Visualization of nucleolar organizer regions in mammalian chromosomes using silver staining. Chromosoma 53:37–50.

Grippo, P., M. Iaccarino, E. Parisi, and E. Scarano. 1968. Methylation of DNA in developing sea urchin embryos. J. Mol. Biol. 36:195–208.

Guntaka, R. V., P. Y. Rao, S. A. Mitsialis, and R. Katz. 1980. Modification of avian sarcoma proviral DNA sequences in nonpermissive XC cells but not in permissive chick cells. J. Virol. 34:569–572.

Harbers, K., B. Harbers, and J. H. Spencer. 1975. Nucleotide clusters in deoxyribonucleic acids. XII. The distribution of 5-methylcytosine in pyrimidine oligonucleotides of mouse L-cell satellite DNA and main band DNA. Biochem. Biophys. Res. Commun. 66:738–746.

Honjo, T., and R. H. Reeder. 1973. Preferential transcription of *Xenopus laevis* ribosomal RNA in interspecies hybrids between *Xenopus laevis* and *Xenopus mulleri*. J. Mol. Biol. 80:217–228.

Howell, W. M., T. E. Denton, and J. R. Diamond. 1975. Differential staining of the satellite regions of human acrocentric chromosomes. Experientia 31:260–262.

Hsu, T. C., B. R. Brinkley and F. E. Arrighi. 1967. The structure and behavior of the nucleolus organizers in mammalian cells. Chromosoma 23:137–153.

Hsu, T. C., S. E. Spirito, and M. L. Pardue. 1975. Distribution of 18+28S ribosomal genes in mammalian genomes. Chromosoma 53:25–36.

Lau, Y. F., W. Wertelecki, R. A. Pfeiffer, and F. E. Arrighi. 1979. Cytological analysis of a 14p+ variant by means of N-banding and combinations of silver staining and chromosome bandings. Hum. Genet. 46:75–82.

Lubit, B. W., R. R. Schreck, O. J. Miller, and B. F. Erlanger. 1974. Human chromosome structure as revealed by an immunoperoxidase staining procedure. Exp. Cell. Res. 89:426–429.

Mandel, J. L., and P. Chambon. 1979. DNA methylation: organ specific variations in the methylation pattern within and around ovalbumin and other chicken genes. Nucleic Acids Res. 7:2081–2103.

Mann, M. B., and H. O. Smith. 1977. Specificity of HpaII and HaeIII DNA methylases. Nucleic Acids Res. 4:4211–4221.

Matsushima, T., and T. H. Yosida. 1971. Change of stemline karyotype in Yoshida sarcoma by appearance of peculiar marker chromosomes. Gann 62:389–394.

McClintock, B. 1934. The relation of a particular chromosomal element to the development of the nucleoli in Zea Mays. Z. Zellforsch. Mikr. Anat. 21:294–328.

McGhee, J. D., and G. D. Ginder. 1979. Specific DNA methylation sites in the vicinity of the chicken β-globin genes. Nature 280:419–420.

Miller, D. A., W. R. Breg, D. Warburton, V. G. Dev, and O. J. Miller. 1978. Regulation of rRNA gene expression in a human familial 14p+ marker chromosome. Hum. Genet. 43:289–297.

Miller, D. A., V. G. Dev, C. Borek, and O. J. Miller. 1972. The quinacrine fluorescent and Giemsa banded karyotype of the rat, *Rattus norvegicus*, and banded chromosome analysis of transformed and malignant rat liver cell lines. Cancer Res. 32:2375–2382.

Miller, D. A., V. G. Dev, R. Tantravahi, and O. J. Miller. 1976a. Suppression of human nucleolus organizer activity in mouse-human somatic hybrid cells. Exp. Cell. Res. 101:235–243.

Miller, D. A., R. Tantravahi, V. G. Dev, and O. J. Miller. 1977. Frequency of satellite association of human chromosomes is correlated with amount of Ag-staining of the nucleolus organizer region. Am. J. Hum. Genet. 29:490–502.

Miller, D. A., R. Tantravahi, B. Newman, V. G. Dev, and O. J. Miller. 1979a. Karyotype of Friend virus-induced mouse erythroleukemia cells. Cancer Genet. Cytogenet. 1:103–113.

Miller, O. J., D. A. Miller, V. G. Dev, R. Tantravahi, and C. M. Croce. 1976b. Expression of human and suppression of mouse nucleolus organizer activity in mouse-human somatic cell hybrids. Proc. Natl. Acad. Sci. USA 73:4531–4535.

Miller, O. J., W. Schnedl, J. Allen, and B. F. Erlanger. 1974. 5-Methylcytosine localised in mammalian constitutive heterochromatin. Nature 251:636–637.

Miller, O. J., R. Tantravahi, D. A. Miller, L.-C. Yu, P. Szabo, and W. Prensky. 1979b. Marked increase in ribosomal RNA gene multiplicity in a rat hepatoma cell line. Chromosoma 71:183–195.

Mullen, V. T., and C. A. Barnett. 1976. Banded karyotypes of H-4-IIE-C3 rat hepatoma cells grown in vitro. In Vitro 12:658–664.

Nawaschin, M. 1934. Chromosome alterations caused by hybridization and their bearing upon certain general genetic problems. Cytologia 5:169–203.

Oshimura, M., M. Sasaki, and S. Makino. 1973. Chromosomal banding patterns in primary and transplanted venereal tumors of the dog. J. Natl. Cancer Inst. 51:1197–1203.

Perry, R. P., D. E. Kelley, U. Schibler, K. Huebner, and C. M. Croce. 1979. Selective suppression of the transcription of ribosomal genes in mouse-human hybrid cells. J. Cell Physiol. 98:553–559.

Procunier, J. D., and K. D. Tartof. 1978. A genetic locus having trans and contiguous cis functions that control the disproportionate replication of ribosomal RNA genes in *Drosophila melanogaster*. Genetics 88:67–79.

Ritossa, F. 1976. The bobbed locus, *in* The Genetics and Biology of Drosophila, vol. 1b, M. Ashburner and E. Novitski, eds. Academic Press, New York, pp. 801–846.

Sasaki, M., M. Oshimura, S. Makino, T. Koike, M. Itoh, F. Watanabe, and N. Tanaka. 1974. Further karyological evidence for contagiousness and common origin of canine venereal tumors. Proc. Jpn. Acad. 50:636–640.

Schreck, R. R., B. F. Erlanger, and O. J. Miller. 1977. Binding of anti-nucleoside antibodies reveals different classes of RNA in the chromosomes of the kangaroo rat *(Dipodomus ordii)*. Exp. Cell Res. 108:403–411.

Sofuni, T., S. Makino, and H. Kobayashi. 1967. A study of chromosomes in Friend virus-induced mouse leukemias. Proc. Jpn. Acad. 43:389–394.

Sutter, D., and W. Doerfler. 1980. Methylation of integrated adenovirus type 12 DNA sequences in transformed cells is inversely correlated with viral gene expression. Proc. Natl. Acad. Sci. USA 77:253–256.

Tantravahi, R., D. A. Miller, G. D'Ancona, C. M. Croce, and O. J. Miller, 1979. Location of rRNA genes in three strains of rat and suppression of rat rRNA gene activity in rat-human somatic cell hybrids. Exp. Cell Res. 119:387–392.

Tantravahi, U., R. V. Guntaka, B. F. Erlanger, and O. J. Miller. 1980. Amplified mammalian ribosomal RNA genes are enriched in 5-methylcytosine. Proc. Natl. Acad. Sci. USA (in press).

Tartof, K. 1974. Unequal mitotic sister chromatid exchange as the mechanism of ribosomal RNA gene magnification. Proc. Natl. Acad. Sci. USA 71:1272–1276.

Tsuchida, R., and M. A. Rich. 1964. Chromosomal aberrations in viral leukemogenesis. I. Friend and Rauscher leukemias. J. Natl. Cancer Inst. 33:33–47.

Vanyushin, B. F., S. G. Tkacheva, and A. N. Belozersky. 1970. Rare bases in animal DNA. Nature 225:948–949.

Waalwijk, C., and R. A. Flavell. 1978. MspI, an isoschizomer of HpaII which cleaves both unmethylated and methylated HpaII sites. Nucleic Acids Res. 5:3231–3236.

Wallace, H. R., and M. L. Birnstiel. 1966. Ribosomal cistrons and the nucleolus organiser. Biochim. Biophys. Acta 114:296–310.

Warburton, D., K. C. Atwood, and A. S. Henderson. 1976. Variation in the number of genes for rRNA among human acrocentric chromosomes: correlation with frequency of satellite association. Cytogenet. Cell. Genet. 17:221–230.

Warburton, D., K. C. Atwood, and A. S. Henderson. 1979. Sequential silver staining and hybridization in situ on nucleolus organizing regions in human cells. Cytogenet. Cell Genet. 24:168–175.

Chromosomal Changes Associated with Neoplasia

Genes, Chromosomes, and Neoplasia, edited by
Frances E. Arrighi, Potu N. Rao, and Elton Stubblefield.
Raven Press, New York © 1981.

Nonrandom Chromosome Changes in Human Leukemia

Janet D. Rowley

Department of Medicine and The Franklin McLean Memorial Research Institute, The University of Chicago, Chicago, Illinois 60637

The role of chromosome changes in the transformation of a normal cell to a cancer cell has been a subject of research and controversy for more than 60 years (Boveri 1914, Levan et al. 1977, Nowell 1976, Sandberg 1979, Rowley 1977). The observations during the 20 years prior to chromosome banding, namely, 1950 to 1970, indicated that, with one exception that I will discuss shortly, the number of chromosomes and the chromosome pattern in leukemic cells seemed to be quite variable, and thus the notion that chromosome changes were merely irrelevant secondary phenomena seemed to be supported by the available data. The application of banding techniques, however, has produced evidence that these views are incorrect. There is now a substantial body of data indicating that nonrandom chromosome changes, particularly consistent translocations, are specifically associated with a number of human tumors (Rowley 1977).

The major support for this statement comes from the study of the chromosome patterns in various hematologic disorders, notably acute and chronic myeloid leukemia. The study of chromosome abnormalities in leukemia serves two functions: one of these is to assist in more accurate diagnosis, and therefore in the more rational choice of the most appropriate treatment for a particular patient. The other function is to provide clues as to the particular chromosome sites that are abnormal so that, in the future, we can understand the alterations in the regulation of genes that are located at these affected sites.

This chapter will provide a general overview of the evidence regarding particular patterns of chromosome change in various human leukemias. The primary data that support these conclusions are presented in a number of current reviews (Sandberg 1979, Mitelman and Levan 1978, Rowley 1978a, b, 1980a).

An analysis of chromosome patterns, to be relevant to a malignant disease, must be based on a study of the karyotype of the tumor cells themselves. In the case of leukemia, the specimen is usually a bone marrow aspirate that is processed immediately or cultured for a short time (Testa and Rowley 1980b). When an abnormal karyotype is found in leukemic cells, it is important to analyze cells from normal tissues; in most instances, cells from these unaffected tissues will have a normal karyotype. The chromosome abnormalities observed in the leukemic cells thus represent somatic mutations in an otherwise normal individual.

The observation of at least two "pseudodiploid" or hyperdiploid cells or three hypodiploid cells, each showing the same abnormality, is considered evidence for the presence of an abnormal clone; patients with such clones are classified as abnormal. Patients whose cells show no alterations, or in whom the alterations involve different chromosomes in different cells, are considered to be normal. Isolated changes may be due to technical artifacts or to random mitotic errors.

In the following discussion, the chromosomes are identified according to the Paris Conference 1971 nomenclature (1972), and the karyotypes are expressed as recommended under this system. The total chromosome number is indicated first, followed by the sex chromosome, and then by the gains, losses, or rearrangements of the autosomes. A + or − sign before a number indicates a gain or loss, respectively, of a whole chromosome; a + or − after a number indicates a gain or loss of part of a chromosome. The letters "p" and "q" refer to the short and long arms of the chromosome, respectively; "i" and "r" stand for isochromosome and ring chromosome. "Mar" is marker, "del" is deletion, "ins" is insertion, and "inv" is inversion. Translocations are identified by "t" followed by the chromosomes involved in the first set of brackets; the chromosome bands in which the breaks occurred are indicated in the second brackets. Uncertainty about the chromosome or band involved is signified by "?".

Ph¹-POSITIVE LEUKEMIA

Chronic Myelogenous Leukemia

Cytogenetic studies on patients with chronic myelogenous leukemia (CML) have been the keystone for karyotype analysis of other human malignancies. New discoveries that have resulted from examination of CML have subsequently been confirmed in other hematologic malignancies, and in many solid tumors as well.

Chromosome Studies of Chronic Myelogenous Leukemia prior to Banding

Nowell and Hungerford, in 1960, reported the first consistent chromosome abnormality in human cancer; they observed an unusually small G-group chromosome in leukemic cells from patients with CML. This chromosome, which appeared to have lost about one half of its long arm, was called the Philadelphia or Ph¹ chromosome in honor of its city of discovery. The question of whether the deleted portion of this long arm was missing from the cell or was translocated to another chromosome could not be answered at that time because it was impossible to identify each human chromosome precisely with the techniques then available. Furthermore, the identity of this chromosome as either a No.

21 or No. 22 could not be established. Despite this uncertainty, the Ph[1] chromosome was a very useful marker in the study of patients with CML.

Bone marrow cells from approximately 85% of patients who have clinically typical CML contain the Ph[1] chromosome (Ph[1] positive) (Whang-Peng et al. 1968, Rowley 1978a, 1980a, Sandberg 1979); the other 15% of patients usually have a normal karyotype (Ph[1] negative), although abnormalities such as an extra C-group chromosome have been observed. Chromosomes obtained from phytohemagglutinin (PHA)-stimulated lymphocytes of patients with Ph[1]-positive CML are normal.

Perplexing observations, still not explained, were that patients with Ph[1]-positive CML had a much better prognosis than those with Ph[1]-negative CML (42- vs. 15-month survival), and that patients with a Ph[1] chromosome and additional chromosome abnormalities did not have a substantially poorer survival than those who had only a Ph[1] chromosome. However, as was shown by Whang-Peng et al. (1968), a change in the karyotype was a grave prognostic sign; the median survival after such a change was about 2 months.

Chromosome Studies of Chronic Myelogenous Leukemia with Banding

Chronic Phase of CML

Chromosome banding techniques were first used in the cytogenetic study of leukemia for identification of the Ph[1] chromosome as a deletion of No. 22 (22q−) (Caspersson et al. 1970, O'Riordan et al. 1971). Since quinacrine fluorescence revealed that the chromosome present in triplicate in Down's syndrome was No. 21, the abnormalities in Down's syndrome and CML were shown to affect different pairs of chromosomes.

The question of the origin of the Ph[1] (22q−) chromosome was answered when Rowley (1973a) reported that the Ph[1] chromosome results from a translocation rather than, as many investigators had previously assumed, a deletion. Additional dully fluorescing chromosomal material was observed at the end of the long arm of one chromosome 9 (9q+) and was approximately equal in length to that missing from the Ph[1] chromosome; it had staining characteristics similar to those of the distal portion of the long arm of chromosome 22 (Figure 1). It was proposed, therefore, that the abnormality of CML was an apparently balanced reciprocal translocation (9;22)(q34;q11). Subsequent measurements of the DNA content of the affected pairs (chromosomes 9 and 22) have shown that the amount of DNA added to chromosome 9 is equal to that missing from the Ph[1] (Mayall et al. 1977); thus, there is no detectable loss of DNA in this chromosome rearrangement. Other studies with fluorescent markers or chromosome polymorphisms have shown that, in a particular patient, the same chromosome 9 or chromosome 22 is involved in each cell (Gahrton et al. 1973). These observations confirm earlier work, based on enzyme markers, indicating that CML cells originate from a single cell (Fialkow 1974).

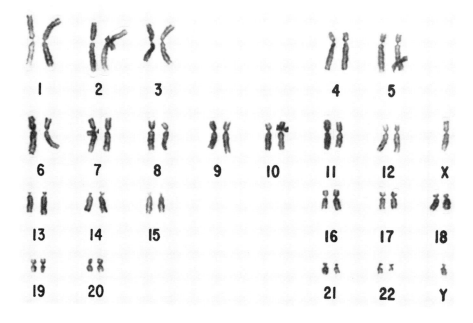

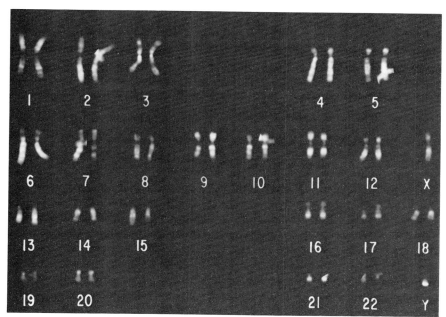

FIG. 1. a (top). Karyotype of a metaphase cell from a bone marrow aspirate obtained from an untreated male patient with CML. The cell has been stained with routine Giemsa prior to banding with quinacrine fluorescence. The Ph[1] chromosome is on the right in pair No. 22. The long arm of the chromosome is much shorter than that of the normal homolog. **b (bottom).** The same cell as in a. The chromosomes were destained, restained with quinacrine mustard,

Karyotypes of 802 Ph[1]-positive patients with CML have been examined with banding techniques by a number of investigators, and the 9;22 translocation has been identified in 739 (92%) (Rowley 1980a). It is now recognized that, in addition to the typical t(9;22), variant translocations may occur (Sandberg 1979, Mitelman and Levan 1978, Rowley 1980a). These appear to be of two kinds: one is a simple translocation involving chromosome 22 and some chromosome other than No. 9, which has been seen in 29 patients. The other is a complex translocation involving three or more different chromosomes; except in two cases, two of the chromosomes involved were found to be No. 9 and No. 22. This type of translocation has been observed in 31 patients.

With two possible exceptions, the break point in the Ph[1] chromosome appears to occur in band 22q11. The two exceptional cases both involve variant translocations between 12p and 22q; in each instance, band 22q12 appears to remain on the Ph[1] (Geraedts et al. 1977, Verma and Dosik 1979). This would suggest either that the break in 22q was in a different band or that more of 12p was translocated to 22q. At least three patients with complex translocations appeared to lack a Ph[1] chromosome. The patient of Engel et al. (1974) had a complex rearrangement that involved translocation of all of chromosome 22 to the recipient, chromosome 17. In the other two patients (Lawler et al. 1976, Tanzer et al. 1977), a complex translocation of material to the end of the Ph[1] chromosome occurred.

A sufficient number of patients has been studied so that one can ask whether any patterns have emerged regarding variant translocations. There are no simple translocations involving chromosomes 1, 4, 5, 7, 8, 18, 20, and Y; and there are no complex translocations involving chromosomes 12, 16, 18, 20, and the Y. At present, only chromosome 18 and the Y have not been observed to be involved in a variant Ph[1] translocation. The chromosome most frequently involved in both simple and complex translocations is No. 17 (four times for each type). These translocations have affected the long or short arm with about equal frequency. On the other hand, 12p is the translocation site in four simple rearrangements, but is not involved in any complex translocations. The significance, if any, of these observations will become clearer in the future as the gene content of these chromosomes and the effects of translocations are understood.

Only three patients have been reported who are said not to have had a translocation. The great specificity of the translocation involving chromosomes 9 and 22 remains an enigma. The survival curves for patients with variant translocations appeared to be the same as those for patients with the standard t(9;22) (Sonta and Sandberg 1977).

---

and photographed with ultraviolet fluorescence. In addition to the Ph[1] (22q−) chromosome, the No. 9 on the right (9q+) has an additional pale band that is not present on the normal No. 9. In a, the No. 9 on the right is longer than that on the left.

Acute Phase of CML

When patients with CML enter the terminal acute phase, about 20% appear to retain the 46,Ph[1]-positive cell line unchanged, whereas other chromosome abnormalities are superimposed on the Ph[1]+ cell line in 80% of patients (Rowley 1978a, 1980a). In a number of cases, the change in the karyotype preceded the clinical signs of blast crisis by 2 to 4 months. In general, if a patient has a clone of Ph[1]-positive cells with a unique marker during the chronic phase, this clone will be the one involved in the transformation.

Bone marrow chromosomes from 242 patients with Ph[1]-positive CML who were in the acute phase have been analyzed with banding techniques (Rowley 1980a). Forty showed no change in their karyotype, whereas 202 patients had additional chromosome abnormalities.

Fifty-six patients (28%) listed in Table 1 had a change in their karyotype at the time of blast crisis that was not reflected in a change in the modal chromosome number. There is thus no substitute for accurate karyotyping, with banding, if one wants to correlate the karyotype with the patient's clinical status. For example, we have one patient who had a −7, +8 karyotype at the time of blast crisis that would not have been detected without banding (Rowley, unpublished results).

In a number of patients, the change in karyotype preceded the clinical signs of blast crisis by 2 to 4 months. There are some reports on patients who showed a change without subsequent development of blast crisis, e.g., two patients reported by Hayata et al. (1975). Since one patient of Van Den Berghe (personal communication) developed blast crisis 2 years after the change in karyotype, this lack of correlation may reflect the slow progression seen in some patients. On the other hand, we cannot state categorically that every patient who shows a karyotypic change will develop acute blast crisis.

The most common changes frequently occur in combination to produce modal numbers of 47 to 52 (Table 1). The gains or structural rearrangements of particular chromosomes observed in 202 patients who had relatively complete analyses

TABLE 1. *Modal chromosome number in aneuploid patients with leukemia whose cells have been studied with banding*

Type of leukemia	Number of patients	Modal chromosome number										
		<44	44	45	46	47	48	49	50	51	52	>53
Ph[1]-positive CML in acute phase	197	0	1	9	56*	60	27	12	11	8	5	8
ANLL	191	7	6	43	63†	58	4	2	1	2	1	4
ALL	54	2	0	2	17†	3	2	4	3	4	4	13

* Cells have the Ph[1] chromosome and other new abnormalities.
† All patients are pseudodiploid.

TABLE 2. *Most common chromosome changes in 202 Ph¹-positive patients in acute phase of CML*

Number of patients with	Chromosome			
	No. 8	No. 17	No. 19	Ph¹
Gain	95	11	38	73
Rearrangement	7	63*	2	5†

* 56 were i(17q).
† All were i(Ph¹).

are summarized in Table 2. Different abnormal chromosomes occur singly or in combinations in a distinctly nonrandom pattern during the acute phase of CML (Rowley 1980a). The significance of these patterns will be understood only when we have more information about the genes carried on these chromosomes. When patients had only a single new chromosome change, this involved a second Ph¹ in 19 and an isochromosome for the Ph¹, i(22q−) in 3 patients, an i(17q) in 18 and a +17 in 1, a +8 in 15, and a +19 in only 1 patient. An extra chromosome 8 and i(17q) occurred together as the only changes in 25 patients. In seven patients, it was possible to determine which chromosome change occurred first; i(17q) was the initial change in 5 of them. An i(17q) was seen with another C group chromosome only in the presence of chromosome 8. A +8 and +17 were never seen as the only changes. On the other hand, if the patient's cells also had a second Ph¹, then a +8,i(17q) was seen in 5 and +8,+17,+Ph¹ was seen in 4 cases. Only 1 patient had a +8,i(17q) and +19, and *no* patient had an i(17q) with +19. An extra chromosome 8 and either a second Ph¹ or a +19 occurred in 10 and 9 patients, respectively. No patients had an i(17q) and a Ph¹, although 1 patient had a +17 and +Ph¹. In patients with more complex patterns, some nonrandom changes were also noted. Of 21 patients with extra C-group chromosomes (excluding the 54 already discussed), 15 had an extra chromosome 8. Thus, only 6 of 75 patients had at least one extra C that did not involve chromosome 8.

As discussed earlier, the frequency of chromosome loss is very low. Of 202 patients, five were lacking one chromosome 7, four were lacking one chromosome 17, three each were lacking one chromosome 19 or a Y, and two each were lacking chromosome 14, 15, 18, or 22, or an X chromosome. Structural rearrangements, other than the Ph¹ translocation and the i(17q), most often involved chromosome 1, usually the long arm (11 patients), chromosome 8 (eight), chromosome 6 (six), chromosome 7 (six), or chromosome 11 (six).

Since virtually all patients who are studied in the acute phase of CML have been treated, usually with busulphan, it is impossible to determine whether this therapy affects the pattern of abnormalities described earlier. Evidence has been presented recently that aggressive chemotherapy in the chronic phase may alter the pattern of chromosome abnormalities seen in the acute phase (Alimena et al. 1979).

Only two reports have been published relating specific chromosome changes in the acute phase with survival of patients, and the results are conflicting. Thus, Prigogina et al. (1978) found that patients who did not show additional abnormalities in the acute phase had a longer survival than those whose karyotypes showed such changes. Sonta and Sandberg (1978), however, found no difference in survival between these two groups.

Ph¹-Positive Acute Leukemia

Our interpretation of the biologic significance of the Philadelphia chromosome has been modified over the course of the last 9 years, as our clinical experience with this marker has widened. Thus, Whang-Peng et al. (1970) proposed that cases of acute myeloblastic leukemia (AML) in which the Ph¹ chromosome was present should be reclassified as cases of CML in blast transformation. This notion, which was broadened to include the cases that appeared to be acute lymphoblastic leukemia (ALL) at diagnosis, was generally accepted until about 1977. More recently, however, the tendency has been to refer to patients who have no prior history suggestive of CML as having Ph¹-positive leukemia (Bloomfield et al. 1977, Rowley 1980a). It is becoming increasingly evident that the observed interrelations of Ph¹-positive leukemias are complex indeed, and that the distinctions between some categories, which are difficult to make, will be determined by the arbitrary judgment of the investigator. Moreover, although the Ph¹ chromosome is used as the marker that defines these leukemias, the cytogenetic studies on patients have often been woefully inadequate. Chromosome banding is an essential requirement not only for establishing whether the "Ph¹" is a 22q— rather than a 21q— chromosome as recently reported by Tosato et al. (1978), but also because there is some evidence that the type and frequency of variant translocations in the Ph¹-positive acute leukemias may differ from those in typical CML. Moreover, most of these patients become chromosomally normal in remission; cytogenetic studies on patients in remission are thus essential.

When one reviews the descriptions of the patients who are first seen with what appears to be acute leukemia and a Ph¹ chromosome, it is quite apparent that a clear demarcation between morphologic types is difficult. Thus, some patients have a high percentage of lymphoblasts, others have a high percentage of myeloblasts, and still others have a mixture of myeloblasts and lymphoblasts. In some instances, cells from the last group have been analyzed for cell surface markers (Janossy et al. 1978).

More recently, levels of terminal deoxynucleotidyl transferase (TdT) have been determined in patients with CML in blast crisis and in patients with Ph¹-positive acute leukemia. It has become apparent that patients in the blast crisis of CML tend to fall into two groups, one with a low and one with a higher level of the enzyme (Marks et al. 1978). When surface marker studies are done on these patients, those with a high TdT level tend to have a lymphoid blast crisis of the null-cell type, whereas patients with low levels of TdT are in myeloid blast crisis (Hoffbrand et al. 1977, Janossy et al. 1978). However, as has been noted by Marks et al. (1978), the correlation is very imprecise.

It is important to determine whether there is any correlation between the karyotype and TdT levels in patients who have elevated blast cell counts. The data available for 21 patients (Rowley 1980a) suggest that CML patients with a lymphoid blast crisis and elevated TdT levels tend to remain only Ph[1]-positive (five of five patients), whereas patients with a myeloid blast crisis may have low or high levels of TdT and tend to show chromosome changes commonly associated with CML blast crisis (four of six patients). Clearly, more patients will have to be examined so that we can determine whether these are consistent differences.

Levels of TdT were elevated in seven of eight patients with Ph[1]-positive acute leukemia, regardless of whether the blasts were predominantly or only partially of lymphoid origin (Rowley 1980a). Two patients remained only t(9;22), whereas other, apparently variable, abnormalities occurred in six. Patients with mixed lymphoid and myeloid blasts seemed to have a mosaic karyotype with some cells having 46,Ph[1]-positive, and with a variable percentage having other abnormalities. Seven of the eight patients were studied in remission, and all had a normal karyotype. Lymphoid cells in five of the eight were of the null-type. Four of the eight patients acheived a complete remission on treatment with vincristine and prednisone, and three had no response to a variety of drug combinations.

ALL Evolving to CML

Twelve patients have been reported on who were first seen with what appeared to be typical ALL that subsequently evolved to CML; these patients have also been discussed recently in an Annotation in *The British Journal of Haematology* (Catovsky 1979) and references to specific patients will be omitted. Two of the 12 were children, 4 and 11 years old; the adults ranged in age from 17 to 59 years (median, 33 years). There were three female and nine male patients. The diagnosis of ALL was made on the basis of morphology and cytochemistry, although a few patients had elevated TdT activity or were anti-ALL antiserum-positive. Cytogenetic studies in the ALL phase were done on only five patients. Four patients had 100% and one had 50% Ph[1]-positive cells; where banding was done, the t(9;22) was found. All of these patients entered complete remission of ALL and then clearly had CML at intervals ranging from the next clinic visit (1 month) to 2 or 3 years (median, 5 months). A cytogenetic study with banding was done on two patients during this period, and bone marrow cells showed only a normal karyotype. When the chronic phase of CML was diagnosed, patients were said to have 75% to 100% Ph[1]-positive cells; four of these patients were studied with banding and had a t(9;22). The chronic phase lasted from 3 weeks to 3 or 4 years. A terminal blast crisis with myeloid morphology was noted in four patients and a blast crisis with lymphoid morphology in five.

The dilemma in our understanding of Ph[1]-positive leukemia has become apparent in recent years as patients with a Ph[1] chromosome in lymphoid-appearing blast cells have been recognized with increasing frequency. The lymphoid blast crisis in a known CML patient can be explained if one assumes that the Ph[1] chromosome is also present in some percentage of the primitive or pluripotent stem cells (PSC). In most patients with CML, this cell remains "silent," but in about 10% to 20% of patients, a Ph[1]-positive PSC undergoes blast transformation and this leads to a lymphoid blast crisis. If such a patient responds to therapy, the blast cells would disappear and the underlying CML would reappear as the dominant hematologic feature.

In patients presenting with Ph[1]-positive acute leukemia, either myeloid or lymphoid, with no prior history of CML, one would have to assume that some portion of a Ph[1]-positive PSC clone underwent rapid expansion to result in acute leukemia. Upon successful therapy, however, this clone would diminish so that the chromosomally normal PSC would repopulate the marrow and the patient would then have a normal karyotype in remission. These uncertainties regarding the interrelationship of the various hematopoietic cell types and the regulation of the differential proliferation of one pool as compared with others will be clarified with increased information about the biology of these cells.

ACUTE NONLYMPHOCYTIC LEUKEMIA

The use of chromosome banding techniques has markedly increased our understanding of the types and frequency of chromosome abnormalities in acute non-lymphocytic leukemia (ANLL). Extra, missing, or rearranged chromosomes previously described on the basis of morphology alone can now be identified precisely in terms of the particular chromosomes or chromosome bands involved.

We and others have also shown that the karyotype pattern of the leukemic cells is correlated with survival. Patients with a normal karyotype have a significantly longer median survival (10 months) than do patients with an abnormal karyotype (4 months) (Golomb et al. 1978b, First International Workshop on Chromosomes in Leukaemia 1978).

Approximately 50% of the patients studied with banding have detectable karyotypic changes; these abnormalities are present prior to therapy and usually disappear when the patient enters remission. The same aberrations reappear in relapse, sometimes showing evidence of further karyotypic change superimposed on the original abnormal clone (Sandberg 1979, Mitelman and Levan 1978, Testa and Rowley 1980a).

Although the karyotypes of patients with ANLL may be variable, examination of the chromosome changes seen in 191 chromosomally abnormal patients are available for analysis (Testa and Rowley 1980a). The modal numbers for the changes are listed in Table 1. The nonrandom distribution of chromosome losses and gains is particularly evident in patients with 45 to 47 chromosomes (Figure 2). A gain of chromosome 8, the most common abnormality in ANLL, was seen in 48 cases, and loss of one chromosome 7, the next most frequent change, was seen in 22 patients. Gains or losses of some chromosomes occurred only in patients with more complex karyotypes; they are likely to represent secondary changes occurring in clonal evolution, rather than primary events. In contrast, structural rearrangements of nearly all of the chromosomes of the complement were usually found in patients with modal numbers of 45 to 47.

We obtained serial samples of leukemic cells from 60 of 90 patients with ANLL for chromosome analysis with banding techniques to determine whether or not a nonrandom pattern of karyotypic evolution is associated with ANLL (Testa et al. 1979). Evolution of the karyotype was observed in 17 of these patients, 7 of whose chromosomes were initially normal and 10, initially abnormal. The most frequent evolutionary change was a gain of one or more chromo-

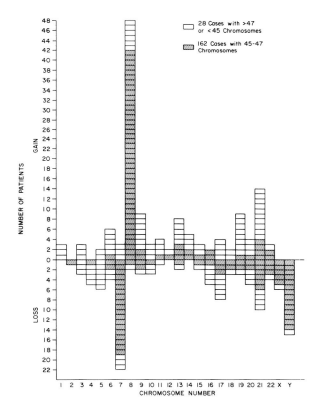

FIG. 2. Histogram of chromosome changes observed in 190 aneuploid patients with ANLL. The cross-hatched portion indicates the changes observed in the 162 patients who had modal chromosome numbers of 45 to 47. (Reproduced from Testa and Rowley 1980a.)

somes (in 12 of the 17 patients). Ten of the 12 patients who acquired one or more additional chromosomes had an extra chromosome 8, and six had an extra chromosome 18. The incidence of karyotypic evolution and the type and frequency of particular evolutionary changes were similar in patients who were initially normal and in those who were initially abnormal. In all but one of the initially abnormal patients, karyotypic evolution involved the original cytogenetically abnormal clone. In no instance was a clone of chromosomally abnormal cells detected when the bone marrow was morphologically normal. ANLL patients whose karyotype evolved tended to have relatively long survival times. However, the median survival after the actual onset of evolution was relatively short (2 months) and the response to further therapy was poor. Thus, karyotypic changes have prognostic significance that may be useful in making decisions on the aggressiveness of further therapy.

The 8;21 Translocation in Acute Myeloblastic Leukemia

Two structural rearrangements seen in ANLL appear to have special significance. The more common of these was first recognized by Kamada et al. (1968) as most likely representing a translocation between a C- and a D-group chromosome. A high incidence of this aberration was noted by Trujillo et al. (1974, 1979), who described it as a complex pattern, −C, +D, +E, −G. The precise nature of the abnormality was resolved by Rowley (1973b), who used the Q-banding technique to determine that it is a balanced translocation between chromosomes 8 and 21 [t(8;21) (q22;q22)] (Figure 3). Chromosomes 8 and 21 can also participate in three-way rearrangements similar to those involving chromosomes 9 and 22 in CML. Lindgren and Rowley (1977) reported on two patients with three-way translocations in whom the third chromosome was either a No. 11 or a No. 17. The 8;21 translocation is frequently associated with loss of a sex chromosome; 32% of the males with the 8;21 translocation are missing a Y, and 36% of the females are missing an X (Second International Workshop on Chromosomes in Leukemia 1980). This association is especially noteworthy since sex chromosome abnormalities are otherwise rarely seen in ANLL. The frequency with which this translocation occurs varies from one laboratory to another, but it probably amounts to about 8% of all cases of ANLL. It appears to be restricted to patients with acute myeloblastic leukemia (M2 in FAB classification).

Slides of bone marrow from 45 patients with a t(8;21) or a variant thereof were examined by hematologists at the Second International Workshop on Chromosomes in Leukemia (1980); every patient had morphologic features that were compatible with a diagnosis of an M2 marrow. It has been reported that patients with the t(8;21) have a low level of leukocyte alkaline phosphatase and a high percentage (64%) of Auer rods (Kamada et al. 1976, Trujillo et al. 1979). They also have a much longer median survival than do patients with other chromosome abnormalities (Sandberg 1979, Second International Workshop on

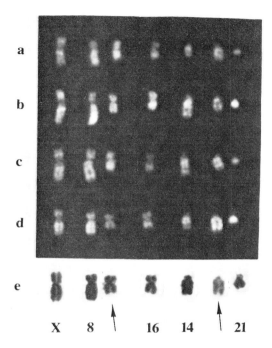

FIG. 3. Partial karyotype of four cells from a female patient with an 8;21 translocation and a missing X chromosome. Rows a–d, chromosomes stained with quinacrine mustard and photographed with ultraviolet fluorescence; row e, standard Giemsa stain of chromosomes in row d. The 8q− chromosome (↑) is broken in band q22 and it resembles a chromosome 16; the 21q+ chromosome (↑) is broken in band q22 and it resembles a chromosome 14.

Chromosomes in Leukemia 1980). Data on survival of 48 patients with t(8;21) were also reviewed at the Second International Workshop (1980); the median survival of the whole group was 11.5 months. For 18 patients who had 46 chromosomes with only the t(8;21), the survival was 14 months; for those with certain other abnormalities superimposed on the translocation, the survival was 15.5 months. In contrast, patients whose cells were missing a sex chromosome, either X or Y, had a median survival of only 5 months.

The 15;17 Translocation and Acute Promyelocytic Leukemia

Another significant structural rearrangement is that observed in acute promyelocytic leukemia (APL), which is a unique form of acute leukemia characterized by hemorrhagic episodes, disseminated intravascular coagulation, and infiltration of the marrow with atypical "hypergranular" promyelocytes. The FAB cooperative study group recently recognized that not all patients have coarse granules and has thus added a category called the M3 variant (French-American-British Cooperative Group 1980). The variant category was identified largely on the basis of the clinical features and a specific chromosome abnormality, namely, a translocation involving the long arms of chromosomes 15 and 17 [t(15; 17)(q25;q22)] (Rowley et al. 1977). We have now studied a total of seven patients with APL; each had the t(15;17) (Golomb et al. 1979). The breakpoint in chromosome 15 appears to be distal to band q24, and in chromosome 17 it appears to be in q22 (Golomb et al. 1979, Second International Workshop on Chromosomes in Leukemia 1980) (Figure 4). The translocation was present in about 40% of patients with APL; it was not seen in patients with other types of acute leukemia (Second International Workshop 1980). Eighty patients with APL were reviewed at the Second International Workshop (1980), and 40 had a normal karyotype, 33 had the t(15;17) alone or with other abnormalities, and 7 had other types of chromosome changes. The median survival of patients with the t(15;17) was 1 month, whereas for those with a normal karyotype or with other changes it was 4 and 5 months, respectively. Two patients with complex translocations involving chromosomes 15 and 17 and either chromosome 2 or 3 have recently been described (Bernstein et al. 1980). Therefore, the same pattern of variation of a specific translocation involves t(15;17) as well as t(9;22) and t(8;21).

There is a curiously uneven frequency in the geographic distribution of this chromosome abnormality. It is seen in 100% of our patients in Chicago, in

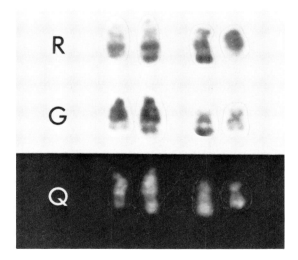

FIG. 4. Partial karyotype of chromosome pairs No. 15 and No. 17 from two patients with APL. The translocation chromosome is on the right in each pair. Top row, R-banding with acridine orange from patient with the M3 variant form of APL. Middle row, modified trypsin G-banding of a cell from the same patient as in top row. Bottom row, Q-banding of cell from another patient with the M3 variant form of APL.

about 70% of patients seen in Belgium, and has not been identified in a single patient in Sweden or Finland. The cause of this variation in frequency is as yet unexplained.

It is important to make a correct early diagnosis of APL because the initial therapy, which includes the use of heparin for control of bleeding, is associated with a significant improvement in survival. Whereas the correct diagnosis should not be difficult in typical cases, the M3 variant, in which granules may be lacking or reduced in number, may cause confusion. Every one of our M3 variant patients also had the typical translocation.

Secondary Acute Leukemia

Acute nonlymphocytic leukemia (ANLL) in patients who have been treated for other diseases is being recognized with increasing frequency. It has been observed in patients treated for Hodgkin's disease, non-Hodgkin's lymphoma, and other solid tumors, as well as nonmalignant diseases. The fact that virtually every one of these patients has a clone of chromosomally abnormal cells in the bone marrow is not well known, nor is the fact that the chromosome changes observed are nonrandom. The observations of nonrandom changes have important implications for accurate clinical diagnosis of treated patients with unexplained cytopenias. In addition, these data are particularly relevant to the identification of patients with ANLL de novo who may have been exposed to potentially mutagenic agents.

We studied the karyotype of 27 patients who developed ANLL either after treatment of a primary malignancy (26 patients) or after a renal transplant (1 patient) (Rowley, Golomb, and Vardiman, manuscript submitted). Fifteen of the patients had previously had both radiotherapy and chemotherapy, eight had had only chemotherapy, and four had had only radiotherapy. The median times from diagnosis of initial disease to the development of ANLL for these treatment groups were 61, 59, and 59 months, respectively. Twenty-six patients had an abnormal karyotype; one or both of two consistent chromosome changes were noted in 23 of 26 patients with aneuploidy. Eleven patients had loss of chromosome 5, and marrow cells from three others were lacking part of the long arm of chromosome 5 (5q−), whereas chromosome 7 was missing from cells of 18 patients and one other had loss of part of the long arm of chromosome 7 (7q−). Although these changes are distinctly different from those seen in lymphomas, they are similar to those seen in 25% of aneuploid patients with ANLL de novo.

In addition to the patients whom we have described, 16 others with secondary ANLL, studied with banding, have been reported (reviewed by Rowley, Golomb, and Vardiman, manuscript submitted); all except one had had a primary malignant disease. Five patients had received primarily radiotherapy, two had received only chemotherapy, and six had received a combination of radiotherapy and chemotherapy. The median time from diagnosis of the initial disease to the leukemic phase was 51 months (range, 26 to 98 months).

Every one of these patients had a clone of chromosomally abnormal cells comprising 50% to 100% of the cells in division. Nine of the 16 patients (56%) had clones with a hypodiploid modal number (Table 3). Moreover, 12 of the 16 (75%) had loss of part or all of chromosome 5 or 7 or both (Table 4). These observations are similar to those that we have reported.

The cytogenetic hallmark of secondary ANLL is thus an abnormal clone of cells, usually with a hypodiploid modal number, that is associated with the nonrandom loss of chromosomes 5 or 7 or both. Our observations regarding the nonrandom abnormalities of chromosomes 5 and 7 are germane to the question of whether the leukemic cells of patients with ANLL de novo contain specific karyotypic changes that allow one to distinguish between patients who have and those who have not been exposed to an environmental mutagen. Although this question cannot be answered at present, several lines of evidence suggest that the answer may be positive. The first of these comes from studies of ANLL in children, and the second from a retrospective study of the correlation of the karyotype with occupational exposure.

Three series describing ANLL de novo in children have recently been published. In one of these, 11 of 22 (50%) untreated patients were aneuploid (Benedict et al. 1979); in the second, 9 of 14 (64%) were aneuploid (Hagemeijer et al. 1979), whereas 10 of 13 (76%) in the third series were aneuploid (Morse et al. 1979). Not one of these 30 aneuploid patients had an abnormal clone with a hypodiploid modal number (Table 3). Moreover, not one of the abnormal clones was missing a chromosome 5, and only one was missing a chromosome

TABLE 3. *Numerical breakdown of patients by modal chromosome numbers of leukemic cells**

Data source	Total	Normal	Modal number for aneuploid patients				
			<45†	45	46	47	>47
ANLL de novo							
Literature review	382	191	13	43	63	58	14
Childhood	49	19†	0	0	16‡	10	4
Sweden—							
exposed	23	4	2	6	4	5	2
nonexposed	33	25	0	1	4	3	0
Chicago	88	44	7	10	18	5	4
ANLL secondary							
Chicago	27	1	8	11	4‡	1	2
Literature review	16	0	6	3	5	2	0

* Modified from Rowley, Golomb, and Vardiman, manuscript submitted.
† Includes one child with Down's syndrome and no abnormality except +21.
‡ One patient with a t(13;14) constitutional abnormality is classified as having 46 chromosomes.

TABLE 4. *Acute leukemia patients with abnormalities of chromosomes 5, 7, and 8**

Data source	Total†	Type of abnormality				
		−5	5q−	−7	7q−	+8
ANLL de novo						
Literature review	190	6	9	22	4	48
Childhood	30	0	0	1‡	5§	3¶
Sweden—						
exposed	19	3	2	3⏐	2	5
nonexposed	8	0	0	1	0	0
Chicago	44	5	5	8**	1	3
ANLL Secondary						
Chicago	26	11¶	3	18¶††	1	5‡‡
Literature review	16	3	4¶	6**	3§§	1¶

* Modified from Rowley, Golomb, and Vardiman, manuscript submitted.
† All studied with banding; these are the patients with aneuploidy from the previous table.
‡ This patient also was +8.
§ Only two patients had the typical del(7)(q22); one other had del(7)(q31).
¶ Includes one patient with this abnormality in evolution of the karyotype, and one who was also −7.
⏐ Includes one patient with −5.
** Includes one patient who was −5 and two who were 5q−.
†† Includes six patients who were −5 and two who were 5q−.
‡‡ Includes two patients who were +8 in karyotype evolution.
§§ Includes the patient who evolved to 5q−.

7 (Table 4). Five patients had deletions involving 7q, but only two had a deletion at 7q22, which is the abnormality seen in most cases of adult ANLL de novo and in secondary ANLL.

A recent study reported by Mitelman and his colleagues (1978) in Sweden provides some additional relevant information. In this retrospective analysis of 56 patients who had ANLL and whose karyotypes and occupations were known, 23 had a history suggesting an occupational exposure to chemical solvents, insecticides, or petroleum products, whereas 33 had no such known exposure. The detailed karyotypic findings showed that there were striking differences between the two groups. Only 24.2% of patients in the nonexposed group had clonal aberrations, compared with 82.6% in the exposed group (Table 3). There was a distinctly nonrandom pattern of abnormalities in the exposed group, with 84.2% of these patients having at least one of four changes, namely, −5, −7, +8, or +21 (Table 4). Only two patients in the nonexposed group had any of these aberrations; one was −7 and the other was +21.

The observation of a missing chromosome 5 or chromosome 7 in patients with secondary leukemia, the occurrence of these same aberrations in patients with de novo ANLL who may have been exposed to mutagenic agents, and the absence of these abnormalities in childhood ANLL all provide support for our proposal that these particular chromosome changes may identify acute leukemia associated with exposure to mutagenic agents (Rowley, Golomb, and Vardiman, manuscript submitted).

Refractory Anemia and Preleukemia

It has been recognized for 30 years that some patients with unexplained cytopenias may develop acute leukemia. These patients may or may not have an increased frequency of blasts in the bone marrow. Data on 244 such patients were analyzed at the Second International Workshop on Chromosomes in Leukemia (1980). As in ANLL, about 50% of patients had a normal karyotype. The types of chromosome changes were very similar to those seen in ANLL. The disease in 52 patients had progressed to overt ANLL at the time of the workshop. Some clear differences emerged between those patients who had and those who did not have a normal karyotype. Only 34 of 118 patients (29%) with a normal karyotype were dead, roughly one half of these having developed ANLL, whereas 66 of 111 patients (60%) with autosomal abnormalities were dead, about one half of them of ANLL. Although eight of 17 patients whose cells had a 5q— abnormality were dead, only two of these died in a leukemic phase. It has been suggested that refractory anemia and the 5q— abnormality may constitute a distinct entity, and that many patients with it do not develop ANLL. The data from the Second International Workshop supports this observation. None of the seven patients with loss of a sex chromosome developed acute leukemia.

Thus, the data indicate that the presence of a chromosome abnormality in patients with cytopenias is a sign of a poor prognosis, since a higher proportion of such patients die compared to those with a normal karyotype. Within each group, about one half of those who died had developed ANLL.

LYMPHOCYTIC LEUKEMIA

Chronic Lymphocytic Leukemia

In the past, chromosome studies of patients with chronic lymphocytic leukemia (CLL) have revealed mainly normal karyotypes. It is generally agreed that this result reflects the absence of dividing malignant cells. The early studies used primarily PHA, which is a T-cell mitogen; more recently, a variety of mitogens have been used with an emphasis on those that stimulate B-cells (Autio et al. 1979, Gahrton et al. 1980). This is appropriate since surface marker studies indicate that the majority (90%) of CLL cells are of B-cell origin. These more recent studies have revealed a variety of chromosome abnormalities, two of which stand out. One of these is an abnormality of chromosome 14 in which material from some other chromosome is translocated to the end of chromosome 14, producing a 14q+ chromosome. This chromosome adnormality was first described in Burkitt's lymphoma (Manolov and Manolova 1972, Zech et al.

1976) and has now been identified in other lymphomas, many of which are derived from B-cells (Rowley and Fukuhara 1980). At least three patients with CLL have been identified whose cells have a 14q+ chromosome (Finan et al. 1978, Gahrton et al. 1979). More recently, Autio et al. (1979) and Gahrton et al. (1980) have identified five patients whose cells have an extra chromosome 12. Various mitogens, including lipopolysaccharide and Epstein-Barr virus, were used in these recent studies. This should be a very productive area for further research.

Other Chronic or Subacute Lymphoid Leukemias

Hairy Cell Leukemia

As in CLL, the study of the chromosome pattern in hairy cell leukemia (HCL) is hampered by the low mitotic rate of the malignant cells and their lack of response to commonly used mitogens. In a study of 26 patients with HCL, we have found clonal chromosome abnormalities in only two of them (Golomb et al. 1978a). These were male patients with a rapidly fatal course whose karyotype showed an extra chromosome 12 with loss of the Y chromosome (46,X−Y,+12).

Plasma Cell Leukemia

Plasma cell leukemia is the leukemic phase of multiple myeloma. Chromosome studies have been described for only a few of these patients, but a 14q+ chromosome has been seen in most of them (Wurster-Hill et al. 1973, Liang et al. 1979). These patients also had very complex structural rearrangements in addition to the 14q+.

Acute Lymphoblastic Leukemia

Chromosome abnormalities have been observed in about one half of the patients with ALL (Mitelman and Levan 1978, Sandburg 1979, Rowley 1980b). Even when banding techniques are used, about 50% of patients appear to have a normal karyotype. It has long been recognized that aneuploid patients with ALL have higher modal chromosome numbers than do patients with acute nonlymphocytic leukemia (Sandberg 1979) (Table 1). Considerably fewer data are available on the types and frequency of chromosome changes in ALL than in ANLL.

Only two unselected series of patients with ALL, each studied with banding, have been reported. Oshimura et al. (1977) described results in 31 patients and Cimino et al. (1979) described results in 16 patients. There have been a number of other reports on one or a few patients, all selected for some unusual cytogenetic abnormalities, most frequently the presence of a Ph[1] chromosome

(Bloomfield et al. 1978, Chessels et al. 1979). (Ph[1]-positive ALL has already been discussed in the section on Ph[1]-positive leukemias.) The small number of patients and the complexity of the karyotypes make the identification of nonrandom patterns in ALL very difficult at present. Despite these handicaps, at least one karyotypic abnormality seems to be consistently associated with one type of ALL classified with the use of immunologic markers. It seems reasonable to assume that other associations will become apparent in the future, as we gain additional information about the karyotypic pattern and about the cell surface markers of subpopulations of lymphoid cells that will be defined with more sophisticated immunologic techniques.

Fifty-three patients with ALL whose karyotypes were abnormal have been reported (Rowley 1980b). The most frequent single change in this group was a gain of one chromosome 21; the second most frequent change was a gain of one chromosome 14, and then a gain of one chromosome 13 (Figure 5). The only chromosome lost with any frequency was one X chromosome. Abnormalities of the Y chromosome have not been described. The most common deletion is that involving the long arm of chromosome 6; the break point in 6q appears

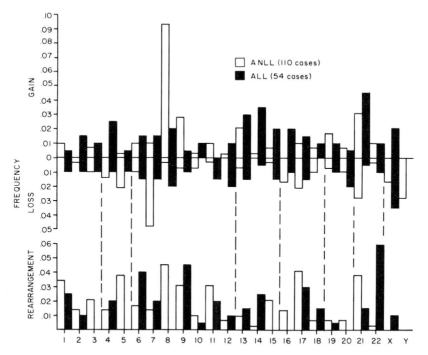

FIG. 5. Histogram of chromosome abnormalities (gains, losses, and rearrangements) in 110 cases of ANLL and in 54 cases of ALL, excluding documented cases of secondary karyotypic evolution. The frequency of each abnormality is calculated as a proportion of all abnormalities. (Reproduced from Cimino et al. 1979.)

to be somewhat variable, involving the region from 6q11 to 6q25. As can be seen in Figure 5, the chromosome changes in ALL differ from those seen in ANLL. These differences probably reflect the action of genes that differ in myeloid and lymphoid cells but that provide proliferative advantages to both.

Patients with B-cell ALL constitute a small percentage (about 4%) of those with ALL, and they are identified because their cells express surface immunoglobulin. Fifteen such patients have been reported in the literature. Every ALL patient whose leukemic cells have been identified as B-cells had an abnormality of chromosome 14 that was the result of a translocation of material from another chromosome to the end of the long arm (14q+). The donor chromosome was identified in 12 cases: it was 11q in the patient of Roth et al. (1979) and 8q in the 10 cases reported by Berger et al. (1979) and the one reported by Mitelman et al. (1979). Some of the cases of Berger et al. had a Burkitt-type solid tumor as well as a leukemic phase, which indicates that at least some B-cell ALLs may represent the leukemic phase of Burkitt's lymphoma. Karyotypic analysis will be useful in clarifying this relationship. Other abnormalities in addition to the 14q+ chromosome have included trisomy for part or all of the long arm of chromosome 6 and an additional chromosome 7 in three patients. Cells from two patients were examined for Epstein-Barr virus and were found to be negative (Mitelman et al. 1979). In one patient, the cells had a characteristically low level of adenosine deaminase (Roth et al. 1979).

CONCLUSIONS

The primary focus in this review has been on the identification of nonrandom chromosome changes in various myeloproliferative and lymphoproliferative disorders. Although the data available for these disorders are quite variable both with regard to the number of patients studied and the quality of banding, patterns of chromosome changes can be discerned that differ among the various groups. Wherever possible, these patterns have been related to structural and functional characteristics of these cells as determined by others, as well as to the clinical correlations of particular chromosome changes. In the future, these correlations must be extended to relate specific chromosome aberrations, particularly translocations and deletions, to alterations of the function of genes located at these sites (Rowley 1977).

ACKNOWLEDGMENTS

This work was supported in part by Grants CA-16910, CA-23954, and CA-19266 from the National Cancer Institute, Department of Health, Education and Welfare, and by the University of Chicago Cancer Research Foundation. The Franklin McLean Memorial Research Institute was operated by the University of Chicago for the U.S. Department of Energy under Contract No. EY-76-C-02-0069.

REFERENCES

Alimena, G., L. Brandt, B. Dallapiccola, F. Mitelman, and P. G. Nilsson. 1979. Secondary chromosome changes in chronic myeloid leukemia: Relation to treatment. Cancer Genet. Cytogenet. 1:79–85.

Autio, K., O. Turunen, O. Penttilä, E. Erämaa, A. de la Chapelle, and J. Schröder. 1979. Human chronic lymphocytic leukemia. Karyotypes in different lymphocyte populations. Cancer Genet. Cytogenet. 1:147–155.

Benedict, W. F., M. Lange, J. Greene, A. Derencsenyi, and O. S. Alfi. 1979. Correlation between prognosis and bone marrow chromosomal patterns in children with acute nonlymphocytic leukemia: Similarities and differences compared to adults. Blood 54:818–823.

Berger, R., A. Bernheim, J. C. Brouet, M. T. Daniel, and G. Flandrin. 1979. t(8;14) Translocation in a Burkitt's type of lymphoblastic leukaemia (L3). Br. J. Haematol. 43:87–90.

Bernstein, R., B. Mendelow, M. R. Pinto, G. Morcom, and W. Bezwoda. 1980. Complex translocation involving chromosomes 15 and 17 in acute promyelocytic leukaemia (APL). Br. J. Haematol. (in press).

Bloomfield, C. D., L. L. Lindquist, R. D. Brunning, J. J. Yunis, and P. F. Coccia. 1978. The Philadelphia chromosome in acute leukemia. Virchow's Arch. B Cell Pathol. 29:81–92.

Bloomfield, C. D., L. C. Peterson, J. J. Yunis, and R. D. Brunning. 1977. The Philadelphia chromosome (Ph¹) in adults presenting with acute leukaemia. A comparison of Ph¹+ and Ph¹− patients. Br. J. Haematol. 36:347–358.

Boveri, T. 1914. Zur Frage der Entstehung maligner Tumoren. Gustav Fischer.

Caspersson, T., G. Gahrton, J. Lindsten, and L. Zech. 1970. Identification of the Philadelphia chromosome as a number 22 by quinacrine mustard fluorescence analysis. Exp. Cell Res. 63:238–244.

Catovsky, D. 1979. Annotation: Ph¹-positive acute leukaemia and chronic granulocytic leukaemia: 1 or 2 diseases. Br. J. Haematol. 42:493–498.

Chessells, J. M., G. Janossy, S. D. Lawler, and L. M. Secker-Walker. 1979. The Ph¹ chromosome in childhood leukaemia. Br. J. Haematol. 41:25–41.

Cimino, M. C., J. D. Rowley, A. Kinnealey, D. Variakojis, and H. M. Golomb. 1979. Banding studies of chromosomal abnormalities in patients with acute lymphocytic leukemia. Cancer Res. 39:227–238.

Engel, E., B. J. McGee, J. M. Flexner, M. T. Russell, and B. J. Myers. 1974. Philadelphia chromosome (Ph¹) translocation in an apparently Ph¹-negative minus G 22, case of chronic myeloid leukemia. N. Engl. J. Med. 291:154.

Fialkow, P. J. 1974. The origin and development of human tumors studied with cell markers. N. Engl. J. Med. 291:26–35.

Finan, J., R. Daniele, D. Rowlands, Jr., and P. Nowell. 1978. Cytogenetics of chronic T cell leukemia, including two patients with a 14q+ translocation. Virchow's Arch. B. Cell Pathol. 29:121–127.

First International Workshop on Chromosomes in Leukaemia. 1978. Chromosomes in acute nonlymphocytic leukaemia. Br. J. Haematol. 39:311–316.

French-American-British (FAB) Co-operative Group. 1980. Bennett, J. M., D. Catovsky, M. T. Daniel, G. Flandrin, D. A. G. Galton, H. R. Gralnick, and C. Sultan. A variant form of hypergranular promyelocytic leukaemia (M3). Ann. Intern. Med. 92:261.

Gahrton, G., J. Lindsten, and L. Zech. 1973. Origin of the Philadelphia chromosomes. Tracing of chromosome 22 to parents of patients with chronic myelocytic leukemia. Exp. Cell Res. 79:246–247.

Gahrton, G., K.-H. Robèrt, K. Friberg, L. Zech, and A. G. Bird. 1980. Extra chromosome 12 in chronic lymphocytic leukaemia. Lancet I:146–147.

Gahrton, G., L. Zech, K.-H. Robèrt, and A. G. Bird. 1979. Mitogenic stimulation of leukemia cells by Epstein-Barr virus. N. Engl. J. Med. 301:438.

Geraedts, J. P. M., A. Mol, G. I. den Ottolander, M. van der Ploeg, and P. L. Pearson. 1977. Variation in the chromosomes of CML patients, *in* Proceedings of the Helsinki Chromosome Conference, p. 194.

Golomb, H. M., V. Lindgren, and J. D. Rowley. 1978a. Hairy cell leukaemia: An analysis of the chromosomes of 26 patients. Virchow's Arch. B. Cell Pathol. 29:113–120.

Golomb, H. M., J. R. Testa, J. W. Vardiman, A. Butler, and J. D. Rowley. 1979. Cytogenetic and ultrastructural features of the de novo acute promyelocytic leukemia: The University of Chicago experience (1973–1978). Cancer Genet. Cytogenet. 1:69–78.

Golomb, H. M., J. W. Vardiman, J. D. Rowley, J. R. Testa, and U. Mintz. 1978b. Correlation of clinical findings with quinacrine-banded chromosomes in 90 adults with acute nonlymphocytic leukemia. N. Engl. J. Med. 299:613–619.

Hagemeijer, A., G. E. Van Zanen, E. M. E. Smit, and K. Hahlen. 1979. Bone marrow karyotypes of children with nonlymphocytic leukemia. Pediat. Res. 13:1247–1254.

Hayata, I., M. Sakurai, S. Kakati, and A. A. Sandberg. 1975. Chromosomes and causation of human cancer and leukemia. XVI. Banding studies of chronic myelocytic leukemia, including five unusual Ph1-translocations. Cancer 36:1177–1191.

Hoffbrand, A. V., K. Ganeshaguru, G. Janossy, M. F. Greaves, D. Catovsky, and R. K. Woodruff. 1977. Terminal deoxynucleotidyl-transferase levels and membrane phenotypes in diagnosis of acute leukaemia. Lancet II:520–523.

Janossy, G., R. K. Woodruff, A. Paxton, M. F. Greaves, D. Capellaro, B. Kirk, E. M. Innes, O. B. Eden, C. Lewis, D. Catovsky, and A. V. Hoffbrand. 1978. Membrane marker and cell separation studies in Ph1-positive leukemia. Blood 51:861–877.

Kamada, N., K. Okada, T. Ito, T. Nakatsui, and H. Uchino. 1968. Chromosome 21–22 and neutrophil alkaline phosphatase in leukemia. Lancet I:364.

Kamada, N., K. Okada, N. Oguma, R. Tanaka, M. Mikami, and H. Uchino. 1976. C-G Translocation in acute myelocytic leukemia with low neutrophil alkaline phosphatase activity. Cancer 37:2380–2387.

Lawler, S. D., F. O'Malley, and D. S. Lobb. 1976. Chromosome banding studies in Philadelphia chromosome positive myeloid leukaemia. Scand. J. Haematol. 17:17–28.

Levan, A., G. Levan, and F. Mitelman. 1977. Chromosomes and cancer. Hereditas 86:15–30.

Liang, W., J. E. Hopper, and J. D. Rowley. 1979. Karyotypic abnormalities and clinical aspects of patients with multiple myeloma and related paraproteinemic disorders. Cancer 44:630–644.

Lindgren, V., and J. D. Rowley. 1977. Comparable complex rearrangements involving 8;21 and 9;22 translocations in leukaemia. Nature 266:744–745.

Manolov, G., and Y. Manolova. 1972. Marker band in one chromosome 14 from Burkitt lymphomas. Nature 237:33–34.

Marks, S. M., D. Baltimore, and R. McCaffrey. 1978. Terminal transferase as a predictor of initial responsiveness to vincristine and prednisone in blastic chronic myelogenous leukemia. N. Eng. J. Med. 298:812–814.

Mayall, B. H., A. V. Carrano, D. H. Moore II, and J. D. Rowley. 1977. Qualification by DNA-based cytophotometry of the 9q+/22q− chromosomal translocation associated with chronic myelogenous leukemia. Cancer Res. 37:3590–3593.

Mitelman, F., M. Anvret-Andersson, L. Brandt, D. Catovsky, G. Klein, G. Manolov, Y. Manolova, E. Mark-Vendel, and P. G. Nilsson. 1979. Reciprocal 8;14 translocation in EBV-negative B-cell acute lymphocytic leukemia with Burkitt-type cells. Int. J. Cancer 24:27–33.

Mitelman, F., L. Brandt, and P. G. Nilsson. 1978. Relation among occupational exposure to potential mutagenic/carcinogenic agents, clinical findings, and bone marrow chromosomes in acute nonlymphocytic leukemia. Blood 52:1229–1237.

Mitelman, F., and G. Levan. 1978. Clustering of aberrations to specific chromosomes in human neoplasms. III. Incidence and geographic distribution of chromosome aberrations in 856 cases. Hereditas 89:207–232.

Morse, H., T. Hays, D. Peakman, B. Rose, and A. Robinson. 1979. Acute nonlymphocytic leukemia in childhood. Cancer 44:164–170.

Nowell, P. C. 1976. The clonal evolution of tumor cell populations. Science 194:23–28.

Nowell, P. C., and D. A. Hungerford. 1960. A minute chromosome in human chronic granulocytic leukemia. Science 132:1197.

O'Riordan, M. L., J. A. Robinson, K. E. Buckton, and H. J. Evans. 1971. Distinguishing between the chromosomes involved in Down's syndrome (trisomy 21) and chronic myeloid leukaemia (Ph1) by fluorescence. Nature (London) 230:167–168.

Oshimura, M., A. I. Freeman, and A. A. Sandberg. 1977. Chromosomes and causation of human cancer and leukemia. XXVI. Banding studies in acute lymphoblastic leukemia (ALL). Cancer 40:1161–1172.

Paris Conference 1971. 1972. Standardization in human cytogenetics. Birth Defects 8:7.

Prigogina, E. L., E. W. Fleischman, M. A. Volkova, and M. A. Frenkel. 1978. Chromosome abnormalities and clinical and morphologic manifestations of chronic myeloid leukemia. Hum. Genet. 41:143–156.

Roth, D. G., M. C. Cimino, D. Variakojis, H. M. Golomb, and J. D. Rowley. 1979. B-cell acute lymphoblastic leukemia (ALL) with a 14q+ chromosome abnormality. Blood 53:235–243.

Rowley, J. D. 1973a. A new consistent chromosomal abnormality in chronic myelogenous leukaemia identified by quinacrine fluorescence and Giemsa staining. Nature (London) 243:290–293.

Rowley, J. D. 1973b. Identification of a translocation with quinacrine fluorescence in a patient with acute leukemia. Ann. de Génét. 16:109–112.

Rowley, J. D. 1977. A possible role for nonrandom chromosomal changes in human hemotologic malignancies, *in* Chromosomes Today, Vol. 6, A. de la Chapelle and M. Sorsa, eds. Elsevier/ North-Holland Biomedical Press, Amsterdam, pp. 345–359.

Rowley, J. D. 1978a. Chromosomes in leukemia and lymphoma. Semin. Hematol. 15:301–319.

Rowley, J. D. 1978b. The cytogenetics of acute leukaemia. Clin. Haematol. 7:385–406.

Rowley, J. D. 1980a. Ph[1] positive leukaemia, including chronic myelogenous leukaemia. Clinics Haematol. 9:55–86.

Rowley, J. D. 1980b. Chromosome abnormalities in acute lymphoblastic leukemia. Cancer Genet. Cytogenet. 1:263–271.

Rowley, J. D., and S. Fukuhara. 1980. Chromosome studies in non-Hodgkin lymphomas. Semin. Oncol. (in press).

Rowley, J. D., H. M. Golomb, J. Vardiman, S. Fukuhara, C. Dougherty, and D. Potter. 1977. Further evidence for a non-random chromosomal abnormality in acute promyelocytic leukemia. Int. J. Cancer 20:869–872.

Sandberg, A. A. 1979. Chromosomes in Human Cancer and Leukemia. Elsevier North-Holland, New York.

Second International Workshop on Chromosomes in Leukemia. 1980. Cancer Genet. Cytogenet. (in press).

Sonta, S., and A. A. Sandberg. 1977. Chromosomes and causation of human cancer and leukemia. XXIV. Unusual and complex Ph[1] translocations and their clinical significance. Blood 50:691– 697.

Sonta, S., and A. A. Sandberg. 1978. Chromosomes and causation of human cancer and leukemia. XXIX. Further studies on karyotypic progression in CML. Cancer 41:153–163.

Tanzer, J., Y. Najean, C. Frocrain, and A. Bernheim. 1977. Chronic myelocytic leukemia with a masked Ph[1] chromosome. N. Engl. J. Med. 296:571.

Testa, J. R., U. Mintz, J. D. Rowley, J. W. Vardiman, and H. M. Golomb. 1979. Evolution of karyotypes in acute nonlymphocytic leukemia. Cancer Res. 39:3619–3627.

Testa, J. R., and J. D. Rowley. 1980a. Chromosomal banding patterns in patients with acute nonlymphocytic leukemia. Cancer Genet. Cytogenet. 1:239–247.

Testa, J. R., and J. D. Rowley. 1980b. Chromosomes in leukemia and lymphoma with special emphasis on methodology, *in* The Leukemic Cell, D. Catovsky, ed. Churchill-Livingston, Edinburgh (in press).

Tosato, G., J. Whang-Peng, A. S. Levine, and D. G. Poplack. 1978. Acute lymphoblastic leukemia followed by chronic myelocytic leukemia. Blood 52:1033–1036.

Trujillo, J. M., A. Cork, M. J. Ahearn, E. L. Youness, and K. B. McCredie. 1979. Hematologic and cytologic characterization of 8/21 translocation acute granulocytic leukemia. Blood 53:695– 706.

Trujillo, J. M., A. Cork, J. S. Hart, S. L. George, and E. J. Freireich. 1974. Clinical implications of aneuploid cytogenetic profiles in adult acute leukemia. Cancer 33:824–834.

Verma, R. S., and H. Dosik. 1979. A case of chronic myelogenous leukemia (CML) with translocation between chromosomes 12 and 22 t(12;22)(p13;q13) resulting in Philadelphia (Ph[1]) chromosome. Cytogenet. Cell Genet. 23:274–276.

Whang-Peng, J., G. P. Canellos, P. P. Carbone, and H. H. Tjio. 1968. Clinical implications of cytogenetic variants in chronic myelocytic leukemia (CML). Blood 32:755–766.

Whang-Peng, J., E. S. Henderson, T. Knutsen, E. J. Freireich, and J. J. Gart. 1970. Cytogenetic studies in acute myelocytic leukemia with special emphasis on the occurrence of Ph[1] chromosome. Blood 36:448–457.

Wurster-Hill, D. H., O. R. McIntyre, G. G. Cornwell III, and L. H. Maurer. 1973. Marker chromosome 14 in multiple myeloma and plasma cell leukaemia. Lancet II:1031.

Zech, L., U. Haglund, K. Nilsson, and G. Klein. 1976. Characteristic chromosomal abnormalities in biopsies and lymphoid-cell lines from patients with Burkitt and non-Burkitt lymphomas. Int. J. Cancer 17:47–56.

Genes, Chromosomes, and Neoplasia, edited by
Frances E. Arrighi, Potu N. Rao, and Elton Stubblefield.
Raven Press, New York © 1981.

Chromosomal Changes in Primary and Metastatic Tumors and in Lymphoma: Their Nonrandomness and Significance

Avery A. Sandberg and Norio Wake

Division of Medicine, Roswell Park Memorial Institute, Buffalo, New York 14263

The status of karyotypic findings in human cancer and lymphoma, the meaning and value of these findings, the vicissitudes and tribulations related to the technical difficulties encountered in determining the chromosomal changes in most cancers, and the advances, however limited, made in this field have been told and discussed in several extensive reviews (Sandberg and Hossfeld 1974, Lawler 1977, Mark 1977, Rowley 1978, Harnden and Taylor 1979) and in books (German 1974, Atkin 1976, Sandberg 1980). Over 4,000 references to the extensive literature of this field may be found in Sandberg (1980). Thus, it would be redundant and presumptuous in this presentation merely to undertake a similar approach or exposition. Instead, we shall concern ourselves with concise descriptions of the accomplishments in the cytogenetics of lymphoma and specific cancers, wherever possible relate these accomplishments to the biology, pathology, and clinical aspects of these neoplasias, and point to those areas which await and beg for accomplishment in the near future. Undoubtedly, the contribution of cytogenetics to the understanding of the etiology of human lymphoma and cancer is in its infancy. However, we do not share the pessimism expressed by some regarding the value of knowing the chromosomal changes in human cancer, for in a little over two decades since 1956 we have seen slow but steady progress in this field. As techniques become available to further expand the level and nature of visualization of chromosomal changes, they will unquestionably broaden our understanding and appreciation of the karyotypic changes in specific cancers and lymphomas.

The third decade of cytogenetic studies in human neoplasia is now in progress. Though banding has unquestionably led to discoveries not readily apparent before the introduction of such techniques, the quality and quantity of karyotypic data obtained on human lymphomas and tumors prior to banding paved the way to appreciation of the old and discovery of the new cytogenetic findings and approaches. In all probability, as new banding and other techniques develop, particularly those revealing sub-band detail, changes characteristic of more tumors will be discovered.

RANDOM AND NONRANDOM KARYOTYPIC CHANGES
IN HUMAN NEOPLASIA

In assessing the chromosomal changes in human neoplasia, we shall concern ourselves primarily with the occurrence of nonrandomness and its significance. Ideal nonrandomness would consist of a characteristic and specific chromosomal change encountered in 100% of the cases of a particular leukemia or cancer. Though no such karyotypic specificity has been found, the presence of the Philadelphia (Ph[1]) chromosome in chronic myelocytic leukemia (CML) in nearly 85% of the cases comes close to meeting the criterion for specificity (Sandberg 1980). Unfortunately, such a specific finding does not characterize the majority of human cancers, and hence a search must be made to establish whether nonrandom changes can be found for each particular neoplasia. Since human cancer includes many diverse diseases, the etiology, biology, clinical presentation, therapy, and prognosis of which differ greatly, it is not surprising that the karyotypic changes show such a wide spectrum and, hence, that the establishment of nonrandomness has not been fulfilled for most of these conditions.

There are some practical aspects to this area that should be mentioned, particularly since they have much to do with the fact that little is known yet regarding the occurrence of specific chromosome changes in human tumors. Thus, a major obstacle to progress in this area resides in the technical difficulties associated with the analysis of solid tumor material and the presence of numerous "random" changes that tend to obscure any specific abnormalities that may be present. This applies in particular to the difficulty in banding a large proportion of solid tumors, thus preventing the establishment of a concise and reliable karyotypic picture for each tumor. Two approaches, however, can be used to possibly uncover nonrandom changes (Harnden and Taylor 1979). The first is to analyze a large number of tumors in order to determine whether any features of the karyotypes are common to the tumors. The second is to search for tumors with minimal changes, since these may represent cancers in which possibly the specific changes are not obscured by other random ones. An attempt will be made in this presentation to address both of these areas.

In deciding upon nonrandomness reliance must be placed on those studies in which adequate banding analyses have been performed, since without these analyses the chromosomal changes cannot be rigorously established, e.g., chromosomes of the same morphology may have different derivations and normal-looking chromosomes may, in fact, be quite abnormal.

The significance of nonrandom changes and their possible application in the classification and diagnosis of certain diseases (lymphoma, leukemia) and the information they yield on the possible biologic behavior of tumors may ultimately turn out to be of value in the treatment of these conditions, with the possibility that chromosomal changes may also shed light upon the etiology of neoplasia. No attempt will be made in this presentation to discuss the possible relation of chromosome changes to the direct causation of neoplasia, since this has

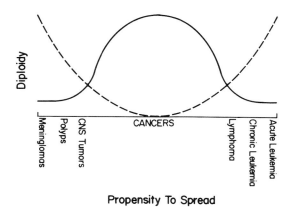

Propensity To Spread

FIG. 1. Schematic presentation of various human neoplastic lesions with the broken line indicating the distribution of diploidy and aneuploidy, i.e., leukemias and meningiomas containing a significant number of diploid cases and the bulk of cancers being aneuploid. The solid line indicates the ability of the cells of the various neoplastic lesions either to metastasize or spread through the body.

been the subject of a number of publications and the area continues to remain controversial (Sandberg 1980). Furthermore, leukemia has been discussed in a preceding presentation, and hence this presentation will address itself to the range of cytogenetic findings in human neoplasms from lymphoma on one end to meningiomas on the other (Figure 1), with the rest of human cancers, exclusive of leukemia, occupying the intervening but large area of human neoplasia.

Under no circumstances do we intend to belittle the so-called random changes in human neoplasia. Their significance and meaning have not been established, though undoubtedly they must reflect basic mechanisms of the tumor and probably have an important bearing on its biologic behavior. However, the complexity of these random changes, their numerical variation and remarkable morphologic variability, preclude at the moment a close application of the findings. We hope that in the future these changes can be correlated with more detailed and meaningful facets of these tumors in order to establish their significance.

SOME DISTRIBUTIONAL ASPECTS OF CHROMOSOMES IN HUMAN NEOPLASIA

Figure 1 is a schematic presentation of the distribution of human neoplasia based on the propensity of the neoplastic cells to metastasize or spread through the body. At one extreme are the meningiomas, which are considered to be benign tumors, though cytogenetic evidence, as well as biologic behavior and histologic aspects point otherwise, and at the other extreme are the leukemias. Adjacent to the former are polyps of various natures, which have been shown to be accompanied by chromosomal changes, possibly indicative of malignant transformation. Next to the leukemias are the lymphomas, which are somewhat intermediate between the leukemias and solid tumors. Between these is the bulk of human cancers subject to more or less propensity to metastasize. Also shown in Figure 1 is the occurrence of chromosome changes in the tumors. As can be seen, at both extremes diploid conditions or tumors are rather common, constituting nearly 50% of all the acute leukemias and about the same percentage of meningiomas and polyps. Diploidy in families of human neoplastic conditions affords the cytogeneticist not only an unusual opportunity to study either totally or partially diploid neoplastic lesions but also to study them at relatively early (if not the earliest) stages of malignant transformation at both extremes of the curve (Figure 1). However, considerable is yet to be learned about the role of gross chromosomal changes in tumor behavior. The bulk of human cancers are invariably aneuploid; convincing evidence that a cancer may be diploid has not been established.

Another interesting facet of the conditions lying at the extremes of the curves shown in Figure 1 is the fact that rather characteristic, if not specific, karyotypic changes have been described both in meningiomas and possibly polyps, as well as in some of the leukemias and lymphomas. On the other hand, the preponderant number of human cancers have to date not been characterized by any nonrandom changes of high frequency, though there is a glimpse of hope in that respect in more recent studies (to be discussed below) dealing with relatively large series of particular tumors and based on detailed analysis of their banding patterns.

THE LYMPHOMAS

Banding made possible the discovery by Manolov and Manolova (1971, 1972) of the 14q+ anomaly in lymphoma (Figures 2–5). The change was apparently too subtle to have been characterized firmly without banding, though when we now examine published metaphases or karyotypes obtained prior to banding, the presence of such extra material on a chromosome 14 can be deduced in certain cases of lymphoma and Burkitt's lymphoma. The 14q+ (and 8q−) anomaly has been seen in almost every form of lymphoma (Table 1), being of highest incidence (probably 100%) in endemic and nonendemic Burkitt's lymphoma and probably least frequent in Hodgkin's disease, among the common lymphomas. In Table 2 are shown the percentages of 14q+ and other karyotypic anomalies present in some lymphomas published in the literature. In Table 3 is shown a list of those chromosomes that serve as donors for the translocation with the long arm of chromosome 14; the most common is due to t(8;14). It should be stressed that, except for Burkitt's lymphoma, a number of donor chromosomes have been found in the various lymphomas. In Burkitt's lymphoma, specificity of the 8q− appears to be well established, and there are some indications that this anomaly may be more specific for the disease than the 14q+, i.e., a 2;8 and an 8;22 translocation in European Burkitt's lymphoma have been described (Van Den Berghe et al. 1979, Miyoshi et al. 1979, Berger et al. 1979). Thus, in our opinion, the diagnosis of this disease should probably not be seriously considered unless an 8q− lesion is present as part of a translocation, usually with the long arm of a chromosome 14.

Since different chromosomes can be donors for the 14q+ in lymphoma (Figure

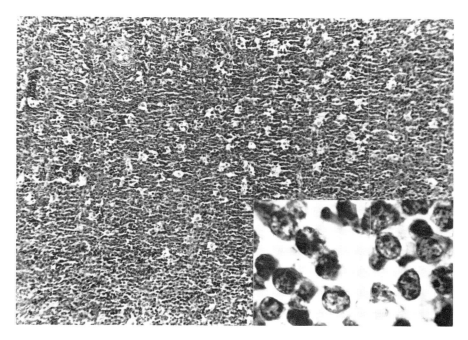

FIG. 2. Burkitt's type undifferentiated lymphoma in a 10-year-old American male. The character-istic morphology of Burkitt's lymphoma, the "starry sky" pattern, makes for easy identification of the disease on histologic basis. In the inset is shown a higher magnification of cells with well-demarcated nuclear membranes, coarsely reticulated and irregularly distributed chromatin, two to five prominent nucleoli, and a moderate amount of cytoplasm. Unfortunately, criteria for the identification on histologic basis of other types of lymphoma appear not to be as rigorous as those in Burkitt's lymphoma.

6), it will be of interest to ascertain in the future whether the nature of the donor chromosome, the amount of material translocated, and the proportion of cells involved by such an anomaly are related to some specific histologic, etiologic, or clinical parameter (Pierre et al. 1980). This question obviously assumes that the receptor chromosome (No. 14) may not play as crucial a role in such lymphomas as the donor chromosome. Why the long arm of chromo-some 14 displays such a consistent and remarkable affinity for receiving material during the process of a translocation with another chromosome remains to be established. Undoubtedly, knowing this process intimately will afford oncologists a somewhat better view of the biology of human lymphoma.

It is natural to ask whether the lymphomas that are not characterized by the presence of 14q+ are in any way different from those that are. At present, the information related to this problem is too scanty and confusing to yield any meaningful conclusions. Thus, it behooves all of those involved and interested in lymphoma, both at the cytogenetic and clinical levels, to study as many lymphomas as possible, not only regarding their chromosome constitution, but

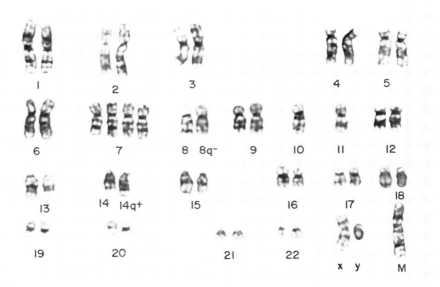

FIG. 3. G-banded karyotype of the case shown in Figure 2, containing 47 chromosomes with a karyotype 47,XY,+7,+7,−10,−11,t(8;14) (q23;q32), plus a marker (M). The type of translocation t(8;14) shown in the figure is characteristic but not invariable in BL.

also the nature of the cells involved (immunologic markers, enzymatic characteristics, etc.) as well as the histology, biology, and clinical behavior of these lymphomas in order to be in a position to correlate these findings with the karyotypic abnormalities. The challenge to cytogeneticists in this area is considerable, though undoubtedly they can contribute much to a more accurate and concise classification of lymphoma, with the possibility that such findings can be of value in determining therapeutic approaches.

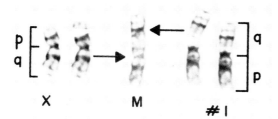

FIG. 4. Partial karyotype based on G-banding showing the genesis of the marker (M) of the previous figure. It is due to a translocation between the X and a No. 1 chromosome.

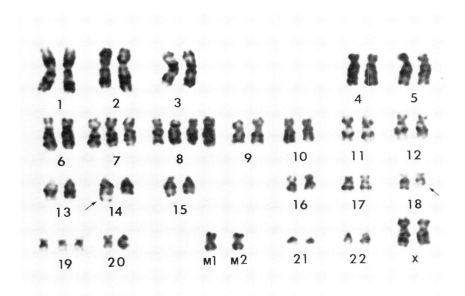

FIG. 5. A 14q+ due to t(14;18) in an established cell line from a B-cell lymphoma. In addition to the translocation, extra chromosomes in groups 7 and 8, an extra chromosome in group 19, and two marker chromosomes whose identity could not be established with certainty are present.

Though much emphasis has been given to the 14q+ anomaly in lymphoma, it should be pointed out that involvement of chromosome 1 occurs in a substantial number of the cases, as it does to a somewhat lesser but nevertheless significant level for chromosomes 6 and 11 (Table 2) (Rowley and Fukuhara 1980, Kakati et al. 1979, 1980).

The demonstration of variant translocations, i.e., t(2;8) and t(8;22), other than the common t(8;14) in nonendemic Burkitt's lymphoma (Van Den Berghe et al. 1979, Miyoshi et al. 1979, Berger et al. 1979) raises the strong possibility of similar variants being present in the endemic disease. The cytogenetic data on the latter condition are based on a relatively small number of cases; furthermore, in most of these the chromosomes were studied after the cells had been established in long-term cell lines. We are now in the strange position of having more karyotypic data on nonendemic than on endemic Burkitt's lymphoma. Thus, it behooves cytogeneticists to study more cases of the endemic disease in order to ascertain whether variant translocations also occur in this disease and close the gap existing now between the cytogenetic data in nonendemic and endemic Burkitt's lymphoma. In addition, it will be important to ascertain whether the cases with variant translocations have features that differentiate them from those with 14q+ histologically, clinically, and immunologically.

TABLE 1. *Incidence (in order of frequency) of 14Q+ in various lymphomas*

1. Burkitt's lymphoma
   Endemic
   Nonendemic
2. Poorly differentiated lymphocytic lymphoma
   (Diffuse = lymphocytic lymphoma; nodular = giant follicle lymphoma)
3. Diffuse histiocytic lymphoma
   (Reticulum cell sarcoma)
4. Well-differentiated lymphocytic lymphoma
5. Mycosis fungoides
6. Hodgkin's disease
7. Multiple myeloma and plasma cell leukemia
8. B-cell ALL
9. T-cell CLL
10. Sézary syndrome(?)

TABLE 2. *Chromosomes most often involved in lymphoma*

Chromosome no.	Percent of cases involved
1	55% (q 31%)
	(p 24%)
6	27% (q 18%)
	(p 9%)
11	26%
14	73%
14q+	64%

TABLE 3. *Donor chromosomes in 14q+ anomaly in lymphoma*

Lymphoma type	Chromosomes involved
Burkitt's lymphoma	
Endemic	8
Nonendemic	8,?1
Variant translocations	t(2;8) (p12;q23) t(8;22) (q24;q11)
Poorly differentiated lymphocytic lymphosarcoma	1,6,8,11,14,18
Diffuse histiocytic lymphoma	1,4,8,10,11,13,14,15,20
Hodgkin's disease	1,8,14
Mycosis fungoides	1,2,8
Multiple myeloma and plasma cell leukemia	1,11
T-cell CLL	11,18
B-cell ALL	8,11

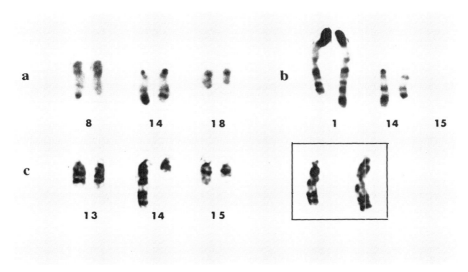

FIG. 6. 14q+ anomaly in several lymphomas in which the donor chromosomes are No. 15 and No. 18. (Figure reproduced courtesy of Janet D. Rowley, M.D.)

Involvement of the long arms of the various chromosomes in lymphoma was three times as frequent as that of the short arms. This may reflect the quantitative distribution of the chromatin in the long and short arms in the human chromosome set. However, nearly 20% of the long arms involved belonged to chromosome 14 and of the short arms to chromosome 1.

It was thought originally that lymphomas originating only from B-cells were characterized by the 14q+ anomaly; recently, however, lymphoproliferative diseases originating from T-cells have been shown to contain the 14q+ (Finan et al. 1978, Nowell et al. 1980). More than 10 chromosomes have been shown to be the donors in the material translocated to chromosome 14. The major feature of these findings is the fact that the deleted long arm of a chromosome 14, characteristically at band 14q32, serves as a "specific" receptor of the translocated material. In addition, the amount of material translocated (and obviously that deleted from the donor chromosomes) appears to vary considerably among the cases. Much work remains to be performed in this area, particularly in correlating the type of lymphoma and the donor chromosome, determining the amount of material translocated, and establishing the consistency of the abnormalities. Since considerable controversy continues to exist among pathologists about the exact histologic classification of lymphomas, it is possible that as more and more cytogenetic findings are established in these diseases the cytogeneticist will be in a unique position to classify the lymphomas.

The other changes related to the lymphomas are concerned with the chromosome numbers in the various diseases.

In Hodgkin's disease a large percentage of normal (diploid) metaphases can be found, and totally diploid tumors are practically nonexistent; chromosomally abnormal clones in Hodgkin's disease are distributed among the near-diploid, near-triploid, and near-tetraploid ranges, with most of the cases belonging to the two last high-ploidy groups. Burkitt's lymphoma, diffuse histiocytic lymphoma, and poorly differentiated lymphocytic lymphosarcoma usually have near-diploid modes. Burkitt's lymphoma is characterized by sharp, prominent modes, whereas all other lymphomas have flat modes, and not infrequently a multimodal chromosome distribution. Abnormal (marker) chromosomes occur most frequently in diffuse histiocytic lymphoma ($> 70\%$), followed by poorly differentiated lymphocytic lymphosarcoma ($\sim 60\%$) and Burkitt's lymphoma ($\sim 50\%$); they are least common in Hodgkin's disease ($\sim 40\%$).

The preponderant number of lymphomas examined have chromosomal changes (Mark et al. 1979, Rowley and Fukuhara 1980, Sandberg 1980). As mentioned previously, the 14q+ anomaly, though it appears to be relatively characteristic for this group of the diseases, does not occur in every tumor. However, other chromosomal changes are invariably present in lymphomas,

TABLE 4. *Involvement of chromosomes in morphological abnormalities as revealed by banding in 102 cases of non-Hodgkin's non-Burkitt's lymphomas* *

Chromosome no.	No. cases	Comments
1	28	13 with 1p—, 10 with 1q—, 4 with i(1q), 1 with 1p+
2	8	7 with 2q—, 1 with 2q+
3	15	7 with 3p—, 6 with 3q—, 3 with 3q+
4	2	
5	3	
6	17	15 with 6q—
7	4	
8	11	10 with 8q—, 1 with 8p—
9	13	8 with 9q—
10	6	5 with 10q—
11	17	all with 11q—
12	6	
13	5	
14	51	48 with 14q+
15	4	
16	3	
17	8	5 with i(17q), 2 with 17p+
18	9	8 with 18q—
21	2	
22	6	
Xq—	3	

* Adapted from Kakati et al. (1980), with cases from Mark et al. (1979) plus additional ones.

and the more common ones are listed in Table 4. As can be seen, though the chromosomal changes are protean, some appear to occur more often than others. Whether these changes carry hidden within the large number of changes a karyotypic anomaly similar to the 14q+ will have to be ascertained not only on the basis of a very large number of studies, but perhaps also after the development of more refined cytogenetic techniques.

Burkitt's lymphoma associated with a reciprocal 8;14 translocation is not limited to the Epstein-Barr (EB) virus-carrying African (endemic) disease. It is also found in EB virus–negative American Burkitt's lymphoma and in rare B-cell forms of acute lymphoblastic leukemia (ALL), believed to represent the neoplastic growth of the same cell type as in Burkitt's lymphoma. The fact that EB virus–transformed cells of non-Burkitt's lymphoma origin do not carry the 8;14 translocation suggests that EB virus is probably not involved in causing the translocation.

It is interesting to note that chromosome 14 anomalies are found with a high frequency in ataxia-telangiectasia, a condition known to be associated with a markedly increased incidence of lymphoma (Sandberg 1980). However, the most frequent break point in chromosomes of patients with ataxia-telangiectasia is in band 14q12, whereas the Burkitt's lymphoma–associated break point is in band 14q32. In other cases, the break point has been shown to be at 14q13 (Fukuhara et al. 1979) with translocations involving chromosomes 1, 2, 4, or 14. It is possible that the 14q31 represents a "receptor" site, and 14q13 a "donor" site. The latter site appears to be identical to that described by Kaiser-McCaw et al. (1977) in translocations involving chromosome 14 in ataxia-telangiectasia.

CHRONIC LYMPHOCYTIC LEUKEMIA

The most recent development in the field of lymphoma is related to observations in chronic lymphocytic leukemia (CLL). This disease is preponderantly of B-cell origin and until recently was characterized by an extremely low mitotic index either in the marrow or involved tissues, thus precluding reliable chromosome analysis. A spuriously normal (diploid) picture was usually obtained in CLL when phytohemmaglutinin (PHA) was used as a mitogen, since it tended to stimulate the normal T-cells to undergo division without apparently affecting the leukemic B-cells. In the case of T-cell CLL, PHA should be used as a mitogen, and, in fact, results based on this approach in T-cell CLL have been published (Finan et al. 1978). The introduction of EB virus and several bacterial extracts as mitotic stimulants of such cells has yielded results that indicate that a substantial number of CLL cases are accompanied by changes of chromosome 2, 12, and 14 (Gahrton et al. 1979, 1980, Autio et al. 1979, Hurley et al. 1980, Morita, M., J. Minowada, and A. A. Sandberg, manuscript in preparation). Undoubtedly, as more cases are examined it will be ascertained how consistent and significant the changes are, in relation to the Ph[1], +8, iso-17, and other karyotypic changes in CML and acute nonlymphocytic leukemia.

MENINGIOMAS

The meningiomas, considered to be one of the most benign of human tumors, offer an unusual opportunity for the study of the role of chromosomal changes in malignancy (Figure 7). The cytogenetic findings in meningioma are of importance in that in a substantial number of such tumors monosomy of chromosome 22 is present (Mark 1974, 1977, Zankl and Zang 1978, 1980). The monosomy apparently develops through several stages (Table 5), ultimately leading to the loss of the chromosome. The loss of chromosome 22 is preceded by deletion of the long arm of this chromosome, but no evidence has been obtained that it is translocated, as it is in the Ph[1] chromosome in CML. A definite correlation between the histologic type of the tumor and cytogenetic findings has been established. Some relationship to the sex of the patient (predominance of the tumor in females) and to the biologic behavior of the tumors has also been established.

The cytogenetic evidence indicates that meningiomas may not be benign tumors, as they have been assumed to be in the past, and that they, in fact, represent the extreme in Figure 1. A substantial number of the tumors may be "benign" and lack chromosomal changes, whereas those with chromosomal changes probably represent one of the earliest stages of malignant transformation. Thus, the cytogenetic findings have led to a reevaluation of the nature of meningiomas and their biologic behavior and to a better understanding of the genesis of the chromosome abnormalities, as well as of the clinical behavior of these tumors.

Chromosome studies in meningioma are usually performed on cultured material, since the original tumor contains too few metaphases for analysis. This presents both an advantage and a problem, i.e., the cells in vitro may not necessarily reflect the in vivo karyotype, but at the same time culture does offer an opportunity to obtain a sufficient number of dividing cells for cytogenetic analysis.

Over 200 meningiomas have been studied to date (Sandberg 1980), and thus we know more about the chromosomes of this tumor than any other. Early

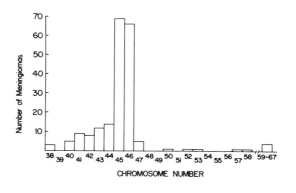

FIG. 7. Distribution of the modal chromosome number in meningiomas.

TABLE 5. *Possible evolution of karyotypic changes in meningioma*

1. Normal diploid stem line
   ↓
2. Normal diploid stem line
   plus
   variant cells with monosomy G
   ↓
3. Normal diploid stem line
   plus
   sideline with monosomy G
   ↓
4. Stem line with monosomy G
   plus
   normal diploid sideline
   ↓
5. Stem line with monosomy G
   plus
   variant cells with diploidy
   ↓
6. Stem line with monosomy G
   ↓
7. Stem line with monosomy G
   plus
   additional karyotypic deviations

in the studies on meningioma attention was drawn to the fact that a strikingly uniform anomaly of the chromosome constitution in these tumors is character-ized by the loss of a G-group chromosome, which has been shown with banding techniques to be consistently a chromosome 22 (Figures 8, 9).

Certain clinical data appear to correlate with the cytogenetic findings in menin-gioma (Zankl and Zang 1980). A higher degree of hypodiploidy and atypical chromosome loss is observed in some of the tumors that had invaded the skull as well as in some of the recurrent meningiomas. A correlation was also found between the site of tumor origin and the karyotype. Tumors located at the convexity of the brain showed mostly an increased hypodiploidy, and those at the base of the brain a normal karyotype. Meningiomas in the spinal canal almost invariably had monosomy of chromosome 22. In the skull, however, the chromosome 22 monosomy was found in a roughly similar frequency to those at the convexity in the base of the brain. Meningiomas with normal karyo-types or monosomy of chromosome 22 predominate in women, whereas tumors with severe hypodiploidy or marker chromosomes were found as frequently in men as in women. Other chromosomes may be involved in numerical or morpho-logic abnormalities in meningiomas, with chromosome 8 being most frequently affected (Figures 8, 9).

An interesting finding has been the increased association between the acrocen-tric chromosomes of meningiomas and the chromosome 22 monosomy. This association has been interpreted as overcompensation for the loss of only one nucleolus organizing region (NOR) (Zankl and Zang 1978).

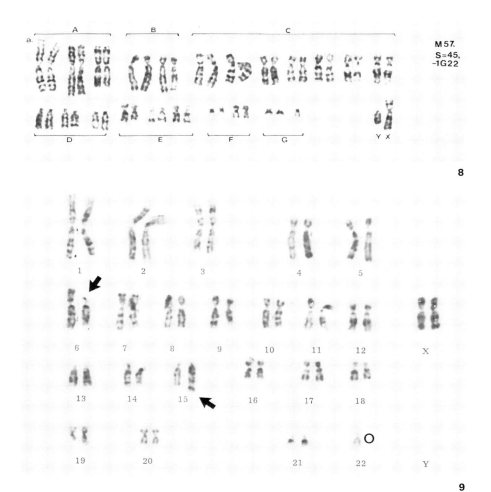

FIGs. 8 and 9. Karyotypes showing the characteristic cytogenetic change in meningioma, i.e., −22. Other changes (shown by arrows in the karyotype of Fig. 9) may also be present in meningioma, these being possibly related to neoplastic transformation of the tumors, usually considered benign. (Figure 8 reproduced courtesy of Dr. J. Mark, Figure 9 of Dr. K. Zang.)

SPECIFIC KARYOTYPIC CHANGE IN OVARIAN PAPILLARY SEROUS CYSTADENOCARCINOMA

The karyotypic findings in ovarian cancer, when obtained from a relatively large group of tumors and when given close scrutiny on banding, serve as an example of likely future results when large series of particular tumors are examined cytogenetically and methodologies allowing reliable banding analysis of the tumors become available. In our hands (Wake et al. 1980), 100% of papillary cystadenocarcinomas of the ovary, though containing a number of karyotypic abnormalities, have been characterized by the presence of 14q+ or 6q−, or both (Figure 10). In five cases, a reciprocal t(6;14) (q21;q24) was established. The donor chromosome in the other cases could not be established with certainty, and the amount of material translocated onto the long arm of chromosome 14 appeared to vary among the cases. We have obtained the impression from these analyses that the presence or absence of the 14q+ anomaly in ovarian cancer may be relatively specific for this group of tumors, i.e., those tumors containing the 14q+ anomaly constitute a rather homogeneous group histologically and clinically. Thus, the findings again demonstrate the value of performing detailed banding analysis in specific human tumors with the aim of defining karyotypic changes that may be specific for, or at least characteristic of, the tumors.

It is interesting to speculate on why the common karyotypic anomalies in ovarian papillary serous cystadenocarcinoma involve chromosomes not infrequently involved in other conditions, e.g., chromosome 6 in ALL and lymphoma (Rowley 1980), chromosome 14 in lymphoma and polyps (Mark 1977). There is the possibility that certain chromosomes are more likely to be affected in human neoplasia regardless of the cause or tissue involved, e.g., chromosome 22 in CML and meningiomas, chromosome 1 in a number of cancers, chromo-

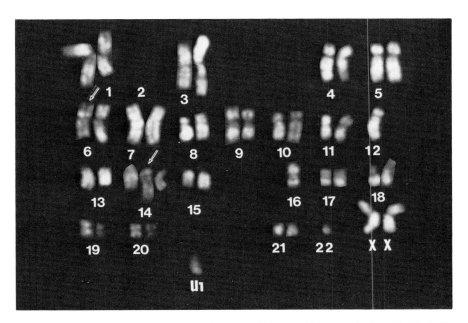

FIG. 10. Karyotype demonstrating the rather specific karyotypic anomaly, i.e., t(6;14), in the cells of a papillary serous cystadenocarcinoma of the ovary. U_1 is a marker chromosome whose origin could not be identified.

some 14 in ovarian cancer, lymphoma, and colonic polyps, chromosome 8 in leukemia, meningioma, and polyps, and chromosome 3 in multiple myeloma and lymphoma (Sandberg 1980).

The recent demonstration (Trent and Davis 1979) with banding techniques that specific involvement of chromosomes 13 and 15 in translocation may occur nonrandomly in endometrial cancer and hyperplasia again indicates the value of detailed karyotypic analysis and the strong possibility for establishing characteristic chromosomal changes for each type of tumor.

KARYOTYPIC OBSERVATIONS ON MIXED SALIVARY GLAND TUMORS

Even though a few parotid gland tumors were examined prior to banding, the results were too complicated to warrant reliable interpretation (Sandberg 1980). However, the recent observations of Mark et al. (1980) on 10 mixed salivary gland tumors are worthy of special note. These tumors are usually considered to be benign, though some of them recur after surgical removal and may undergo pleomorphic changes suggestive of malignancy. Mark et al. (1980) examined cultured cells from 10 such tumors and found chromosomal changes of interest in four. The authors felt that structural changes in the chromosomes of these and other tumors are more frequent and of more importance than numerical deviations. The most striking observation was the occurrence of an identical (3;8) (p25;q21) translocation in two different pleomorphic adenomas. This finding gains interest by the occurrence of 3p— and 8p—q— in a third adenoma with signs of early cancerous transformation. The breakpoints in the short arm of chromosome 3 and the long arm of chromosome 8 in the last case are close to those observed in the cases with the 3;8 translocation.

Though the 3p;8q involvement is a very suggestive feature, the authors (Mark et al. 1980) felt that great caution must be used in evaluating the results, since the number of cases with this particular structural involvement was small and such a karyotypic change was not seen in a fourth adenoma with abnormal stem lines, indicating that several evolutionary karyotypic patterns exist in such mixed tumors of the parotid gland; particularly significant was the occurrence of two related and a third unrelated abnormal cell line in consecutive preparations of this fourth adenoma. In this connection, of interest is the recent description of a constitutional reciprocal 3;8 translocation [t(3;8) (p21;q24)] in an Italian-American family predisposing to the development of renal carcinoma (Cohen et al. 1979). The tumors tended to develop at an earlier age than is usual for renal carcinoma and were often bilateral and multifocal. Though no such hereditary change was seen in the blood lymphocytes of the patients with parotid tumors examined by Mark et al. (1980), the two reports indicate that losses or changes in position or character of loci in the short arm of chromosome 3 and in the long arm of chromosome 8 may predispose to the development of neoplasia in different tissues. The unexplained but well-established common involvement of chromosome 8 in various disorders such as meningioma, CML, acute myeloblastic leukemia, and colonic polyps and of chromosome 3 in non-Burkitt's lymphomas and multiple myeloma are observations in line with the above idea.

Cytogenetic studies in meningiomas and mixed tumors of the parotid gland, in which transformation to a maligant state probably occurs slowly and can be observed at early stages, point to specific chromosomal changes in such cancerous transformations. In all probability, involvement of chromosome 22 in meningioma and chromosomes 3 and 8 in the mixed tumors is associated intimately with malignant transformation, with the subsequent karyotypic changes possibly being a reflection of either further progression of the neoplastic state or change in the biologic behavior of the tumor. Unfortunately, such early cytogenetic studies are usually not available for the preponderant number of human cancers, and the cytogeneticist is usually faced with a kaleidoscopic karyotypic picture defying rational interpretation. However, as was shown by us in ovarian cancer and by Mark et al. (1980) in mixed tumors of the parotid, some nonrandom changes can be established, with the hope that as more and more data are collected, ideally on early tumors, the original and specific karyotypic change can be established.

POLYPS (ADENOMAS) OF LARGE BOWEL

The generally held view is that most cancers of the large bowel arise from polyps. However, although polyps are common, the incidence of their malignant transformation is not too high and varies with the histological type. In adenomatous polyps, the occurrence of cancer has been estimated to be not more than 1%, whereas in villous adenomas it may range from 35% to 70%; some polyps have the features of both adenomatous and villous types (Mark 1977).

Polyposis coli is a familial disorder due to an autosomal dominant gene in which a large number of small polyps and multiple cancers usually develop. These lesions arise at a lower age than large bowel cancer generally, the maximum prevalence occurring at 35 years of age.

Chromosomes in polyps that were surgically removed incidentally with a cancer or because they produced symptoms were studied. Prior to banding, a number of reports of aneuploidy in polyps, with or without diploid cells and with or without evidence of a clone, were reported. Generally, extra chromosomes in groups C and D (Mark et al. 1973, Mitelman et al. 1974) were reported in colonic polyps of both sporadic and hereditary types. G-banding revealed that the deviations preferentially affected chromsomes 8 and 14.

An interesting observation is that of Danes and Alm (1979), who confirmed previously observed cytogenetic results in cell cultures of material obtained from patients with adenomatosis of the colon and rectum (ACR), i.e., an increased incidence of tetraploidy in epithelium-containing cultures from colonic mucosa but not of dermis derived from the same patients. Thus, increased tetraploidy was not observed in all cultured cells with the ACR gene, but only in colonic mucosa that, although appearing normal in vivo, was known from clinical phenotypes and family histories to undergo malignant transformation in vivo. Since not all cases with ACR showed such an increased tetraploidy in vitro, the authors (Danes and Alm 1979) suggested that genetic heterogeneity exists within this clinically defined group.

Recent developments in techniques for examining large bowel tumors, both benign and malignant, should go a long way towards increasing the efficacy of cytogenetic examinations in these tumors (Martin et al. 1979).

It appears to us that cytogenetic analysis of polyploid lesions of the large bowel have revealed that early malignant transformation can be established on the basis of karyotypic analysis. They point to the possibility that most large bowel cancers may develop from polyploid lesions. It is also possible that the cytogenetic analysis can be of considerable help to the pathologist and surgeon in deciding upon the histologic status of polyps removed from the colon, where criteria for their malignant transformation are often not clear-cut.

INVOLVEMENT OF CHROMOSOME 1

In all probability, the most commonly involved chromosome in human cancer (and leukemia) is chromosome 1, particularly its long arm (Figures 11 and 12). This was first established in our laboratory with banding of chromosomes

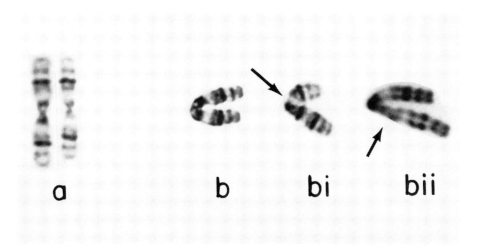

a   b   bi   bii

FIG. 11. Isochromosome markers of similar morphology in different tumors and involving the long arm of chromosome 1. a, A pair of markers, i(1q), from a single metaphase spread originating from a highly undifferentiated cancer. b, An identical i(1q) marker from an ovarian tumor. bi and bii, Further changes in the previous marker in the case with ovarian cancer in which loss of the dark-banded region q12 in one of the arms is shown in bi and insertion of a light band at q21 in one of the arms in bi. This figure illustrates the occurrence of markers of identical derivation and morphology in tumors of different nature.

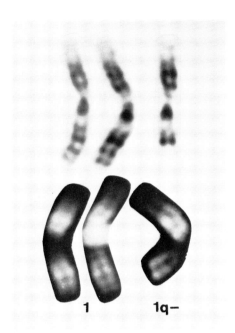

FIG. 12. Partial Q- and G-banded karyotypes from an adenocarcinoma of the rectum in which three No. 1 chromosomes were present in the cells, with one chromosome being morphologically abnormal due to deletion of the long arm, shown as 1q- (q32→qter). This figure demonstrates the not infrequent involvement of chromosome 1 in human cancer, leading to the presence of markers. Though in this case the long arm was deleted, a more frequent finding is trisomy of that arm and sometimes the whole of chromosome 1.

from cancer of the ovary, lung, breast, and malignant melanoma (Kakati et al. 1975, 1976a, 1976b, 1977) and subsequently confirmed in these and other tumors (Sandberg 1980). Trisomy, either partial or complete, of that arm is a frequent accompaniment of karyotypic changes in a number of human tumors (breast, cervix, lung, malignant melanoma, ovary, etc.) (Atkin and Pickthall 1977, Atkin and Baker 1978, 1979, van der Riet-Fox et al. 1979, Sandberg 1980) and though no specific change in this chromosome has been correlated to date with a particular tumor, it will be important to establish whether the breakpoint or sub-band involvement in chromosome 1 has some specificity. Obviously, such a demonstration requires the development of more refined techniques allowing sub-band analysis in human cancers, something that has not been possible to accomplish rigorously to date. Thus, even though the involvement of chromosome 1 lacks the specificity and nonrandomness of other chromosomes, e.g., chromosome 14 in lymphoma and ovarian cancer and chromosome 22 in meningioma, close examination of the changes involving this particular chromosome should be tabulated and reported in order to establish whether some characteristic change of chromosome 1 may not ultimately prove to be specific for one tumor or another.

MARKERS

Marker chromosomes (morphologically abnormal chromosomes) (Figures 13–17) are present in the majority of human cancers. It has been shown with

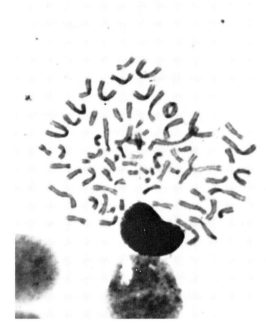

FIG. 13. Unbanded cancerous metaphase containing a ring chromosome, constituting almost invariable cytogenetic evidence of cancer. Even though this metaphase is unbanded, the presence of the ring chromosome and the large number of chromosomes, are indisputable evidence of a cancerous cell.

banding techniques that markers of identical origin may be present in tumors of diverse nature and that tumors of the same organ may be characterized by markers of differing origin. These markers range from minute units to chromosomes much larger than any in the normal set. Ring chromosomes are usually

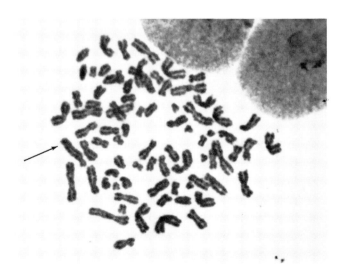

FIG. 14. Unbanded cancerous metaphase in which a marker chromosome (arrow) could be readily identified, even without banding. Though abnormal chromosomes, not readily identified without banding, may be present in the set, in the era before banding only those markers with definitely abnormal morphology could be identified.

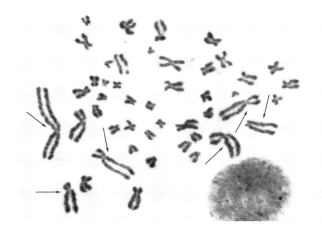

FIG. 15. A pseudodiploid metaphase containing marker chromosomes (arrows).

invariable evidence of malignancy (Bertrand et al. 1978), and though they often occur following radiation, they may occur spontaneously in a small proportion of human cancers. Since it has been shown that transposable elements in chromosomes can cause large-scale rearrangements of adjacent DNA sequences, and since markers constitute markedly displaced chromatin material, such changes may lead to alteration of the normal mechanism for genetic regulation of the cell and possibly contribute to the biologic behavior of a tumor.

That the genesis of marker chromosomes in human neoplasia may either be limited by the nature of the chromosomes or preferentially involve others is indicated by a description of identical chromosomes occurring in the effusion

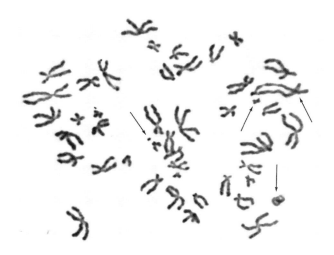

FIG. 16. Pseudodiploid metaphase containing several abnormal markers (arrows), i.e., ring chromosome, minute chromosome, and large acrocentric. This metaphase, as well as that in Figure 15, even though examined without banding, leaves little doubt about the neoplastic nature of the cells.

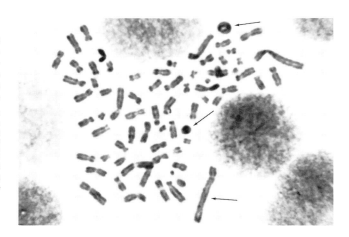

FIG. 17. Metaphase from an ovarian cancer containing a number of abnormal chromosomes that could be identified as definitely abnormal (arrows), even without banding, and indicative of the malignant nature of the cell. The abnormal chromosomes include two ring chromosomes and a very large marker (arrows).

of a patient with breast cancer (examined without culture) and those observed in the well-established cell line HeLa (Pathak et al. 1979).

Banding has been of inestimable value in establishing the origin of marker chromosomes (Figures 18–21). Thus, prior to banding, testicular tumors were thought to be characterized by a rather consistent marker, which, however, has not stood up to examination with banding techniques. On the other hand, markers that were unrecognized prior to banding because of their seemingly normal morphology have been shown with these techniques to be of abnormal origin (Oshimura et al. 1977).

ADVANCES IN THE CYTOGENETIC STUDIES OF OVARIAN TERATOMAS AND TROPHOBLASTIC TUMORS

Teratomas of the ovary are classified into the cystic and solid varieties. The former is the common form (dermoid cyst) and is predominantly composed of ectodermal elements, with at times considerable mesodermal admixture, though evidence of all three germ layers may occasionally be found. Sebaceous material and hair may fill the cysts. These and all other structures within the cysts arise from a nodular growth nidus. It has been known that the cystic type of teratoma has a normal female karyotype, 46, XX (Galton and Benirschke 1959, Corfman and Richart 1964). Because of their location in the ovary and the presence of cellular and tissue elements that are inappropriate to the ovary, it had been speculated for a number of years that these tumors were of germ cell derivation.

Linder (1969) and Linder and Power (1970) attempted to test the theory that dermoid cysts arise from germ cells at or after meiosis by investigations of electrophoretic variants between the tumor and the host. The findings were

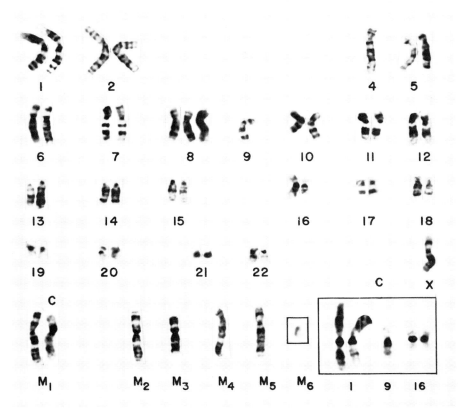

FIG. 18. G-banded karyotype of a cell from the effusion of a patient with ovarian carcinoma. Six markers (M_1-M_6) are present in this cell, whose origin could be established (Kakati et al. 1975). The chromosomes in the bracketed box were obtained from another metaphase of the same case. C-banding (indicated by the letter C over the appropriate chromosomes) of M_1 and chromosomes 1, 9, and 16 is also shown. This case demonstrates the feasibility of identifying with banding all the chromosomes, including abnormal ones, in a cancer cell.

as follows: (1) There was no change in the phenotype of multiple samples from the same tumor and replicate cell cultures; (2) Tumors obtained from heterozygous hosts frequently had a homozygous phenotype for that particular gene product (loss of gene effect); (3) In hosts heterozygous for PGM 1 (phosphoglucomutase) and PGM 3 the tumors lost a gene effect at one but not the other locus; (4) Within the same host, multiple tumors may have different phenotypes; this could be explained by loss of a gene, suppression of gene activity, or homozygosity incidental to meiosis. The last explanation was favored, but could not be proved.

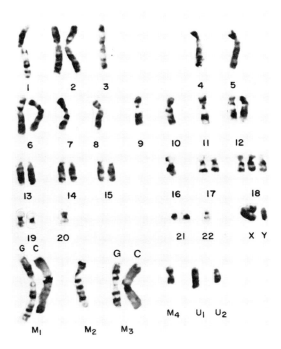

FIG. 19. G-banded karyotype of a malignant melanoma cell with missing chromosomes in 12 autosomal pairs and an extra chromosome in pair No. 18. The cell contained six marker chromosomes, four of which (M_1-M_4) could be identified with banding and two of which remain unidentified (U_1 and U_2). (G = G-banding, C = C-banding). C-banded markers M_1 and M_2 from another cell are also shown in the karyotype. This case illustrates a condition not infrequently encountered in human cancer, in which the origins of some of the abnormal (marker) chromosomes can be identified with certainty and some cannot, even when based on a number of banding methods.

Comparison of chromosomal polymorphism between the cells of the tumor and host was helpful in clarifying which of the three mechanisms mentioned above fit the biochemical data best (Linder et al. 1975a). This comparison showed that five dermoid cysts, of which the hosts were found to be heterozygous for a total of 17 chromosomal polymorphisms, were uniformly homozygous. This result ruled out the possibility that the tumors arose from a somatic cell or an oogonium.

The cytological and biochemical differences between the host and tumor indicated that the latter could originate from a germ cell after the first meiotic division. The likely mechanism was failure of meiosis II or, equivalently, reentry of the second polar body. Dermoid cysts, therefore, contain a haploid chromosome complement in duplicate derived only from their hosts; the possible presence of both homozygosis and heterozygosis was indicated by gene products that depended upon the number and nature of the cross-over events.

It has been demonstrated in four studies (Linder 1969, Linder and Power 1970, Linder et al. 1975b, McCaw et al. 1977) that in a total of 45 cases, dermoid cysts were of parthenogenetic origin on the basis of electrophoretic and chromosomal polymorphism. These two findings could be utilized for mapping of gene-centromere distances. The mathematical analysis of the chromosome

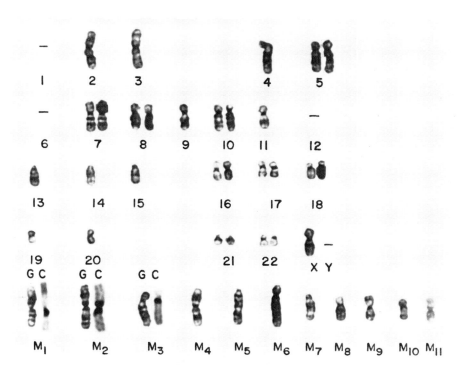

FIG. 20. G-banded karyotype of a malignant melanoma cell with missing chromosomes in most of the groups and containing 11 marker chromosomes (M_1-M_{11}), shown at the bottom of the figure (G = G-banding; C = C-banding). The C-banded chromosomes are from another cell of the same tumor with an identical karyotype and are shown for comparison. Note the missing Y chromosome. Even though the cells of this case contained a large number of marker chromosomes, their origin and identity could be established on the basis of their characteristic banding patterns.

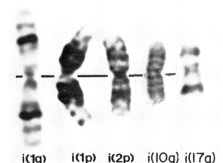

i(1q)   i(1p) i(2p) i(10g) i(17g)

FIG. 21. An example of G-banded isochromosomes originating from various chromosomes in human cancers and leukemia observed in our laboratory. The occurrence of isochromosomes in human neoplasia is not infrequent and the involvement of chromosome 1 in isochromosome formation is probably the commonest of all chromosomes in the set.

and gene marker data in ovarian teratomas has been utilized for the estimation of the distance between the centromere and the PGM 3 locus. Undoubtedly, this approach will provide much information about the genetic constitution of chromosomes in the near future.

Solid teratomas of the ovary differ from simple dermoid cysts not only in that they are solid tumors, but also in their being malignant and containing elements derived from all three of the fetal layers. However, there appear to be no data related to the origin of these tumors based on chromosomal and enzyme markers.

Linder et al. (1975a) examined single-gene products and chromosomes in extragonadal teratomas and reported that these tumors developed in a manner different from that of dermoid cysts. The cell of origin could either be a somatic cell or a misplaced germ cell that had failed to undergo meiosis and proceeded directly to mitosis. Heterozygous expression of glucose-6-phosphate-dehydrogenase (G-6-PD) in extragonadal teratomas suggested that they could possibly originate from several cells, because the enzyme is. X-linked, and in female cells one of the two alleles on the X chromosome undergoes random inactivation.

Hydatidiform moles are lesions of chorionic villi. One of the most characteristic macroscopic features is the presence of swollen vesicles, which can fill the uterine cavity. On the basis of anatomopathologic and cytogenetic studies in specimens with gross swelling of the chorionic villi, Vassilakos et al. (1977) and Szulman and Surti (1978a, 1978b) classified hydatidiform moles into partial and complete moles. Most cases of the former had chromosome abnormalities, including triploidy and trisomies, whereas the latter showed exclusively a normal female karyotype (46,XX). Aside for the macroscopic morphology of the villi, these two entities were distinct from each other; complete moles with villi consisting of pronounced hyperplastic and anaplastic trophoblasts were never accompanied by a fetus, cord, or amniotic membranes, whereas partial moles with villi consisting of slightly hyperplastic to hypoplastic trophoblasts were always associated with cord or amniotic membranes, or both. Complete moles have attracted deserved attention, primarily because of possible malignant transformation into choriocarcinoma.

Recent cytogenetic investigations (Kajii and Ohama 1977, Wake et al. 1978a, 1978b, Jacobs et al. 1978) have revealed the following findings in complete moles: (1) Homologous chromosomes of the mole were completely homozygous for banding polymorphism; (2) Neither maternal homologue was transmitted to the mole; (3) A paternal chromosome was inherited in duplicate by the mole. These observations clearly ruled out host cells, as well as a normally fertilized ovum, as the origin of complete moles, but strongly pointed to androgenesis as a cause of this type of chorionic lesion. Androgenesis is the development of an egg under the influence of a spermatozoan nucleus, the egg nucleus being either absent or inactivated. The presence of paternally derived marker chromosomes in duplicate suggests that the doubling of a paternal haploid set occurred either after fertilization or at the second meiotic division. However, which of the two possibilities is responsible for the mechanism of androgenesis could not be determined by chromosomal analysis.

Studies based on HL-A absorption in molar tissues also showed that all of these tissues expressed homozygous HL-A-A and -B specificities that were inherited exclusively from the father and none from the mother (Yamashita et al. 1979). These results suggest that androgenesis is responsible for the pathogenesis of most, if not all, cases of complete moles.

The human histocompatibility complex (HL-A) is located on the short arm of chromosome 6. It is composed of several genetic loci (A, B, C, and D) that are closely linked and inherited as a genetic unit. By determining the expression of HL-A on molar tissues it is possible to clarify whether a diploid or a haploid sperm took part in fertilization; the moles are uniformly homozygous in expression, indicating postmeiotic duplication, but occasionally one may be heterozygous if its origin was meiotic.

Two HL-A specificities (one from A and the other from B locus) of 21 molar tissues were determined (Yamashita et al. 1979). They were uniformly homozygous and consistent with the haplotype of the father except for one case. In no case did the molar tissue absorb HL-A antiserum that was unique only for the mother. HL-A-A and -B specificities in the exceptional case were not homozygous but completely identical with the ones of the father. Studies by Otto et al. (1976) on ovarian teratomas suggested that cross-over events between the centromere and the PGM 3 locus occurred in a third of the cases. Both the HL-A region and the PGM 3 locus are located on the short arm of chromosome 6, and the latter locus is proximal to the centromere. It is possible to assume, therefore, that the frequency of crossing-over between the centromere and the HL-A region could be higher than a third. However, only one case of a heterozygote has been detected for each locus in a total of 23 cases and two cases of invasive moles (Tsuji et al. 1978). Furthermore, this heterozygote might have been an XY mole resulting from the fertilization by an XY diploid sperm due to abnormal first meiotic division, though the karyotype could not be determined.

The above findings suggest that most complete moles with a 46,XX karyotype arise from an androgenetic ovum fertilized by a haploid sperm. In such a case, the duplication of a paternal haploid set might have occurred either at or after the first cleavage division. A mosaic, consisting of haploid and diploid cells, may be expected in the latter case, though the haploid cells may be eliminated during the course of growth.

This sequence leads to the preponderance of 46,XX karyotypes in moles, inasmuch as YY counterparts are probably lost during early cleavage stages. A unique case with a 48,XXYY karyotype reported by Shinohara et al. (1971) seems to be compatible with an androgenetic origin. Exclusive of apparently partial moles, however, three XY cases (Sasaki et al. 1962, Bourgoin et al. 1965, Shinohara et al. 1971) out of 94 cases with complete moles were reported in the literature. Kajii and Ohama (1977) also observed three X chromatin–negative moles; two were Y chromatin negative, and one was Y chromatin positive in 54% of its stromal cells. The moles with a 46,XY karyotype were of special interest since they must be different from the 46,XX variety, in that they cannot develop by the mechanism postulated for the former. Recently, Surti et al. (1979) reported on the chromosomal origin of a complete mole with a 46,XY karyotype. A study of polymorphic chromosomal and enzyme markers of the molar and paternal genomes revealed that the molar chromosomes were entirely of paternal origin derived from both haploid sets, there being no maternal contribution. Thus, XY moles turn out to be androgenetic in origin, analogous to the 46,XX moles. However, it remains to be elucidated whether the actual total paternal genome contained in one diploid sperm (meiosis I error) or an XY condition with 46 chromosomes resulting from dispermy represents the origin of 46,XY moles, in contrast to the 46,XX variety. Further investigations by the use of chromosomal polymorphisms, HL-A, and enzyme markers for the former are necessary to determine which of the two mechanisms is more likely to have occurred.

Androgenetic ova (fertilized through the process of androgenesis) were occasionally observed in rat and mice at the pronuclear stage (Austin and Braden 1954). Such eggs may occur in other species. Various attempts have been made in mice to develop a fertilized egg in which only the maternal (gynogenesis) or paternal (androgenesis) genome is retained. Efforts with microsurgical techniques either to remove one pronucleus shortly after fertilization (Markert and Petters 1977) or to bisect the fertilized egg at the pronuclear stage (Tarkowski 1977) rarely have resulted in the production of blastocysts. Nevertheless, the experiment by Hoppe and Illmensee (1977) has shown that viable homozygous-uniparental diploid mice can be generated by microsurgically removing one pronucleus from the fertilized egg and diploidizing the residual one in the presence of cytochalasin B. These mice, furthermore, have been fertile. However, ordinarily there is no trace of a fetus in complete moles. It remains to be explored why the differentiation pattern of androgenetic ova is different in men and mice.

Kajii and Ohama (1977) postulated the existence of a recessive mutation of a gene (or genes) controlling cell growth for the frequent neoplastic transformation of moles into choriocarcinoma. Exclusive transmission of a paternal genome in choriocarcinoma cells presents a serious question: why should such an allogeneic tumor grow progressively in the maternal body? One of the possible reasons for this is that major histocompatibility antigens are not expressed on the trophoblasts, though Tsuji et al. (1978) recognized that the expression of HL-A in the trophoblasts is much the same as in other cellular components of the mole. Recently, we investigated the chromosomes of two cases of metastatic chorionic tumors and found that they were characterized by a high incidence of cells in the triploid and tetraploid ranges and that homologous chromosomes of certain pairs with polymorphic chromosome variants were not morphologically identical. Thus, the above data seem not to favor the view of cellular continuity between moles and choriocarcinoma, though choriocarcinoma has been clinically assumed to be preceded by molar conception.

SOME COMMENTS ON THE CHROMOSOMAL CHANGES IN HUMAN CANCER

A major challenge to cancer cytogeneticists is to ascertain whether existing within the "jungle" of karyotypic changes characterizing most human cancers is a nonrandom karyotypic path indicative of and possibly leading to the underlying mechanisms causing these malignancies.

Even though cytogenetic examinations prior to banding often failed to reveal the nature of karyotypic abnormalities present in cancer cells, much information was obtained regarding the modal chromosome number in various cancers (Figures 22 and 23) and in primary tumors vs. their metastases (Figure 24). A panoramic view of the range and extent of morphologic and numerical changes (Figures 25 and 26) associated with human neoplasia was obtained. Since some cancers had modal chromosome numbers in the hypodiploid range and others in the near-triploid or hypotetraploid range, it is inconceivable that these do not affect the behavior of the tumor. We feel that little attention has been paid to the possible role of the chromosome number in cancer genesis and biology.

Cytogenetic experience indicates that, generally, once a karyotype is established in a tumor, it tends to imprint itself on that tumor as reflected in recur-

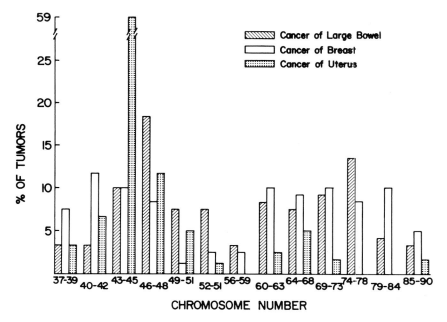

FIG. 22. Distribution of the modal chromosome numbers in three human cancers. The modal number varies among these cancers, with some being diploid and others being hyperdiploid. The exact significance of these differences has yet to be established.

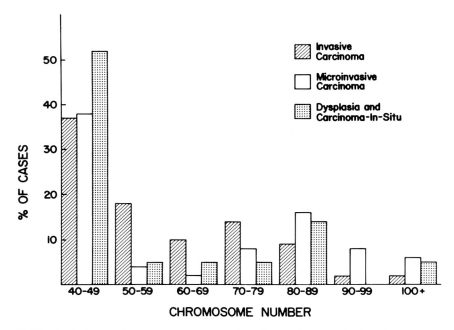

FIG. 23. Distribution of chromosome numbers in invasive carcinoma, microinvasive carcinoma, and dysplasia and carcinoma in situ of the cervix.

rences and metastases (which are more accessible and, in effusions, more readily analyzed). The latter, though often containing a higher number or a multiple of the modal chromosome number, are characterized by karyotypes that are merely variations on the cytogenetic theme of the modal karyotype in the primary tumor. Of course, exceptions and deviations from the above occur, but generally human cancers prefer to retain their original cytogenetic constitution since they appear to offer tumors definite growth advantages unique to each cancer. The challenge to cytogeneticists is to decipher which of the karyotypic changes are essential to tumor growth and spread, how each tumor generates these chromosomal changes, and what factors lead to these changes.

It is our strong belief that the establishment of nonrandom and, particularly, specific chromosomal changes in various human tumors will ultimately prove to be of definite aid in the etiologic determination, diagnosis, classification, and therapy of the diseases. However, even though the chromosomal changes may serve as messengers for the above areas, we must be careful not to ascribe to these changes authorship of these messages. The origin of the chromosomal changes, however specific, may reside in a much more subtle chromosomal change, far removed from the gross microscopic changes observed by cytogeneticists, and be merely phenotypically manifested in the karyotypic changes.

In reviewing the chromosome changes in specific human tumors it is apparent

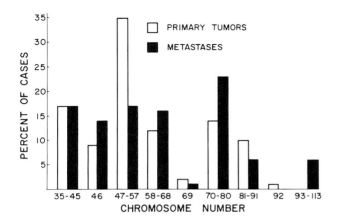

FIG. 24. Distribution of the modal chromosome numbers in human primary and metastatic tumors. Most of the primary tumors tend to have a mode in the diploid range, whereas the metastatic tumors tend to have a distribution in both the diploid and hypertriploid ranges. Generally, higher numbers are observed in the metastatic cells than in those of the primary tumor.

that the information supplied by each raises questions that at the moment may be difficult to spin into the total yarn of the role of chromosomal changes in human neoplasia. Yet, each such specific tumor sheds light on an area, however narrow, related to some facets of karyotypic changes characterizing the tumors. It is hoped that ultimately the cytogenetic pieces supplied by each tumor can be fit into the general puzzle of the exact place of the chromosomal changes

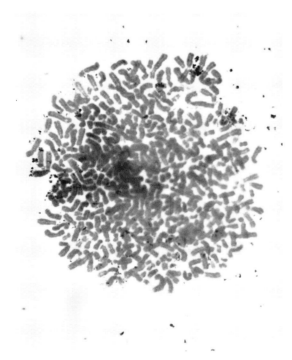

FIG. 25. Unbanded cancerous metaphase containing a very large number of chromosomes.

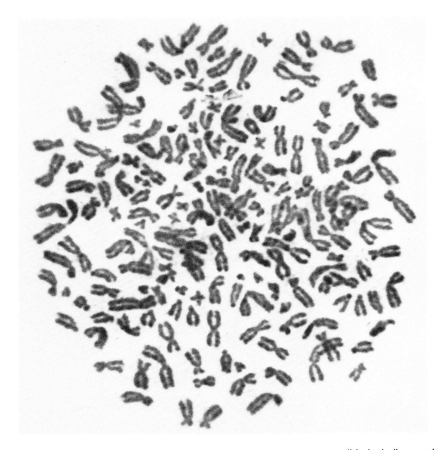

FIG. 26. Metaphase containing a large number of chromosomes, possibly including marker chromosomes, which could not be definitely ascertained with the staining procedures not relying on banding. However, the large number of chromosomes per se is definitely indicative of the neoplastic nature of the cell.

in the causation and biology of human cancers. These changes must be reconciled with the fact that advanced human neoplasia, particularly acute leukemia, may develop without any accompanying karyotypic changes, at least at the level of banding presently utilized, which thus puts the contention that such changes cause the tumor in a somewhat doubtful position.

The usefulness of chromosomal changes in evaluating a clinical condition may be demonstrated in bladder cancer. In two separate studies it has been shown that the presence of marker chromosomes in noninvasive or microinvasive bladder cancer carries with it a very high risk of recurrence, whereas the absence of such markers indicates less than a 5% chance of such recurrence (Falor and Ward 1976, 1978, Sandberg 1977, 1980). The nature of the marker did not seem to play as much a role as its presence or absence, and apparently recurrence was not related to the modal chromosome number, though most of these tumors had chromosome numbers in the diploid range.

The concept has been advanced that the same causative agent may lead to similar nonrandom karyotypic changes in various tissues. Though undoubtedly this is true to some extent, exceptions do occur and indicate that it is the tissue involved that may decide to a large extent the type of nonrandom karyotypic change that will be observed. Furthermore, the fact that such nonrandom changes can develop after a neoplastic condition has become well established points to the possibility that such nonrandom changes are not essential to the development of the neoplastic process. The fact that such nonrandom changes are not seen in every tumor of a particular tissue points also to the possibility that the genome of the host may play a key role in affecting the presence or absence of such nonrandom changes. Thus, it is our belief that the occurrence of nonrandom (or random) changes probably reflects a combination of factors, i.e., the causative agent, the nature of the tissue involved, and the genome of the host.

The cytogeneticist's ability to examine early karyotypic changes in both meningioma and leukemia, two conditions at the extremes of the curves shown in Figure 1, is probably due to the slow growth and production of very early symptoms in meningioma and the rapid spread of cells and, generally, early diagnosis in leukemia. Because of these considerations, it is usually found that the modal chromosome number is rather sharp in these conditions and that wide karyotypic variation within a single case is relatively rare. The bulk of the other cancers, which are probably observed at a relatively late stage of development, are accompanied by many variations on the karyotypic theme of the modal cell of a particular tumor.

Klein (1979) has restated the view that chemical or physical carcinogens as well as viruses play essentially the same role of initiators in tumor progression. Their major effect is the establishment of long-lived preneoplastic cells. Specific genetic changes are responsible for the transition of the preneoplastic cells to frankly malignant ones. In some systems they are expressed as cytogenetically detectable chromosomal changes, characteristic of the majority of the tumors that originate from the same target cell. The changes may arise by random mechanisms and are selectively fixed due to the increased growth advantages of the clone that carries such changes. This advantage is based on a decreased responsiveness to growth-controlling or differentiation-inducing host signals. This selection process, rather than any specific induction mechanism, is responsible for the "cytogenetic convergence" of preneoplastic cell lineages, initiated ("caused") by widely diverse agents, towards the same nonrandom chromosomal change. The cytogenetic changes act by shifting the balance between genes that favor progressive growth in vivo and genes that counteract it. Changes in effective gene dosage are brought about by nonrandom duplication of a whole chromosome, as in trisomy, or by reciprocal translocation that may affect gene expression on the donor or recipient chromosome.

A cytogenetic feature of most human cancers is the range of karyotypic variation within a tumor, often reflected in a flat distribution of the modal chromosome number. However, the chromosome consititutions of the cells within such a tumor often consist of variations on the modal karyotypic theme, e.g., multiples of markers present in the modal cell or missing chromosomes in certain groups. In this respect, cancers differ in their cytogenetic picture from that usually seen in leukemic cells, where the modal number is usually sharp. It is possible that these differences between tumors and leukemia reflect a more concise control, at least cytogenetically, of the leukemic cell versus that of cancer cells.

ACKNOWLEDGMENTS

The work performed in our laboratory and cited in this paper has been supported in part by a grant CA-14555 from the National Cancer Institute. We wish to thank Miss Anne Marie Conti for clerical help.

REFERENCES

Atkin, N. B. 1976. Cytogenetic Aspects of Malignant Transformation. Karger, Basel, München, Paris, London, New York, Sydney.

Atkin, N. B., and M. C. Baker. 1978. Duplication of the long arm of chromosome 1 in a malignant vaginal tumour. Br. J. Cancer 38:468–471.

Atkin, N. B., and M. C. Baker. 1979. Chromosome 1 in 26 carcinomas of the cervix uteri. Structural and numerical changes. Cancer 44:604–613.

Atkin, N. B., and V. J. Pickthall. 1977. Chromosomes 1 in 14 ovarian cancers. Hum. Genet. 38:25–33.

Austin, C. R., and A. W. Braden. 1954. Anomalies in rat, mice, and rabbit eggs. Aust. J. Biol. Sci. 7:537–542.

Autio, K., O. Turunen, O. Penttilä, E. Erämaa, and A. de la Chapelle. 1979. Human chronic lymphocytic leukemia: Karyotypes in different lymphocyte populations. Cancer Genet. Cytogenet. 1:147–155.

Berger, R., A. Bernheim, H.-J. Weh, G. Flandrin, M. T. Daniel, J.-C. Brouet, and N. Colbert. 1979. A new translocation in Burkitt's tumor cells. Hum. Genet. 53:111–112.

Bertrand, S., M. R. Branger, and F. Cheix. 1978. Use of G banding technique in the cytogenetic study of metastatic breast cancer effusion. A case report. Eur. J. Cancer 15:737–743.

Bourgoin, P., R. Baylet, C. Ballon, and H. Grattepanche. 1965. Exploration d'une hypothese sur l'etiopathogenie des moles hydatidiformis. Etude chromosomique. Dakar-Rev. Franc. Gynec. 60:673–684.

Cohen, A. J., F. P. Li, S. Berg, D. J. Marchetto, S. Tsai, S. C. Jacobs, and R. S. Brown. 1979. Hereditary renal-cell carcinoma associated with a chromosomal translocation. N. Engl. J. Med. 301:592–595.

Corfman, P. A., and R. M. Richart. 1964. Chromosome number and morphology of benign ovarian cystic teratomas. N. Engl. J. Med. 271:1241–1244.

Danes, B. S., and T. Alm. 1979. In vitro studies on adenomatosis of the colon and rectum. J. Med. Genet. 16:417–422.

Falor, W. H., and R. M. Ward. 1976. Cytogenetic analysis. A potential index for recurrence of early carcinoma of the bladder. J. Urol. 115:49–52.

Falor, W. H., and R. M. Ward. 1978. Prognosis in early carcinoma of the bladder based on chromosomal analysis. J. Urol. 119:44–48.

Finan, J., R. Daniele, D. Rowlands, Jr., and P. Nowell. 1978. Cytogenetics of chronic T cell leukemia, including two patients with a 14q+ translocation. Virchows Arch. (Cell Pathol.) 29:121–127.

Fukuhara, S., J. D. Rowley, D. Variakojis, and H. M. Golomb. 1979. Chromosome abnormalities in poorly differentiated lymphocytic lymphoma. Cancer Res. 39:3119–3128.

Gahrton, G., K.-H. Robèrt, K. Friberg, L. Zech, and A. G. Bird. 1980. Extra chromosome 12 in chronic lymphocytic leukaemia. Lancet I: 146–147.

Gahrton, G., L. Zech. K.-H. Robèrt, and A. G. Bird. 1979. Mitogenic stimulation of leukemia cells by Epstein-Barr virus. N. Engl. J. Med. 301:438–439.

Galton, M., and K. Benirschke. 1959. Forty-six chromosomes in an ovarian teratoma. Lancet II: 761–762.

German, J., Ed. 1974. Chromosomes and Cancer. John Wiley & Sons, New York.

Harnden, D. G., and A. M. R. Taylor. 1979. Chromosomes and neoplasia, *in* Advances in Human Genetics. Plenum Press, New York and London, pp. 1–70.

Hoppe, P. C., and K. Illmensee. 1977. Microsurgically produced homozygous-diploid uniparental mice. Proc. Natl. Acad. Sci. USA 74:5657–5661.

Hurley, J. N., S. Man Fu, H. G. Kunkel, R. S. K. Chaganti, and J. German. 1980. Chromosome abnormalities of leukaemic B lymphocytes in chronic lymphocytic leukaemia. Nature 283:76–78.

Jacobs, P. A., T. J. Hassold, A. M. Matsuyama, and I. M. Newlands. 1978. Chromosome consitution of gestational trophoblastic disease. Lancet II:49.

Kaiser-McCaw, B., A. L. Epstein, K. M. Overton, H. S. Kaplan, and F. Hecht. 1977. The cytogenetics of human lymphomas: Chromosome 14 in Burkitt's, diffuse histiocytic and related neoplasms. Chromosomes Today 6:383–390.

Kajii, T., and K. Ohama. 1977. Androgenetic origin of hydatidiform mole. Nature 268:633–634.

Kakati, S., M. Barcos, and A. A. Sandberg. 1979. Chromosomes and causation of human cancer and leukemia. XXXVI. 14q+ anomaly in an American Burkitt lymphoma and its value in the definition of lymphoproliferative disorders. Med. Pediatr. Oncol. 6:121–129.

Kakati, S., M. Barcos, and A. A. Sandberg. 1980. Chromosomes and causation of human cancer and leukemia. XLI. Cytogenetic experience with non-Hodgkin, non-Burkitt lymphomas. Cancer Genet. Cytogenet. (in press).

Kakati, S., I. Hayata, M. Oshimura, and A. A. Sandberg. 1975. Chromosomes and causation of human cancer and leukemia. X. Banding patterns in cancerous effusions. Cancer 36:1729–1738.

Kakati, S., I. Hayata, and A. A. Sandberg. 1976a. Chromosomes and causation of human cancer and leukemia. XIV. Origin of a large number of markers in a cancer. Cancer 37:776–782.

Kakati, S., M. Oshimura, and A. A. Sandberg. 1976b. Chromosomes and causation of human cancer and leukemia. XIX. Common markers in various tumors. Cancer 38:770–777.

Kakati, S., S. Y. Song, and A. A. Sandberg. 1977. Chromosomes and causation of human cancer and leukemia. XXII. Karyotypic changes in malignant melanoma. Cancer 40:1173–1181.

Klein, G. 1979. Lymphoma development in mice and humans: Diversity of initiation is followed by convergent cytogenetic evolution. Proc. Natl. Acad. Sci. USA 76:2442–2446.

Lawler, S. D. 1977. Chromosomes in haematology. Br. J. Haematol. 36:455–460.

Linder, D. 1969. Gene loss in human teratomas. Proc. Natl. Acad. Sci. USA 63:699–704.

Linder, D., and J. Power. 1970. Further evidence of postmeiotic origin of teratomas in the human female. Ann. Hum. Genet. 34:21–30.

Linder, D., F. Hecht, B. K. McCaw, and J. R. Campbell. 1975a. Origin of extragonadal teratomas and endodermal sinus tumors. Nature 254:597–598.

Linder, D., B. K. McCaw, and F. Hecht. 1975b. Parthenogenetic origin of benign ovarian teratomas. N. Engl. J. Med. 292:63–66.

Manolov, G., and Y. Manolova. 1971. A marker band in one chromosome 14 in Burkitt lymphomas. Hereditas 69:300.

Manolov, G., and Y. Manolova. 1972. Marker band in one chromosome 14 from Burkitt lymphomas. Nature 237:33–34.

Mark, J. 1974. The human meningioma: A benign tumor with specific chromosome characteristics, *in* Chromosomes and Cancer, J. German, ed. John Wiley, New York, pp. 497–517.

Mark, J. 1977. Chromosomal abnormalities and their specificity in human neoplasms: An assessment of recent observations by banding techniques. Adv. Cancer Res. 24:165–222.

Mark, J., R. Dahlenfors, and C. Ekedahl. 1979. Recurrent chromosomal aberrations in non-Hodgkin and non-Burkitt lymphomas. Cancer Genet. Cytogenet. 1:39–56.

Mark, J., R. Dahlenfors, C. Ekedahl, and G. Stenman. 1980. The mixed salivary gland tumor—a usually benign human neoplasm frequently showing specific chromosomal abnormalities. Cancer Genet. Cytogenet. (in press).

Mark, J., F. Mitelman, H. Dencker, C. Norryd, and K.-G. Tranberg. 1973. The specificity of the chromosomal abnormalities in human colonic polyps. A cytogenetic study of multiple polyps in a case of Gardner's syndrome. Acta Pathol. Microbiol. Scand. 81A:85–90.

Markert, C. L., and R. M. Petters. 1977. Homozygous mouse embryos produced by microscopy. J. Exp. Zool. 201:295–302.

Martin, P., B. Levin, H. M. Golomb, and R. H. Riddell. 1979. Chromosome analysis of primary large bowel tumors. Cancer 44:1656–1664.

McCaw, B. K., F. Hecht, D. Linder, E. W. Lovrien, H. Wyandt, D. Bacon, B. Clark, and N. Lea. 1977. Ovarian tetratomas: Cytologic data. Cytogenet. Cell Genet. 18:391–395.

Mitelman, F., J. Mark, P. G. Nilsson, H. Dencker, C. Norryd, and K.-G. Tranberg. 1974. Chromosome banding pattern in human colonic polyps. Heriditas 78:63–68.

Miyoshi, I., S. Hiraki, I. Kimura, K. Miyamoto, and J. Sato. 1979. 2/8 translocation in a Japanese Burkitt's lymphoma. Experientia 35:742–743.

Nowell, P., R. Daniele, D. Rowlands, and J. Finan. 1980. Cytogenetics of chronic B cell and T cell leukemia. Cancer Genet. Cytogenet. 1:273–280.

Oshimura, M., S. Kakati, and A. A. Sandberg. 1977. Chromosomes and causation of human cancer and leukemia. XXVII. Possible mechanisms for the genesis of common chromosome abnormalities, including isochromosomes and the Ph¹. Cancer Res. 37:3501–3507.

Otto, J., D. Linder, B. K. McCaw, E. W. Lovrien, and F. Hecht. 1976. Estimating distances from the centromere by means of benign ovarian teratomas in man. Ann. Hum. Genet. 40:191–196.

Pathak, S., M. J. Siciliano, R. Cailleau, C. L. Wiseman, and T. C. Hsu. 1979. A human breast adenocarcinoma with chromosomes and isoenzyme markers similar to those of the HeLa line. J. Natl. Cancer Inst. 62:263–271.

Pierre, R. V., G. W. Dewald, and P. C. Banks. 1980. Cytogenetic studies in malignant lymphoma: Possible role in staging studies. Cancer Genet. Cytogenet. 1:257–261.

Rowley, J. D. 1978. Chromosomes in leukemia and lymphoma. Semin. Hematol. 15:301–319.

Rowley, J. D. 1980. Chromosome abnormalities in acute lymphoblastic leukemia. Cancer Genet. Cytogenet. 1:263–271.

Rowley, J. D., and S. Fukuhara. 1980. Chromosome studies in non-Hodgkin lymphomas. Semin. Oncol. (in press).

Sandberg, A. A. 1977. Chromosome markers and progression in bladder cancer. Cancer Res. 37:2950–2956.

Sandberg, A. A. 1980. Chromosomes in Human Cancer and Leukemia. Elsevier-North Holland, New York.

Sandberg, A. A., and D. K. Hossfeld. 1974. Chromosomal changes in human tumors and leukemias, in Handbuch der allgemeinen Pathologie, Vol. VI, H. W. Altmann, F. Büchner, H. Cottier, E. Grundmann, G. Holle, E. Letterer, W. Masshoff, H. Meessen, F. Roulet, G. Seifert, and G. Siebert, eds. Springer-Verlag, Berlin, Heidelberg, New York, pp. 141–287.

Sasaki, M., T. Fukushima, and S. Makino. 1962. Some aspects of the chromosomal constitution of hydatidiform moles and normal chorionic villi. Gann 53:101–106.

Shinohara, T., M. S. Sasaki, A. Tonomura, T. Shimamine, T. Yokoyama, and T. Hasegawa. 1971. Cytogenetic studies in human chorionic lesions. Jpn. J. Hum. Genet. 16:111–112.

Surti, U., A. E. Szulman, and S. O'Brien. 1979. Complete (classic) hydatidiform mole with 46,XY karyotype of paternal origin. Hum. Genet. 51:153–155.

Szulman, A. E., and U. Surti. 1978a. The syndrome of hydatidiform mole. I. Cytogenetic and morphologic correlations. Am. J. Obstet. Gynecol. 131:665–671.

Szulman, A. E., and U. Surti. 1978b. The syndrome of hydatidiform mole. II. Morphologic evolution of the complete and partial mole. Am. J. Obstet. Gynecol. 132:20–27.

Tarkowski, A. K. 1977. In vitro development of haploid mouse embryos produced by bisection of one-cell fertilized eggs. J. Embryol. Exp. Morph. 38:187–202.

Trent, J. M., and J. R. Davis. 1979. D-Group chromosome abnormalities in endometrial cancer and hyperplasia. Lancet II: 361.

Tsuji, T., S. Matsuda, T. Hirai, H. Fukiwara, and T. Hamaoka. 1978. Selective expression of paternal HLA on the surface of hydatidiform mole cells. Gann 69:849.

Van Den Berghe, H., C. Parlois, S. Gosseye, V. Englebienne, G. Cornu, and G. Sokal. 1979. Variant translocation in Burkitt lymphoma. Cancer Genet. Cytogenet. 1:9–14.

van der Riet-Fox, M. F., A. E. Retief, and W. A. van Niekerk. 1979. Chromosome changes in 17 human neoplasms studied with banding. Cancer 44:2108–2119.

Vassilakos, P. G. Riotton, and T. Kajii. 1977. Hydatidiform mole: Two entities. Am. J. Obstet. Gynecol. 127:167–170.

Wake, N., M. M. Hreshchyshyn, S. M. Piver, S. Matsui, and A. A. Sandberg. 1980. Specific chromosome change in ovarian cancer. Cancer Genet. Cytogenet. 2:87.

Wake, N., Y. Shiina, and K. Ichinoe. 1978a. A further cytogenetic study of hydatidiform mole, with reference to its androgenetic origin. Proc. Jpn. Acad. 54:533–537.

Wake, N., N. Takagi. and M. Sasaki. 1978b. Androgenesis as a cause of hydatidiform mole. J. Natl. Cancer Inst. 60:51–53.

Yamashita, K., N. Wake, T. Araki, K. Ichinoe, and M. Kuroda. 1979. Human lymphocyte antigen expression in hydatidiform mole: Androgenesis following fertilization by a haploid sperm. Am. J. Obstet. Gynecol. 135:597–600.

Zankl, H., and K. D. Zang. 1978. Quantitative studies on the arrangement of human metaphase chromosomes. V. The association pattern of acrocentric chromosomes in human meningiomas after the loss of G and D chromosomes. Hum. Genet. 40:149–155.

Zankl, H., and K. D. Zang. 1980. Correlations between clinical and cytogenetical data in 180 human meningiomas. Cancer Genet. Cytogenet. 1:351–356.

Genes, Chromosomes, and Neoplasia, edited by
Frances E. Arrighi, Potu N. Rao, and Elton Stubblefield.
Raven Press, New York © 1981.

Tumor Etiology and Chromosome Pattern— Evidence from Human and Experimental Neoplasms

Felix Mitelman

Department of Clinical Genetics, Lund University Hospital, Lund, Sweden

During the past two decades the study of chromosomal aberrations and their significance in the development and progression of tumors became a rapidly expanding branch of cancer research. The introduction of chromosome banding techniques 10 years ago has no doubt led to a great expansion of our knowledge in this field. Chromosome analysis using these modern techniques have already brought a clearer understanding of such dynamic processes as cell competition, selection, and adaptation, all of which operate within a malignant cell population, and may help our understanding of basic tumor biology. The information available is, however, in many respects still fragmentary and permits only certain generalizations to be advanced concerning the role of chromosome changes in carcinogenesis. Thus, it is generally accepted that the chromosome aberrations found in the majority of experimental and human neoplasms are an integral part of tumor evolution, but chromosome abnormality is certainly not a *sine qua non* of neoplastic change, and the fundamental question, the initiating role of chromosome changes, is far from clarified. The aim of this paper is to review the information that has been obtained from recent cytogenetic studies of tumor cells in relation to neoplastic development in experimental animals and man.

NONRANDOM CHROMOSOME CHANGES IN EXPERIMENTAL TUMORS

Evidence for nonrandom karyotypic aberrations was obtained in experimental neoplasms even before the introduction of chromosome banding techniques. Thus, a considerable number of primary mouse leukemias, both spontaneous ones (Wakonig and Stich 1960) and those induced with chemicals (Stich 1960) or viruses (Mark 1967), as well as primary virus-induced rat sarcomas (Levan 1961, Nichols 1963) were characterized by trisomic stem lines. Although individual identification of the extra chromosomes was impossible, the results were compatible with the presence of a pattern. Sugiyama et al. (1967) first demon-

335

strated a consistent karyotypic abnormality—trisomy of the longest telocentric chromosome—in rat leukemias induced by 7,12-dimethyl-benz (α) anthracene (DMBA). The same abnormality was also present in rat leukemias induced by the closely related hydrocarbon carcinogens 6,8,12- and 7,8,12-trimethylbenz (α) anthracene (TMBA) (Sugiyama and Brillantes 1970).

During this period an even stronger indication of a nonrandom pattern in chromosome variation was found in the Chinese hamster. In this species, each abnormal clone could be characterized in terms of gains and losses of individual chromosomes, and Kato (1968), studying the chromosomes of primary sarcomas induced by Rous sarcoma virus (RSV), found that the most common change in these tumors was the addition of one or, less often, two or three specific chromosomes, *viz.* chromosomes 5, 6, and 10. The most striking nonrandom pattern hitherto observed was, however, found in RSV-induced sarcomas of the rat (Mitelman 1971, 1972a). Three specific chromosomes were regularly involved in aberrations, most characteristically as trisomy. It was also demonstrated that the chromosomal additions occurred in a given sequence, indicating a predetermined karyotypic evolution, the first step being gain of the telocentric chromosome earlier noted by Levan (1961) and Nichols (1963). Overall, there was a close similarity between the chromosomes of primary tumors and their metastases, although the latter tended to show an accelerated chromosomal progression (Mitelman 1972b). In both primary and metastatic tumors there was an inverse relationship between the proportion of normal diploid cells and tumor age, and the loss of diploid cells was accompanied by a histologic dedifferentiation.

The results of these early studies, supplemented with analyses of sarcomas induced by DMBA in rats (Mitelman and Levan 1972) and the Chinese hamster (Mitelman et al. 1972a) led to the conclusion that karyotype evolution was determined somehow by the inducing agent (Mitelman et al. 1972b). This hypothesis, developed further by Rowley (1974), offered a possible explanation for the variable chromosome picture often seen in neoplasms of man; quite likely histologically defined tumor types are composed of tumors with many different etiologies.

With the advent of banding techniques, identification of individual mouse and rat chromosomes became possible, and new possibilities for a highly refined analysis of the different chromosome aberrations involved in malignant cells became available. Results from such recent cytogenetic studies utilizing modern banding techniques are summarized in Table 1 and will briefly be described below for each species analyzed.

Rat

RSV-induced sarcomas were studied in G-banding by Levan and Mitelman (1975a, 1976). The extra medium-sized telocentric chromosome, repeatedly found in previous studies, was consistently identified as No. 7; the second and

TABLE 1. *Characteristic chromosome aberrations in experimental neoplasms*

Species	Tumor type	Inducing factor									References
		RSV	DMBA	BP	MC	NBU	Mineral oil	X ray	RadLV	"Spontaneous"	
Rat	Sarcoma	+7,+13 (+12)	+2	+2	+2						Mitelman 1971, 1972a Levan 1974 Levan et al. 1974 Levan and Levan 1975 Levan and Mitelman 1975a, 1976
	Leukemia					+2					Sugiyama et al. 1978 Uenaka et al. 1978 Ahlström 1974
Chinese hamster	Carcinoma		+2								Kato 1968
	Sarcoma	+5,+6, (+10)	+11								Mitelman et al. 1972a
Mouse	Lymphocytic leukemia		+15	+15				+15	+15 (+17)	+15	Dofuku et al. 1975 Chang et al. 1977 Chan 1978 Kodama et al. 1978 Wiener et al. 1978a,b Hayata et al. 1979
	Myelocytic leukemia							del(2) (±Y)			
	Plasmocytoma						del(15)				Shepard et al. 1974, 1976 Yoshida and Moriwaki 1975 Yoshida et al. 1978 Ohno et al. 1979
	Carcinoma									+13 (−X)	Dofuku et al. 1979

RSV = Rous sarcoma virus, DMBA = 7,12,dimethylbenz (α) anthracene, BP = 3,4-benzpyrene, MC = 20-methylcholanthrene, NBU = N-nitroso-N-butylurea, RadLV = radiation leukemia virus.

third step in the karyotypic evolution of these tumors was identified as trisomy 13 and trisomy 12, respectively.

Trisomy 2 was confirmed as a characteristic aberration in DMBA-induced leukemias (Sugiyama et al. 1978), carcinomas (Ahlström 1974), and sarcomas (Levan et al. 1974). It could also be demonstrated that in several instances apparently without trisomy 2, additional material of chromosome 2 was in fact revealed as part of different marker chromosomes, i.e., partial trisomy 2. By detailed structural analysis of the marker chromosomes containing extra segments of chromosome 2, the segment of critical significance was tentatively located on certain bands in the middle region of chromosome 2 (Levan et al. 1974).

Cytogenetic studies have been performed on three other chemically induced primary rat neoplasms: sarcomas produced by 20-methylcholanthrene (MC) and 3,4-benzpyrene (BP) (Levan 1974, Levan and Levan 1975) and leukemias induced by N-nitroso-N-butylurea (NBU) (Uenaka et al. 1978). In general, the chromosome pattern of all three inductions was similar to the DMBA pattern, but differed sharply from the RSV pattern. All three tumor types showed a high frequency of trisomy 2 or marker chromosomes involving chromosome 2.

There were, however, some notable differences in the karyotypic patterns of the different chemical inductions: (1) The frequency of normal diploid cells was higher in neoplasms induced by DMBA and NBU than in sarcomas induced by BP and MC. (2) Trisomy 2 was more common in DMBA-induced neoplasms than in sarcomas and leukemias induced by MC, BP, and NBU. On the other hand, "hidden" trisomy 2 was more common in sarcomas induced by MC and BP than in neoplasms induced by DMBA and NBU.

Mouse

Dofuku et al. (1975) and more recently Kodama et al. (1978) found that spontaneous thymic lymphomas of AKR mice had a characteristic aberration: trisomy 15. A similar nonrandom predominance of trisomy 15 has also been identified in leukemias and lymphomas induced by X rays in C57 BL (Chang et al. 1977) and CFW/D mice (Chan 1978), leukemias induced by DMBA in C57 BL and CFW/D mice (Chan 1978, Wiener et al. 1978b), leukemias induced by BP in CFW/D mice (Chan 1978), and leukemias induced by two different substrains of the radiation leukemia virus (RadLV) in C57 BL mice (Wiener et al. 1978a). It is noteworthy than the RadLV-induced leukemias were characterized also by a second nonrandom event, trisomy 17, which was much less frequent than trisomy 15 and never found without the latter.

In contrast to the regular trisomy of chromosome 15 in most T-cell leukemias and lymphomas, Abelson virus–induced murine leukemias were purely diploid, and remained diploid for several consecutive passages in vivo (Klein et al. 1980). Trisomy 15 has not been found in other mouse lymphomas of non-T-cell origin induced by the Rauscher, Friend, Graffi, and Duplan viruses (Klein 1979), nor in mineral oil–induced plasmocytomas, X-ray–induced myelocytic leuke· mias, or spontaneous mammary tumors.

A deletion of chromosome 15 seems to be a consistent pattern in mineral oil–induced plasmocytomas of the mouse (Shepard et al. 1974, 1976, Yoshida and Moriwaki 1975, Yoshida et al. 1978, Ohno et al. 1979). In most cases the segment missing from the deleted chromosome 15 was translocated to chromosome 12, but in some cases chromosomes 6 and 10 were the recipient chromosomes. This finding indicates that the chromosomal events leading to formation of the deleted chromosome 15 are more important in the pathogenesis of plasmocytoma than changes in any particular recipient chromosome.

A deletion of chromosome 2 was consistently observed in six of seven myelocytic leukemias of irradiated C3H/He and RFM mice (Hayata et al. 1979). Another consistent abnormality was an addition or a loss of the Y chromosome in a fraction of cells in all six males. Although the deleted chromosome 2 varied in size in the six mice studied, one common characteristic was noted in all these deletions: a segment lying between a certain band in region 2C and a band in region 2E, including the whole region 2D, was missing.

Trisomy 13 was a consistent chromosomal abnormality in spontaneous mammary tumors of GR, C3H, and non-inbred Swiss mice studied by Dofuku et al. (1979). Loss of the X chromosome was also frequently observed, but the incidence was relatively low in each tumor compared to that of trisomy 13.

Chinese Hamster

Chromosome banding studies have not been reported from primary tumors or leukemias of any species other than the mouse and the rat. Information is, however, available on two types of tumors studied with conventional staining in the Chinese hamster. Since the chromosome complement of this species permits detailed karyotype analysis even with conventional staining technique, the results are highly informative.

Kato (1968) studied the chromosomes of 42 primary sarcomas induced by RSV. As mentioned above, the chromosome pattern was clearly nonrandom. Where chromosome aberrations were present, the most common change was the addition of one to three specific chromosomes, i.e., chromosomes, 5, 6, or 10.

In contrast, Mitelman et al. (1972a) found a different pattern in histologically indistinguishable sarcomas induced by DMBA in the same inbred Chinese hamster strain. The main cytogenetic event in these sarcomas was trisomy for chromosome 11.

AGENT VERSUS TARGET CELL SPECIFICITY OF NONRANDOM
ABERRATIONS IN EXPERIMENTAL NEOPLASMS

Some of the results presented above are, no doubt, suggestive and indicate a relationship between the etiologic agent and the karyotypic pattern. This is mainly supported by the following three observations:

1. The same etiological agent may produce the same sequence of karyotypic changes in histologically different neoplasms. Thus, DMBA-induced sarcomas, leukemias, and carcinomas of the rat display the same characteristic chromosomal aberration.

2. Different etiological agents may produce different karyotypic patterns in histologically indistinguishable neoplasms. Thus, sarcomas induced by RSV and DMBA, respectively, in rats and the Chinese hamster, are characterized by completely different nonrandom karyotypic changes.

3. Chromosome 2, which is frequently present in trisomic state in DMBA-induced neoplasms of the rat, is specifically vulnerable to the chromosome-damaging action of DMBA, as evidenced by a selective increase in chromosome breakage frequency (Sugiyama 1971) and sister chromatid exchange (Ueda et al. 1976). It has been demonstrated by these authors that DMBA and TMBA, after intravenous injections into rats, induces chromosome breakage in bone marrow cells and that the breaks are nonrandomly distributed; chromosome 2 was found to contain five times more breaks per unit length than the other chromosomes. Furthermore, the breaks were distributed over the length of the chromosome in a distinctly nonrandom fashion, indicating that there are "hot spots" for the interaction between the oncogen and the DNA of chromosome 2. Also, administration of DMBA or TMBA in vitro to bone marrow cells induces an increase in nonrandomly distributed sister chromatid exchanges along the length of chromosome 2. The locations of the most frequent sister chromatid exchanges roughly coincide with those most susceptible to chromosome breakage in vivo after intravenous administration of these carcinogens (Ueda et al. 1976).

In contrast, there is now also evidence showing that different etiological agents may produce the same karyotypic pattern in the same target cell type. Thus, rat leukemias induced by DMBA and NBU, as well as rat sarcomas induced by DMBA, MC, and BP, were all characterized by trisomy 2. MC and BP are both polycyclic hydrocarbons related to DMBA, whereas NBU has a completely different chemical structure. Some differences in the karyotypic patterns of the different chemical inductions were, however, discernible. The situation in induced murine leukemias is particularly remarkable in this context. All T-cell murine leukemias so far studied, whether spontaneous or induced by X ray, RadLV, BP, or DMBA, display the same characteristic aberration, trisomy 15. On the other hand, murine leukemias of non-T-cell origin induced by different viruses do not have this particular aberration. Whether they have other types of distinctive changes is not yet known. Interestingly mineral oil–induced plasmocytomas do not have trisomy 15, but are characterized by a deletion of chromosome 15.

The results from the murine leukemia/lymphoma system seem at first glance to contradict the results obtained in induced neoplasms of the rat and the Chinese hamster and support the idea that the karyotypic changes are dependent on the target cell rather than the etiological agent. It could, of course, be argued that all murine T-cell lymphomas are due to the activation of latent type C viruses, but several facts seem to make this interpretation unlikely (Klein 1979). According to Klein (1979), the initiation process in lymphoma development creates long-lived preneoplastic cells "frozen in their state of differentiation and capable of continued division. These cells constitute the raw material for the subsequent cytogenetic evolution that converges towards a common distinctive pattern. The nature of this pattern, as it appears in overt lymphoma, depends on the subclass of the target lymphocyte, rather than on the initiating ('etiological') agent." As will be discussed below, this hypothesis does not contradict the hypothesis proposed by us (Mitelman and Levan 1978) that the *primary* change is the result of a direct interaction between the oncogenic factor(s) and the genetic material of the target cell. According to this hypothesis, such primary changes lead to a "transformed" cell, released from host control and with the potential of indefinite life. Such a cell could be highly malignant in itself, but more often it would have a very slow rate of proliferation and constitute no immediate hazard to the host. Due to the transformed state, however, the developing cell population can accumulate secondary chromosome changes, passively and at random but subjected to cell selection, so that only aberrations resulting in cells with proliferative advantage would persist and later dominate the cell population. We speculate that secondary aberrations that amplify the effect of the primary lesion often would be at a selective advantage, and the nature of such aberrations might very well depend on the type of target cell (see section on "Tumor Etiology and Chromosome Pattern in Neoplasms of Man," this chapter).

Apparently, the two hypotheses—agent specificity versus target cell specificity—need not be mutually exclusive. As indicated above, both types of aberrations may occur, and both may together contribute to the final cytogenetic event: nonrandom karyotypic changes. Further experimental work is urgently needed to clarify the relative importance of the target cell and the etiological agent for the appearance of nonrandom chromosome change in different neoplasms.

NONRANDOM CHROMOSOME CHANGES IN HUMAN NEOPLASMS

About 10 years ago, the confusing variety of karyotypic changes in human tumors seemed to refute any possibility that specific chromosome changes were associated with specific tumor types. The tremendous variability in the karyotypic changes and the fact that no two similar karyotypes had been found in any human neoplasm certainly indicated that the chromosomal abnormalities observed were secondary phenomena to the neoplastic state and that at least visible aberrations could not be connected with any important parameters of the disease. The discovery of the Ph[1]-chromosome in chronic myeloid leukemia (CML) 20 years ago by Nowell and Hungerford (1960) had suggested that such correlations might exist and also raised the possibility that chromosome changes might be causally related to malignant disease. Apparently, however, this association between a specific aberration and a specific malignant disorder was an exception; no consistent karyotypic picture emerged for any other neoplasm or group of neoplasms. The enthusiasm for tumor cytogenetics, to a large extent aroused by the discovery of the Ph[1] in CML, gradually faded.

The first indication of nonrandom karyotypic changes in human tumors was detected independently in 1966 by Levan and van Steenis in materials from the literature consisting mainly of ascitic forms of gastric, mammary, uterine, and ovarian carcinomas. The authors analyzed the chromosome changes in about 40 published cases and found that certain chromosome types tended to increase in number, others to decrease. Soon afterwards evidence for nonrandom patterns was demonstrated also in specific types of solid tumors such as meningiomas (Zang and Singer 1967, Mark 1970), gliomas (Mark 1971), and cervical carcinomas (Granberg 1971), as well as in leukemia (Whang-Peng et al. 1969, Hart et al. 1971) and other myeloproliferative disorders, e.g., polycythemia vera (Lawler et al. 1970).

In the 10 years since modern techniques of human cytogenetics first were applied to the study of human neoplasms, a sufficient number of cases have been investigated to permit certain generalizations to be advanced concerning the types of chromosome changes present in the majority of these disorders. The information presently available clearly shows that all tumor types that have been studied in a number sufficient to permit conclusions are characterized by various nonrandom chromosome patterns. We have tried systematically to collect and survey data on banded tumor chromosomes in man (Levan and Mitelman 1975b, 1977, Mitelman and Levan 1976a, 1978, unpublished observations, Mitelman 1980). The bulk of the material represents published cases, which we have retrieved by the aid of three separate computer-based literature scans; the remainder consists of unpublished cases from our laboratory and from laboratories of numerous colleagues who have kindly contributed their cases in response to our requests (Mitelman and Levan 1976b, 1979). By the end of 1979, our survey comprised more than 4,000 human neoplasms submitted to karyotype analysis with any of the banding techniques, the majority being leukemias, preleukemias, or other myeloproliferative disorders. In the total material, 1,613 had clonal karyotypic changes; cases with CML and the Ph[1] chromosome produced by the standard 9;22 translocation as the sole abnormality were not included in the survey.

Table 2 shows the distribution of the tumors studied classified into 15 tumor types. The distribution is heavily skewed: thus, the group myeloproliferative disorders constitutes about 70% of the cases, whereas all solid tumors, including

TABLE 2. *Subdivision of 1,613 cases of human neoplasms with chromosome aberrations identified by banding techniques*

Myeloproliferative disorders	Chronic myeloid leukemia (CML)	
	t (9;22) and other aberrations	338
	Aberrant translocations	76
	Acute myeloid leukemia (AML)	391
	Polycythemia vera (PV)	77
	Various myeloproliferative disorders (MD)	263
Lymphoproliferative disorders	Malignant lymphomas (ML) (non-Burkitt)	105
	Burkitt's lymphoma (BL)	22
	Acute lymphocytic leukemia (ALL)	122
	Chronic lymphocytic leukemia (CLL)	24
	Monoclonal gammopathies (MG)	21
Solid tumors	Meningiomas (MA)	56
	Benign epithelial tumors (BET)	9
	Carcinomas (CA)	86
	Malignant melanoma (MEL)	8
	Neurogenic tumors (NT)	7
	Sarcomas (SA)	8

all benign and malignant epithelial tumors, constitute only about 10% of the material. Apparently, much more information is needed on solid tumors before any definite conclusions can be drawn concerning the types of chromosome variation in human neoplasia. Nevertheless, it is already at this stage possible to make the generalization that the aberrations in each tumor type that have been studied in a sufficient number are strictly nonrandom. Table 3 lists the chromosomes involved in aberrations in more that 20% of each particular tumor type. In several instances the aberrations are highly consistent and show the same degree of specificity as the Ph^1 chromosome in CML, e.g., the 8;14 translocation in Burkitt's lymphoma, the 15;17 translocation in acute promyelocytic leukemia (APL), the 5q— marker in refractory anemia, and the loss or deletion of one chromosome 22 in meningiomas (Mitelman 1980). In other instances, the nonrandom involvement of certain chromosomes is less specific but still in a statistical sense highly reproducible. For example, one of chromosomes 5, 7, 8, and 21 is involved in aberrations in more than 80% of patients with acute nonlymphocytic leukemia (ANLL) and in CML about 80% of patients show an evolution characterized by three particular aberrations: an additional Ph^1-chromosome, trisomy 8, and the acquisition of an isochromosome 17. In our opinion, both types of changes—the consistent and specific aberrations as well as the nonrandom involvement of certain chromosomes—lends support to the hypothesis that chromosomal changes play an important role in tumor development and may contribute to the direct causation of the malignant state.

Another interesting observation is that in the total material only certain chro-

TABLE 3. *Chromosomes preferentially engaged in 15 tumor types*
(1,613 cases)

Tumor type	Chromosome nos.
Myeloproliferative Disorders	
Acute myeloid leukemia	5, 7, 8, 17, 21
Chronic myeloid leukemia	8, 9, 17, 22
Polycythemia vera	1, 8, 9, 20
Various myeloproliferative disorders	5, 7, 8
Lymphoproliferative Disorders	
Malignant lymphomas	1, 3, 9, 14
Burkitt's lymphoma	7, 8, 14
Acute lymphocytic leukemia	1, 8, 9, 14, 17, 21, 22
Chronic lymphocytic leukemia	1, 14, 17
Monoclonal gammopathies	1, 3, 14
Solid Tumors	
Meningiomas	8, 22
Benign epithelial tumors	8, 14
Carcinomas	1, 3, 5, 7, 8
Malignant melanoma	1, 9
Neurogenic tumors	1, 22
Sarcomas	1, 3, 5, 13, 14, 22

mosomes participate in the variation of different tumor types. In Figure 1, the abbreviated designations of the 15 tumor classes have been listed under the chromosomes most often involved in aberrations in each tumor type. It is striking that all tumor classes cluster to 12 of the 24 chromosome types, and three chromosomes, Nos. 1, 8, and 14, are involved in at least seven disorders each. It should be mentioned that no chromosome has been added to those subject to aberration in this considerably extended material, amounting to 1,613 cases, compared to our previously published survey of 856 cases (Mitelman

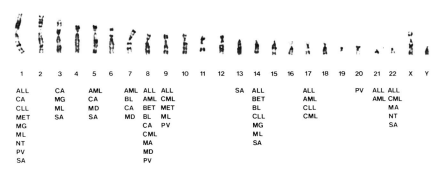

FIG. 1. Survey of the most common chromosome changes in 1,613 human neoplasms. The abbreviated designations of the 15 groups of malignant or premalignant disorders (see Table 2) have been listed below the chromosome types preferentially involved (F. Mitelman and G. Levan, unpulblished observation).

and Levan 1978); only now more tumors have been assigned to each of the chromosomes preferentially involved in aberrations.

The fact that chromosome aberrations in human tumors are nonrandom, and that in different tumor types aberrations tend to cluster in specific chromosomes, may suggest that only certain chromosomes carry genes of special importance for malignant development.

TUMOR ETIOLOGY AND CHROMOSOME PATTERN IN NEOPLASMS OF MAN

Extensive evidence obtained by recent cytogenetic studies implies that the chromosome aberrations observed in experimental and human neoplasms are essential for the initiation of malignant development and not merely secondary phenomena in tumor progression. However, this interpretation does not mean that all chromosome aberrations contribute to the direct causation of the malignant state. Some aberrations are undoubtedly random events, but may still be of great importance. According to this hypothesis, elaborated in detail in Mitelman and Levan (1978), there are two essentially different kinds of chromosome aberration: primary nonrandom changes and secondary random events. We believe that the first type is the result of a direct interaction between oncogenic factors and the genetic material of specific loci. Such actively induced gene changes could be expressed as small structural aberrations, such as translocations, deletions, or duplications, but probably more often as microscopically undetectable point mutations. Sometimes these aberrations may directly lead to a highly malignant cell, but more often to a "premalignant" cell in which further secondary changes may take place. The latter, passive changes are random and most of the cells resulting are eliminated; only aberrations leading to cells with a selective advantage persist. We believe that secondary aberrations that enhance the effect of the primary lesion often would be at a selective advantage. An increase in the relative amount of the aberrant gene product can be obtained in two ways: (1) amplification of the chromosome region in which the primary change has occurred, e.g., through nondisjunction producing trisomy, by duplication, or by translocation; (2) elimination of the unaffected homologous segment, e.g., through nondisjunction producing monosomy, by deletion and by translocation.

It should also be noted that translocation between homologous chromosomes can lead to duplication of the affected segment and simultaneous removal of the homologous unaffected segment without any visible karyotypic change. Such a mechanism would explain the fact that a substantial number of malignant neoplasms, e.g., 50% of all ANLL, have a completely normal karyotype.

This working hypothesis is, as we have pointed out, clearly vulnerable to criticism but the assumption of two essentially different types of chromosomal change accords well with data now available from human and experimental malignancies.

The initial change—the primary nonrandom change—is supposed to be caused by the direct action of the inducing factors on specific gene loci. This is, of course, an unproved hypothesis, but is supported by several results from experimental studies reviewed above, although the relationship between etiologic factor and chromosome pattern may be more complex than originally anticipated. Is there evidence obtained from studies in human neoplasia that any of the nonrandom or specific aberrations are tissue specific or agent specific?

The catalogue of the nonrandom changes in various tumors presented in Table 3 and Figure 1 provides clear evidence that the same chromosomes may be affected in a variety of tumors; chromosomes 1, 8, and 14 are good examples. These chromosomes are nonrandomly affected in myeloproliferative and lymphoproliferative disorders as well as in epithelial, mesenchymal, and neurogenic tumors. On the other hand, some chromosomes seem to be preferentially involved in neoplasia affecting a particular tissue; the involvement of chromosome 14 in various lymphoproliferative disorders is a good example. The question of tissue specificity of specific chromosome changes was recently reviewed in some detail (Mitelman 1980). Additional information now available supports the conclusion presented that although some specific changes undoubtedly are more common in certain types of neoplasia, e.g., the 8;14 translocation in B-cell leukemia and lymphoma, no specific aberration so far described is exclusive for any particular tumor or cell type.

Unfortunately, it is presently almost impossible to put the alternative hypothesis, i.e., agent specificity of chromosome aberrations, to a direct test in human neoplasms. Tumors associated with known environmental carcinogens are uncommon and, so far, none has been subjected to chromosome analysis. However, two indirect methods to approach this question are available and will be briefly commented on.

Our recently published survey of 856 cases with aberrations (Mitelman and Levan 1978) revealed certain hints that differences in chromosome pattern may exist among geographic regions. The same trend is also obvious in our present extended material. A few examples are chosen for brief consideration.

In AML the incidence of monosomy 5 or deletion of the long arm of chromosome 5 is about 25% in the U.S.A., which is about twice the incidence in Europe, whereas no single case has been reported from Japan. Similarly, monosomy 7 or a deletion of the long arm of chromosome 7 has so far only been reported in one case from Japan, whereas the frequency in both Europe and the U.S.A. is about 25%. On the other hand, the 8;21 translocation is rather uncommon in Europe, but this aberration is present in more than 25% of the cases from Japan.

In APL, an uneven geographic distribution of the characteristic 15;17 translocation was reported by Teerenhovi et al. (1979). The Second International Workshop on Chromosomes in Leukemia (1980) reviewed 67 cases of APL from different regions of the world. Again, an unusual geographic distribution was noted: in Chicago, 6 of 6 patients had the translocation; in Buffalo, 1 of 18; in Leuven, 16 of 18; in Paris, 8 of 12; and in Scandinavia, 0 of 13.

In CML the notable differences were that trisomy 19 was more frequent in Europe than in the U.S.A. and about six times more common in Japan; also, the presence of an extra Ph^1 seemed to be much more common in Japan than in other parts of the world, whereas an iso (17q) was decidedly less common in Japan than in Europe and the U.S.A.

In polycythemia vera, the characteristic 20q− marker, present in about 20% of the cases from the U.S.A. and about 40% of the cases from Europe, has not been reported in one single case from Israel. It may be noteworthy that the most common aberration in Israel was trisomy 8, while in Europe and the U.S.A. it is the most uncommon one.

In meningiomas, about one third of the tumors with aberrations display a loss of one chromosome 8 in addition to the characteristic and consistent monosomy or deletion of one chromosome 22. It is striking that monosomy 8 has been noted in 65% of cases from Sweden, whereas this aberration was only found in 3% of cases studied in Germany.

These data admittedly suffer from certain weaknesses. The results are in some instances based on small numbers of materials and may also be biased since many published cases represent selections from larger analyzed series. Nevertheless, some of the correlations are suggestive and indicate that geographic differences in chromosome aberrations may exist. Such a geographic heterogeneity may be taken to indicate heterogeneity in the distribution of etiologic factors.

Another approach to elucidate possible agent dependence of chromosome aberrations in human neoplasms was used in a study by our group (Mitelman et al. 1978), and recently extended in collaboration with G. Alimena and coworkers in Rome, Italy (Mitelman et al. 1979). In this investigation, the case records of 156 patients with ANLL were reviewed and the patients divided into two groups: 50 patients occupationally exposed to potential mutagenic/carcinogenic agents and 106 patients with no history of occupational exposure to such agents. The exposed individuals were divided into three groups according to exposure to chemical solvents, insecticides, and petrol products. In the total material, about 50% of the patients had clonal chromosomal aberrations at diagnosis, and these were nonrandom, in perfect agreement with previously published data on ANLL (e.g., Van Den Berghe et al. 1978). There were, however, striking differences between the chromosomal changes in the unexposed group and the exposed group: in the unexposed group only 32% of the patients had chromosome aberrations in their bone marrow cells, whereas 76% of the exposed patients had chromosome aberrations. Among patients exposed to chemical solvents, 92% had an abnormal bone marrow karyotype. Furthermore, in the exposed group about 90% showed one or more of four particular clonal changes: −5 or 5q−, −7 or 7q−, +8, or +21; in the unexposed group only about 55% of the patients had any of these changes (F. Mitelman, P. G. Nilsson, L. Brandt, G. Alimena, B. Dallapiccola, and R. Gastaldi, unpublished observation).

Our data were obtained retrospectively from case records. Therefore, the information on environmental hazards is incomplete and may also be biased, since a considerable number of patients had to be excluded from the study because the occupational data were insufficient. The fact that differences in karyotypic aberrations between unexposed and exposed patients still were discernible indicate that the karyotypic pattern of the leukemic cells were, in fact, influenced by the exposure. The possible relation between exposure to potential mutagenic/carcinogenic agents and the presence of clonal chromosomal abnormalities should be persued by prospective studies. Such a study is in progress in our laboratory. In this context, it might be of interest that intensive chemotherapy during the chronic phase of CML has been found to produce new stable clones with a preferential engagement of chromosome 1 (Alimena et al. 1979). Thus, although only indirect evidence, the results support the concept that nonrandom chromosome changes may be influenced by external factors.

Even though there is no direct evidence from human materials that specific chromosome changes are induced by any particular agent, some of the relationship of chromosomal deviations to geographic area on the one hand, and to exposure to potential mutagenic/carcinogenic agents, on the other, undoubtedly suggest that such correlations may exist.

ACKNOWLEDGMENTS

Original studies described in this paper were supported by the Swedish Cancer Society and the John and Augusta Persson Foundation for Medical Scientific Research.

REFERENCES

Ahlström, U. 1974. Chromosomes of primary carcinomas induced by 7,12-dimethylbenz (α) anthracene in the rat. Hereditas 78:235–244.

Alimena, G., L. Brandt, B. Dallapiccola, F. Mitelman and P. G. Nilsson. 1979. Secondary chromosome changes in chronic myeloid leukemia: Relation to treatment. Cancer Genet. Cytogenet. 1:79–85.

Chan, F. P. 1978. Chromosome studies in induced murine thymomas. Diss. Abstr. Intern., Sect. B, 38:3994–3995.

Chang, T. D., J. L. Biedler, E. Stockert, and L. J. Old. 1977. Trisomy of chromosome 15 in X-ray-induced mouse leukemia. Cancer Res. 18:225.

Dofuku, R., J. L. Biedler, B. A. Spengler, and L. J. Old. 1975. Trisomy of chromosome 15 in spontaneous leukemia of AKR mice. Proc. Natl. Acad. Sci. USA 72:1515–1517.

Dofuku, R., T. Utakoji, and A. Matsuzawa. 1979. Trisomy of chromosome no. 13 in spontaneous mammary tumors of GR, C3H, and noninbred Swiss mice. J. Natl. Cancer Inst. 63:651–656.

Granberg, I. 1971. Chromosomes in preinvasive, microinvasive and invasive cervical carcinoma. Hereditas 68:165–218.

Hart, J. S., J. M. Trujillo, E. J. Freireich, S. L. George, and E. Frei. 1971. Cytogenetic studies and their clincal correlates in adults with acute leukemia. Ann. Intern. Med. 75:353–360.

Hayata, I., T. Ishihara, K. Hirashima, T. Sado, and J. Yamagiwa. 1979. Partial deletion of chromosome no. 2 in myelocytic leukemias of irradiated C3H/He and RFM mice. J. Natl. Cancer Inst. 63:843–848.

Kato, R. 1968. The chromosomes of forty-two primary Rous sarcomas of the Chinese hamster. Hereditas 59:63–119.

Klein, G. 1979. Lymphoma development in mice and humans: Diversity of initiation is followed by a convergent cytogenetic evolution. Proc. Natl. Acad. Sci. USA 76:2442–2446.

Klein, G., S. Ohno, N. Rosenberg, F. Wiener, J. Spira, and D. Baltimore. 1980. Cytogenetic studies on Abelson-virus induced mouse leukemias. Int. J. Cancer 25:805–811.

Kodama, Y., M. C. Yoshida, and M. Sasaki. 1978. A note on banded karyotypes in spontaneous leukemias of the AKR mouse. Proc. Jpn. Acad. 54:222–227.

Lawler, S. D., R. E. Millard, and H. E. M. Kay. 1970. Further cytogenetical investigations in polycythemia vera. Eur. J. Cancer 6:223–233.

Levan, A. 1961. Preliminary chromosome data on Rous sarcomas in rats. First Scandinavian Symposium on Carcinogenesis, Oslo, 1961.

Levan, A. 1966. Non-random representation of chromosome types in human tumor stemlines. Hereditas 55:28–38.

Levan, G. 1974. The detailed chromosome constitution of a benzpyrene-induced rat sarcoma. A tentative model for G-band analysis in solid tumors. Hereditas 78:273–289.

Levan, G., U. Ahlström, and F. Mitelman. 1974. The specificity of chromosome A2 involvement in DMBA-induced rat sarcomas. Hereditas 77:263–280.

Levan, G., and A. Levan. 1975. Specific chromosome changes in malignancy: Studies in rat sarcomas induced by two polycyclic hydrocarbons. Hereditas 79:161–198.

Levan, G., and F. Mitelman. 1975a. G-band analysis in serially transplanted Rous rat sarcoma. Hereditas 80:140–145.

Levan, G., and F. Mitelman. 1975b. Clustering of aberrations to specific chromosomes in human neoplasms. Hereditas 79:156–160.

Levan, G., and F. Mitelman. 1976. G-banding in Rous rat sarcomas during serial transfer: Significant chromosome aberrations and incidence of stromal mitoses. Hereditas 84:1–14.

Levan, G., and F. Mitelman. 1977. Chromosomes and the etiology of cancer, in Chromosomes Today, A. de la Chapelle and M. Sorsa, eds. Elsevier/North Holland Biomedical Press, Amsterdam, pp. 363–371.

Mark, J. 1967. Chromosomal analysis of ninety-one primary Rous sarcomas in the mouse. Hereditas 57:23–82.

Mark. J. 1970. Chromosomal patterns in human meningiomas. Eur. J. Cancer 6:489–498.

Mark, J. 1971. Chromosomal characteristics of neurogenic tumours in adults. Hereditas 68:61–100.

Mitelman, F. 1971. The chromosomes of fifty primary Rous rat sarcomas. Hereditas 69:155–186.

Mitelman, F. 1972a. Predetermined sequential chromosome changes in serial transplantation of Rous rat sarcomas. Acta. Pathol. Microbiol. Scand. [A] 80:313–328.

Mitelman, F. 1972b. Comparative chromosome analysis of primary and metastatic Rous sarcomas in rats. Hereditas 70:1–14.

Mitelman, F. 1980. Cytogenetics of experimental neoplasms and non-random chromosome correlations in man. Clin. Haematol. 9:195–219.

Mitelman, F., L. Brandt, and P. G. Nilsson. 1978. Relation among occupational exposure to potential mutagenic/carcinogenic agents, clinical findings, and bone marrow chromosomes in acute nonlymphocytic leukemia. Blood 52:1229–1237.

Mitelman, F., and G. Levan. 1972. The chromosomes of primary 7,12-dimethylbenz (α) anthracene-induced rat sarcomas. Hereditas 71:325–334.

Mitelman, F., and G. Levan. 1976a. Clustering of aberrations to specific chromosomes in human neoplasms. II. A survey of 287 neoplasms. Hereditas 82:167–174.

Mitelman, F., and G. Levan. 1976b. Do only a few chromosomes carry genes of prime importance for malignant transformation? Lancet II:264.

Mitelman, F., and G. Levan. 1978. Clustering of aberrations to specific chromosomes in human neoplasms. III. Incidence and geographic distribution of chromosome aberrations in 856 cases. Hereditas 89:207–232.

Mitelman, F., and G. Levan. 1979. Chromosomes in neoplasia: An appeal for unpublished data. Cancer Genet. Cytogenet. 1:29–32.

Mitelman, F., J. Mark, and G. Levan. 1972a. The chromosomes of six primary sarcomas induced in the Chinese hamster by 7,12-dimethylbenz (α) anthracene. Hereditas 72:311–318.

Mitelman, F., J. Mark, G. Levan, and A. Levan. 1972b. Tumor etiology and chromosome pattern. Science 176:1340–1341.

Mitelman, F., P. G. Nilsson, L. Brandt, G. Alimena, A. Montuoro, and B. Dallapiccola. 1979. Chromosomes, leukaemia, and occupational exposure to leukaemogenic agents. Lancet II:1195–1196.

Nichols, W. W. 1963. Relationship of viruses, chromosomes and carcinogenesis. Hereditas 50:53–80.

Nowell, P. C., and D. A. Hungerford. 1960. A minute chromosome in human chronic granulocytic leukemia. Science 132:1497.

Ohno, S., M. Babonits, F. Wiener, S. Spira, G. Klein, and M. Potter. 1979. Nonrandom chromosome changes involving the Ig gene-carrying chromosomes 12 and 6 in pristane-induced mouse plasmocytomas. Cell 18:1001–1007.

Rowley, J. D. 1974. Do human tumors show a chromosome pattern specific for each etiologic agent? J. Natl. Cancer Inst. 52:315–320.

Second International Workshop on Chromosomes in Leukemia. 1980. Chromosomes in acute promyelocytic leukemia. Cancer Genet. Cytogenet. (in press).

Shepard, J. S., O. S. Pettengill, D. H. Wurster-Hill, and G. D. Sorenson. 1976. Karyotype, marker formation, and oncogenicity in mouse plasmocytomas. J. Natl. Cancer Inst. 56:1003–1011.

Shepard, J. S., D. H. Wurster-Hill, O. S. Pettengill, and G. D. Sorenson. 1974. Giemsa-banded chromosomes of mouse myeloma in relationship to oncogenicity. Cytogenet. Cell Genet. 13:279–304.

Stich, H. F. 1960. Chromosomes of tumor cells. I. Murine leukemias induced by one or two injections of 7,12-dimethylbenz (α) anthracene. J. Natl. Cancer Inst. 25:649–661.

Sugiyama, T. 1971. Specific vulnerability of the largest telocentric chromosome of rat bone marrow cells to 7,12-dimethylbenz (α) anthracene. J. Natl. Cancer Inst. 47:1267–1275.

Sugiyama, T., and F. P. Brillantes. 1970. Cytogenetic studies of leukemia induced by 6,8,12- and 7,8,12-trimethylbenz (α) anthracene. J. Exp. Med. 131:331–341.

Sugiyama, T., Y. Kurita, and Y. Nishizuka. 1967. Chromosome abnormality in rat leukemia induced by 7,12-dimethylbenz (α) anthracene. Science 158:1058–1059.

Sugiyama, T., H. Uenaka, N. Ueda, S. Fukuhara and S. Maeda. 1978. Reproducible chromosomal changes of polycyclic hydrocarbon-induced rat leukemia; incidence and chromosome banding pattern. J. Natl. Cancer Inst. 60:153–160.

Teerenhovi, L., G. H. Borgström, F. Mitelman, L. Brandt, P. Vuopio, T. Timonen, A. Almqvist, and A. de la Chapelle. 1978. Uneven geographical distribution of 15;17-translocation in acute promyelocytic leukaemia. Lancet II:797.

Ueda, N., H. Uenaka, T. Akematsu, and T. Sugiyama. 1976. Parallel distribution of sister chromatid exchanges and chromosome aberrations. Nature 262:581–583.

Uenaka, H., N. Ueda, S. Maeda, and T. Sugiyama. 1978. Involvement of chromosome No. 2 in chromosome changes in primary leukemia induced in rats by N-nitroso-N-butylurea. J. Natl. Cancer Inst. 60:1399–1404.

Van Den Berghe, H., G. H. Borgström, L. Brandt, D. Catovsky, A. de la Chapelle, H. Golomb, D. K. Hossfeld, S. Lawler, J. Lindsten, A. Louwagie, F. Mitelman, B. R. Reeves, J. D. Rowley, A. A. Sandberg, L. Teerenhovi, and P. Vuopio. 1978. Chromosomes in acute non-lymphocytic leukemia. Br. J. Haematol. 39:311–316.

van Steenis, H. 1966. Chromosomes and cancer. Nature 209:819–821.

Wakonig, R., and H. F. Stich. 1960. Chromosomes in primary and transplanted leukemias of AKR mice. J. Natl. Cancer Inst. 25:295–303.

Whang-Peng, J., E. J. Freireich, J. J. Oppenheim, E. Frei, and J. H. Tjio. 1969. Cytogenetic studies in 45 patients with acute lymphocytic leukemia. J. Natl. Cancer Inst. 42:881–897.

Wiener, F., S. Ohno, J. Spira, N. Haran-Ghera, and G. Klein. 1978a. Chromosomal changes (trisomy 15 and 17) associated with tumor progression in leukemias induced by radiation leukemia virus (Rad LV). J. Natl. Cancer Inst. 60:227–237.

Wiener, F., J. Spira, S. Ohno, N. Haran-Ghera, and G. Klein. 1978b. Chromosome changes (trisomy 15) in murine T-cell leukemia induced by 7,12-dimethylbenz (α) anthracene (DMBA). Int. J. Cancer, 22:447–453.

Yoshida, M. C., and K. Moriwaki. 1975. Specific marker chromosome involving a translocation (12;15) in a mouse myeloma. Proc. Jpn. Acad. 51:588–592.

Yoshida, M. C., K. Moriwaki, and S. Migita. 1978. Specificity of the deletion of chromosome No. 15 in mouse plasmacytoma. J. Natl. Cancer Inst. 60:235–238.

Zang, K. D., and H. Singer. 1967. Chromosomal constitution of meningiomas. Nature 216:84–85.

Genes, Chromosomes, and Neoplasia, edited by
Frances E. Arrighi, Potu N. Rao, and Elton Stubblefield.
Raven Press, New York © 1981.

The Chromosome Changes in Bloom's Syndrome, Ataxia-Telangiectasia, and Fanconi's Anemia

James H. Ray and James German

*Laboratory of Human Genetics, The New York Blood Center,
New York, New York 10021*

The relationship between chromosome changes and the onset of neoplasia remains obscure in spite of numerous investigations. The best-documented examples of specific chromosome changes associated with human tumors are the Philadelphia chromosome (Ph[1]) in chronic granulocytic leukemia, the 14q+ chromosome in Burkitt's lymphoma, and the numerically or structurally deleted No. 22 chromosome in meningiomas (Mark 1977). Structurally abnormal chromosomes as well as extra or missing chromosomes are also found in other types of human cancer, but their constancy is not nearly that of the associations just mentioned. Compounding the problem of assigning nonrandom chromosome changes a causative role in neoplasia is the fact that more than one karyotypic change may be evident in the cells of a given cancer. Because the cells having the nonrandom chromosome changes themselves originated from neoplastic tissue, it has been impossible to determine whether they arose during the period of unrestricted growth following the onset of cancer. The lack of understanding stems in large measure from our being unable to monitor the early stages of human cancer evolution, particularly in solid tumors. We glimpse only a small segment, the final segment, in what may well be a process that has been under way for a long time. Therefore, much speculation and deduction have necessarily been employed in the interpretation of the results of cancer cytogenetic studies.

A more direct approach for assessing the potential role of chromosome change in cancer development is the study of chromosome changes in persons predisposed to cancer but who have yet to manifest clinical symptoms of the disease. Cancer predisposition can be on the basis of both environmental and genetic factors. Bloom's syndrome, ataxia-telangiectasia, and Fanconi's anemia—three so-called chromosome-breakage syndromes—represent conditions in which affected individuals are predisposed on a genetic basis. In all three syndromes, an increased risk of cancer, as well as increased chromosome instability, exists. Numerous cytological studies of these three rare conditions have been made during the past 15 years (see Schroeder and Kurth 1971, Polani 1976, for earlier reviews). We summarized these studies here because of the possibility that a broad and comparative view of all three may help clarify the significance of chromosome change in the etiology of human cancer.

BLOOM'S SYNDROME

General

Bloom's syndrome (BS) is an autosomal recessive disorder (German et al. 1965) found relatively commonly among Ashkenazic Jews (German et al. 1977b, German 1979a). Its major clinical features are pre- and postnatal growth retardation, a sun-sensitive facial skin lesion, and a defective immune response (Hütteroth et al. 1975, Weemaes et al. 1979). Of the 90 individuals known to have had BS, 73 are still living, and their progress is being monitored by our laboratory. A drastically increased risk of developing cancer at an early age exists in BS (German et al. 1977a, German 1979b); one of every four persons known to have been affected has developed one or more primary cancers by the average age of 23 (German et al. 1979; J. German, unpublished results). Leukemia is the predominant cancer, but lymphomas and tumors of the respiratory and alimentary tracts and of the kidney have also occurred.

Cytogenetic Studies

Chromosome Aberrations

The cytogenetic disturbance in BS was reported a decade after the clinical syndrome itself had been described (Bloom 1954, German 1964, German et al. 1965). BS cells exhibit an increased incidence of chromosome aberrations and an increased frequency of sister chromatid exchange (SCE). Chromosome instability has been observed in all cell types examined—phytohemagglutinin (PHA)-stimulated blood lymphocytes, uncultured and cultured bone marrow, dermal fibroblasts, and Epstein-Barr virus–transformed lymphoblastoid cells. The aberrations observed in BS cells include chromatid and isochromatid gaps and breaks, acentric fragments, transverse centromeric separations, sister chromatid reunions, dicentric and rearranged monocentric chromosomes, and triradial (Tr) and quadriradial (Qr) interchange configurations. Until the discovery of increased SCE (see below), Qr configurations were considered to be the most characteristic cytogenetic feature of BS (German et al. 1974). The typical BS Qr (Figure 1) is symmetrical, with the centromeres of the two affected homologous chromosomes positioned in opposite arms of the interchange figure (German et al. 1974, Schroeder and German 1974, Schroeder and Stahl Mauge 1976). Asymmetrical Qr configurations involving either homologous or nonhomologous chromosomes and Tr configurations are also occasionally encountered in BS. Qr configurations occur in 2.4% (range 0–14%) of BS T-lymphocytes (German 1974, German et al. 1974) and in approximately 1% of BS dermal fibroblasts

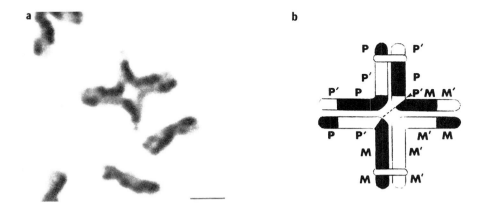

FIG. 1. a, Differentially stained BS Qr interchange configuration between two apparently homologous chromosomes. b, Diagrammatic representation of the exchange event that lead to the Qr formation in a. Chromosomes have been arbitrarily designated PP' (paternal) and MM' (maternal) to distinguish the chromosomes and the exchanged chromosome segments. The presence of P segments on P' chromatids and P' segments on P chromatids indicates that sister chromatid exchanges have occurred. The above is also true for M and M' chromatids. The presence of M and M' segments on P and P' chromatids and P and P' segments on M and M' chromatids signifies that an exchange has occurred between homologous but non-sister chromatids. The points of exchange between the nonsister chromatids are denoted by the dashed line. Bar in a represents 2 μm.

(German and Crippa 1966, Ved Brat 1979). A recently reported abnormality in BS cells is the terminal association (TA) of homologous chromosomes (Schonberg et al. 1978). Whether TA represents an unusual chromatid-interchange configuration or some other phenomenon remains to be determined. The sites of breakage and exchange in Qr configurations are reported to be in interband regions or at band-interband junctions in G-banded (Shiraishi and Sandberg 1977) and Q-banded chromosomes (Kuhn 1976).

Whether the Qr configurations typical of cultured BS cells occur in vivo is of considerable general biological interest because their demonstration there would represent direct cytogenetic evidence that somatic recombination occurs in vivo. Adjacent hypo- and hyperpigmented spots on the skin of persons with BS are compatible with its occurrence in vivo (Szalay and Weinstein 1972, German 1973, Festa et al. 1979). However, Qr configurations were not detected in five reported cytogenetic studies of BS bone marrow cells (Landau et al. 1966, Schoen and Shearn 1967, Kiossoglou et al. 1976, Shiraishi et al. 1976, Shiraishi and Sandberg 1977, Festa et al. 1979). We have studied 800 cells from uncultured marrow preparations of three BS individuals and have not found a single Qr configuration, although other chromatid aberrations were sometimes seen (J. H. Ray and J. German, unpublished results). Of the 800 cells studied, 650 were from one patient whose lymphocyte Qr frequency ranged from 1–4%. Therefore, if Qr configurations occur in vivo, their frequency is greatly reduced from that in cultured cells.

One of the main questions remaining unanswered for all the chromosome breakage syndromes, and one which we are currently addressing experimentally with respect to BS, is to what extent the chromosome instability observed in cultured cells is representative of the in vivo condition. The possibility exists that the increased aberration frequencies observed in cultured cells might be the result of some nutritional deficiency of the culture medium. Data from the BS marrow analyses just mentioned suggest that an increase in at least some aberrations occurs in vivo. Also arguing against BS chromosome instability being the result of in vitro incubation conditions is the observation of Shiraishi et al. (1976) that similarly increased numbers of chromosome aberrations occur in marrow cells examined either directly after aspiration or after incubation for 46 hours in tissue culture medium. We, in an effort to test further the possibility that chromosome aberrations are induced in vitro, cultured BS lymphocytes from two persons in the presence of bromodeoxyuridine (BrdU) so that first- and second-division cells could be recognized following differential staining (Crossen and Morgan 1977, 1978, Scott and Lyons 1979). No significant differences were noted between the aberration frequencies of first- and second-division lymphocytes (J. H. Ray and J. German, unpublished results). Therefore, if the observed chromosome instability was induced in vitro, it occurred prior to the first metaphase. During this study we found an interesting cell among the 200 first-division cells examined that appears to be informative with respect to in vivo chromosome instability. The cell has a chromosome complement that includes a dicentric chromosome along with 45 other apparently normal chromosomes, but it has no acentric fragments. Such an abnormal chromosome complement could not have resulted from disruptive events in the first in vitro division cycle, i.e., it must have arisen in vivo. Observations such as these, fragmentary though they are at present, suggest that in vivo instability does exist in BS.

Sister Chromatid Exchange

In the early 1970s it became possible to differentiate between sister chromatids on the basis of their staining patterns following growth for two cell cycles in the presence of the thymidine analogue BrdU (Latt 1973). The chromosomes in cells so treated have one chromatid bifilarly substituted and one unifilarly substituted with BrdU. The bifilarly substituted chromatid shows relatively dimmer fluorescence following 33258 Hoechst staining (Latt 1973) and relatively paler staining with Giemsa (Perry and Wolff 1974). This method of chromosome examination permitted the direct demonstration that Qr configurations in BS represent exchanges between homologous nonsister chromatids (Chaganti et al. 1974) as originally suggested (German 1964). BS T-lymphocytes (Figure 2) were also found to have a striking increase (10- to 14-fold) in the SCE frequency

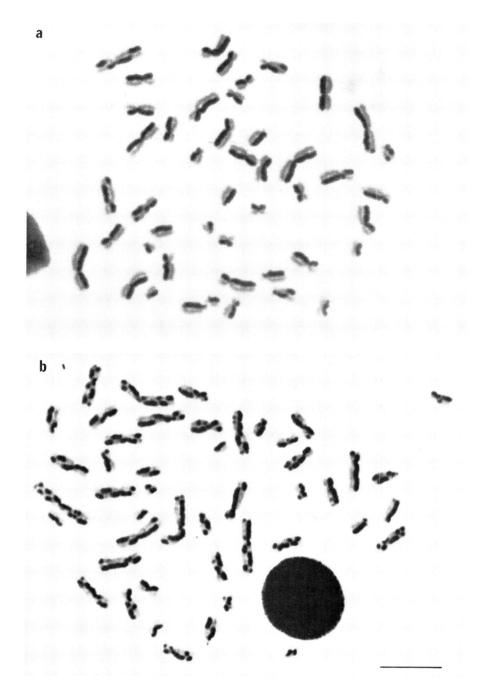

FIG. 2. Differentially stained PHA-stimulated lymphocytes from a normal individual (a) and a BS patient (b) demonstrating the significantly increased SCE frequency of BS cells. The cell in a is from the mother of the patient in b. Bar represents 10 μm.

(Chaganti et al. 1974, Bartman et al. 1976, German et al. 1977c). An elevated SCE frequency is now considered diagnostic of BS (Sperling et al. 1976, Hustinx et al. 1977, Dicken et al. 1978). Elevated SCE frequencies have been observed not only in PHA-stimulated lymphocytes but also in cultured bone marrow cells (Shiraishi and Sandberg 1977, German and Schonberg 1980), dermal fibroblasts (Ved Brat 1979), and lymphoblastoid cells (Henderson and German 1978). Shiraishi et al. (1976) tested the possibility that the high SCE frequency in BS cells was due to BrdU hypersensitivity by using ^{3}H-deoxycytidine to label the chromatids of BS lymphocyte chromosomes differentially. The autoradiographs were interpreted to indicate that the incidence of SCE in ^{3}H-deoxycytidine-labeled chromosomes is elevated just as in BrdU-labeled chromosomes and that, whatever the cause of the increased SCE frequency in BS, it is not related specifically to BrdU hypersensitivity.

A puzzling and highly provocative finding in blood lymphocyte cultures from some BS patients is the presence of two types of cells with respect to SCE frequency, one with the characteristically elevated SCE frequency and one with a frequency roughly in the range of that of non-BS cells (German et al. 1977c, Hustinx et al. 1977). The presence of two cell populations is unexpected in view of the recessive transmission of the disorder and appears to indicate that homozygosity for the BS mutation may be expressed differently in different cells. The presence of two populations of cells in the blood also probably accounts for the unexpected finding that BS blood can give rise to cloned lymphoblastoid cell lines (LCLs) of two types, one with high and one with low frequencies of SCE (Henderson and German 1978, unpublished results). Of four BS LCLs studied to date, two had increased SCE frequencies, and two had low SCE frequencies; the two LCLs with high SCE frequencies also exhibited many chromatid aberrations. The finding of elevated SCE frequencies and increased numbers of chromatid aberrations in some BS LCLs and low SCEs and low numbers of aberrations in others suggests that SCE and chromosome aberrations ultimately have the same molecular basis in BS cells.

Abnormal Clones

Clones of cells with abnormal chromosome complements are occasionally detectable in BS. Rauh and Soukup (1968) found a t(1q+;Dq−) chromosome in approximately 50% of the cells of a BS dermal fibroblast culture. Cohen et al. (1979) found clones with a Dq+ chromosome in 2% and 7% of the blood lymphocytes of two affected individuals and in 4% of the fibroblasts of one of them. However, since similar mutant cells are sometimes present in lymphocyte cultures derived from normal persons (Welch and Lee 1975, Beatty-DeSana et al. 1975, Hecht et al. 1975), studies comparing Dq+ clone frequencies in cell lines derived from BS versus normal people are needed before their significance in BS is understood. An apparent isochromosome of the long arm of the Y chromosome has been found repeatedly in our laboratory in a small proportion of PHA-stimulated short-term blood cultures from each of two males with BS (S. Schonberg and J. German, unpublished results). The repeated presence of such clones in short-term blood cultures is evidence for in vivo chromosome mutation in BS.

ATAXIA-TELANGIECTASIA

General

Ataxia-telangiectasia (AT, or the Louis-Bar syndrome) is a rare autosomal recessive disorder (Tadjoedin and Fraser 1965) that is relatively common in only one group, Jews from the Atlas Mountains (Levin and Perlov 1971), probably on the basis of founder effect. Three independent descriptions of AT were made before it became widely recognized as a specific clinical entity (Syllaba and Henner 1926, Louis-Bar 1941, Boder and Sedgwick 1957, 1958). The predominant clinical features are cerebellar ataxia and oculocutaneous telangiectasia (Boder 1975), but immunodeficiency and hypogonadism may also be prominent. Abnormally low concentrations of one or more immunoglobulins, particularly IgA, are characteristic, as is decreased lymphocyte transformation following PHA stimulation. Frequent and severe sinopulmonary infections occur, often causing death in childhood or adolescence. Cancer, which accounts for 10% of AT deaths (Paterson and Smith 1979), mainly affects lymphoreticular tissues, but solid tumors of the gastrointestinal tract, ovaries, and central nervous system also occur. Cancer has also been reported by one group (Swift 1976, Swift et al. 1976) to occur more commonly in AT heterozygotes than in the normal population.

Cytogenetic Studies

Chromosome Aberrations

Many early reports of AT cytogenetics stated that the chromosomes were "normal" (Zellweger and Khalifeh 1963, Utian and Plit 1964, Ferák et al. 1965, Schuster et al. 1966). The existence of AT chromosome instability was first detected by Hecht et al. (1966), who found abnormalities in 20–30% of the PHA-stimulated lymphocytes from three affected persons. Other laboratories have in general concurred with the findings of Hecht et al., but persons with AT lacking increased breakage have also been reported (Pfeiffer 1970, Schuler et al. 1971, Passarge 1972, Hayashi and Schmid 1975a, Nelson et al. 1975, Jean et al. 1979, Saxon et al. 1979). Some AT heterozygotes also have an increased aberration frequency in cultured blood and dermal fibroblasts (Cohen et al. 1975, Oxford et al. 1975).

Chromosome aberrations found in AT cells include chromatid and isochromatid gaps and breaks, acentric fragments, structurally rearranged chromosomes, and Tr and Qr interchange figures. Increased numbers of endoreduplicated and polyploid cells have been found in some individuals (Bochkov et al. 1974). The frequency of dicentric chromosomes is increased (Pfeiffer 1970, Schmid and Jerusalem 1972, Hecht et al. 1973, Harnden 1974, Hayashi and Schmid 1975a), and in the view of Hayashi and Schmid (1975a) they result from the fusion of two essentially complete chromosomes without the loss of significant amounts of chromosome material. AT Qr configurations are of the nonhomologous asymmetrical type rather than the homologous, equal, and symmetrical type characteristic of BS. Cells with rearranged monocentric chromosomes occur with a significantly increased frequency in AT. Some of the cells with rearranged chromosomes belong to clones (i.e., the rearranged chromosome is present in multiple cells), while other chromosome rearrangements appear to be sporadic. The presence of clones of cells with mutant karyotypes is the most characteristic cytogenetic feature of AT (see below).

In cultures of AT dermal fibroblasts the incidence of chromosome abnormalities is typically 2–3 times greater than in PHA-stimulated blood lymphocytes (Cohen et al. 1973, 1975, 1979, Oxford et al. 1975, Webb and Harding 1977), although some fibroblast cultures lack aberrations (Schmid and Jerusalem 1972, Oxford et al. 1975). AT LCLs have been reported to show no increased chromosome breakage (Cohen et al. 1979, Jean et al. 1979); however, Jean et al. (1979) found all the cells of the AT LCL used in their study to contain a specific translocation chromosome, t(8;14). One AT LCL studied in our laboratory had chromosome aberrations in 15% of the cells, which is comparable to one of the 11 studied by Cohen et al. (1979) in which 11% of the cells had chromosome abnormalities. These studies raise the possibility that some AT LCLs exhibit increased breakage while others do not, as has been shown to be the case in BS. AT bone marrows have been studied in only a few instances (Lisker and Cobo 1970, Hecht et al. 1973, Al Saadi et al. 1975, Cohen et al. 1975, Levitt et al. 1978, Saxon et al. 1979). Of the eight marrows studied, only two exhibited increased abnormalities (Lisker and Cobo 1970, Levitt et al. 1978); one of the marrows was from a leukemia patient (Levitt et al. 1978), and in the other marrow only 12 cells were analyzed (Lisker and Cobo 1970). Therefore, the question of whether the increased breakage observed in cultivated AT cells is also present in vivo remains unanswered.

The SCE frequency of AT cells is similar to that of normal cells (Chaganti et al. 1974, Galloway and Evans 1975, Hayashi and Schmid 1975b, Hatcher et al. 1976).

Abnormal Clones

The most characteristic cytogenetic feature of AT is, as was stated above, the presence of clones with aberrant chromosome complements. The frequency of abnormal clones varies from 1% to 100%. A striking specificity for the involvement of one, sometimes both, No. 14 chromosomes in clone formation has been observed (see below). Clones with rearranged chromosome complements are found in cultured AT blood lymphocytes, dermal fibroblasts, and LCLs and in a leukemic bone marrow preparation. Some AT heterozygotes also have clones in cultured skin fibroblasts (Cohen et al. 1975, Oxford et al. 1975).

Abnormal clones have been detected among the lymphocytes from 61 AT individuals. Group D chromosomes are involved in 56 of the 70 reported lymphocyte clones (Table 1). The most common lymphocyte clones reported are those with Dq+ chromosomes; the remaining chromosomes in the cells with the Dq+ chromosomes were normal in the cases where complete chromosome analyses were made. In most cases the origin of the extra material on the long arm of the group D chromosome could not be determined even when G-banding was employed; however, in at least one case the extra material resulted from tandem duplication of the long arm of chromosome 14 (Schmid and Jerusalem 1972).

Clones having a translocation chromosome involving not just one but two group D chromosomes (t(Dq−;Dq+)) are also significantly increased in AT lymphocytes. Although many of the clones were studied before the advent of G-banding techniques, recent studies utilizing G-banding have shown the Nos. 14 to be the group D chromosomes involved. In the usual rearrangement, one chromosome 14 has a break in the region q11 or q12 while the other No. 14 has a break near q31 or q32; the long arm (q11 or q12→qter) is then translocated onto the end of the long arm of the other chromosome 14, producing one abnormally long and one abnormally short chromosome 14, the latter about the size of the Ph[1] chromosome.

Translocations involving groups C and D chromosomes are also common in AT PHA-stimulated lymphocytes. The group C chromosomes reported to be involved are the X, No. 6, and No. 7. Again, breakage in chromosome 14 occurs at q11 or q12 and in the X, chromosome 6, or chromosome 7 toward the ends of the chromosomes.

A striking array of clones with rearrangements involving chromosome 7 has recently been detected by three groups independently, studying either G-banded or R-banded PHA-stimulated lymphocytes from AT. The initial report was that of Hustinx et al. (1979), who studied metaphases from a boy in whom the clinical diagnosis was left tentative because he had neither ataxia nor telangiectasia. Fibroblasts from the boy have been shown to be moderately hypersensitive to gamma-irradiation (T. W. J. Hustinx, personal communication), which suggests that the patient represents a previously undescribed variation in the already notoriously variable AT phenotype. The chromosome rearrangement found most often was a pericentric inversion in chromosome 7. A t(7;7) translocation chromosome was also present in some PHA-stimulated lymphocytes as was the previously described t(7;14). More recently, two additional AT patients studied in the same laboratory were found to have a low frequency of cells with an inversion of chromosome 7 (Scheres et al., in press). B. Scott de Martinville (unpublished results from our laboratory) and Aurias et al. (1980) also found similar inversions in chromosome 7 as well as t(7;7) translocations in

TABLE 1. *Clones in ataxia-telangiectasia PHA-stimulated lymphocytes, dermal fibroblasts, lymphoblastoid cells, and bone marrow*

Cell type / Clones	Number of clones	References
Lymphocytes		
Dq+	20*	Pfeiffer 1970, Schmid and Jerusalem 1972, Cohen et al. 1973, 1975, 1979, Hatcher et al. 1974, Hayashi and Schmid 1975a, Hook et al. 1975, McCaw et al. 1975
t(Dq−;Dq+)	18†	Goodman et al. 1969, Hecht et al. 1973, Hecht and McCaw 1973, Bochkov et al. 1974, Harnden 1974, Rary et al. 1974, 1975, Al Saadi et al. 1975, McCaw et al. 1975, Oxford et al. 1975, Levitt et al. 1978, Cohen et al. 1979
t(C;D)‡	14	Hecht and McCaw 1973, Bochkov et al. 1974, Harnden 1974, McCaw et al. 1975, Nelson et al. 1975, Cohen et al. 1979, Hustinx et al. 1979, Aurias et al. 1980
t(?;D)	1§	Hatcher et al. 1974
+C	3§	Bochkov et al. 1974
r14	1	McCaw et al. 1975
Dq−	1§	Bochkov et al. 1974
inv(7)	9	Hustinx et al. 1979, Aurias et al. 1980, Scheres et al. in press
inv(14)	1	Aurias et al. 1980
t(7;7)	2	Aurias et al. 1980
Fibroblasts		
Dq+	7	Cohen et al. 1973, 1975, 1979
+marker,del(11)(q13q25)	1	Webb and Harding 1977
+marker,+E/D marker, −D,−C,−C,+marker, +long D-like marker	1	Webb and Harding 1977
−2,+long B-like	1	Webb and Harding 1977
t(1;2)(q42;q37)	1	Webb and Harding 1977
−D,+A-like	1	Cohen et al. 1979
t(D;F)	1	Oxford et al. 1975, Cohen et al. 1979
Cp+	1	Oxford et al. 1975
Lymphoblastoid Cell Lines		
t(8;14)(8q−;14q+),+21,+X-like	1	Jean et al. 1979
t(2;14)	1	Cohen et al. 1979
Bone Marrow		
t(14q−;14q+)	1	Levitt et al. 1978

* Of the Dq+ chromosomes, 9 were positively identified as 14q+ by G-banding.

† Of the t(Dq−;Dq+) chromosomes, 14 were positively identified as t(14q−;14q+) by G-banding.

‡ The 14 t(C;D) clones include 1 t(6;14), 7 t(7;14), 3 t(X;14), and 3 t (C;D) clones of unknown composition.

§ These clones were found prior to development of G-banding; therefore, specific chromosomes were not identifiable.

AT lymphocytes. An additional rearrangement found by the latter group was a chromosome 14 that had arisen by breakage at band q12 and at qter with inversion of the intervening segment. The significance of these newly discovered chromosome rearrangements in AT remains to be determined. It is possible that they have been present and simply overlooked in previous studies.

Abnormal clones are also readily detectable in AT fibroblasts; however, the rearrangements are often quite complex, suggesting an in vitro origin and evolution for some of them (Table 1). Among 14 dermal fibroblast clones reported, at least one group D chromosome (often identifiable as a chromosome 14) was involved in 10 of them; seven of these clones had the characteristic 14q+ chromosome. AT LCL and bone marrow clones have also contained altered No. 14 chromosomes.

The frequency of cells belonging to specific clones among PHA-stimulated blood lymphocytes changes with time (Schmid and Jerusalem 1972, Hecht et al. 1973, Al Saadi et al. 1975, McCaw et al. 1975, Nelson et al. 1975, Oxford et al. 1975, Rary et al. 1974, 1975, Levitt et al. 1978). Thus, Hecht et al. (1973) studied a patient over a 52-month period and found the frequency of his 46,XY,t(14;14)(q12;q31) clone to increase from 1.7% to 25–46% and finally, before the patient's death from infection, to 57–68%. The cells that contained the translocation chromosome were said to show less chromosome breakage than cells without it (1.7% vs. 4.3%) and to respond better to PHA stimulation. Oxford et al. (1975) studied two sibs with AT over a 29-month period; one sib had a t(14;14) clone while the other had a t(X;14) clone. The frequency of t(14;14) cells increased from 1% to 38% during the sampling period, whereas the frequency of t(X;14) cells did not increase from the initial frequency of 80%. Such data suggest that the presence of some translocations involving chromosome 14 may confer a proliferative advantage on a cell, permitting it eventually to become the predominant, if not the only, cell type in the blood to respond to PHA.

Three AT patients with lymphocytic leukemia have been analyzed cytogenetically (McCaw et al. 1975, Levitt et al. 1978, Saxon et al. 1979). In all three cases the leukemias were shown to be T-cell in origin. A clone was present in the circulating lymphocytes in all three individuals (100% in two cases, 98% in the other). Each of the clones had a translocation chromosome derived from both No. 14 chromosomes, the breakpoints having been at q11-q12 and q31-q34 (Levitt et al. 1978). Additional chromosome changes, including the loss of the 14q− chromosome, were also observed. In all of these individuals cytogenetic studies had been made prior to the diagnosis of leukemia, and clones having a t(14;14) chromosome had been detected; in two of the leukemia patients (McCaw et al. 1975, Levitt et al. 1978) the clones were originally present at a reduced frequency—21% and 1.7%, respectively, while in the third patient (Saxon et al. 1979) the abnormal clone originally accounted for 100% of the PHA-stimulated lymphocytes (Goodman et al. 1969). While it is interesting that a t(14;14) chromosome was present in virtually all dividing T-lymphocytes in each of the three cytogenetically well-documented cases of AT leukemias, it is too early to attempt to establish a cause and effect relationship between t(14;14) and the leukemias of AT. Many AT patients are known to have clones of lymphocytes bearing the t(14;14) rearrangements circulating in their blood, but have no clinical indication of cancer.

FANCONI'S ANEMIA

General

Fanconi's anemia (FA) is a rare genetic disorder believed to follow autosomal recessive transmission (Reinhold et al. 1952, Schroeder et al. 1976b). Its clinical features include bone marrow failure, hyperpigmentation of the skin, and developmental anomalies of the kidney and skeleton (Alter et al., in press). The first signs of anemia, leukopenia, or thrombocytopenia usually occur before age 10, and without treatment, marrow failure leads to death in a few months or years. While the majority of FA deaths result from hemorrhage or infection, a significant proportion of those with FA develop cancer (Swift 1971, German 1979b), particularly acute nonlymphocytic leukemia and hepatic tumors. The widely quoted evidence that purported to show that FA heterozygotes also had an increased risk of developing cancer (Swift 1976) has not been supported by more extensive studies (Caldwell et al. 1979).

Cytogenetic Studies

Chromosome Aberrations

The cytogenetic disturbance in FA was reported 37 years after the clinical syndrome was described (Fanconi 1927, Schroeder et al. 1964). Schroeder et al. (1964) reported that 25% of the lymphocytes from two affected brothers showed chromosome breakage. Cytogenetic analyses of lymphocytes from more than 100 persons with FA have now been published, and few individuals have been found who do not exhibit significantly increased aberration frequencies (Bloom et al. 1966, Germain et al. 1968, Nathanson et al. 1968, Lieber et al. 1972, von Koskull and Aula 1973, Bushkell et al. 1976). The cytogenetic abnormalities appear to be unrelated to the stage of the disease or the age of the patient, because some studies showing increased breakage have utilized cells from individuals in the preanemic stage (Schroeder et al. 1964, German et al. 1965, Schroeder 1966b, Varela and Sternberg 1967, Perkins et al. 1969). Varela and Sternberg (1967) found striking abnormalities in cells from blood and skin of a newborn lacking thumbs who was to develop fatal hematological manifestations only several years later (M. Varela, personal communication). Further evidence that increased chromosome aberrations in FA result from an inherent metabolic defect of the cells rather than from the response to some extracellular component is derived from the observation that incubation of FA lymphocytes with serum from normal individuals failed to alter the aberration frequency from that observed when FA lymphocytes were incubated in FA serum (Bloom et al. 1966, Perkins et al. 1969); conversely, incubation of normal lymphocytes in FA serum did not increase the amount of breakage in the normal cells.

The frequency of aberrant cells varies widely both among FA individuals (Bloom et al. 1966, Perkins et al. 1969, de Grouchy et al. 1972, Beard et al. 1973, Sasaki and Tonomura 1973, Schmid and Fanconi 1978, Zaizov et al. 1978) and among repeated cultures from the same individual (Bloom et al. 1966, Gmyrek et al. 1968, Hirschman et al. 1969, Crossen et al. 1972, Lisker and de Gutiérrez 1974, Schroeder et al. 1976a, Berger et al. 1977, Bourgeois and Hill 1977, Zaizov et al. 1978). Thus, Berger et al. (1977) conducted eight cytogenetic studies on PHA-stimulated lymphocytes from a single FA individual over a 3-year period and found the aberration frequencies to vary from a low of 5% to a high of 42%. A similar variation in the aberration frequencies (11%–52%) was reported by Schroeder et al. (1976a) in 19 blood cultures from a man with FA studied over a 6-year period. The wide variation in the aberration frequencies prompted Schroeder et al. (1976a) to suggest that repeated cytogenetic analyses be undertaken to examine large numbers of cells before ruling out increased breakage when FA is suspected clinically.

Chromatid and isochromatid gaps and breaks are the major aberrations found in FA cells, but rings, translocations, acentric fragments, and interchange (Tr and Qr) configurations are also increased. The breaks in FA chromosomes have been reported by some to occur exclusively in the lightly stained regions of G-banded chromosomes (von Koskull and Aula 1973, 1977, Berger et al. 1977), whereas Dutrillaux et al. (1977), on the basis of an elaborate analysis combining G-, Q-, and R-banding, suggested that the breaks occur in the interband regions between G- and R-bands. Detailed analyses of 536 Qr configurations from FA lymphocyte cultures have been made (Schroeder and German 1974, Schroeder and Stahl Mauge 1976). In contrast to the situation found in BS cells in which the majority of the Qr configurations represent symmetrical exchanges between homologous chromosomes, 95% of the exchanges in FA cells occur between nonhomologous chromosomes (Figure 3). Such chromosome exchanges are potentially much more damaging than those that occur between homologues at homologous sites because there is a potential for the deletion or duplication of genetic material. Endomitoses also occur with an increased frequency in FA cells according to some reports (Schroeder et al. 1964, Bloom et al. 1966, Schroeder 1966a, 1966b, Varela and Sternberg 1967, Germain et al. 1968, Perkins et al. 1969, Guanti et al. 1971, Beard et al. 1973, von Koskull and Aula 1973, Schroeder and German 1974).

Chromosome aberrations are also increased in FA dermal fibroblasts (German and Crippa 1966, Swift and Hirschhorn 1966, Varela and Sternberg 1967, McDougall 1971, Meisner et al. 1978), although exceptions have been reported

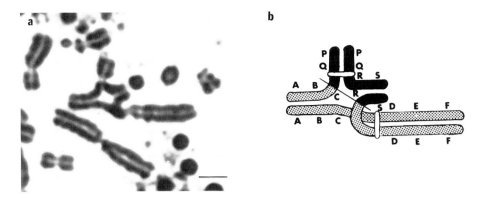

FIG. 3. a, FA PHA-stimulated lymphocyte Qr interchange configuration between two nonhomologous chromosomes. b, Diagrammatic representation of the exchange event that occurred in the formation of the Qr interchange in a. Segments of one chromosome have been arbitrarily designated ABCDEF while those of the other chromosome have been designated PQRS. The nonhomologous chromatid segments ABC and RS were involved in the exchange event between the two chromosomes. Bar in a represents 2 μm.

(Schmid et al. 1965, Beard et al. 1973). The aberration frequency in fibroblasts is generally lower than in lymphocytes. Here too, aberrations consist predominantly of chromatid and isochromatid gaps and breaks, but chromosome aberrations are also found. FA LCLs also exhibit an increased aberration frequency (Cohen et al. 1979, J. German and E. Louie, unpublished results).

Of particular interest in determining whether in vivo chromosome instability exists are studies of FA bone marrows. Of 28 marrows analyzed, 8 had no breakage (Schmid et al. 1965, Bloom et al. 1966, Schmid 1967, Lieber et al. 1972, Beard et al. 1973, Bargman et al. 1977), 13 had breakage in 1–10% of the cells (Hoefnagel et al. 1966, Swift and Hirschhorn 1966, Varela and Sternberg 1967, Hirschman et al. 1969, Dosik et al. 1970, Crossen et al. 1972, Shahid et al. 1972, Wolman and Swift 1972, Lisker and de Gutiérrez 1974, Berger et al. 1977, Bourgeois and Hill 1977), and 7 had breakage in more than 10% of the cells (Guanti et al. 1971, Bersi and Gasparini 1973, Meisner et al. 1978, Puligandla et al. 1978, Zaizov et al. 1978). Four FA marrows have been analyzed in our laboratory after just 90 minutes of incubation in colchicine-containing medium (J. German and J. H. Ray, unpublished results); two showed no breakage, and two showed breakage in 3–4% of the cells. Because so much variability exists in chromosome breakage between different FA marrows and because the incidence of breakage is low even in those in which breakage seems to be increased, definite conclusions are not yet possible about whether increased chromosome instability exists in vivo in FA. Reports that micronuclei (Schroeder 1966a, 1966b, Gmyrek et al. 1968) and anaphase bridges (Schroeder 1966a, 1966b, Shahid et al. 1972) have been found in smears of bone marrow also favor the existence of in vivo chromosome instability in FA. In strong support of this are the observations of clones of cells with aberrant chromosome complements in freshly aspirated marrows (Hirschman et al. 1969, Crossen et al. 1972, Lisker and de Gutiérrez 1974, Bargman et al. 1977, Berger et al. 1977, Bourgeois and Hill 1977, Meisner et al. 1978, Zaizov et al. 1978). Certain types of chromosome aberrations that have been found in FA bone marrow preparations also favor in vivo chromosome instability, viz., dicentric chromosomes (Shahid et al. 1972, J. German and J. H. Ray, unpublished results), ring chromosomes (Shahid et al. 1972), and chromosome interchange figures (Lisker and de Gutiérrez 1974, J. German and J. H. Ray, unpublished results).

SCE frequencies in untreated FA lymphocytes (Chaganti et al. 1974, Hayashi and Schmid 1975b, Latt et al. 1975, Sperling et al. 1975, Kato and Stich 1976, Burgdorf et al. 1977, Novotná et al. 1979) and fibroblasts (Latt et al. 1975) are similar to those of control cultures.

Abnormal Clones

Clones with abnormal chromosome complements (Table 2) have been found in PHA-stimulated lymphocytes (Schroeder et al. 1976a, Berger et al. 1977, Harnden 1977, Zaizov et al. 1978, J. German and D. Warburton, unpublished re-

TABLE 2. *Clones in Fanconi's anemia PHA-stimulated lymphocytes, dermal fibroblasts, and bone marrow*

Cell type / Clones	Clone frequency (%)	Reference
Lymphocytes		
1p−	<1	Schroeder et al. 1976a
t dicentric (1;D/G)	<1	
t dicentric (B;18)	<1	
t dicentric (1;C)	<1	
t dicentric (C;D)	<1	
+21	3	Berger et al. 1977
3q+	1	
3q+	17	
3q+,12q+	90	
1p+,3q+,7p+,12q+	10	
t(1;6)(1q−;6p+)	5–85	Harnden 1977
t(1;6)(1q−;6p+),11q+	?	
Aq+	?	Zaizov et al. 1978
t(4;9)(4p−;9q+)	8	J. German and D. Warburton, unpublished results

Table 2 (continued)

Cell type / Clones	Clone frequency (%)	Reference
Fibroblasts		
−A,+2 abnormal Bs	100	Beard et al. 1973
+F	78	
Bone Marrow		
t(B;E)	50	Bourgeois and Hill 1977
Dq+	60	Lisker and de Gutiérrez 1974
−2 Cs,+1,+1 large submetacentric	100	Hirschman et al. 1969
+C	8–13	Crossen et al. 1972
−C	100	Bargman et al. 1977, Meisner et al. 1978
−C,+"2"	35	Meisner et al. 1978
Bq+	9	
−C,+"2"	30	
−C,+"B"	30	
−C,+"1"	25	
+21	39	Berger et al. 1977
3q+,12q+	99	
12q+	1	
3q+,12q+	100	
3q+,12q+	81	
12q+	13	
Dq+	4	
3q+,12q+	39	
12q+	3	
1p+,3q+,7p+,12q+	45	
7p+,12q+	12	
t(1;6)(1q−;6p+)	5–85	Harnden 1977
t(1;6)(1q−;6p+),11q+	?	

Braces denote multiple clones found in single individuals at different times. Sometimes more than one clone per sample was found. (Double spacing between entries indicates separate samples.)

sults), dermal fibroblasts (Beard et al. 1973), and freshly aspirated bone marrow preparations (Hirschman et al. 1969, Crossen et al. 1972, Lisker and de Gutiérrez 1974, Bargman et al. 1977, Berger et al. 1977, Bourgeois and Hill 1977, Harnden 1977, Meisner et al. 1978). Group A and C chromosomes are involved in many of these clones, although the consistency of the involvement of these chromosomes is far less than that found for group D chromosomes in AT clones. In their 6-year study of one man with FA, Schroeder et al. (1976a) found five clones in PHA-stimulated blood lymphocyte cultures. One clone bore an abnormal chromosome 1 (1p−), while the other four had dicentric chromosomes. Each clone was present at a low frequency and probably would have been overlooked had not serial chromosome analyses over a long period of time been made. The finding of clones with abnormal chromosome complements in such a small proportion of cells indicates that chromosome rearrangements may occur with a much greater frequency in vivo than is currently recognized and are simply being overlooked because so few cells (50–100) are usually analyzed. Beard et al. (1973) cultured FA dermal fibroblasts from two separate biopsies and found all the cells from the first biopsy to lack a group A chromosome and to have two abnormal group B chromosomes. In this mutant clone, additional chromosome abnormalities, including an extra group F chromosome and the absence of a group C or G chromosome, were also present in varying proportions of the cells. In the second biopsy, 78% of the cells, presumably representing a clone, had an extra group F chromosome as their only abnormality.

Also significant for in vivo events are studies of direct preparations of bone marrow. Clones have been observed in eight of 28 FA bone marrows reported. In six of the eight, group C chromosomes were involved. In three, a loss of one or more group C chromosomes had occurred (Hirschman et al. 1969, Bargman et al. 1977, Meisner et al. 1978), in one case an extra group C chromosome was present (Crossen et al. 1972), and in another, cells were found containing various combinations of chromosomes with partial duplications of the short arm of chromosome 1 (1p+), the long arm of chromosome 3 (3q+), the short arm of chromosome 7 (7p+), and the long arm of chromosome 12 (12q+) (Berger et al. 1977). Harnden (1977) found a major clone with a translocation between chromosomes 1 and 6 (t(1;6) (1q−;6p+)) in both PHA-stimulated lymphocytes and direct marrow preparations; a subclone having an 11q+ chromosome was also present. Within a 6-week period, the clone's frequency increased from 5% to 85%.

Proliferative advantage and evolution of some abnormal clones are also suggested from the results of two other FA marrow studies. Meisner et al. (1978) studied a patient who had been treated with prednisone and testosterone 6 years earlier. In the initial marrow study 35% of the metaphases had the complement 46,XX,−C,+2, and 9% had 46,XX,Bq+; other individual cells were observed that had the karyotypes 46,XX,−C,+B and 46,XX,−C,+1. When a second marrow sample was analyzed 8 months later, the 46,XX,−C,+2 karyotype was found in 30% of the cells, but the 46,XX,Bq+ clone was not detected; clones with the abnormal complements 46,XX,−C,+B and 46,XX,−C,+1, detected only as single cells in the earlier study, were present in 30% and 25% of the dividing marrow population, respectively.

Berger et al. (1977), as mentioned earlier, studied marrow from one individual with FA on five occasions over a 3-year period; the final occasion was after leukemia had been diagnosed. In the initial study, 39% of the cells had the karyotype 46,XX,+21; this clone was not observed in any later marrow samples.

In the second marrow sample >99% of the cells had the karyotype 46,XX,3q+,12q+, and a single cell had the karyotype 46,XX,12q+; these clones were also found in subsequent marrow specimens. In the third sample 100% of the cells had the karyotype 46,XX,3q+,12q+. In the fourth, 81% had the karyotype 46,XX,3q+,12q+, and 13% the karyotype 46,XX,12q+; in addition, 4% of the cells had a Dq+ chromosome. In the fifth marrow sample, taken after clinical leukemia had supervened, the karyotypes 46,XX,3q+,12q+ and 46,XX,12q+ were found again, now in 39% and 3% of the cells, respectively; however, two previously undetected clones were now present, one with the karyotype 46,XX,1p+,3q+,7p+,12q+ and another with 46,XX,7p+,12q+, in 45% and 12% of the cells, respectively.

DISCUSSION

The central concern in the study of BS, AT, and FA is directly related to the theme of this symposium—the relationship between genes, chromosomes, and neoplasia. Of particular interest to us and others in this field is the role, if any, that the increased chromosome instability observed in each of the syndromes may play in the cancer predisposition of affected individuals. If chromosome changes do indeed play a role in the neoplastic transformation that occurs in persons with these disorders, not only should chromosome changes of the type leading to long-term genomic modification be observable in vitro and in vivo, but chromosome alterations similar to those observable in vitro should also be demonstrable in the cancers themselves. Evidence pertaining to these questions has been difficult to obtain but is accumulating slowly.

Chromosome Changes Observed in the Syndromes

In vitro Chromosome Changes

In populations of cultured BS, AT, and FA cells, two types of chromosome lesions occur. The first and major type is the chromosome breakage that presents itself as chromatid and isochromatid gaps and breaks, acentric fragments, reunions between sister chromatids, and bizarre interchange figures. In spite of their prevalence in BS, AT, and FA, it seems probable that these aberrations are inconsequential in terms of long-lived alterations of the genomes, because cells having them will be eliminated from the body if further mitotic divisions occur. Of much greater potential significance are the stable chromosome rearrangements that occur as evidenced by the presence of certain types of Qr configurations and translocation chromosomes. These stable rearrangements produce such subtle changes in the genomes of the affected cells that survival probably is not curtailed; in fact, growth may be enhanced, and mutant clones may assume numerical predominance (Hirschman et al. 1969, Pfeiffer 1970, Schmid and Jerusalem 1972, Beard et al. 1973, Hecht et al. 1973, Harnden

1974, 1977, Al Saadi et al. 1975, Hayashi and Schmid 1975a, McCaw et al. 1975, Oxford et al. 1975, Nelson et al. 1975, Bargman et al. 1977, Berger et al. 1977, Bourgeois and Hill 1977, Levitt et al. 1978, Meisner et al. 1978, Jean et al. 1979, Saxon et al. 1979).

The stable chromosome rearrangements resulting in interchanges and translocations may have diverse effects on the genome. The symmetrical homologous Qr configurations found in BS cells signify the interchange of identical segments of paternal and maternal chromosomes (Figure 1). Depending on the segregation patterns of the affected chromatids at anaphase, the potential exists for converting cells heterozygous for a given genetic trait to the homozygous state. Asymmetrical nonhomologous Qr configurations as seen in AT and FA signify the exchange of chromosome segments bearing different genes (Figure 3). Again, depending on the chromatid segregation patterns, the potential exists for the production of cells bearing duplications of some chromosome segments and deficiencies of others. In addition to the interchange configurations just mentioned, cultured AT lymphocytes and dermal fibroblasts also contain an increased number of cells with abnormal chromosome complements that are the result of exchange events at an earlier cell division cycle. Sometimes clones of such cells are found. The chromosome rearrangements found in clones of AT cells involve translocations between specific chromosomes. One of the chromosomes is No. 14; the other chromosome may be the X, No. 6, No. 7, or the other No. 14. Clones apparently arising from the duplication of part of the long arm of chromosome 14 and the pericentric inversion of chromosome 7 may also be present. The end result of all these stable rearrangements is the alteration of the cell's genetic material.

In vivo Chromosome Changes

As stated above, if the stable chromosome rearrangements observed in vitro just discussed are to have a role in the etiology of neoplasia, they quite obviously must occur in vivo. Although detailed in vivo BS, AT, and FA cytogenetic studies are lacking, the available data do tend to support the existence of in vivo chromosome instability. Studies of direct BS marrows have shown occasional cells with interchange figures or dicentric chromosomes as well as increased numbers of gaps and breaks (Landau et al. 1966, Kiossoglou et al. 1976, Shiraishi et al. 1976, Shiraishi and Sandberg 1977). Some BS blood lymphocyte cultures have contained abnormal clones that were probably of in vivo origin (Cohen et al. 1979, S. Schonberg and J. German, unpublished results).

Further support for the existence of in vivo chromosome instability in BS comes from our inability to demonstrate a difference in the aberration frequencies of first- and second-division PHA-stimulated lymphocytes and from our finding of the above-mentioned first-division cell bearing a dicentric chromosome, no fragments, and 45 other normal chromosomes, a cell that could only have been derived from in vivo chromosome rearrangement (J. H. Ray and J. German, unpublished results). Clones of cells have been found in AT and FA bone marrows and peripheral blood lymphocytes (Tables 1 and 2). In vivo chromosome instability in FA is also suggested by the existence of rearranged chromosomes (Hirschman et al. 1969, Crossen et al. 1972, Lisker and de Gutiérrez 1974, Bargman et al. 1977, Berger et al. 1977, Bourgeois and Hill 1977, Harnden 1977, Meisner et al. 1978) and increased breakage in some direct bone marrow preparations (Guanti et al. 1971, Bersi and Gasparini 1973, Meisner et al. 1978, Puligandla et al. 1978, Zaizov et al. 1978) and by the presence of anaphase bridges and micronuclei in bone marrow smears (Schroeder 1966a, 1966b, Gmyrek et al. 1968, Shahid et al. 1972).

Cytogenetic Studies of Cancers in the Syndromes

If, given the above information, it is accepted that at least some in vivo chromosome instability of the type leading to stable rearrangements does exist in BS, AT, and FA, what, if any, is the relationship between the chromosome rearrangements and cancers found in individuals with these disorders? To answer this question, descriptions of the chromosome complements in the cancers that have occurred in BS, AT, and FA are needed, but few are available; those few are mentioned below.

No published cytogenetic information is available on BS cancers.

In AT, chromosome studies have been conducted on 10 cancer patients (Haerer et al. 1969, Bochkov et al. 1974, Goldsmith and Hart 1975, Al Saadi et al. 1975, McCaw et al. 1975, Buyse et al. 1976, Levitt et al. 1978, Saxon et al. 1979). It is unlikely that appropriate tissues were analyzed in seven of the 10 cytogenetic studies. Haerer et al. (1969) reported PHA-stimulated lymphocyte chromosomes to be normal in a patient having a gastric adenocarcinoma, as did Goldsmith and Hart (1975) and Buyse et al. (1976) in two patients with gonadoblastoma. Al Saadi et al. (1975) found a t(14;14) clone in 85% of the PHA-stimulated lymphocytes of an AT patient with a hepatoma, a finding probably unrelated to the existence of the hepatoma. Finally, Bochkov et al. (1974) stated that three patients with AT had unspecified lymphoreticular neoplasms; of these three, only one had a clone of cells in PHA-stimulated lymphocytes, and the clone was present in only 5% of the cells. In three AT T-cell leukemias, the malignancy has been sampled (McCaw et al. 1975, Levitt et al. 1978, Saxon et al. 1979). In each case, essentially all of the PHA-stimulated lymphocytes contained a clone bearing a t(14;14) translocation chromosome; additional chromosome abnormalities were present as well.

In FA, eight patients with cancer have been analyzed cytogenetically (Gymrek et al. 1968, Swift et al. 1971, Mulvihill et al. 1975, Berger et al. 1977, Bourgeois and Hill 1977, Meisner et al. 1978, Prindull et al. 1979). In two patients with solid tumors (Swift et al. 1971, Mulvihill et al. 1975), no chromosome abnormalities were found in peripheral blood samples. A clone of cells containing a translocation chromosome (t(B;E)) accounting for 50% of the dividing bone marrow cells was found in a patient with acute myelomonocytic leukemia (Bourgeois and Hill 1977). In one of two other acute leukemia patients no chromosome abnormalities were present in peripheral blood cells (Gmyrek et al. 1968); in the other patient several clones with different combinations of abnormalities of chromosomes 1, 3, 7, and 12 (1p+, 3q+, 7p+, 12q+) were found in both unstimulated blood cultures and direct bone marrow preparations (Berger et al. 1977). Two FA patients with erythroleukemia have been studied (Bargman et al. 1977, Meisner et al. 1978, Prindull et al. 1979); in one, no abnormalities were found in peripheral blood (Prindull et al. 1979), while in the other patient, 100% of both unstimulated peripheral blood lymphocytes and dividing bone marrow cells lacked a group C chromosome.

The available information suggests that chromosome rearrangements figure prominently in neoplasms of chromosome-breakage syndrome patients, as they do in cancers in general. However, with the exception of the three T-cell leukemias in AT, detailed and repeated cytogenetic analyses have not been conducted on appropriate tissues of BS, AT, and FA, i.e., on the cancerous tissues themselves. Only as additional BS, AT, and FA cytogenetic data are accumulated from studies made before, during, and after the onset of neoplasia will associations between specific chromosome changes and cancer development in persons with the chromosome-breakage syndromes become more apparent.

The Role of Chromosome Change in Neoplastic Transformation

The existence of clones of cells with abnormal chromosome complements in individuals with the chromosome-breakage syndromes both before and after the clinical diagnosis of cancer constitutes some of the strongest evidence presently available that chromosome change plays an etiological role in human neoplastic transformation. Cancers in humans (de Grouchy and Turleau 1974, Nowell 1974) as well as in certain other species (Klein 1979) can be demonstrated in general to be clones of cells with mutated chromosome complements. The few appropriately studied cancers in persons with one of the chromosome-breakage syndromes also have been clones with mutant chromosome complements. The exciting information that is emerging from the cytogenetic study of persons with the chromosome-breakage syndromes, particularly AT, is that the chromosome rearrangements in the neoplastic clones resemble those found in ostensibly benign clones that can populate the tissues of affected persons for long periods of time. Additionally, it can be stated that chromosome rearrangements that tend to occur frequently in benign clones in persons with AT resemble those in lymphoid neoplasms that occur in the general population (Mark 1977).

That most cancers are clones with mutated chromosome complements and that clones with abnormal chromosome complements are observed in persons with the chromosome-breakage syndromes and in certain populations predisposed to cancer because of previous exposure to environmental carcinogenic agents prior to the clinical diagnosis of cancer raise certain fundamental questions pertaining to our current concepts of neoplasia. The concepts of benign and malignant growth are useful in clinical medicine, but no compelling reason exists for biologists to employ them. In fact, the growth disturbances that have become apparent during the study of cancer-prone populations lead us to suspect that a spectrum of neoplasia may exist. Thus, it is known from observation of irradiated populations such as A-bomb survivors that clones with Ph[1] chromosomes can proliferate for long periods of time, during which they are represented in only a low percentage of dividing marrow cells and the person is asymptomatic (Kamada and Uchino 1978, Ohkita and Kamada 1979). In time, the cells bearing the Ph[1] chromosome may essentially completely replace those with normal chromosome complements, and clinical cancer—chronic granulocytic leukemia— will be said to have supervened. In the chromosome-breakage syndromes, particularly AT in which the tissue type most predisposed to cancer is relatively easily sampled, we can also begin to discern a spectrum concerning disturbed growth. Sometimes a single cell or two constitute the only detectable evidence of a mutant clone, while at other times the entire population of PHA-responsive cells has a mutated genome. In the latter instance clinical evidence of cancer may or may not be present. In the three cytogenetically well-documented T-cell leukemias in AT, the leukemic clones had the same chromosome rearrangements, perhaps with additional mutations, that had been found in what had been considered benign cells earlier. That the clones just mentioned bearing the Ph[1] and the t(14;14) chromosomes were detectable at all among dividing cells suggests that the cells did not just have a proliferative advantage over cells with normal karyotypes but were, in fact, dividing relatively autonomously. This would indicate that neoplasia can be said to have existed in such instances when the first chromosome rearrangement occurred. Two questions pertaining to the concept of neoplasia are raised by such observations: (1) When can a mutant clone be said to have become neoplastic? (2) When can a benign neoplastic clone be said to have become malignant? Although the current cytogenetic information available from the study of the chromosome-breakage syndromes does not permit us to answer these questions, few if any areas of biology exist with a greater potential for elucidation of the step or steps taken as cells transform from "normal" to "neoplastic" and from "benign" to "malignant."

ACKNOWLEDGMENTS

This investigation was supported partially by grants HD 04134 and HL 09011 from the National Institutes of Health and grant No. CD-10J from the American Cancer Society.

REFERENCES

Alter, B. P., R. Parkman, and J. M. Rappeport. 1980. Bone marrow failure syndromes, *in* Hematology of Infancy and Childhood, D. G. Nathan and F. A. Oski, eds. W. B. Saunders, Philadelphia, (in press).

Al Saadi, A., M. Paluke, and K. Kumar. 1975. Cytogenetic and immunologic studies in ataxia telangiectasia. Am. J. Hum. Genet. 27:78A.

Aurias, A., B. Dutrillaux, D. Buriot, and J. Lejeune. 1980. High frequencies of inversions and translocations of chromosomes 7 and 14 in ataxia telangiectasia. Mutat. Res. 69:369–374.

Bargman, G. J., N. T. Shahidi, E. F. Gilbert, and J. M. Opitz. 1977. Studies of malformation syndromes in man XLVII: Disappearance of spermatogonia in the Fanconi anemia syndrome. Eur. J. Pediatr. 125:163–168.

Bartman, C. R., T. Koske-Westphal, and E. Passarge. 1976. Chromatid exchanges in ataxia telangiectasia, Bloom's syndrome, Werner syndrome, and xeroderma pigmentosum. Ann. Hum. Genet. (Lond.) 40:79–86.

Beard, M. E. J., D. E. Young, C. J. T. Bateman, G. T. McCarthy, M. E. Smith, L. Sinclair, A. W. Franklin, and R. B. Scott. 1973. Fanconi's anemia. Quart. J. Med. 42:403–422.

Beatty-DeSana, J. W., M. J. Hoggard, and J. W. Cooledge. 1975. Non-random occurrence of 7—14 translocations in human lymphocyte cultures. Nature (Lond.) 255:242–243.

Berger, R., A. Bussel, and C. Schenmetzler. 1977. Somatic segregation in Fanconi anemia. Clin. Genet. 11:409–415.

Bersi, M., and C. Gasparini. 1973. Anomalie cromosomiche in un caso di anemia di Fanconi. Minerva Med. 64:1633–1637.

Bloom, D. 1954. Congenital telangiectatic erythema resembling lupus erythematosus in dwarfs. Am. J. Dis. Child. 88:754–758.

Bloom, G. E., S. Warner, P. S. Gerald, and L. K. Diamond. 1966. Chromosome abnormalities in constitutional aplastic anemia. N. Engl. J. Med. 274:8–14.

Bochkov, N. P., Y. M. Lopukhin, N. P. Kuleshov, and L. V. Kovalchuk. 1974. Cytogenetic study of patients with ataxia-telangiectasia. Humangenetik 24:115–128.

Boder, E. 1975. Ataxia-telangiectasia: Some historic, clinical, and pathologic observations. Birth Defects 11:255–270.

Boder, E., and R. P. Sedgwick. 1957. Ataxia-telangiectasia. A familial syndrome of progressive cerebellar ataxia, oculocutaneous telangiectasia and frequent pulmonary infection. Univ. South. Calif. Med. Bull. 9:15–27.

Boder, E., and R. P. Sedgwick. 1958. Ataxia-telangiectasia: A familial syndrome of progressive cerebellar ataxia, oculocutaneous telangiectasia and frequent pulmonary infection. Pediatrics 21:526–553.

Bourgeois, C. A., and F. G. H. Hill. 1977. Fanconi anemia leading to acute myelomonocytic leukemia: Cytogenetic studies. Cancer 39:1163–1167.

Burgdorf, W., K. Kurvink, and J. Cervenka. 1977. Sister chromatid exchange in dyskeratosis congenita lymphocytes. J. Med. Genet. 14:256–257.

Bushkell, L. L., J. H. Kersey, and J. Cervenka. 1976. Chromosomal breaks in T and B lymphocytes in Fanconi's anemia. Clin. Genet. 9:583–587.

Buyse, M., C. T. Hartman, and M. G. Wilson. 1976. Gonadoblastoma and dysgerminoma with ataxia-telangiectasia. Birth Defects 12:165–169.

Caldwell, R., C. Chase, and M. Swift. 1979. Cancer in Fanconi anemia families. Am. J. Hum. Genet. 31:132A.

Chaganti, R. S. K., S. Schonberg, and J. German. 1974. A manyfold increase in sister chromatid exchanges in Bloom's syndrome lymphocytes. Proc. Natl. Acad. Sci. USA 71:4508–4512.

Cohen, M. M., G. Kohn, and J. Dagan. 1973. Chromosomes in ataxia-telangiectasia. Lancet 2:1500.

Cohen, M. M., M. Sagi, Z. Ben-Zur, T. Schaap, R. Voss, G. Kohn, and H. Ben-Bassat. 1979. Ataxia-telangiectasia: Chromosomal stability in continuous lymphoblastoid cell lines. Cytogenet. Cell Genet. 23:44–52.

Cohen, M. M., M. Shaham, J. Dagan, E. Shmueli, and G. Kohn. 1975. Cytogenetic investigations in families with ataxia-telangiectasia. Cytogenet. Cell Genet. 15:338–356.

Crossen, P. E., J. E. L. Mellor, A. C. Adams, and F. W. Gunz. 1972. Chromosome studies in Fanconi's anaemia before and after treatment with oxymetholone. Pathology 4:27–33.

Crossen, P. E., and W. F. Morgan. 1977. Analysis of human lymphocyte cell cycle time in culture measured by sister chromatid differential staining. Exp. Cell Res. 104:453–457.

Crossen, P. E., and W. F. Morgan. 1978. Occurrence of 1st division metaphases in human lymphocyte cultures. Hum. Genet. 41:97–100.

de Grouchy, J., C. de Nava, J. C. Marchand, J. Feingold, and C. Turleau. 1972. Études cytogénétique et biochimique de huit cas d'anémie de Fanconi. Ann. Génét. 15:29–40.

de Grouchy, J., and C. Turleau. 1974. Clonal evolution in the myeloid leukemias, *in* Chromosomes and Cancer, J. German, ed. John Wiley and Sons, New York, pp. 287–311.

Dicken, C. H., G. Dewald, and H. Gordon. 1978. Sister chromatid exchanges in Bloom's syndrome. Arch. Dermatol. 114:755–760.

Dosik, H., L. Y. Hsu, G. J. Todaro, S. L. Lee, K. Hirschhorn, E. S. Selirio, and A. A. Alter. 1970. Leukemia in Fanconi's anemia: Cytogenetic and tumor virus susceptibility studies. Blood 36:341–352.

Dutrillaux, B., J. Couturier, E. Viegas-Péquignot, and G. Schaison. 1977. Localization of chromatid breaks in Fanconi's anemia using three consecutive stains. Hum. Genet. 37:65–71.

Fanconi, G. 1927. Familiäre infantile perniziosaartige Anämie (perniziöses Blutbild und Konstitution). Jahrb. für Kinderheil. 117:257–280.

Ferák, V., J. Benko, and E. Čajková. 1965. Genetik der Ataxie Teleangiectasie. Acta Genet. Med. Gemellol. 14:57–72.

Festa, R. S., A. T. Meadows, and R. A. Boshes. 1979. Leukemia in a black child with Bloom's syndrome: Somatic recombination as a possible mechanism for neoplasia. Cancer 44:1507–1510.

Galloway, S. M., and H. J. Evans. 1975. Sister chromatid exchange in human chromosomes for normal individuals and patients with ataxia telangiectasia. Cytogenet. Cell Genet. 15:17–29.

Germain, D., C. Requin, J. Robert, J.-J. Viala, and F. Freycon. 1968. Les anomalies chromosomiques dans l'anémie de Fanconi (A propos de 6 observations personnelles). Pediatrie 23:153–167.

German, J. 1964. Cytological evidence for crossing-over in vitro in human lymphoid cells. Science 144:298–301.

German, J. 1973. Genetic disorders associated with chromosomal instability and cancer. J. Invest. Dermatol. 60:427–434.

German, J. 1974. Bloom's syndrome. II. The prototype of human genetic disorders predisposing to chromosome instability and cancer, *in* Chromosomes and Cancer, J. German, ed. John Wiley and Sons, New York, pp. 601–617.

German, J. 1979a. Bloom's syndrome. VIII. Review of clinical and genetic aspects, *in* Genetic Diseases among Ashkenazi Jews, R. M. Goodman and A. G. Motulsky, eds. Raven Press, New York, pp. 121–139.

German, J. 1979b. The cancers in chromosome-breakage syndromes, *in* Radiation Research, Proceedings of the Sixth International Congress of Radiation Research, Tokyo, S. Okada, M. Imamura, T. Terashima, and H. Yamaguchi, eds. Toppan Printing Co., Tokyo, pp. 496–505.

German, J., R. Archibald, and D. Bloom. 1965. Chromosomal breakage in a rare and probably genetically determined syndrome of man. Science 148:506–507.

German, J., D. Bloom, and E. Passarge. 1977a. Bloom's syndrome. V. Surveillance for cancer in affected families. Clin. Genet. 12:162–168.

German, J., D. Bloom, and E. Passarge. 1979. Bloom's syndrome. VII. Progress report for 1978. Clin. Genet. 15:361–367.

German, J., D. Bloom, E. Passarge, K. Fried, R. M. Goodman, I. Katzenellenbogen, Z. Laron, C. Legum, S. Levin, and J. Wahrman. 1977b. Bloom's syndrome. VI. The disorder in Israel and an estimation of the gene frequency in the Ashkenazim. Am. J. Hum. Genet. 29:553–562.

German, J., and L. Crippa. 1966. Chromosomal breakage in diploid cell lines from Bloom's syndrome and Fanconi's anemia. Ann. Génét. 9:143–154.

German, J., L. P. Crippa, and D. Bloom. 1974. Bloom's syndrome. III. Analysis of the chromosome aberration characteristic of this disorder. Chromosoma (Berl.) 48:361–366.

German, J., and S. Schonberg. 1980. Bloom's syndrome. IX. Review of cytological and biochemical aspects, *in* Genetic and Environmental Factors in Experimental and Human Cancer, H. V. Gelboin, B. MacMahon, T. Matsushima, T. Sugimura, S. Takayama, and H. Takebe, eds. University of Tokyo Press, Tokyo, in press.

German, J., S. Schonberg, E. Louie, and R. S. K. Chaganti. 1977c. Bloom's syndrome. IV. Sister-chromatid exchanges in lymphocytes. Am. J. Hum. Genet. 29:248–255.

Gmyrek, D., R. Witkowski, I. Syllm-Rapoport, and G. Jacobasch. 1968. Chromosomal aberrations and abnormalities of red-cell metabolism in a case of Fanconi's anemia before and after development of leukaemia. Germ. Med. Monthly 13:105–111.

Goldsmith, C. I., and W. R. Hart. 1975. Ataxia-telangiectasia with ovarian gonadoblastoma and contralateral dysgerminoma. Cancer 36:1838–1842.

Goodman, W. N., W. C. Cooper, G. B. Kessler, M. S. Fischer, and M. B. Gardner. 1969. Ataxia-telangiectasia. A report of two cases in siblings presenting a picture of progressive spinal muscular atrophy. Bull. Los Angeles Neurol. Soc. 34:1–22.

Guanti, G., P. Petrinelli, and F. Schettini. 1971. Cytogenetical and clinical investigations in aplastic anaemia (Fanconi's type). Humangenetik 13:222–243.

Haerer, A. F., J. F. Jackson, and C. G. Evers. 1969. Ataxia-telangiectasia with gastric adenocarcinoma. JAMA 210:1884–1887.

Harnden, D. G. 1974. Ataxia-telangiectasia syndrome: Cytogenetic and cancer aspects, *in* Chromosomes and Cancer, J. German, ed. John Wiley and Sons, New York, pp. 619–636.

Harnden, D. G. 1977. The relationships between induced chromosome aberrations and chromosome abnormality in tumour cells, *in* Human Genetics, S. Armendares and R. Lisker, eds. Excerpta Medica, Amsterdam, pp. 355–366.

Hatcher, N. H., P. S. Brinson, and E. B. Hook. 1976. Sister chromatid exchanges in ataxia-telangiectasia. Mutat. Res. 35:333–336.

Hatcher, N. H., B. Pollara, and E. B. Hook. 1974. Chromosome breakage in two siblings with ataxia-telangiectasia: A search for intrafamilial similarities. Am. J. Hum. Genet. 26:39A.

Hayashi, K., and W. Schmid. 1975a. Tandem duplication q14 and dicentric formation by end-to-end chromosome fusions in ataxia telangiectasia (AT): Clinical and cytogenetic findings in 5 patients. Humangenetik 30:135–141.

Hayashi, K., and W. Schmid. 1975b. The rate of sister chromatid exchanges parallel to spontaneous chromosome breakage in Fanconi's anemia and to Trenimon-induced aberrations in human lymphocytes and fibroblasts. Humangenetik 29:201–206.

Hecht, F., R. D. Koler, D. A. Rigas, G. S. Dahnke, M. P. Case, V. Tisdale, and R. W. Miller. 1966. Leukemia and lymphocytes in ataxia-telangiectasia. Lancet 2:1193.

Hecht, F., and B. K. McCaw. 1973. Evidence for a consistent chromosomal abnormality in ataxia-telangiectasia (A-T) lymphocytes. Am. J. Hum. Genet. 25:32A.

Hecht, F., B. K. McCaw, and R. D. Koler. 1973. Ataxia-telangiectasia—Clonal growth of translocation lymphocytes. N. Engl. J. Med. 289:286–291.

Hecht, F., B. K. McCaw, D. Peakman, and A. Robinson. 1975. Nonrandom occurrence of 7—14 translocations in human lymphocyte cultures. Nature (Lond.) 255:243–244.

Henderson, E., and J. German. 1978. Development and characterization of lymphoblastoid cell lines (LCLs) from "chromosome breakage syndromes" and related genetic disorders. J. Supramolec. Struc. (Suppl.) 2:83.

Hirschman, R. J., N. R. Shulman, J. G. Abuelo, and J. Whang-Peng. 1969. Chromosomal aberrations in two cases of inherited aplastic anemia with unusual clinical features. Ann. Intern. Med. 71:107–117.

Hoefnagel, D., M. Sullivan, O. R. McIntyre, J. A. Gray, and R. C. Storrs. 1966. Panmyelopathy with congenital anomalies (Fanconi) in two cousins. Helv. Paediat. Acta 21:230–238.

Hook, E. B., N. H. Hatcher, and O. J. Calka. 1975. Apparent "in situ" clone of cytogenetically marked ataxia-telangiectasia lymphocytes. Humangenetik 30:251–257.

Hustinx, T. W. J., J. M. J. C. Scheres, C. M. R. Weemaes, B. G. A. ter Haar, and A. J. Janssen. 1979. Karyotype instability with multiple 7/14 and 7/7 rearrangements. Hum. Genet. 49:199–208.

Hustinx, T. W. J., B. G. A. ter Haar, J. M. J. C. Scheres, F. J. Rutten, C. M. R. Weemaes, R. L. E. Hoppe, and A. H. Janssen. 1977. Bloom's syndrome in two Dutch families. Clin. Genet. 12:85–96.

Hütteroth, T. H., S. D. Litwin, and J. German. 1975. Abnormal immune responses of Bloom's syndrome lymphocytes in vitro. J. Clin. Invest. 56:1–7.

Jean, P., C.-L. Richer, M. Murer-Orlando, D. H. Luu, and J. H. Joncas. 1979. Translocation 8;14 in an ataxia-telangiectasia-derived cell line. Nature (Lond.) 277:56–58.

Kamada, N., and H. Uchino. 1978. Chronologic sequence in appearance of clinical and laboratory findings characteristic of chronic myelocytic leukemia. Blood 51:843–850.

Kato, H., and H. F. Stich. 1976. Sister chromatid exchanges in ageing and repair-deficient human fibroblasts. Nature (Lond.) 260:447–448.

Kiossoglou, K., A. Moschos, K. Mantalenaki-Lambrou, S. Haïdas, and J. German. 1976. Acute lymphoblastic leukemia in Bloom syndrome. Clin. Genet. 10:362.

Klein, G. 1979. Lymphoma development in mice and humans: Diversity of initiation is followed by convergent cytogenetic evolution. Proc. Natl. Acad. Sci. USA 76:2442–2446.

Kuhn, E. M. 1976. Localization by Q-banding of mitotic chiasmata in cases of Bloom's syndrome. Chromosoma (Berl.) 57:1–11.

Landau, J. W., M. S. Sasaki, V. D. Newcomer, and A. Norman. 1966. Bloom's syndrome: The syndrome of telangiectatic erythema and growth retardation. Arch. Dermatol. 94:687–694.

Latt, S. A. 1973. Microfluorometric detection of deoxyribonucleic acid replication in human metaphase chromosomes. Proc. Natl. Acad. Sci. USA 70:3395–3399.

Latt, S. A., G. Stetten, L. A. Juergens, G. R. Buchanan, and P. S. Gerald. 1975. Induction by alkylating agents of sister chromatid exchanges and chromatid breaks in Fanconi's anemia. Proc. Natl. Acad. Sci. USA 72:4066–4070.

Levin, S., and S. Perlov. 1971. Ataxia-telangiectasia in Israel: With observations on its relationship to malignant disease. Isr. J. Med. Sci. 7:1535–1541.

Levitt, R., R. V. Pierre, W. L. White, and R. G. Siekert. 1978. Atypical lymphoid leukemia in ataxia-telangiectasia. Blood 52:1003–1011.

Lieber, E., L. Hsu, L. Spitler, and H. H. Fudenberg. 1972. Cytogenetic findings in a parent of a patient with Fanconi's anemia. Clin. Genet. 3:357–363.

Lisker, R., and A. Cobo. 1970. Chromosome breakage in ataxia-telangiectasia. Lancet 1:618.

Lisker, R., and A. C. de Gutiérrez. 1974. Cytogenetic studies in Fanconi's anemia. Description of a case with bone marrow clonal evolution. Clin. Genet. 5:72–76.

Louis-Bar, D. 1941. Sur un syndrome progressif comprenant des télangiectasies capillaires cutenées et conjonctivales symmétriques, à disposition naevoïde et des troubler cérébelleux. Confin. Neurol. (Basel) 4:32–42.

Mark, J. 1977. Chromosomal abnormalities and their specificity in human neoplasms: An assessment of recent observations by banding techniques. Adv. Cancer Res. 24:165–222.

McCaw, B. K., F. Hecht, D. G. Harnden, and R. L. Teplitz. 1975. Somatic rearrangement of chromosome 14 in human lymphocytes. Proc. Natl. Acad. Sci. USA 72:2071–2075.

McDougall, J. K. 1971. Spontaneous and adenovirus type 12-induced chromosome aberrations in Fanconi's anemia fibroblasts. Int. J. Cancer 7:526–534.

Meisner, L. F., A. Taher, and N. T. Shahidi. 1978. Chromosome changes and leukemic transformation in Fanconi's anemia, *in* Aplastic Anemia, S. Hibino, F. Takaku, and N. T. Shahidi, eds. University Park Press, Baltimore, pp. 253–271.

Mulvihill, J. J., R. L. Ridolfi, F. R. Schultz, M. S. Borzy, and P. B. T. Haughton. 1975. Hepatic adenoma in Fanconi anemia treated with oxymetholone. J. Pediatr. 87:122–124.

Nathanson, S. D., S. M. van Biljon, and J. Kallmeyer. 1968. Constitutional aplastic anemia (Fanconi type): Case presentation and review of the literature. S. Afr. Med. J. 42:1159–1161.

Nelson, M. M., A. Blom, and L. Arens. 1975. Chromosomes in ataxia-telangiectasia. Lancet 1:518–519.

Novotná, B., P. Goetz, and N. I. Surkova. 1979. Effects of alkylating agents on lymphocytes from controls and from patients with Fanconi's anemia. Studies of sister chromatid exchanges, chromosome aberrations, and kinetics of cell division. Hum. Genet. 49:41–50.

Nowell, P. C. 1974. Chromosome changes and the clonal evolution of cancer, *in* Chromosomes and Cancer, J. German, ed. John Wiley and Sons, New York, pp. 267–285.

Ohkita, T., and N. Kamada. 1979. Leukemia among atomic bomb survivors, *in* Radiation Research, Proceedings of the Sixth International Congress of Radiation Research, Tokyo, S. Okada, M. Imamura, T. Terashima, and H. Yamaguchi, eds. Toppan Printing Co., Tokyo, pp. 59–68.

Oxford, J. M., D. G. Harnden, J. M. Parrington, and J. D. A. Delhanty. 1975. Specific chromosome aberrations in ataxia telangiectasia. J. Med. Genet. 12:251–262.

Passarge, E. 1972. Spontaneous chromosomal instability. Humangenetik 16:151–157.

Paterson, M. C., and P. J. Smith. 1979. Ataxia-telangiectasia: An inherited human disorder involving hypersensitivity to ionizing radiation and related DNA-damaging chemicals. Annu. Rev. Genet. 13:291–318.

Perkins, J., J. Timson, and A. E. H. Emery. 1969. Clinical and chromosome studies in Fanconi's aplastic anemia. J. Med. Genet. 6:28–33.

Perry, P., and S. Wolff. 1974. New Giemsa method for the differential staining of sister chromatids. Nature (Lond.) 251:156–158.

Pfeiffer, R. A. 1970. Chromosome abnormalities in ataxia-telangiectasia (Louis Bar's syndrome). Humangenetik 8:302–306.

Polani, P. E. 1976. Cytogenetics of Fanconi anaemia and related chromosome disorders, *in* Congenital Disorders of Erythropoiesis, Ciba Foundation Symposium 37, Elsevier/Excerpta Medica, North Holland, Amsterdam, pp. 261–306.

Prindull, G., E. Jentsch, and I. Hansmann. 1979. Fanconi's anaemia developing erythroleukemia. Scand. J. Haematol. 23:59–63.

Puligandla, B., S. A. Stass, H. R. Schumacher, T. P. Keneklis, and F. J. Bollum. 1978. Terminal deoxynucleotidyl transferase in Fanconi's anaemia. Lancet 2:1263.

Rary, J. M., M. A. Bender, and T. E. Kelly. 1974. Cytogenetic studies of ataxia telangiectasia. Am. J. Hum. Genet. 26:70A.

Rary, J. M., M. A. Bender, and T. E. Kelly. 1975. A 14/14 marker chromosome lymphocyte clone in ataxia telangiectasia. J. Hered. 66:33–35.

Rauh, J. L., and S. W. Soukup. 1968. Bloom's syndrome. Am. J. Dis. Child. 116:409–413.

Reinhold, J. D. L., E. Neumark, R. Lightwood, and C. O. Carter. 1952. Familial hypoplastic anemia with congenital abnormalities (Fanconi's syndrome). Blood 7:915–926.

Sasaki, M. S., and A. Tonomura. 1973. A high susceptibility of Fanconi's anemia to chromosome breakage by DNA cross-linking agents. Cancer Res. 33:1829–1836.

Saxon, A., R. H. Stevens, and D. W. Golde. 1979. Helper and suppressor T-lymphocyte leukemia in ataxia telangiectasia. N. Engl. J. Med. 300:700–704.

Scheres, J. M. J. C., T. W. J. Hustinx, and C. M. R. Weemaes. 1980. Chromosome 7 in ataxia-telangiectasia. J. Pediatr. (in press).

Schmid, W. 1967. Familial constitutional panmyelocytopathy, Fanconi's anemia (F.A.). II. A discussion of the cytogenetic findings in Fanconi's anemia. Semin. Hematol. 4:241–249.

Schmid, W., and G. Fanconi. 1978. Fragility and spiralization anomalies of the chromosomes in three cases, including fraternal twins, with Fanconi's anemia, type Estren-Dameshek. Cytogenet. Cell Genet. 20:141–149.

Schmid, W., and F. Jerusalem. 1972. Cytogenetic findings in two brothers with ataxia-telangiectasia (Louis Bar's syndrome). Arch. Genet. (Zur.) 45:49–52.

Schmid, W., K. Schärer, T. Baumann, and G. Fanconi. 1965. Chromosomenbrüchigkeit bei der familiären Panmyelopathie (Typus Fanconi). Schweiz. Med. Wochenschr. 95:1461–1464.

Schoen, E. J., and M. A. Shearn. 1967. Immunoglobulin deficiency in Bloom's syndrome. Am. J. Dis. Child. 113:594–596.

Schonberg, S., J. German, and R. S. K. Chaganti. 1978. A new cytogenetic finding in Bloom's syndrome: Terminal association of homologous chromosomes at metaphase. Genetics (Suppl.) 88:s88–s89.

Schroeder, T. M. 1966a. Cytogenetischer Befund und Ätiologie bei Fanconi-Anämie Ein Fall von Fanconi-Anämie ohne Hexokinase-defekt. Humangenetik 3:76–81.

Schroeder, T. M. 1966b. Cytogenetische und cytologische Befunde bei enzymopenischen Panmyelopathien und Pancytopenian Familiäre Panmyelopathie Typus Fanconi, Glutathionreduktasemangel-Anämie und megaloblastäre Vitamine B$_{12}$-Mangel-Anämie. Humangenetik 2:287–316.

Schroeder, T. M., F. Anschütz, and A. Knopp. 1964. Spontane Chromosomenaberrationen bei familiärer Panmyelopathie. Humangenetik 1:194–196.

Schroeder, T. M., P. Drings, P. Beilner, and G. Buchinger. 1976a. Clinical and cytogenetic observations during a six-year period in an adult with Fanconi's anemia. Blut 34:119–132.

Schroeder, T. M., and J. German. 1974. Bloom's syndrome and Fanconi's anemia: Demonstration of two distinctive patterns of chromosome disruption and rearrangement. Humangenetik 25:299–306.

Schroeder, T. M., and R. Kurth. 1971. Spontaneous chromosomal breakage and high incidence of leukemia in inherited disease. Blood 37:96–112.

Schroeder, T. M., D. Tilgen, J. Krüger, and F. Vogel. 1976b. Formal genetics of Fanconi's anemia. Hum. Genet. 32:257–288.

Schroeder, T. M., and C. Stahl Mauge. 1976. Spontaneous chromosome instability, chromosome reparation and recombination in Fanconi's anemia and Bloom's syndrome, *in* DNA Repair and Late Effects, International Symposium of the "IGEGM," H. Altmann, ed. Institut für Biologie, Forschungszentrum Seibersdorf/Wien, Eisenstadt, pp. 35–50.

Schuler, D., L. Schöngut, E. Cserháti, J. Siegler, and G. Gács. 1971. Lymphoblastic transformation, chromosome pattern, and delayed-type skin reaction in ataxia telangiectasia. Acta Paediat. Scand. 60:66–72.

Schuster, J., Z. Hart, C. W. Stimson, A. J. Brough, and M. D. Poulik. 1966. Ataxia telangiectasia with cerebellar tumor. Pediatrics 37:776–786.

Scott, D., and C. Y. Lyons. 1979. Homogeneous sensitivity of human peripheral blood lymphocytes to radiation-induced chromosome damage. Nature (Lond.) 278:756–758.

Shahid, M. J., F. P. Khouri, and S. K. Ballas. 1972. Fanconi's anaemia: Report of a patient with significant chromosomal abnormalities in bone marrow cells. J. Med. Genet. 9:474–478.

Shiraishi, Y., A. I. Freeman, and A. A. Sandberg. 1976. Increased sister chromatid exchange in bone marrow and blood cells from Bloom's syndrome. Cytogenet. Cell Genet. 17:162–173.

Shiraishi, Y., and A. A. Sandberg. 1977. The relationship between sister chromatid exchanges and chromosome aberrations in Bloom's syndrome. Cytogenet. Cell Genet. 18:13–23.

Sperling, K., U. Goll, J. Kunze, E.-K. Lüdtke, M. Tolksdorf, and G. Obe. 1976. Cytogenetic investigations in a new case of Bloom's syndrome. Hum. Genet. 31:47–52.

Sperling, K., R. D. Wegner, H. Riehm, and G. Obe. 1975. Frequency and distribution of sister-chromatid exchanges in a case of Fanconi's anemia. Humangenetik 27:227–230.

Swift, M. 1971. Fanconi's anaemia in the genetics of neoplasia. Nature (Lond.) 230:370–373.

Swift, M. 1976. Malignant disease in heterozygous carriers. Birth Defects 12:133–144.

Swift, M., and K. Hirschhorn. 1966. Fanconi's anemia: Inherited susceptibility to chromosome breakage in various tissues. Ann. Intern. Med. 65:496–503.

Swift, M., L. Scholman, M. Perry, and C. Chase. 1976. Malignant neoplasms in the families of patients with ataxia-telangiectasia. Cancer Res. 36:209–215.

Swift, M., D. Zimmerman, and E. R. McDonough. 1971. Squamous cell carcinomas in Fanconi's anemia. JAMA 216:325–326.

Syllaba, L., and K. Henner. 1926. Contribution a l'indépendance de l'athétose double idiopathique et congénitale. Rev. Neurol. 1:541–562.

Szalay, G. C., and E. D. Weinstein. 1972. Questionable Bloom's syndrome in a Negro girl. Am. J. Dis. Child. 124:245–248.

Tadjoedin, M. K., and F. C. Fraser. 1965. Heredity of ataxia-telangiectasia (Louis-Bar syndrome). Am. J. Dis. Child. 110:64–68.

Utian, H. L., and M. Plit. 1964. Ataxia telangiectasia. J. Neurol. Neurosurg. Psychiat. 27:38–40.

Varela, M. A., and W. H. Sternberg. 1967. Preanaemic state in Fanconi's anaemia. Lancet 2:566–567.

Ved Brat, S. 1979. Sister chromatid exchange and cell cycle in fibroblasts of Bloom's syndrome. Hum. Genet. 48:73–79.

von Koskull, H., and P. Aula. 1973. Nonrandom distribution of chromosome breaks in Fanconi's anemia. Cytogenet. Cell Genet. 12:423–434.

von Koskull, H., and P. Aula. 1977. Distribution of chromosome breaks in measles, Fanconi's anemia, and controls. Hereditas 87:1–10.

Webb, T., and M. Harding. 1977. Chromosome complement and SV40 transformation of cells from patients susceptible to malignant disease. Br. J. Cancer 36:583–591.

Weemaes, C. M. R., J. A. J. M. Bakkeren, B. G. A. ter Haar, T. W. J. Hustinx, and P. J. J. van Munster. 1979. Immune responses in four patients with Bloom syndrome. Clin. Immunol. Immunopathol. 12:12–19.

Welch, J. P., and C. L. Y. Lee. 1975. Non-random occurrence of 7—14 translocations in human lymphocyte cultures. Nature (Lond.) 255:241–242.

Wolman, S. R., and M. Swift. 1972. Bone marrow chromosomes in Fanconi's anemia. J. Med. Genet. 9:473–474.

Zaizov, R., Y. Matoth, and Z. Mamon. 1978. Long-term observations in children with Fanconi's anemia, in Aplastic Anemia, S. Hibino, F. Takaku, and N. T. Shahidi, eds. University Park Press, Baltimore, pp. 243–251.

Zellweger, H., and R. R. Khalifeh. 1963. Ataxia-telangiectasia. Report of two cases. Helv. Paediatr. Acta 18:267–279.

Genes, Chromosomes, and Neoplasia, edited by
Frances E. Arrighi, Potu N. Rao, and Elton Stubblefield.
Raven Press, New York © 1981.

Premature Chromosome Condensation Studies in Human Leukemia

Walter N. Hittelman, Potu N. Rao, and Kenneth B. McCredie

*Department of Developmental Therapeutics, The University of Texas System Cancer Center
M. D. Anderson Hospital and Tumor Institute, Houston, Texas 77030*

The title *Genes, Chromosomes, and Neoplasia* brings to mind chromosomes that are normally visualized at mitosis when the interphase chromatin becomes condensed and packaged into discrete units. The work to be described in this paper is the result of looking at chromosomes in an unconventional manner. Rather than waiting for the cells to reach mitosis in order to see the chromosomes, we artificially induce the interphase chromatin to condense prematurely in whatever phase of the cell cycle the cell is at the time of the experiment. In this paper, we describe how this phenomenon of premature chromosome condensation (PCC) is useful in the study of malignancy, and in particular, human leukemia. The early finding that is basic to all these studies is that the PCC technique allows one to distinguish normal from malignant cells in resting phase, regardless of the morphology of the cell. Applying this basic notion to the study of human leukemia, we have found that the PCC technique is useful in the determination of response at various stages of the disease.

The paper begins with a brief description of the phenomenon of PCC and how it is useful in determining the cytogenetic and cell kinetic characteristics of cell populations. The next section presents the evidence that malignant cells accumulate in a different stage in the cell cycle than normal cells both in vitro and in vivo. The third section begins the focus on human leukemia and describes the pretreatment PCC characteristics of bone marrow populations of patients with different forms of leukemia. The fourth section enumerates the early changes after the initiation of chemotherapy as viewed by the PCC technique and illustrates when this technique is useful in determining prognosis after therapy. The fifth section focuses on patients apparently in complete remission and describes how the PCC technique can predict relapse an average of four months prior to any clinical signs of disease relapse. The sixth section describes our early efforts at understanding the cell biology behind our findings, and the final section considers both the basic and clinical implications of this work.

USE OF PREMATURE CHROMOSOME CONDENSATION IN CYTOGENETIC AND CYTOKINETIC STUDIES OF CELL POPULATIONS

In an earlier paper in this symposium, Dr. Potu Rao described his and Dr. Robert Johnson's observation that when mitotic cells were fused together with interphase cells by Sendai virus, the interphase chromatin was induced to condense into visible chromosomes within minutes (Rao et al. 1981, see pages 49 to 60, this volume). This phenomenon has been termed "chromosome pulverization," prophasing (Kato and Sandberg 1968a, b, Tagaki et al. 1969, Sandberg et al. 1970, Matsui et al. 1972), and "premature chromosome condensation" (Johnson and Rao 1970) (Figure 1). When the fusion mixture is subjected to hypotonic treatment, fixed in methanol:glacial acetic acid, and placed on wet glass slides, the resulting chromosome spreads can be stained and observed with the light microscope. The morphology of the PCC reflects the stage of the cell in the cell cycle at the time of fusion. Cells from G1 phase yield PCC with a single chromatid per chromosome. Cells fused from S phase exhibit a pulverized appearance, with both single and double elements visible. Cells from G2 phase exhibit two closely aligned, extended chromatids per chromosome (Figure 2). The PCC technique can therefore be used to determine the cell cycle distribution of cells in any cell population based on the frequency of G1, S, and G2 PCC obtained after the fusion reaction (Rao et al. 1977).

Since the chromosomes of the cell can be induced to condense from any cell cycle phase, the PCC technique can be useful in cytogenetic studies. The PCC of G1 and G2 cells can be analyzed both for karyotype and for the presence of chromosome aberrations. This adds a new dimension to cytogenetic studies since it allows the cytogenetic analysis of nondividing cell populations. In addition, it allows comparisons to be made between interphase and metaphase cells in mixed cell populations. This might be very useful in studying the clonal evolution of various malignancies. In one study, for example, DNA content determinations by flow cytometry had indicated that a patient with leukemia had a hypodiploid amount of DNA, whereas conventional mitotic analyses only revealed diploid cells reaching mitosis. PCC analysis of the interphase bone marrow cells revealed a group of cells in G1 phase with only 44 chromosomes, thus confirming the DNA content observation (Barlogie et al. 1977). Apparently, this hypodiploid population of cells was resting in G1 phase and not cycling to mitosis where it could be detected by conventional means.

The PCC technique is also useful in visualizing chromosome damage in interphase cells. In the 1979 M. D. Anderson Symposium on Fundamental Cancer Research (Hittelman et al. 1980c), our work in this area was reviewed. One can measure the immediate and delayed accumulation of chromosome damage in cells after exposure to a variety of agents including radiation (Hittelman and Rao 1974a, b, Waldren and Johnson, 1974), alkylating agents (Hittelman and Rao 1974b), bleomycin (Hittelman and Rao 1974c), Adriamycin (Hittelman and Rao 1975), and neocarzinostatin. The PCC technique has proved to be

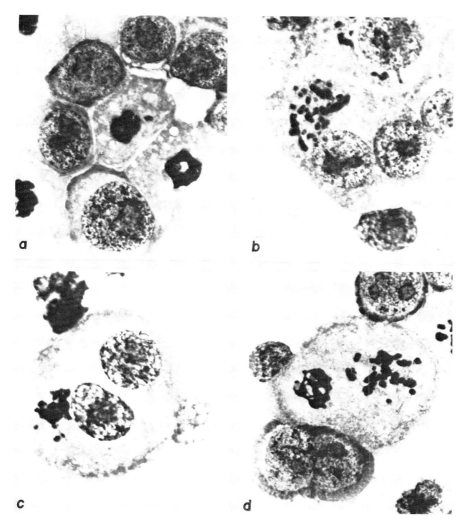

FIG. 1. The phenomenon of PCC by cell fusion via Sendai virus. a, Mixture of mitotic and interphase cells prior to fusion. b, Early stages of PCC formation. Note the condensing chromatin of the interphase component. c, Intermediate stage of PCC formation. Note increased chromatin condensation and breakdown of the nuclear membrane. d, Late stage of PCC formation. The interphase chromatin is condensed and the nuclear membrane has disappeared.

uniquely sensitive in detecting chromosome damage, since damaged cells often do not reach mitosis and may accumulate in G2 phase (Hittelman and Rao 1975, Rao and Rao 1976). By delaying the fusion until a period of time after clastogenic treatment, one can also directly determine the extent of chromosome repair that has taken place (Hittelman and Rao 1974c, Sognier et al. 1979).

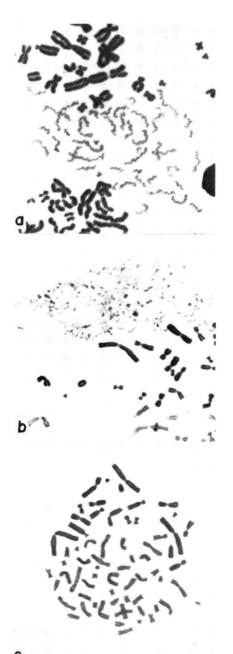

FIG. 2. PCC of human bone marrow cells from various stages of the cell cycle. a, G1 PCC exhibiting a single chromatid per chromosome. b, S PCC exhibiting a pulverized appearance with both single (pre-replicative) and double (post-replicative) elements. c, G2 PCC exhibiting two chromatids per chromosome.

DISTINGUISHING MALIGNANT AND NORMAL RESTING CELLS

Mazia suggested a number of years ago that a chromosome condensation cycle exists within the cell cycle (Mazia 1963). Studies with a variety of biophysical and chemical probes have shown that the chromatin is most condensed at mitosis and least condensed in the early part of S phase (Pederson 1972, Pederson and Robbins 1972, Niccolini et al. 1975). Early studies showed that the PCC technique might be useful in visualizing the transition from a condensed to a decondensed state at the chromosomal level (Schor et al. 1975, Hittelman and Rao 1976, Rao et al. 1977). Using synchronized Chinese hamster ovary (CHO) G1 cells, we found that early G1 cells exhibit highly condensed G1 PCC, while late G1 cells yield highly extended G1 PCC (Hittelman and Rao 1978a). Thus, a progressive decondensation of chromatin during G1 occurs sequentially as the cell progresses toward S phase (Figure 3). Similarly, unstimulated peripheral blood lymphocytes yield condensed G1 PCC upon fusion with mitotic cells. After phytohemagglutinin (PHA) stimulation, however, the cell population exhibits a higher fraction of extended G1 PCC as the cells move toward S phase (Hittelman and Rao 1976). These early studies, therefore, indicated that the PCC technique was useful not only in determining whether a cell was in G1, S, or G2 phase, but also in mapping the location of a cell within a phase, e.g., early, mid, or late G1 phase.

Upon entering a quiescent phase (whether due to population limitations, nutritional deprivation, or differentiation), most cell populations appear to accumulate cells in the G1 phase of the cell cycle. We therefore decided to employ the PCC technique to determine where in the G1 phase cell populations come to rest. Early studies concentrating on in vitro cell cultures grown to a plateau phase showed that normal and transformed cell populations exhibit completely distinct patterns. Nearly all the cells in plateau phase cultures of the normal cell lines tested (human PA2 and WI38 and mouse 3T3) gave rise to early G1 PCC morphologies, i.e., condensed G1 PCC, with few or no cells in S or G2 phase. In contrast, plateau phase transformed cell populations (HeLa, SV3T3, CHO) gave rise to G1 PCC with predominately mid to late G1 PCC morphologies, i.e., extended G1 PCC (Hittelman and Rao 1978a).

This difference between quiescent normal and transformed cells with regard to their location in G1 phase persists in vivo. For example, in a mouse fibrosarcoma grown in the mouse, nearly all the G1 PCC observed from the solid tumor were highly extended, suggesting that most of these cells accumulate in late G1 phase. In some cases, condensed G1 PCC, typical of early G1 cells, were observed; however, the majority of these condensed G1 PCC exhibited 40 chromosomes. Since the mouse fibrosarcoma cells used in these experiments have a modal number of 62 chromosomes per cell, these G1 PCC must have originated from normal mouse cells that had invaded the tumor (Grdina et al. 1977). Similar observations have been made on human breast cancer tumors grown in the nude mouse (Hittelman, unpublished observations) and in studies with normal tissues of the mouse and even fish. Thus, as a general rule, normal resting cells appear to accumulate both in vitro and in vivo in early G1 phase, whereas transformed or tumor cells appear to accumulate in late G1. These differences can be shown by the PCC technique.

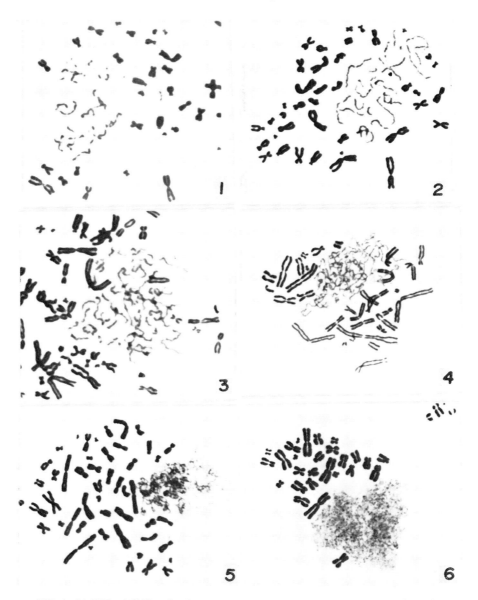

FIG. 3. G1 PCC of CHO cells showing varying degrees of chromosome condensation.

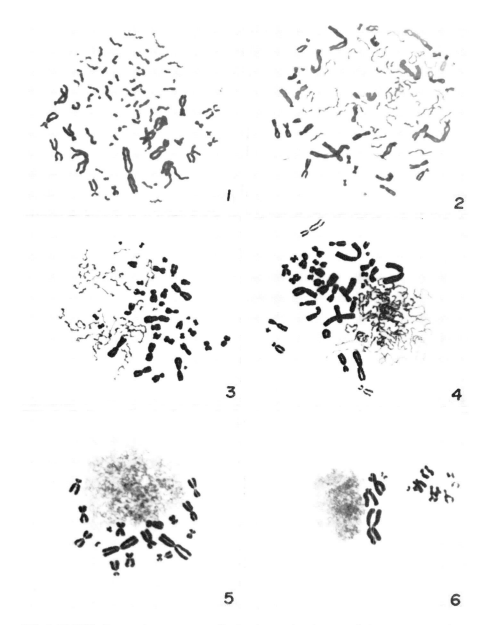

FIG. 4. G1 PCC of human bone marrow cells showing varying degrees of chromosome condensation. The G1 PCC have been arbitrarily divided into six stages according to the degree of chromosome condensation.

In an attempt to quantitate the distribution of cells in G1 phase, we arbitrarily divided G1 into six categories according to PCC morphology, with a value of 1 representing the most highly condensed G1 PCC (Figure 4). Since stimulation of normal resting cells (e.g., fibroblasts, lymphocytes) results in an increase in the frequency of G1 PCC belonging to classes 4 to 6, we defined the term proliferative potential index (PPI) as the fraction of G1 PCC that are highly extended (classes 4,5,6/classes 1–6). Thus the higher the PPI, the higher the fraction of cells in late G1 phase. By this definition, normal resting cells would have a low PPI while resting tumor cells would have a high PPI value.

PCC ANALYSIS OF BONE MARROW CELLS FROM LEUKEMIA PATIENTS BEFORE THERAPY

Our studies in human leukemia derived from a project designed to use the PCC technique to predict sensitivity of a patient's tumor cells to chemotherapy. The original idea was to obtain bone marrow cells from leukemia patients, concentrate mononuclear cells by Ficoll-Hypaque (Boyum 1968), treat these cells with potentially useful chemotherapeutic agents in vitro, and then induce PCC in the treated cells to detect the degree of chromosome damage produced. Since most chemotherapeutic agents are clastogens, if no chromosome damage were induced, the leukemic cells would most likely be resistant to the chemotherapeutic agents. We had hoped that this technique might be useful in providing a basis for the individualization of remission-induction therapy. However, while beginning this project, we observed that PCC of leukemic bone marrows exhibited unusual and varied patterns in the distribution of cells within G1 phase.

Early in these studies, we found that the G1 PCC morphologies varied from patient to patient. Bone marrow populations from untreated patients with leukemia exhibited an average PPI value of 27.5%, ranging from a low of 6% to a high of 76%. In contrast, bone marrow populations from normal individuals or from solid tumor patients without bone marrow involvement yielded an average of 11.7%, ranging from 4.0% to 18.2% (Hittelman and Rao 1978b). The above results were in agreement with the notion that resting cells from normal cell populations tend to accumulate in early G1 phase while those from tumor populations tend to accumulate in late G1 phase (their S phase fractions were similar). However, our early study grouped all types of leukemia patients together, including patients with chronic and those with active disease. It was of interest, therefore, to determine whether patients with different leukemic disease categories yielded distinct PCC characteristics.

In one study (Hittelman et al. 1979), 58 bone marrow aspiration samples from untreated patients with documented acute myelogenous leukemia (AML), acute monomyelocytic leukemia (AMML), or acute promyelocytic leukemia (ApML) were analyzed by the PCC technique. As shown in Figure 5, the average PPI for this class of patients was 35.5% (range 10–79.5%), nearly three times that found for normal marrow. The average percentage of S PCC was 7.4%

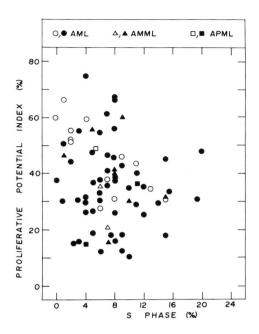

FIG. 5. Scatter diagram showing the relationship between PPI values and fraction of cells in S phase for patients with AML, AMML, or ApML. Closed symbols, patients who achieved CR; open symbols, patients who did not achieve CR.

(range 0–20%). The average PPI value for 23 untreated patients with acute lymphocytic leukemia (ALL) or acute undifferentiated leukemia (AUL) was higher (42%, range 11.3–92.9%), as was the average percent S phase cells (13.1%, range 1–30.7%). Thus, ALL and AUL appear to be more actively proliferating diseases than myelogenous disease when studied by the PCC technique. Interestingly, however, there appeared to be little correlation between the PPI values and the fraction of cells in S phase, or the PPI and the blast count or blast infiltrate (blast percentage × cellularity) in the bone marrow.

Since the variability in PPI values was great in active leukemia, we were interested in determining whether the pretreatment PPI value was useful in predicting response to therapy. All patients with AML, AMML, or ApML who had initial PPI values of less than 20% achieved complete remission, whereas only half of the patients with PPI values greater than 50% achieved complete remission with remission induction therapy. Intermediate PPI values were not as useful in predicting response.

The PPI values observed in patients with chronic myelogenous leukemia (CML) reflected the activity of their disease in an interesting way. Untreated patients with CML not obviously in blast crisis gave rise to an average PPI value of 25.8% (range 11.7–54.3%). Patients with benign CML (which is easily controlled with therapy such as hydroxyurea or busulfan) exhibited nearly normal PPI values (average 14%). In contrast, patients with CML in blast crisis yielded an average PPI value of 34.4%, similar to that observed in acute leukemia. Interestingly, one patient showed a low PPI value (14.3%) during the benign phase and a high PPI value (38.3%) at the blastic phase of disease. Another patient showed a high PPI value of 35.7% during the controlled phase yet showed clinical evidence of blast crisis within 5 months of the high PPI value. These results suggest that the transition from the more benign to the blastic phase of CML is preceded by an increased PPI; thus the PCC technique may be useful in predicting the onset of blast crisis.

Only a few patients with chronic lymphocytic leukemia (CLL) have been studied, and the average PPI value for them was 59.4% (range 42–70%). Interestingly, although most of the cells in the bone marrow of a CLL patient at an active stage of disease appear to be morphologically differentiated lymphocytes, they nevertheless yield the PCC characteristics of malignant cells (i.e., resting in late G1 phase).

While it is not understood why there should be such a great variability in PPI values from patient to patient within each disease category, these results do suggest that each patient's disease is biologically distinct. In CML, the PPI value may reflect the activity of the disease. In acute leukemia, this hypothesis is difficult to support. For example, some patients show low blast infiltrates yet high PPI values. Thus, some cells that appear morphologically normal give rise to late G1 PCC characteristic of resting malignant cells. This enigma will be discussed further in a later section.

BONE MARROW PCC CHANGES DURING REMISSION INDUCTION THERAPY

PCC analysis of pretreatment bone marrow populations showed that leukemic populations could be distinguished from normal marrow populations on the basis of raised PPI values. In early studies of a few patients after treatment, complete remission (CR) was correlated with PPI values intermediate between normal and pretreatment figures (i.e., PPI<35%). Based on this observation, we decided to ask whether PCC changes observed during remission induction therapy might be useful for the early prediction of response or progression. For this purpose, we studied the PCC characteristics of sequential bone marrow aspiration samples obtained from 21 patients with AML or AMML during primary remission induction therapy (Hittelman et al. 1980b). Treatment generally included a combination chemotherapy program of vincristine, cytosine arabinoside, and prednisone with an anthracycline (Adriamycin or rubidizone) (AdOAP or ROAP, respectively).

Prior to treatment, the average PPI and percentage of S PCC for this group of patients was 34.0% and 6.7%, respectively, with little or no chromosome damage in the PCC. One of the first notable changes observed in the PCC after initiation of therapy was extensive chromosome damage. For example, prior to therapy, the patient of Figure 6a, b exhibited a ring chromosome that could be visualized in both G1 and G2 PCC. Within 48 hours of the initiation of therapy, extensive chromosome damage could be detected in both G1 and

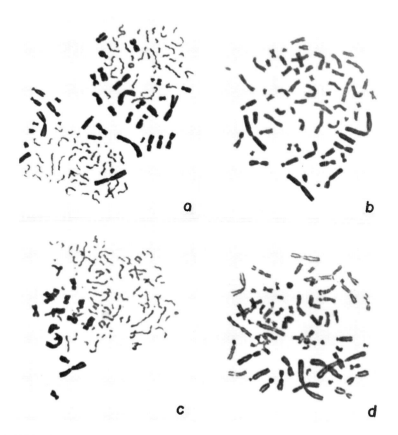

FIG. 6. PCC of bone marrow cells obtained from a patient undergoing remission-induction therapy. a, G1 PCC and b, G2 PCC obtained prior to initiation of therapy. Note the presence of a ring chromosome. c, G1 PCC and d, G2 PCC obtained 2 days after initiation of chemotherapy. Note extensive damage in the PCC.

G2 cells (Figure 6c, d). This damage was typical of that observed after treatment of cells in vitro with Adriamycin (i.e., a high frequency of chromatid exchanges) (Hittelman and Rao 1975).

Figure 7 illustrates a typical pattern observed for patients achieving a complete remission. During therapy, patients showed an initial drop in PPI value (if their initial PPI value was high) with evidence of moderate to extensive chromosome damage. This was followed by a rise in the PPI and the fraction of S PCC, just prior to and accompanying early regeneration of the bone marrow. In patients who achieved CR, the PPI dropped to values below 35% as normal peripheral counts began to recover. Continued PPI values below 35% at the end of each course of therapy correlated with continued CR.

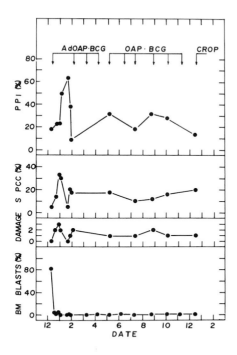

FIG. 7. Serial PCC-derived characteristics and bone marrow blast counts obtained from a patient with leukemia achieving complete remission after one course of remission-induction therapy. Note the increase in PPI, fraction of cells in S phase, and chromosome damage during the early regenerative period and the return of PPI values to intermediate values by the end of a course of therapy.

Patients showing either a partial remission or leukemic progression tended to exhibit one of two distinct patterns. For some patients failing remission induction therapy, little change in the PPI or percentage S PCC and little or no chromosome damage were observed (Figure 8). The second pattern observed for non-responding patients was continued high PPI values despite bone marrow recovery.

PCC patterns appeared to be predictive of therapeutic outcome in some patients. The patient of Figure 9 achieved a CR in the first course of therapy, which was accompanied by low PPI values. However, during the second course of therapy, the bone marrow blast count was observed to increase to greater than 40%. At this time, conventional bone marrow morphology studies were not able to determine whether these represented normal blasts associated with active regeneration or, alternatively, a return of the leukemic blasts. PCC analysis at this time showed low PPI values and high S PCC values accompanied by significant chromosome damage. These PCC patterns were suggestive of normal bone marrow regeneration. In fact, this patient's marrow regenerated normally prior to the next course of therapy. As shown in Table 1, this pattern (i.e. low PPI values despite blast counts >10% during regeneration) was observed in seven patients in this study, and all seven patients achieved complete remission. Table 1 also shows other situations in which the PCC technique can be used to predict therapeutic outcome.

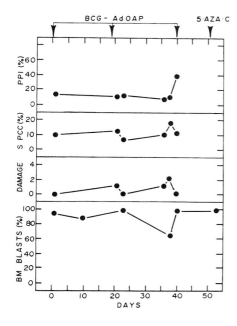

FIG. 8. PCC-derived characteristics and bone marrow blast counts of a patient failing remission-induction therapy. Note only slight PCC changes with therapy.

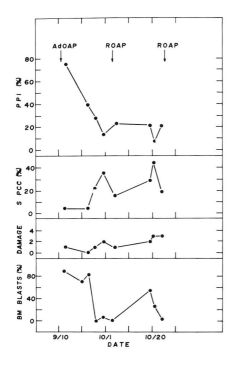

FIG. 9. PCC-derived characteristics and blast counts of a patient achieving a complete remission in the first course of therapy yet exhibiting a high blast percentage during the second course. Note the low PPI values observed during the regeneration period.

TABLE 1. *Correlations between proliferative potential index and clinical prognosis*

Patient's clinical setting	PPI	Prognosis	Frequency of observation
Not in CR after therapy	Low	CR	5/6
Not in CR after therapy	High	PR or failure	3/4
Blast >10% during re-generative period	Low	CR	7/7
In CR after therapy	High	Progressive disease	2/4

In general, continued high PPI values (>35%) correlated with progression of disease, while low PPI values correlated with achievement of remission. These results suggest that the PCC technique is a useful tool with which to monitor the pattern of response of patients with AML to remission induction therapy. This technique is especially helpful during the regenerative phase when it is difficult by bone marrow morphology and cellularity studies alone to predict the future course of the disease. It will be of interest to determine whether other forms of leukemia show similar patterns during therapy.

PCC CHANGES DURING MAINTAINED REMISSION—PREDICTION OF RELAPSES

In our earliest study of patients with leukemia (Hittelman and Rao 1978b), we had obtained single PCC measurements on 13 patients who were clinically in complete remission. The average PPI value for this group was close to 24% compared to average PPI values of >40% for patients who had obviously progressive disease despite continued therapy. Four patients in the CR group showed PPI values >35% and three of these patients relapsed within 6 months. On the other hand, of the nine patients who exhibited low PPI values while in CR, only one relapsed within 6 months. These early results suggested, therefore, that a high PPI value during complete remission might predict for the onset of relapse. As shown in Figure 10, this pattern was also observed in some patients followed sequentially during remission. In this particular patient, a rise in PPI value above 35% preceded clinically detectable relapse by 3 months.

Based on the early observations mentioned above and those observed during remission induction therapy, we decided to test the hypothesis that a rise in PPI values above 35% during complete remission was predictive for relapse of disease. If the hypothesis proved true, we were also interested in determining the time interval between a rise in PPI value and clinical evidence of relapse. For this purpose, we serially monitored 19 patients clinically in remission for more than 6 months who were entered on a late-intensification protocol in an attempt to prevent relapse (Hittelman et al. 1980). Bone marrow specimens were obtained serially during and after late-intensification CROP therapy (cyclo-

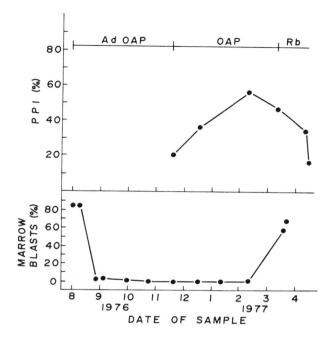

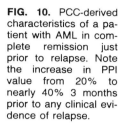

FIG. 10. PCC-derived characteristics of a patient with AML in complete remission just prior to relapse. Note the increase in PPI value from 20% to nearly 40% 3 months prior to any clinical evidence of relapse.

phosphamide, rubidizone, vincristine, and prednisone) and were evaluated by the PCC technique.

Of the 19 patients, 14 have relapsed so far, and an increase in PPI values to above 35% preceded relapse in 11 of the 14. In the remaining 3 relapsing patients, the PPI value rose above 35% at the time of relapse. The average time from an elevation of PPI to above 35% to clinical evidence of relapse was 4.6 months (range 1.3 to 11 months). While a prediction interval of 11 months is difficult to believe, this patient exhibited a low PPI followed by 3 high PPI values prior to relapse. If this patient were not included in the analysis, the average prediction time would still be 3.9 months. Figure 11 shows a typical example. In this case, the PPI rose above 35% 1.8 months prior to clinical evidence of relapse.

The incidence of false-positive (a rise in PPI without a subsequent relapse) and false-negative (a low PPI value just prior to relapse) measurements were low (13.3% and 5.1%, respectively). Only one false-positive measurement was observed in the five patients who have not relapsed, and this measurement was associated with an early regenerating marrow. In fact, half of the false-positive measurements were associated with actively regenerating marrows, for which the PPI is expected to be high.

It was of interest in this study to determine whether the PCC technique might be useful in measuring the effectiveness of the late intensification protocol; i.e., did late-intensification prolong remission duration in patients who were

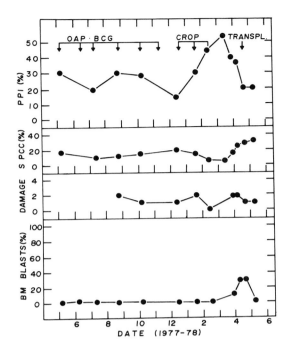

FIG. 11. PCC-derived characteristics of a patient relapsing after late-intensification CROP therapy. Note the rise in PPI value to >35% 1.8 months prior to clinical evidence of relapse.

predicted to relapse at the initiation of therapy? Two patients coincidentally exhibited high PPI values (not associated with active bone marrow regeneration) at the initiation of late-intensification therapy. Figure 12 illustrates the results obtained for one of these patients. This patient exhibited a high PPI value (65.6%) just prior to initiation of LI therapy. With CROP therapy, the PPI values dropped back below 35%, which is consistent with continued complete remission. Subsequently, however, the PPI values rose to levels greater than 35% and remained high. Clinical evidence of relapse was observed 3½ months after the second PPI rise. Since this patient relapsed 6 months after late intensification therapy was initiated, it is estimated that remission was prolonged by 2½ months (6 − 3½ months) by this regimen. The other patient who showed a high PPI value at the initiation of late-intensification therapy showed a similar pattern, and by similar calculation, remission was estimated to be prolonged by 5½ months.

These results suggest that the PCC technique can be very useful in the prediction of relapse in patients with leukemia who are clinically in complete remission. It still remains to be determined whether this technique will be useful in the prediction of response and progression in patients with other forms of leukemia. For example, we are interested in determining whether a high PPI value in a patient with CML in the benign phase is predictive of the blast crisis phase.

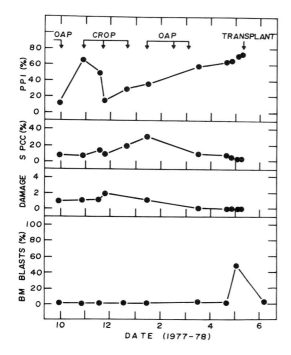

FIG. 12. PCC-derived characteristics of a patient with high PPI values upon initiation of late intensification therapy. Note the transient drop in PPI values with CROP therapy, followed by a rise in PPI values and subsequent relapse (3½ months after the second rise in PPI values).

Similarly, are high PPI values in patients with smoldering or oligoblastic leukemia predictive of progressive disease?

THE BIOLOGICAL BASIS FOR THE PCC FINDINGS IN HUMAN LEUKEMIA

The results discussed in the previous three sections raise an interesting question: in a heterogenous cell population such as the bone marrow, which cell types are giving rise to cells in late G1 phase? On presentation, patients with AML and ALL showed a wide variability in PPI values that did not correlate with the percentage of blasts in the marrow or the fraction of cells in S phase. Thus, some patients exhibited a high blast count yet a low PPI value. The converse posed a similar enigma: a high PPI value for the patient clinically in complete remission was prognostic of impending relapse. In these cases, a leukemic pattern (i.e., high PPI value) was observed in bone marrow populations where, by morphological criteria, few or no leukemia cells were present. Yet this value occurred just prior to relapse. While we have only begun to investigate the biological reasons for these apparent paradoxes, some preliminary experiments have already shed some light on the question.

Two alternative possibilities might explain the cases in which high PPI values were observed in apparently normal bone marrow populations just prior to relapse. First, the late G1 cells that give rise to high PPI values might reflect an effect of a rising leukemic burden on the normal cells, i.e., leukemic cells might cause normal bone marrow cells to be held up in late G1 phase without having an effect on normal cell differentiation. That leukemic cells affect normal hematopoiesis has been shown both in vitro and in vivo by a variety of investigators (Broxmeyer et al. 1978a,b, 1979, Knudtzen and Mortensen 1976, Chiyoda et al. 1975, 1976, Miller et al. 1977). This effect is generally observed as a block in the ability of normal cells to grow into colonies in agar and differentiate under the influence of colony-stimulating activity. No one has yet determined where in the cell cycle the normal cells are blocked by the leukemic factors, although S phase cells seem to be uniquely sensitive to extracts from leukemic cells (Broxmeyer et al. 1978b).

An alternative explanation for a high PPI value in a remission marrow just prior to relapse is that these late G1 cells may represent a growing population of leukemic cells, which in that period of remission are capable of differentiating into mature-looking cells. Thus, while these leukemic cells can mature morphologically and enzymatically, they retain the malignant characteristic of resting in late G1 phase. Studies with mouse myeloid leukemia cell lines, whereby particular leukemic clones can be induced to differentiate in vitro by treatment with a variety of agents (for review, see Sachs 1978), support this notion. A second piece of supporting data for this possibility is that bone marrow cells from many AML patients in complete remission immunologically react positively with hetero-antisera raised against AML blasts (Baker et al. 1979). Thus, even though these cells appear to be maturing normally, they exhibit a cell surface property of AML blast cells. Interestingly, these AML cell surface–positive cells appear on the average of 3–4 months prior to relapse; this correlates well with our findings that the PPI rises to high levels an average of 4 months before relapse.

To determine which of the above two biological hypotheses explain our PCC findings, we have sought to determine which cells give rise to late G1 PCC in normal and leukemic bone marrow populations both in remission and during an active leukemia state. In collaboration with the bone marrow transplantation team at M. D. Anderson Hospital (Drs. K. Dicke, A. Zander, and L. Vellekoop), bone marrow aspirations were fractionated by albumin density sedimentation, and aliquots from each of five fractions were characterized by the PCC technique.

In bone marrow fractionations obtained from normal donors and solid tumor patients with no evidence of disease in the marrow, a characteristic pattern was observed (Figure 13a). PPI values throughout the gradient were generally below 30%, and the most dense fractions containing the most differentiated cells (late myelocytes, polymorphonucleocytes, and late nucleated red cells) always showed the lowest PPI values. A similar pattern was observed in six fractionations from leukemia patients who were in definite complete remission.

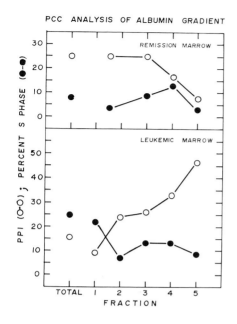

FIG. 13. PCC characteristics of albumin-fractionated bone marrow populations. **Top,** Normal bone marrow fractionation. Note that the densest fractions with the most mature cells yield the lowest PPI values. **Bottom,** Leukemic bone marrow fractionation. Note that the most dense fractions yield the highest PPI values.

PCC analysis of bone marrow fractionations obtained from patients with active leukemic disease (i.e., untreated active disease or clinically defined relapse) showed a quite different pattern from that obtained with normal marrows. In general, the PPI values were much higher than in the normal setting, and more interestingly, the densest fractions representing the most differentiated cells yielded the highest PPI values (Figure 13b). In addition, two patients in early relapse (<10% blasts) showed a similar pattern, with the most differentiated fractions yielding the highest PPI values. In some cases, the PPI values remained uniformly high throughout the gradient.

Combining the findings that (1) high PPI values predict for relapse, and (2) the high PPI arises from apparently differentiating cells suggests that for some time prior to, or at the time of, apparent active leukemic disease, the differentiating bone marrow cells rest in late G1, rather than in early G1 as they do normally. However, these results do not tell us whether the differentiated cells were derived from normal or leukemic cells. To help distinguish between these two possibilities, we asked where leukemic cells stop in the cell cycle when induced to differentiate in vitro.

A human promyelocytic leukemia cell line, HL60, has been recently established from the peripheral blood of a patient with leukemia (Collins et al. 1977). These cells have been shown to be capable of limited maturation to mature elements including myelocytes, metamyelocytes, and banded and segmented neutrophils when induced with dimethylsulfoxide (DMSO) (Collins et al. 1978). Other inducing agents can be employed, including retinoids and phorbol esters

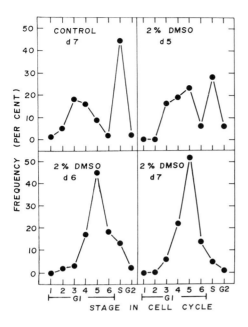

FIG. 14. PCC-derived characteristics of human promyelocytic leukemia HL-60 cells induced to mature with 2% DMSO. Note the drop in fraction of cells in S phase and the accumulation of cells in late G1 phase with time as the cells mature.

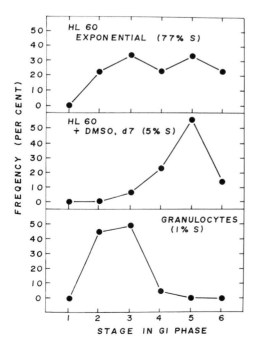

FIG. 15. PCC-derived characteristics of leukemic and normal granulocytes. **Top,** Undifferentiated HL-60 cells in exponential phase. **Center,** DMSO-induced mature HL-60 cells. **Bottom,** Normal peripheral blood granulocytes. Note the late G1 phase accumulation in differentiated leukemia cells and the early G1 phase accumulation in normal differentiated granulocytes.

(Huberman and Callaham 1979, Rovera et al. 1979a,b), and in these cases, the cells resemble monocytes. The differentiation process is accompanied by a cessation of cell growth. We treated HL60 populations with 2% DMSO in medium with 0.5% serum and analyzed the cell populations using the PCC technique at various time intervals after DMSO treatment. Exponentially growing HL60 cells were found to have up to 70% of cells in S phase at any one time, and G1 cells were distributed evenly throughout the G1 phase. As shown in Figure 14, after 2% DMSO is added to the cell populations, the fractions of cells in S phase decrease with time, and cells accumulate in late G1 phase. Thus, as these leukemic cells mature, they become arrested in late G1 phase. This result contrasts with that for normal differentiated cells, such as normal human peripheral blood polymorphonuclear cells, which rest in early G1 phase (Figure 15).

These results suggest that while leukemic cells can be induced to mature, they still retain the malignant characteristic of resting in late G1 phase. This does not, however, solve the dilemma raised earlier because it has not yet been determined where normal cells are arrested when under the influence of leukemic factors. Experiments to do so are under way.

BASIC AND CLINICAL IMPLICATIONS—FUTURE DIRECTIONS

The preceding sections have described how the PCC technique can be used to perform a cell cycle analysis of cell populations, including a means with which to map the location of cells within G1 phase. While this has proved to be useful for predicting the response of patients with leukemia, the implications of these findings are only now beginning to be discovered.

At the basic level of cell regulation, these results support the notion that normal cells can recognize a restriction point early in G1 phase while malignant cells fail to do this (Pardee 1974). More interestingly, the process of differentiation of malignant cells, e.g., of HL60 cells, does not cause an arrest in early G1 phase but in late G1 phase. Thus, malignant cells might arrest upon differentiation at one of the other growth restriction points in G1 phase as described by the experiments of Scher with fibroblasts (Scher et al. 1979). From another perspective, these results suggest that leukemic cells may differentiate in their morphology and enzyme patterns, but their chromatin retains malignant characteristics. Alternatively, in vivo, leukemia cells may influence normal hematopoietic cells to pass beyond the restriction point in G1 phase before arresting yet still retain the normal differentiation ability. Since the PCC technique can be used to detect chromatin differences between cells that look morphologically and enzymatically similar, this technique will be very useful in further studies of the regulation of cell growth and differentiation.

In order to be more useful at the clinical level, the PCC technique either needs to be simplified or alternative techniques need to be developed to determine the same parameters detected by the PCC technique. This hope is not far off. For example, where the PCC technique is useful in the prediction of relapse by detecting the presence of maturing cells arrested in late G1 phase, it may be possible to detect the same cells in the peripheral blood since the bone marrow cells, once mature, are released into the blood. The blood cells could then be analyzed by other microscopic techniques that can distinguish between early and late G1 phase. This would allow any clinical laboratory to make such predictions of relapse in leukemia.

If the PCC technique turns out to be a useful predictor for continued response or progressive disease, it might raise some new clinical considerations. For example, if the PCC technique predicts for progression of disease in a patient with leukemia, it might be helpful to change the therapeutic regimen in order to attack the resistant disease while the tumor burden is still apparently low. At the other end of the spectrum, for the patient in complete remission with continued low PPI values, it might be feasible to discontinue therapy until the PPI values begin to rise again. This latter approach might save the patient unneeded therapy and perhaps allow a period of immunological restoration.

A further interaction between the basic and clinical levels of this research might also lead to an alternative approach to the treatment of leukemia. If it turns out that a high PPI value in the face of a morphologically normal marrow reflects leukemic cell maturation, leukemic cells might be induced to differentiate in the patient with proper manipulation. This phenomenon has been demonstrated in mouse myeloid leukemia model systems (see Sachs 1978), and it has also been shown that induction of differentiation of leukemia cells in vivo can increase the survival of mice injected with differentiation-sensitive leukemia cell lines (Honma et al. 1979). Thus, a thorough study of the patients predicted to relapse by the PCC technique might allow a determination of conditions by which the patient's residual leukemic cells can be induced to mature. In addition, the PCC technique might be useful in determining which particular patients would benefit by a differentiation therapy approach.

CONCLUSION

The work described here arose out of a basic finding that chromosomes can be artificially induced to prematurely condense. When applied to human leukemia, it was found that this phenomenon might be useful in the study of malignancy, and further, that it could be utilized in the individualization of treatment for each patient. At the same time, the clinical findings provided a whole host of new basic questions to be considered. Thus, this research alternates between the basic and clinical levels of research, each level providing new questions to be asked at the other level while still providing immediately useful information in the management of each patient's disease.

ACKNOWLEDGMENTS

We thank Lowanda C. Broussard for excellent technical assistance and Jobie Carrillo for typing the manuscript. We also thank the clinical staff and bone marrow nurses of the Department of Developmental Therapeutics at M. D. Anderson Hospital for patient material and consultation. Supported in part by NIH Grants CA-14528, CA-16480, CA-11520, and CA-5831 from The National Cancer Institute and by Grant GM-23252 from The National Institute of General Medical Sciences.

REFERENCES

Baker, M. A., J. A. Falk, W. H. Carter, R. N. Taub, and the Toronto Study Group. 1979. Early diagnosis of relapse in acute myeloblastic leukemia. Serological detection of leukemia-associated antigens in human marrow. N. Engl. J. Med. 301:1353–1357.

Barlogie, B., W. Hittelman, G. Spitzer, J. Trujillo, J. Hart, L. Smallwood, and B. Drewinko. 1977. Correlation of DNA distribution abnormalities with cytogenetic findings in human adult leukemia and lymphoma. Cancer Res. 37:4400–4407.

Boyum, A. 1968. Separation of leukocytes from blood and bone marrow. Scand. J. Clin. Lab. Invest. 21(Suppl. 97):77–89.

Broxmeyer, H. E., N. Jacobsen, J. Kurland, N. Mendelsohn, and M. A. S. Moore. 1978a. In vitro suppression of normal granulocytic stem cells by inhibitory activity derived from human leukemia cells. J. Natl. Cancer Inst. 60:497–511.

Broxmeyer, H. E., E. Grossband, N. Jacobsen, and M. A. S. Moore. 1978b. Evidence for proliferative advantages of human leukemia colony-forming cells in vitro. J. Natl. Cancer Inst. 60:513–521.

Broxmeyer, H. E., E. Grossband, N. Jacobson, and M. A. S. Moore. 1979. Persistence of inhibitory factor against normal bone-marrow cells during remission of acute leukemia. N. Engl. J. Med. 301:346–351.

Chiyoda, S., H. Mizoguchi, S. Asano, F. Takaku, and Y. Miura. 1976. Influence of leukaemic cells on the colony formation of human bone marrow cells in vitro. II. Suppressive effects of leukaemic cell extracts. Br. J. Cancer 33:379–384.

Chiyoda, S., H. Mizoguchi, K. Kosaka, F. Takaku, and Y. Miura. 1975. Influence of leukaemic cells on the colony formation of human bone marrow cells in vitro Br. J. Cancer 31:355–358.

Collins, S. J., R. C. Gallo, and R. E. Gallagher. 1977. Continuous growth and differentiation of human myeloid leukaemic cells in suspension culture. Nature 270:347–349.

Collins, S. J., F. W. Ruscetti, R. E. Gallagher, and R. C. Gallo. 1978. Terminal differentiation of human promyelocytic leukemia cells induced by dimethyl sulfoxide and other polar compounds. Proc. Natl. Acad. Sci. USA 75:2458–2462.

Grdina, D., W. N. Hittelman, R. A. White, and M. L. Meistrich. 1977. Relevance of density, size, and DNA content of tumour cells to the lung colony assay. Br. J. Cancer 36:659–669.

Hittelman, W. N., L. C. Broussard, G. Dosik, and K. McCredie. 1980a. Predicting relapse of human leukemia by means of premature chromosome condensation. N. Engl. J. Med. 303:479–484.

Hittelman, W. N., L. C. Broussard, and K. McCredie. 1979. Premature chromosome condensation studies in human leukemia. I. Pretreatment characteristics. Blood 54:1001–1014.

Hittelman, W. N., L. C. Broussard, K. McCredie, and S. G. Murphy. 1980b. Premature chromosome condensation studies in human leukemia: II. Proliferative potential changes after induction therapy for AML patients. Blood 55:457–465.

Hittelman, W. N., and P. N. Rao. 1974a. Visualization of X-ray induced chromosome damage in interphase cells. Mutat. Res. 23:251–258.

Hittelman, W. N., and P. N. Rao. 1974b. Premature chromosome condensation. II. The nature of chromosome gaps produced by alkylating agents and ultraviolet light. Mutat. Res. 23:259–266.

Hittelman, W. N., and P. N. Rao. 1974c. Bleomycin-induced damage in prematurely condensed chromosomes and its relationship to cell cycle progression in CHO cells. Cancer Res. 34:3433–3439.

Hittelman, W. N., and P. N. Rao. 1975. The nature of Adriamycin-induced cytotoxicity in Chinese hamster cells as revealed by premature chromosome condensation. Cancer Res. 35:3027–3035.

Hittelman, W. N., and P. N. Rao. 1976. Premature chromosome condensation: Conformational changes of chromatin associated with phytohemagglutinin-stimulation of peripheral lymphocytes. Exp. Cell Res. 100:219–222.

Hittelman, W. N., and P. N. Rao. 1978a. Mapping G1 phase by the structural morphology of the prematurely condensed chromosomes. J. Cell Physiol. 95:333–341.

Hittelman, W. N., and P. N. Rao. 1978b. Predicting response or progression of human leukemia by premature chromosome condensation of bone marrow cells. Cancer Res. 38:416–423.

Hittelman, W. N., M. A. Sognier, and A. Cole. 1980c. Direct measurement of chromosome damage and its repair by premature chromosome condensation, *in* Radiation Biology in Cancer Research (The University of Texas System Cancer Center 32nd Annual Symposium on Fundamental Cancer Research) R. E. Meyn and H. R. Withers, eds. Raven Press, New York, pp. 103–123.

Honma, Y., T. Kasukabe, J. Okabe, and M. Hozumi. 1979. Prolongation of survival time of mice inoculated with myeloid leukemia cells by inducers of normal differentiation. Cancer Res. 39:3167–3171.

Huberman, E., and M. F. Callaham. 1979. Induction of terminal differentiation in human promyelocytic leukemia cells by tumor-promoting agents. Proc. Natl. Acad. Sci. USA 76:1293–1297.

Johnson, R. T., and P. N. Rao. 1970. Mammalian cell fusion: Induction of premature chromosome condensation in interphase nuclei. Nature 226:717–722.

Kato, H., and A. A. Sandberg. 1968a. Chromosome pulverization in human cells with micronuclei. J. Natl. Cancer Inst. 40:165–179.

Kato, H., and A. A. Sandberg. 1968b. Chromosome pulverization in Chinese hamster cells induced by Sendai virus. J. Natl. Cancer Inst. 41:1117–1123.

Knudtzen, S., and B. T. Mortensen. 1976. Interaction between normal and leukaemic human cells in agar culture. Scand. J. Haematol. 17:369–378.

Matsui, S., H. Yoshida, H. Weinfeld, and A. A. Sandberg. 1972. Induction of prophase in interphase nuclei by fusion with metaphase cells. J. Cell Biol. 54:120–132.

Mazia, D. 1963. Synthetic activities leading to mitosis. J. Cell Comp. Physiol. 62(Suppl.):123–140.

Miller, A., P. Page, B. Hartwell, and S. Robinson. 1977. Inhibition of growth of normal, murine granulocytes by cocultured acute leukemic cells. Blood 50:799–809.

Nicolini, C., K. A. Ajiro, T. W. Borun, and R. Baserga. 1975. Chromatin changes during the cell cycle of HeLa cells. J. Biol. Chem. 250:3381–3385.

Pardee, A. B. 1974. A restriction point for control of normal animal cell proliferation. Proc. Natl. Acad. Sci. USA 71:1286–1290

Pederson, J. 1972. Chromatin structure and the cell cycle. Proc. Natl. Acad. Sci. USA 69:2224–2228.

Pederson, T., and E. Robbins. 1972. Chromatin structure and the cell division cycle. J. Cell Biol. 55:322–327.

Rao, A. P., and P. N. Rao. 1976. The cause of G2-arrest in Chinese hamster ovary cells treated with anticancer drugs. J. Natl. Cancer Inst. 57:1139–1143.

Rao, P. N., P. S. Sunkara, and D. A. Wright. 1981. Chromosome condensation factors of mammalian cells, *in* Genes, Chromosomes, and Neoplasia (The University of Texas System Cancer Center 33rd Annual Symposium on Fundamental Cancer Research) F. E. Arrighi, E. Stubblefield, and P. N. Rao, eds. Raven Press, New York, pp. 49–60.

Rao, P. N., B. Wilson, and T. Puck. 1977. Premature chromosome condensation and cell cycle analysis. J. Cell Physiol. 91:131–142.

Rovera, G., T. G. O'Brien, and L. Diamond. 1979a. Induction of differentiation in human promyelocytic leukemia cells by tumor promoters. Science 204:868–870.

Rovera, G., D. Santoli, and C. Damsky. 1979b. Human promyelocytic leukemia cells in culture differentiate into macrophage-like cells when treated with a phorbol ester. Proc. Natl. Acad. Sci. USA 76:2779–2783.

Sachs, L. 1978. Control of normal cell differentiation and the phenotypic reversion of malignancy in myeloid leukaemia. Nature 274:535–539.

Sandberg, A. A., T. Aya, T. Ikeuchi, and H. Weinfeld. 1970. Definition and morphologic features of chromosome pulverization: A hypothesis to explain the phenomenon. J. Natl. Cancer Inst. 45:615–623.

Scher, C. P., R. C. Shepard, H. N. Antoniades, and C. D. Stiles. 1979. Platelet-derived growth factor and the regulation of the mammalian fibroblast cell cycle. Biochim. Biophys. Acta 560:217–241.

Schor, S. L., R. T. Johnson, and C. A. Waldren. 1975. Changes in the organization of chromosomes during the cell cycle: Response to ultraviolet light. J. Cell Sci. 17:539–565.

Sognier, M. A., W. N. Hittelman, and P. N. Rao. 1979. Effect of DNA repair inhibitors on the induction and repair of bleomycin-induced chromosome damage. Mutat. Res. 60:61–72.

Takagi, N., T. Aya, H. Kato, and A. A. Sandberg. 1969. Relation of virus-induced cell fusion and chromosome pulverization to mitotic events. J. Natl. Cancer Inst. 43:335–347.

Waldren, C. A., and R. T. Johnson. 1974. Analysis of interphase chromosome damage by means of premature chromosome condensation after X- and ultraviolet irradiation. Proc. Natl. Acad. Sci. USA 71:1137–1141.

Genes, Chromosomes, and Neoplasia, edited by
Frances E. Arrighi, Potu N. Rao, and Elton Stubblefield.
Raven Press, New York © 1981.

Nonrandom Chromosome Abnormalities in Transformed Syrian Hamster Cell Lines

S. Pathak, T. C. Hsu, J. J. Trentin,* J. S. Butel,† and B. Panigrahy‡

*Department of Cell Biology, The University of Texas System Cancer Center, M. D. Anderson Hospital and Tumor Institute, *Division of Experimental Biology and †Department of Virology and Epidemiology, Baylor College of Medicine, Houston, Texas 77030; ‡Department of Veterinary Microbiology, College of Veterinary Medicine, Texas A & M University, College Station, Texas 77840*

The phenomenon of cell transformation in which normal cells can be converted to malignant cells by chemicals, viruses, and physical agents has largely aided in understanding the complex process of malignancy. Transformation, in vitro as well as in vivo, of several rodent cell lines, including Syrian and Chinese hamsters, mouse, rat, and guinea pig, has been well documented. The concept that neoplastic transformation of clones arises from chromosomally unbalanced cell or cells dated from the time of Boveri (1912). A large body of information on chromosome constitutions of both spontaneous and experimentally induced tumors has subsequently been accumulated.

The development of chromosomal banding techniques has facilitated the recognition of individual chromosome or chromosome segments and thereby provides suitable opportunity to examine in detail the Boveri hypothesis. A number of investigators have examined the chromosomal characteristics of chemically and virally transformed Syrian hamster embryo cells. Using a Giemsa banding procedure and cloned cell populations, Yamamoto et al. (1973) and Bloch-Shtacher and Sachs (1976) demonstrated specific chromosome involvement with the expression and suppression of malignancy in transformed Syrian hamster cells. These observations have been further supported by the results obtained on Ara-C–transformed Syrian hamster cells (Benedict et al. 1975). However, DiPaolo and co-workers have reported karyological instability and random chromosome involvement in neoplastic transformation of Syrian hamster embryo cells (DiPaolo et al. 1973, Popescu et al. 1974, DiPaolo 1975). Most recently, Kessous et al. (1979) demonstrated trisomy of an acrocentric chromosome in two clones of Syrian hamster cells transformed by herpes simplex virus type 2.

In the present communication we describe the results of karyological analyses of virally transformed Syrian hamster cell lines and tumor-derived cultures. Our data demonstrate the nonrandom involvement of a specific Syrian hamster chromosome in the majority of these transformed lines and tumor-derived cultures.

MATERIALS AND METHODS

Tissue Culture and Cell Transformation

Syrian hamster *(Mesocricetus auratus)* cells were transformed in vivo with simian adeno 7 and chicken embryo lethal orphan (CELO) viruses and in vitro with simian adeno 7, CELO, herpes simplex, and simian papova viruses following the techniques described elsewhere (Panigrahy et al. 1976, Asch et al. 1979, Butel et al. 1975). The tumors were induced in vivo in the newborn baby hamsters, less than 24 hours of age, by inoculating viruses intraperitoneally or subcutaneously following the procedure of Larson et al. (1965) and Trentin et al. (1968). Characteristics are summarized in Table 1.

Chromosome Preparation

Transformed cell lines and tumor-derived cultures were grown in T-60 flasks in Eagle's or Dulbecco's modified medium supplemented with 10% fetal bovine serum. Metaphase cells were accumulated by treating cultures with Colcemid (0.04 μg/ml) for 3–4 hours during the exponential growth phase. Cells were dislodged with a rubber policeman, treated with sodium citrate solution (1.0%)

TABLE 1. *Characteristics of the transformed Syrian hamster cell lines and tumor-derived cultures*

Identification no.	Passage no.	Transforming agent	Modal chromosome no.	Monosomy C15
A	22	SA7	44	+
B	22	SA7	42	+
I-A	22	SA7	42	+
III-A	22	SA7	43	+
TU-26	5	SA7	40	+
HEL1	32	HSV-2	45	+
HEL2	5	HSV-2	46	+
HEL4	10	HSV-2	45	+
J19L	25	SV40	53	−
Ha4A28	75	SV40	44	+
IV	42	CEL0	42	+
IX	42	CEL0	44	+
X	21	CEL0	56	−
VI	8	spontaneous	43	−
VII	N.A.	MSV	41	+
XI	N.A.	MSV	78	−
T-71-30	5	AD7	50	+
AD-7	N.A.	AD7	48	+
AD-7 (TU)	N.A.	AD7	46	+

N.A. = Information not available; for other abbreviations, see text.
+ = Monosomy C15 present
− = Monosomy C15 not present

for 15–20 minutes at room temperature, fixed in aceto-methanol (1:3), and finally air-dried on acetone-cleaned slides. Some slides were stained directly with Giemsa while others were used for C- and G-bandings.

A minimum of 50 metaphases from conventionally stained preparations were counted to determine the stem line number. Individual chromosomes were identified in 10 to 15 G-banded metaphase spreads. Five additional G-banded metaphase cells were photographed to prepare the karyotypes. The chromosomes were arranged following the nomenclature system published for Syrian hamster karyotype (Popescu and DiPaolo 1972) with minor modifications. The entire hamster karyotype was divided into A, B, C, and D groups only. Chromosomes with minor alterations in their morphology were placed by the side of the normal homologs. Those with complex rearrangements or unidentified ones were placed on the left hand side of each karyotype. In certain cases, only partial karyotypes containing monosomy, trisomy, tetrasomy, and other altered chromosome morphologies were presented. The sex chromosomes (XX and XY), whenever identified positively, were marked and also placed on the left hand side of each karyotype.

RESULTS

Simian Adenovirus (SA7)–Transformed Cell Lines

Five cell lines were analyzed for their chromosomal characteristics. Line A, transformed in vitro, had a stem line chromosome number of 44. The proportion between 1s and 2s elements was approximately 92:8 in 100 metaphases counted. The G-banded karyotype (Figure 1a) revealed that one A2 chromosome had a terminal deletion in the long arm, and a small submetacentric chromosome, with darkly stained long arm, was present. The morphology of this chromosome was very similar to that of C15. However, C-banding showed that the long arm of this chromosome was made of constitutive heterochromatin (Figure 1a, inset). This chromosome, therefore, is not C15, and C15 is represented by only one chromosome. Other autosomes, i.e., B9 and C11, showed heteromorphism between the short arms of their homologs.

Line B was transformed in vivo. The ratio of 1s to 2s in our sample was approximately 62:38 in 53 metaphases counted. The karyotype contained monosomy of B7, C13, and C15 (Figure 1b). A new, unidentified chromosome was placed on the left side of the karyotype. Both lines A and B were derived from male hamsters as indicated by the presence of XY sex chromosomes.

Line III-A, transformed in vitro, was apparently derived from a female hamster cell (Figure 2a). In many cells three to five altered chromosomes were present. Chromosomes B7, C15, and D16 were present in monosomic form. Two extra chromosomes (on the left side of the karyotype) resembled C15 at the distal end of the long arm but not the remaining segment. Such altered chromosomes were not present in other cells, but monosomy C15 was evident in all cases.

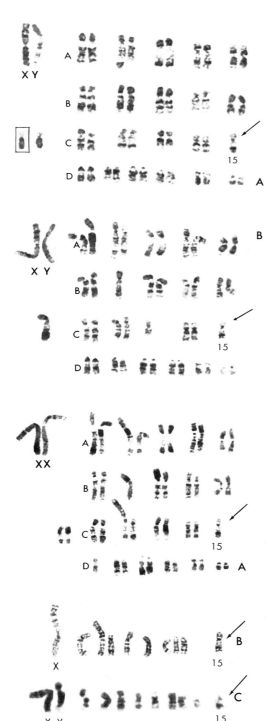

FIG. 1. Giemsa-banded karyotypes of SA7-transformed male Syrian hamster cell lines. a. Line A. Note the presence of C15 in single copy (arrow). An unidentified small submetacentric chromosome is arranged on the left side in group C. The same chromosome (from another cell) has a C-band positive long arm (inset). One homolog of pair A2 has deletion in the long arm. b. Line B. C15 is present in monosomic form (arrow). B7 and C13 chromosomes are also monosomic. An unidentified chromosome is placed on the left side of the karyotype.

FIG. 2. G-banded complete and partial karyotypes of SA7-transformed hamster cell lines. a. Line III-A, derived from a female hamster. Chromosomes B7, C15, D16 are present in single copy. Two extra chromosomes (arranged on extreme left side in row C) have some resemblance to C15 in their distal long arm segments. One homolog of each pair A2 and C12 has added material in the short arm. b. Partial karyotype of I-A. Note the presence of a single C15 chromosome. Only one sex chromosome (X) could be identified. c. Partial karyotype of a tumor clone TU-26. Note the presence of a single copy of C15, XY, and other altered chromosomes.

Cell line I-A was transformed in vivo. Partial karyotypes from lines I-A and TU-26 are shown in Figures 2b and 2c, respectively. Several morphologically altered, monosomic, and trisomic chromosomes are present. Monosomy of C15 is evident in both clones. Only one X chromosome was present in line I-A, whereas in TU-26 both XY chromosomes were observed.

Herpes Simplex Virus-Transformed Clones

These clones belong to the human embryonic lung (HEL) series. In vitro transformed HEL cells were injected into the newborn hamsters. HEL #1, #2, and #4 are cultures derived from primary hamster tumors that developed in the hamsters following inoculation of transformed HEL cells. These hamster tumor cells contained HSV antigens as detected by indirect immunofluorescence. HEL #1 has a stem line chromosome number of 45. A typical G-banded karyotype is shown in Figure 3a. Chromosomes A3, C15, and D18 were present in monosomy, whereas B8 was trisomic. The presence of XY chromosomes indicated a male origin for the tumor. Three unidentified chromosomes were placed on the left side of the karyotype. Monosomy of C15 is indicated by an arrow (Figure 3a). HEL #2 and HEL #4 had stem line numbers of 46 and 45 chromosomes, respectively. Partial karyotypes of HEL #2 (Figure 3b) and HEL #4 (Figure 3c indicated the presence of XY chromosomes, thus confirming the origin of these tumors as male hamsters. In addition to the monosomy

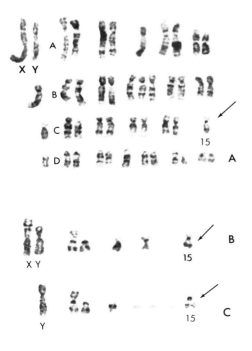

FIG. 3. Giemsa-banded karyotypes of herpes simplex virus type 2–induced hamster tumors. a. HEL #1. Note the presence of three altered chromosomes arranged on extreme left hand side of rows B, C, and D. Presence of XY sex chromosomes, monosomy of A3, C15, D18, and trisomy of B8 is evident. b. Partial karyotype of HEL #2. Note the presence of three altered chromosomes, XY sex chromosomes, and monosomy of C15. c. Partial karoytype of HEL #4. Note the presence of two or three minute chromosomes and monosomy of C15 (arrow). The X chromosome is apparently missing in this cell.

of C15, several other marker chromosomes were present. Two to three minute chromosomes were observed in HEL #4 cells. Other altered chromosomes were randomly distributed between cells.

Simian Papovavirus (SV40)–Transformed Clones

Clone J19L is derived from a tumor induced by hamster cells transformed in vitro. A typical G-banded karyotype is shown in Figure 4a. The stem line

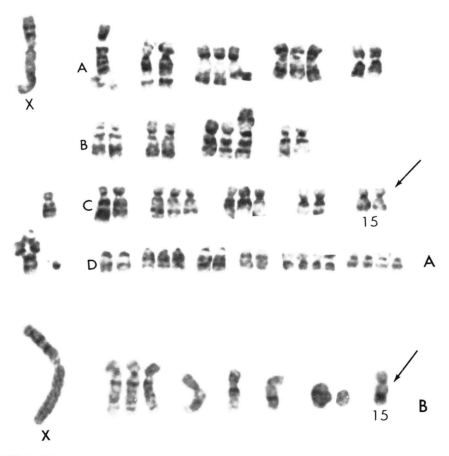

FIG. 4. G-banded karyotypes of simian papovavirus (SV40)–induced tumors in baby hamsters. a. J19L clone. Autosomes A3, A4, B8, C12, C13, and D17 are present in trisomic, A1 in monosomic, D20 and D21 in tetrasomic forms. One homolog in B8 has an enlarged short arm and B10 is apparently missing. A single X and three altered chromosomes are arranged on the left side of the karyotype. Note the presence of exchange figures in the karyotype. b. Partial karyotype of a Ha4A28 cell. Trisomy of B9, monosomy of X, C13, and C15 are evident. Note the presence of two altered and two variable-size ring chromosomes.

number of the majority of metaphases was in the hyperdiploid range, and the chromosomes were present in monosomic, trisomic, and tetrasomic forms. B10 was apparently missing in a good number of cells. Minute chromosomes and exchange figures were not uncommon. Only one X chromosome could be identified positively. Chromosome C15 was present in two copies (Figure 4a, arrow).

Clone Ha4A28 was transformed in vitro. A partial karyotype of this clone is shown in Figure 4b. The stem line chromosome number was 44, but many altered chromosomes were present. Ring chromosomes were also noticed in some metaphase spreads. A large majority of chromosomes were present in monosomic and disomic forms, whereas others were in the trisomic and tetrasomic state. Chromosome C15 was present in monosomic form. Again in this clone only one X chromosome could be identified. These cells were maintained at 33°C because of their temperature-sensitive nature.

Chicken Embryo Lethal Orphan (CELO) Virus-Transformed Cell Lines

The line X is from a hepatocellular carcinoma induced in vivo. In the majority of metaphases the chromosome numbers were within the hyperdiploid range, with a peak at 56. A typical G-banded karyotype is shown in Figure 5a. Many autosomes were present in tetrasomic and others in monosomic and disomic forms. In the present cell (Figure 5a) chromosome B9 was apparently missing. Several altered chromosomes were noted in the majority of spreads. Chromosome C15 was present in disomic form. The sex chromosomes were so much altered that it was difficult to identify them in some cells. In only a few metaphases it was possible to identify two X chromosomes.

The IV tumor cell line was obtained from a tumor induced in vitro. The majority of metaphases had pseudodiploid chromosome numbers with a peak at 42. A partial G-banded karyotype is shown in Figure 5b. Many chromosomes were present in monosomic and tetrasomic forms. Chromosome C15 was again present in a single copy. The presence of two X chromosomes indicated its female origin.

The line IX was obtained from hamster embryo cells transformed by a large plaque-forming mutant of CELO virus. The stem line chromosome number was determined to be 44. A partial G-banded karyotype is shown in Figure 5c. Three to four altered autosomes were present per metaphase plate. Some autosomes were present in monosomic and others in tetrasomic forms. Chromosome C15 was present in monosomic form. The presence of XY chromosomes indicated its male origin.

Tumor Lines Induced by RNA Tumor Viruses

The cell line XI was derived from a tumor induced in vivo by murine sarcoma virus. These hamster tumor cells produce sarcoma C-type viruses. The stem line chromosome numbers varied between 62 and 82 with a peak at 78. Figure

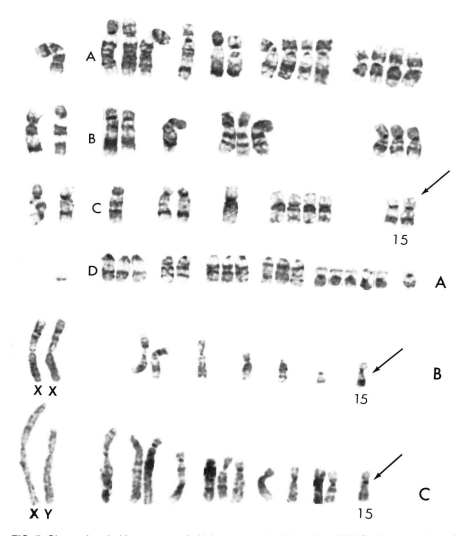

FIG. 5. Giemsa-banded karyotypes of chicken embryo lethal orphan (CELO) virus-transformed hamster cells. a. Line X. Note the presence of autosomes in variable copies. B9 and sex chromosomes are apparently missing. Six altered chromosomes are placed on left side of the karyotype. C15 (arrow) is present in two copies. b. Partial karyotype of line IV. Note the presence of two X chromosomes, monosomy of some autosomes including C15, and two altered chromosomes. c. Partial karyotype of IX. Note the presence of altered chromosomes and variable copies of normal chromosomes. Chromosome C15 is present in single copy (arrow).

6a shows a G-banded karyotype of line XI. Most of the autosomes were present in four copies, but A1, B10, C11, C13, C14, and C15 were present in disomic form. Six or seven altered chromosomes were present in the majority of cells analyzed. Apparently this tumor was induced in a male hamster because the

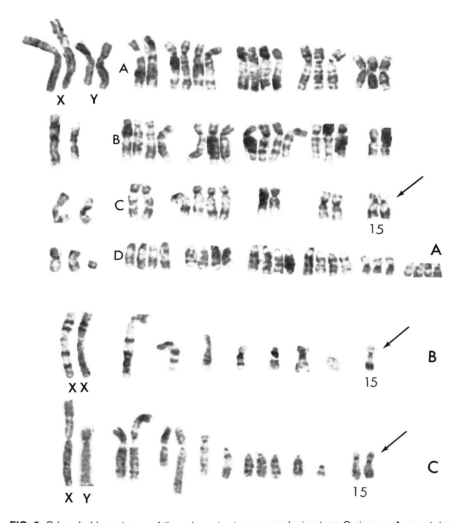

FIG. 6. G-banded karyotypes of three hamster tumors producing type-C virus. a. A near tetra-ploid cell from XI. Two X, two Y, and seven altered chromosomes are evident. Only A1, B10, C11, C13, C14, and C15 are present in disomic forms. b. Partial karyotype of VII. Note the presence of two X chromosomes, five altered chromosomes, and single copies of B8, B10, and C15 autosomes. The long arm of one of the X chromosomes (left) appeared to have a translocation. c. Partial karyotype of VI. The XY sex chromosomes and several altered autosomes are present. Some D group autosomes are present in variable copies. C15 is present in disomic form (arrow).

sex chromosomes were of XY type. There were two X and two Y chromosomes, indicating that it was a tetraploid cell (Figure 6a).

Tumor VII was induced by murine sarcoma virus. The chromosome numbers in the majority of metaphases were within the hypodiploid range, with a peak at 41 chromosomes. A G-banded partial karyotype is shown in Figure 6b. Two or three large, altered chromosomes were noticed in each cell. The majority of the autosomes were present in disomic form. Small submetacentric and acrocentric chromosomes were also present in some cells. Monosomic state of C15 was observed in all cells karyotyped. The presence of two X chromosomes indicated the female origin of this tumor.

Spontaneous Lymphoma

Tumor line VI was established from a lymphoma which developed spontaneously in an old hamster. This tumor produces abundant type C virus particles. The chromosome numbers in different metaphases ranged from 40 to 56 with a peak at 43 chromosomes. A G-banded partial karyotype is shown in Figure 6c. Two to three new biarmed chromosomes were noticed in some cells. Chromosome C15 was present in disomic form in all cells karyotyped. Apparently this tumor arose in a male hamster cell because it exhibited X and Y chromosomes.

Adenovirus Type 7–Induced Tumor Cell Lines

The cell line T-71-30 was obtained from a tumor induced in vivo by AD7. In the majority of metaphases the chromosome distribution ranged between 45 and 56 with a peak at 50 chromosomes. A typical G-banded karyotype is shown in Figure 7a. Three to four large biarmed altered chromosomes were present per metaphase plate. Only a few autosomes were present in trisomic and tetrasomic forms. In the cell shown in Figure 7a, chromosome B10 was apparently missing. The most consistent chromosomal characteristic was the presence of C15 chromosome in monosomic form. The presence of two X chromosomes indicated the female origin of this tumor.

The second tumor designated as AD-7 has a modal chromosome number of 48. A G-banded partial karyotype is shown in Figure 7b. The altered chromosomes varied in number from cell to cell. Monosomy of C15 was again the most characteristic feature of this tumor karyotype. In some cells the heterochromatic long arm of the X chromosome was totally missing. A medium-sized acrocentric chromosome present in such metaphases is the euchromatic arm of the X because it has two characteristic bands of the mammalian X chromosome (Pathak and Stock 1974).

The third tumor line designated as AD-7 (TU) had a modal chromosome number of 46. A G-banded partial karyotype is shown in Figure 7c. Four to five altered chromosomes were present in each metaphase plate examined. Chro-

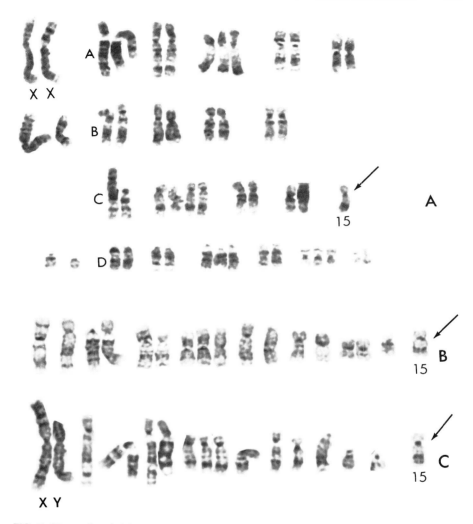

FIG. 7. Giemsa-banded karyotypes of type 7 adeno virus–induced hamster tumors. a. Tumor T-71-30. Note the absence of B10, monosomy of C15, and trisomy of some autosomes. Chromosome C12 is present in four copies. Two X chromosomes and four unidentified elements are placed on left side of the karyotype. One C11 homolog has translocation in the short arm. b. Partial karyotype of AD-7. Note the presence of structural and numerical anomalies in many chromosomes. A large-sized acrocentric represents the euchromatic short arm of the X chromosome. Chromosome C15 is present in single copy. c. Partial karyotype of AD-7 (TU) line. Note the presence of XY sex chromosomes, altered autosomes, and variable copies of some chromosomes. C15 is present in single copy (arrow).

mosome C15 was present in monosomic form (Figure 7c, arrow). The presence of the X and Y chromosomes indicated the male origin of this tumor.

DISCUSSION

Specific chromosome changes have been discovered in several types of human and murine neoplasms. The change may be numerical, structural, or both. In T cell leukemias of the mouse, both spontaneous and induced, trisomy 15 is the dominant alteration. In the well-known human chronic myelogenous leukemias, the distal segment of a chromosome 22 is translocated to another chromosome (Berghe et al. 1978). In human retinoblastomas, an interstitial deletion in chromosome 13 appears to be the specific anomaly (Wilson et al. 1973), and in human meningiomas, a partial or complete monosomy of chromosome 22 has been discovered (Zankl and Zang 1972). In Burkitt's lymphomas, human chromosome 8 is involved in translocation to chromosome 14 and other chromosomes (Zech et al. 1976).

It appears that genetic alterations of a target cell are responsible for neoplastic transformation and in many cases such genetic alterations are cytologically demonstrable. This concept has been supported by the recent work of Ohno et al. (1979) on the chromosomes of pristane-induced murine plasmacytomas. Instead of trisomy 15 (T cell leukemias), the tumor cells exhibited translocations between 15:6 and 15:12. Genetic data of the mouse indicated that the immunoglobulin genes of the mouse are located in chromosomes 6 and 12. The translocations may have affected the Ig genes, the principal function of plasma cells. Perhaps the finding of a No. 14 translocation in human plasmacytomas (Rowley 1974) can be explained in a similar manner, since chromosome 14 has been shown to contain the heavy-chain gene for immunoglobulin (Croce et al. 1979).

Unfortunately, cytogenetic analyses of many tumors are confronted with an array of chromosomal aberrations. If a primary change did occur in the target cell, numerous additional chromosome changes as the result of cancer progression may mask the original lesion. Some of these additional changes may be nonrandom and others may be random. Because of such complications in the chromosomal constitutions of tumors in vivo, cell transformation in vitro may offer a better system for detecting the primary change(s), since the clones can be analyzed as soon as a sufficient supply of cells becomes available.

Our data on the Syrian hamster cells transformed by a variety of oncogenic viruses appear to suggest that monosomy C15 is the primary chromosome aberration. DiPaolo and his associates (DiPaolo et al. 1973, DiPaolo and Popescu 1976, Popescu et al. 1974) reported monosomy of C15 in Syrian hamster cells transformed both by virus and by chemical carcinogens. However, these investigators did not attach much significance to monosomy C15 because not all the transformed clones and not all cells within a clone exhibited this property. In our opinion, the preponderance of monosomy C15 in transformed Syrian hamster cell lines and tumors, from data of our group and DiPaolo's group, is too great to be a random cytological anomaly. In a number of human and murine neoplasms, the specific anomalies (e.g., 8:14 translocation in Burkitt's lymphoma and the trisomy 15 of mouse T cell leukemias) are not present in 100% of the cells examined. The missing C15 of the Syrian hamster may be either lost or broken up in the process of translocation. The genes of that chromosome, being altered, may be responsible for the transformation.

ACKNOWLEDGMENT

This investigation was a joint project of The University of Texas System Cancer Center, M. D. Anderson Hospital and Tumor Institute, and the John S. Dunn Research Foundation of Houston, Texas and Baylor College of Medicine. It was also supported in part by grants CA-16672, CA-22555, CA-3367, and K 06-14219 from the National Cancer Institute.

REFERENCES

Asch, B. B., K. J. McCormick, and J. J. Trentin. 1979. Unusual features of the oncogenicity of chicken embryo lethal orphan (CELO) virus in hamsters, in Progress in Experimental Tumor Research, Vol. 23, J. J. Trentin and F. Homburger (eds.). S. Karger, Basel, pp. 58–88.

Benedict, W., N. Rucker, C. Mark, and R. Kouri. 1975. Correlation between balance of specific chromosomes and expression of malignancy in hamster cells. J. Natl. Cancer Inst. 54:157–162.

Berghe, H. Van Den, G. H. Borgström, L. Brandt, D. Catovsky, A. DeLa Chapelle, H. Golomb, D. K. Hossfeld, S. Lawler, J. Lindsten, A. Louwagie, F. Mitelman, B. R. Reeves, J. D. Rowley, A. A. Sandberg, L. Teerenhovi, and P. Vuopio. 1978. Chromosomes in Ph¹-positive chronic granulocytic leukemia. Br. J. Haematol. 39:305–309.

Bloch-Shtacher, N., and L. Sachs. 1976. Chromosome balance and the control of malignancy and cell transformation. J. Cell Physiol. 87:89–100.

Boveri, T. 1912. Beitrag zum studium des chromatins in den epithelzellen der carcinome. Beitr Pathol. 14:249.

Butel, J. S., M. Tálas, J. Ugur, and J. L. Melnick. 1975. Demonstration of infectious DNA in transformed cells. III. Correlation of detection of infectious DNA-protein complexes with persistence of virus in simian adenovirus SA7-induced tumor cells. Intervirology 5:43–56.

Croce, C. M., M. Shander, J. Martinis, L. Cicurel, G. G. D'Ancona, T. W. Dolby, and H. Koprowski. 1979. Chromosomal location of the genes for immunoglobulin heavy chains. Proc. Natl. Acad. Sci. USA 76:3416–3419.

DiPaolo, J. A. 1975. Karyological instability of neoplastic somatic cells. In Vitro 11:89–96.

DiPaolo, J. A., and N. C. Popescu, 1976. Relationship of chromosome changes to neoplastic cell transformation. Am. J. Pathol. 85:709–738.

DiPaolo, J. A., N. C. Popescu, and R. L. Nelson. 1973. Chromosomal banding patterns and in vitro transformation of Syrian hamster cells. Cancer Res. 33:3250–3258.

Kessous, A., V. Bibor-Hardy, M. Suh, and R. Simard. 1979. Analysis of chromosomes, nucleic acids, and polypeptides in hamster cells transformed by herpes simplex virus type 2. Cancer Res. 39:3225–3234.

Larson, V. M., A. J. Girardi, M. R. Hilleman, and R. E. Zwickey. 1965. Studies of oncogenicity of adenovirus type 7 viruses in hamsters. Proc. Soc. Exp. Biol. Med. 118:15–24.

Ohno, S., M. Babonits, F. Wiener, J. Spira, G. Klein, and M. Potter. 1979. Nonrandom chromosome changes involving the Ig gene-carrying chromosomes 12 and 6 in Pristane-induced mouse plasmacytomas. Cell 18:1001–1007.

Panigrahy, B., K. J. McCormick, and J. J. Trentin. 1976. In vitro transformation of rodent cells by simian adenovirus 7 and bovine adenovirus type 3 (strain WBR-1). Am. J. Vet. Res. 37:1503–1504.

Pathak, S. and A. D. Stock. 1974. The X chromosomes of mammals: Karyological homology as revealed by banding techniques. Genetics 78:703–714.

Popescu, N. C., and J. A. DiPaolo. 1972. Identification of Syrian hamster chromosomes by acetic-saline-Giemsa (ASG) and trypsin techniques. Cytogenetics 11:500–507.

Popescu, N. C., C. D. Olinici, B. C. Casto, and J. A. DiPaolo. 1974. Random chromosome changes following SA7 transformation of Syrian hamster cells. Int. J. Cancer 14:461–472.

Rowley, J. 1974. Do human tumors show a chromosome pattern specific for each etiologic agent? J. Natl. Cancer Inst. 52:315–320.

Trentin, J. J., G. L. VanHoosier, and L. Samper. 1968. The oncogenicity of human adenoviruses in hamsters. Proc. Soc. Exp. Biol. Med. 127:683–687.

Wilson, M. G., J. W. Towner, and A. Fujimoto. 1973. Retinoblastoma and D-chromosome deletions. Am. J. Hum. Genet. 25:57–61.

Yamamoto, T., Z. Rabinowitz, and L. Sachs. 1973. Identification of the chromosomes that control malignancy. Nature (New Biol.) 243:247–250.

Zankl, H., and K. D. Zang. 1972. Cytological and cytogenetical studies on brain tumors. IV. Identification of the missing G chromosome in human meningiomas as No. 22 by fluorescence technique. Humangenetik 14:167–169.

Zech, L., U. Haglund, K. Nilsson, and G. Klein. 1976. Characteristic chromosomal abnormalities in biopsies and lymphoid-cell lines from patients with Burkitt and non-Burkitt lymphomas. Int. J. Cancer 17:47–56.

Genes, Chromosomes, and Neoplasia, edited by
Frances E. Arrighi, Potu N. Rao, and Elton Stubblefield.
Raven Press, New York © 1981.

Relationships between Chromosome Complement and Cellular DNA Content in Tumorigenic Cell Populations

Larry L. Deaven,* L. Scott Cram, Robert S. Wells, and Paul M. Kraemer

Experimental Pathology Group, Los Alamos Scientific Laboratory, University of California, Los Alamos, New Mexico 87545

Elucidation of the relationship between chromosome changes and malignancy has been one of the most challenging areas of biological research for the past 60 years. During this period of time, attention has been focused on two major aspects of this problem: (1) studies of the chromosomal changes at the first stages of malignant transformation and (2) studies of the chromosomal rearrangements characteristic of highly evolved tumorigenic cell populations. Chromosome changes have been found in association with both of these stages of malignancy, but it has been difficult to determine whether the changes are primary or secondary to the neoplastic state. A major reason for this difficulty has been the inadequacy of chromosome resolution and the inefficiency of conventional methods. New developments in chromosome band staining and in flow cytometry have improved the sensitivity of detection of chromosome alterations and may permit a critical evaluation of Boveri's (1914) cancer hypothesis. In this manuscript, we will discuss the application of these techniques to the analysis of cellular DNA–chromosome relationships in long-term cell lines and in tumors grown in nude mice.

LONG-TERM CELL LINES

Although the Boveri hypothesis that chromosome imbalance may lead to malignancy was first proposed in 1914, it was not until the 1950s that advances were made that could lead to its proof or disproof. During this period, ascites tumor systems were developed (Klein 1951), methods for chromosome preparation were improved (Hsu 1952), and a number of cultured cell lines were established (Gey et al. 1952, Chang 1954, Eagle 1955). Chromosome analysis of

---

* Present Address: Health Effects Research Division, Office of Health and Environmental Research, Office of Environment, United States Department of Energy, Washington, D.C. 20545

these in vitro cell populations led to some unexpected results. Cancer cytologists had known for some time that tumor cells contained abnormal numbers of chromosomes; however, many investigators were surprised to find extensive aneuploidy in cultured cells derived from normal tissues. These numerical and structural chromosome changes have been extensively documented and confirmed and are believed to be similar, if not identical, to the chromosome rearrangements found in tumor cells (Hsu 1961).

An attempt was made during this early period to draw the many published observations together into a central theme called the stem line theory. The original stem line concept was derived from studies of ascites tumors, but later was extended to include evolved in vitro cell populations (Hsu 1961). This concept defined stem cells as that portion of the cell population having a specific karyotype and as being the principal progenitors of tumor growth (Makino 1952). The cells with altered karyotypes, compared to stem cells, were considered to be the products of aberrant mitotic events, to be less competitive in growth rate, and hence to be minor or transient components of the tumor cell population.

Further studies demonstrated that this view is an oversimplification. For example, improved cytogenetic analyses showed that karyotype deviations exist within a stem line (structural rather than numerical changes) (Tjio and Levan 1956), and microspectrophotometric analyses suggested that some tumors have multiple stem lines or no distinct stem lines. Therefore, more recent observations have recognized that tumor stem lines are in a state of flux, with changing forms and modes, and that stem line chromosome constancy is not absolute (Hsu 1961, Hauschka and Levan 1958).

Early studies of cellular DNA content in aneuploid cells suggested that chromosomally aneuploid cells also contain "aneuploid amounts of DNA" (Hsu 1961). Some of these studies are difficult to interpret because investigators failed to consider cell cycle progression or because of relatively imprecise staining methodology (Hsu and Moorhead 1957, Bader 1959). Again, more recent studies with improved instrumentation and stain technology have changed the original stem line concepts. Atkin (1969, 1970) was the first investigator to emphasize the relatively small variation in cell-to-cell DNA content in tumor populations. Pooled Feulgen-DNA values from several tumors (about 1,500 cells in the G1 mode) were compared with DNA values for normal cells (sample sizes of 116, 302, and 839 cells). The coefficient of variation (CV) for the tumor cells was about 10% while that for normal cells ranged from 6.2% to 8.2%. The development of flow cytometry, which permitted the analysis of large numbers of cells (typical sample size, 10^5 cells) with relative ease, resulted in a two- to three-fold improvement in CV. Using this technology, highly evolved cell populations of tumor or normal cell origin were found to have CVs as low as or lower than those of their euploid counterparts (Kraemer et al. 1972). These studies on cultured cells have been repeated on a variety of human tumors with similar results (Barlogie et al. 1979).

Chromosome and DNA Content Measurements in Cultured Cells of Human Tumor Origin

Examples of the results mentioned above are illustrated in Figure 1, in which DNA distributions of human euploid (WI-38) cells are compared with two cell lines derived from human tumors. The WI-38 cells had a well-defined modal number of 46 chromosomes and a CV of 4.87% for the G1 mode of the DNA distribution. The tumor-derived lines each showed an elevated dispersion of chromosome number but without any corresponding increase in cell-to-cell variability of DNA content (melanoma CV = 4.76%, HeLa CV = 3.16%). These results have been extensively confirmed and validated (Kraemer et al. 1972), and no exceptions have been found to the original observation of DNA constancy in highly evolved, exponentially growing, heteroploid* cell populations. We have

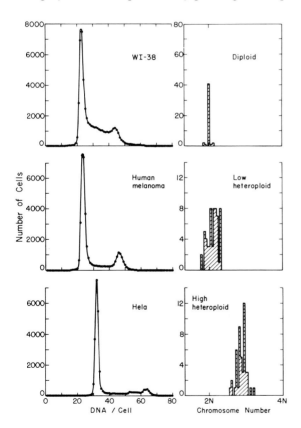

FIG. 1. Flow cytometric DNA distributions and chromosome number histograms of cultured diploid and heteroploid human cells.

---

* In this paper *heteroploid* refers to mammalian cell populations with abnormal variability of chromosome number and structure.

examined 40–50 different heteroploid cell lines, either of tumor origin or by in vitro evolution and two conclusions may be drawn: (1) all contain some degree of elevated dispersion in chromosome number without a corresponding dispersion of cellular DNA content and (2) the DNA content of these cell lines tends to cluster in approximate multiples of the haploid genome (Deaven and Petersen 1974). One example of this latter trend is illustrated in Figure 2. Two HeLa lines selected for different sensitivities to human enterovirus infection were found to contain different amounts of DNA per cell and different modal numbers of chromosomes. One line, designated HeLa-4, had a chromosome mode of 65 and, like six other HeLa lines examined, contained 1.5 times as much DNA as euploid mammalian cells. The second line, designated HeLa-229, contained a chromosome number mode of 83 and had a DNA/cell value

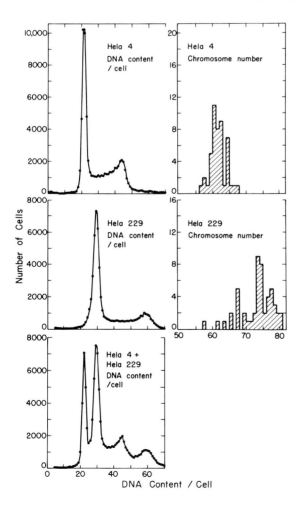

FIG. 2. Acriflavine-Feulgen DNA distributions for two HeLa sublines selected for different sensitivities to human enterovirus infection.

of two times mammalian euploid values. This comparison of two HeLa sublines indicates that the DNA content and chromosome number of stem lines in highly evolved cell populations can change as a result of selective pressure. However, when they stabilize as a consequence of continuous exponential growth they tend to contain invariant DNA values close to multiples of the haploid genome. Both lines had CVs in the 3–4% range in contrast to the chromosome number variabilities illustrated.

Detailed descriptions of various types of validation experiments for the flow data have been described previously (Kraemer et al. 1972). One of these experiments is critical to a proper interpretation of these data and will be described here in a brief manner. This involves a comparison of the effects of induced nondisjunction on flow distributions. When Chinese hamster cells are blocked with Colcemid for short periods of time, a portion of the blocked population experiences nondisjunctive events. The number of nondisjunctive events is proportional to the time of Colcemid exposure and is a transitory or lethal event for most affected cells. Only those cells that have either gained or lost a small chromosome continue to grow and form clones (Cox and Puck 1969, Kato and Yosida 1970). When Chinese hamster cells were treated with Colcemid and the blocked population was allowed to divide in drug-free medium, the effect of nondisjunction on the G1 peak was readily apparent (Figure 3). The G1 peaks of the distributions shown in Figure 3 consist of two populations of cells: (1) the cells that did not undergo nondisjunction and formed the bulk of the G1 peak and (2) the cells that lost or gained chromosomes and, therefore, broadened the DNA per cell distribution at the base of the curve. This demonstrates that when cells gain or lose DNA the flow technique accurately reflects those changes.

An alternative way of obtaining cell populations with a broad or a narrow range of chromosome numbers per cell is to clone a heteroploid population and select clones with a reduced chromosome number distribution (Chu and Giles 1958). Accordingly, we cloned a HeLa line and selected a clone that had cells with either 66 or 68 chromosomes (parental dispersion of 62–74 chromosomes). When this cloned line was compared with the parental line, no significant difference could be discerned between them (Figure 4). Taken together, these experiments indicate that when changes occur in DNA content per cell, the flow method detects them readily, and that the karyotypic number variability in HeLa cells contributes very little to the low CVs characteristic of heteroploid lines.

Chromosome analysis of HeLa cells showed that cellular chromosome complements could be divided conveniently into three groups: (1) normal unaltered chromosomes, (2) altered chromosomes common to many cells in the population, and (3) altered chromosomes unique to one or a minority of cells in a population. There were indications that the HeLa cells were basically triploid; the DNA content was about 1.5 times the human diploid value, and in some groups of chromosomes triploid complements could be found. The origin of chromosomal

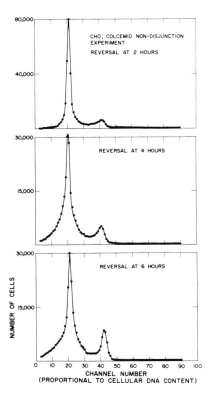

FIG. 3. Flow cytometric DNA distributions of Chinese hamster cells following treatment with Colcemid. Monolayer cultures were given 0.06 μg/ml of Colcemid and shaken at 2,4, and 6 hours to remove mitotic cells. These cells were then grown in drug-free medium for 8 hours and prepared for flow analysis by staining with the acriflavine-Feulgen technique. The broadening at the base of the G1 peaks reflects the loss or grain of chromosomes as a result of Colcemid-induced nondisjunction.

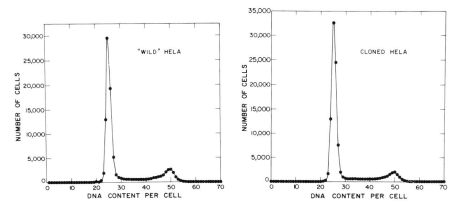

FIG. 4. Flow cytometric DNA distributions of a "wild" type and a cloned population of HeLa cells. The coefficient of variation of the G1 peak for the wild type is 3.52%, and for the clone it is 3.75%.

material of some of the rearranged chromosomes could be identified. However, because there also were many small abnormal chromosomes with complex band rearrangements, a complete determination of chromosomal origin of all chromatin was not possible. All of the HeLa lines analyzed had some common marker chromosomes. A striking difference between some lines was in the ratios of normal to abnormal chromosomes. Figures 5 and 6 compare karyotypes of typical cells from HeLa S-3 and HeLa-4 populations. Both of these lines had DNA values at 1.5 times diploid human cells and both had similar chromosome number distributions. However, HeLa S-3 cells contained about 40% abnormal chromosomes while there were only about 24% in HeLa-4. The increased percentage of normal chromosomes in HeLa-4 resulted in many of the individual numbers having triploid groups. Another unexpected observation was the large cell-to-cell differences in karyotype within HeLa lines. Because many individual chromosomes had variable numbers of copies and because marker complements also varied in number from one cell to another, it was difficult to find two HeLa cells with identical karyotypes. This cell-to-cell variability in specific chromosome content was as large among cells having the modal chromosome number as it was among the nonmodal cells (Deaven et al. 1975).

An attempt was made to estimate the cell-to-cell variability in chromosome arm length measurements. Previous measurements have demonstrated that total arm length is approximately proportional to DNA content and therefore can

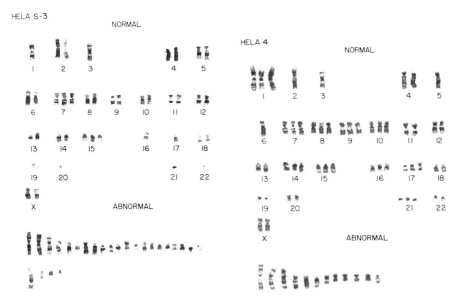

FIG. 5. Giemsa bands of a HeLa S-3 cell. Chromosome number, 65.

FIG. 6. Giemsa bands of a HeLa 4 cell. Chromosome number, 64.

be used to provide an independent measure of cell-to-cell variability in DNA content (B. Mayall, personal communication). Each normal chromosome and each identifiable portion of an altered chromosome in three HeLa lines was assigned a length value from pooled measurements (Bergsma 1966). When an abnormal chromosome could not be positively identified, it was assigned to an appropriate Denver position on the basis of length alone and given that arm length regardless of its banding pattern. When cells from the three lines were compared to each other on this basis, the difference between lines S-3 and 229 was about 28%, and the difference between 229 and 4 was 32%. The latter is in good agreement with flow cytometry data that indicated a difference of 33%. When cells within a given line were analyzed (8 from S-3, 7 from 229, and 11 from 4), cell-to-cell dispersion varied from less than 1% to 4.8%. These values also agree with flow data, which indicate that these populations vary at most from 3% to 5%.

Chromosome and DNA Content Measurements in Heteroploid Chinese Hamster Cells

Because of the complexity of the karyotype of human heteroploid cells, a system originating from a relatively simple karyotype was selected for further studies. This sytem consists of a heteroploid line of Chinese hamster cells (derived from CHO cells) with a modal chromosome number of 35 and a perimodal dispersion of 10 chromosomes, and a series of clones derived from it, selected on the basis of chromosome number.

Cell populations derived from the selected clones were examined to determine the stability of chromosome number, the cellular DNA content, and the specific chromosome complement by G-banding. Some clones had relatively tight chromosome number distributions (perimodal dispersion of 2 or 3), while others appeared to revert rapidly to dispersions as large as that of the parent line (10). Although the modal chromosome number of the clones varied from 29 to 40, they all contained amounts of DNA almost identical to that of the CHO 38 line (Figure 7a,b). In each case, the DNA content was approximately 1.5 times the DNA content of diploid Chinese hamster cells, with CVs of 6%.

Karyologic analysis of the clones showed some trends that were not apparent in the HeLa studies. All cells analyzed, regardless of clone or chromosome number, contained the same complement of large chromosomes (No. 1 and 2). In the medium sized chromosomes (No. 4 through 8 and X) some variability in numbers of copies was found, but the greatest amount of variability was in the small chromosomes (No. 9 to 11 and small unidentifiable chromosomes and minutes).

When perimodal cells were compared with those containing the modal chromosome number, few, if any, systematic differences could be found between them. That is, two modal cells each containing 36 chromosomes often had as many differences between them as were found between a cell with 30 chromo-

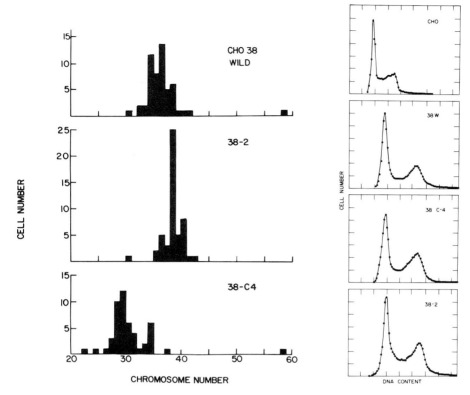

FIG. 7. A (left), Chromosome number counts of metaphase cells from CHO 38 "wild type" (parental line) and clones CHO 38–2 and CHO 38C–4. **B (right),** Flow cytometric DNA distributions of CHO 38W, CHO 38–2, and CHO 38C–4. Each of the high heteroploid lines contains almost identical amounts of DNA per cell (1.5 times the diploid Chinese hamster).

somes and one with 36 chromosomes. Two cells from the CHO-38 parental line are illustrated in Figure 8. The chromosomes are organized with normal chromosomes on the top row and rearranged chromosomes on the two lower rows. Each cell has 36 chromosomes, but variable numbers of copies in several classes of chromosomes make each karyotype unique.

This variability in numbers of copies of a particular chromosome is often believed to be due to high rates of nondisjunction. As mentioned earlier, studies of nondisjunction in diploid or near diploid Chinese hamster cells indicate that the addition or deletion of most chromosomes is not compatible with cell survival. Only those cells that gain or lose a small chromosome (10 or 11) form viable clones (Cox and Puck 1969, Kato and Yosida 1970). A comparable experiment with heteroploid cells would be very difficult if not impossible to perform. The chromosome heterogeneity would tend to mask the induced nondisjunction.

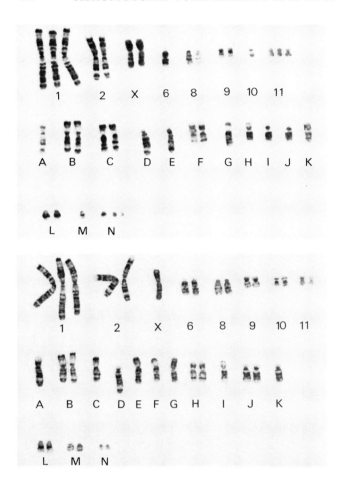

FIG. 8. Trypsin-induced Giemsa bands in two CHO 38 "wild type" cells. Unaltered Chinese hamster chromosomes are on the top line, and the rearranged chromosomes are on the bottom lines. The chromosome number for each cell is 36; however, there are compensatory substitutions in several classes of chromosomes that make each karyotype unique.

However, the use of flow analysis on large numbers of cells can easily reveal the effects of nondisjunction as seen in Figure 3. Accordingly, cultures of a euploid Chinese hamster cell strain (LA-CHE♀) and the CHO-38 heteroploid line were treated with low doses of Colcemid (0.06 μg/ml and 0.12 μg/ml, respectively). After 3 hours the blocked cells were selectively detached and plated in new flasks with normal medium. Samples were collected every 24 hours for 8 days for flow DNA analysis. The resulting distributions are illustrated in Figure 9a,b. In each case at day 1 the G1 modes showed evidence of DNA gains and losses; however, by day 5 both the euploid and the heteroploid cells returned to normal G1 modes. This suggests that heteroploid cells are as sensitive to the effects of induced nondisjunction as euploid cells.

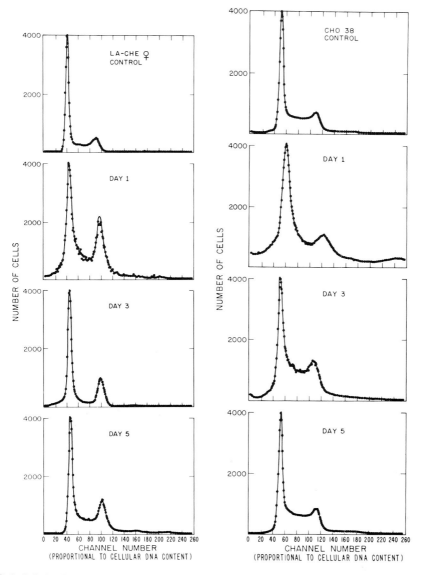

FIG. 9. A (left), Flow cytometric DNA distributions of euploid Chinese hamster cells (LA-CHE+) showing the transitory effects of Colcemid-induced nondisjunction (see text). **B (right),** DNA distributions of heteroploid Chinese hamster cells (CHO-38) following the induction of nondisjunction by a brief Colcemid treatment (see text).

INITIAL CHROMOSOME CHANGES FOLLOWING MALIGNANT
TRANSFORMATION

Early studies on the relationships between chromosome changes and malignancy were addressed primarily to analyses of heteroploid cells either from experimental tumor systems or from highly evolved in vitro populations (Hsu 1961). In spite of intensive and meticulous studies, progress in understanding the basic mechanisms of chromosome control in these cells seems disappointing. We now believe that the available tools were inadequate for the task. However, certain new experimental tumor systems now permit detailed analyses of chromosome changes in the early stages of malignant progression. Therefore, we have shifted our attention to these early changes in karyotype because heteroploid cell lines seem to be too complex to support or reject the chromosome imbalances postulated by Boveri.

For this purpose, we selected an experimental tumor system developed in the laboratory of Dr. Ruth Sager (Sager and Kovac 1978, Howell and Sager 1978). Briefly, the system consists of two doubly cloned cell populations designated as CHEF/18–1 and CHEF/16–2. Both of these lines have been reported to have a diploid chromosome mode and a euploid karyotype as determined by G-banding. The lines have similar morphological features and growth rates, but marked differences in anchorage dependence, colonial morphology, and tumorigenicity. As reported (Sager and Kovac 1978), CHEF/18–1 is not tumorigenic with cell inocula as high as 1×10^7, whereas line CHEF/16–2 cells formed tumors from as few as 10 cells. Sager's studies have been involved with the expression or suppression of tumor-forming ability following cell hybridization or the formation of cybrids between the two lines. We believed that the CHEF/16–2 line could provide an excellent model system for studies of the primary changes in karyotype that accompany malignant growth.

The CHEF/16–2 and CHEF/18–1 lines were maintained in alpha-MEM (minimal essential medium) (GIBCO) supplemented with 10% fetal calf serum (Irvine Scientific Co). The cells were grown as monolayers in 75 cm² culture flasks and subcultured twice a week at a 1:10 dilution for CHEF/18–1 and 1:20 for CHEF/16–2. Chromosome preparations and banding were according to Deaven and Petersen (1974). For tumor production, 10^5 cells from CHEF/16–2 were inoculated into the flanks of nude mice. For some tumors, the cells were implanted subcutaneously in Gelfoam^R gelatin discs (Wells et al. 1980). When palpable tumors (fibrosarcomas) appeared (3–4 weeks), the tumors were excised, passed through a steel mesh screen, and plated in triplicate in 75 cm² culture flasks. In 1 or 2 days, the cells were harvested for flow DNA analysis, for chromosome slide preparations, and for freeze preservation.

Chromosome counts on 50 cells from CHEF/16–2 and CHEF/18–1 were performed at passage 2 following receipt of the cells from Dr. Sager. Both clones had a tight distribution with a modal number of 22 (Figure 10). Chromosome counts from five tumors are compared with the parental CHEF/16–1

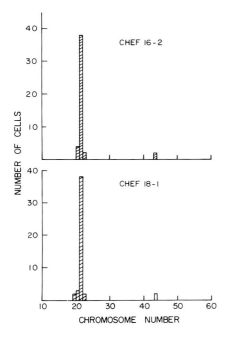

FIG. 10. Chromosome number per cell for CHEF 16–2 (tumorigenic) and CHEF 18–1 (nontumorigenic) Chinese hamster cells.

clone in Figure 11. Tumors 1,2, and 3 each had a modal number of 23 chromosomes. Tumor 4 had a modal number of 40 chromosomes with a range of 35 to 42. Tumor 5 had a modal number at the diploid value of 22 but had significant numbers of cells with 21 and 23 chromosomes. Each of the tumors had from 2% to 5% of the cells at the double stem line number and tumor 4 had 8% at 25 to 30 chromosomes.

DNA measuremenis for each of the tumors are shown in Figures 12 and 13. In each figure the tumor cells are compared with CHO cells, previously shown to contain about 4% less DNA than euploid Chinese hamster cells (Deaven and Petersen 1973). As expected, tumors 1,2,3, and 5 each have cellular DNA contents close to the diploid value. Tumor 4 is 78% higher than the other tumors. The DNA content distribution for tumor 5 suggests the existence of a second stem line, not detected by chromosome counts, at the subtetraploid level. Coefficients of variation for the G1 peaks varied from 6% for tumors 1,2,3, and 4 to 8% for tumor 5.

Karyotype analysis was done by photographing 20 cells from each tumor and parental line. Ten of these photographs were selected for analysis on the basis of band quality. These cells were karyotyped and the resulting data are given in Tables 1–9. In some cases, additional cells were karyotyped from photographs or by microscopic examination. The numerical chromosome assignments are according to the standard Chinese hamster idiogram as described by Hsu and Zenzes (1964).

Tables 1 and 2 contain individual karyotypes for CHEF/16–2 and 18–1 cells as they appeared two subcultures after receipt. Each of the clones was predominately euploid. In the CHEF/18–1 clone, one cell contained an extra chromosome 11 and in CHEF/16–2 one cell had a deletion of the long arm of

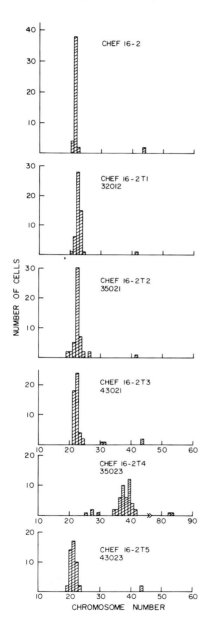

FIG. 11. Chromosome number per cell for CHEF 16–2 and five primary tumors derived from it. Tumors T1, T2, and T3 have a modal number of 23; T4 has a modal number of 39–40; T5 has a modal number of 22 chromosomes.

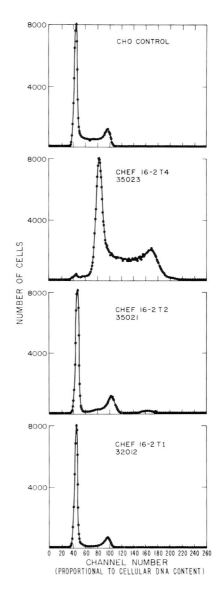

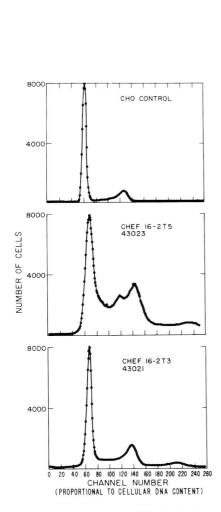

FIG. 12. Flow cytometric DNA distributions for CHEF 16–2 tumors 3 and 5. Both tumors have a near diploid DNA content.

FIG. 13. Flow cytometric DNA distributions for CHEF 16–2 tumors 1, 2, and 4. Tumors 1 and 2 have a near diploid DNA content per cell and tumor 4 is 78% higher.

TABLE 1. *Individual karyotypes for CHEF 16–2 (early passage)*

						Number of chromosomes in each group							
1	2	X	Y	4	5	6	7	8	9	10	11	Marker	Total
2	2	1	1	2	2	2	2	2	2	2	2		22
2	2	1	1	2	2	2	2	2	2	2	2		22
2	2	1	1	2	2	2	2	2	2	2	2		22
1	2	1	1	2	2	2	2	2	2	2	2	1p—	22
2	2	1	1	2	2	2	2	2	2	2	2		22
2	2	1	1	2	2	2	2	2	2	2	2		22
2	2	1	1	2	2	2	2	2	2	2	2		22
2	2	1	1	2	2	2	2	2	2	2	2		22
2	2	1	1	2	2	2	2	2	2	2	2		22
2	2	1	1	2	2	2	2	2	2	2	2		22

TABLE 2. *Individual karyotypes for CHEF 18–1 (early passage)*

						Number of chromosomes in each group							
1	2	X	Y	4	5	6	7	8	9	10	11	Marker	Total
2	2	1	1	2	2	2	2	2	2	2	2		22
2	2	1	1	2	2	2	2	2	2	2	2		22
2	2	1	1	2	2	2	2	2	2	2	3		23
2	2	1	1	2	2	2	2	2	2	2	2		22
2	2	1	1	2	2	2	2	2	2	2	2		22
2	2	1	1	2	2	2	2	2	2	2	2		22
2	2	1	1	2	2	2	2	2	2	2	2		22
2	2	1	1	2	2	2	2	2	2	2	2		22
2	2	1	1	2	2	2	2	2	2	2	2		22
2	2	1	1	2	2	2	2	2	2	2	2		22

chromosome 1. After approximately 32 subcultures (16 weeks of continuous growth), the clones were again karyotyped and these data appear in Tables 3 and 4. In each cell line, euploid karyotypes had given way to aneuploid cells; in CHEF/16–2 this tendency was dramatic. CHEF/18–1 had one cell with a trisomy for the long arm of chromosome 7 and two cells contained small, unidentifiable chromosomes or chromosome fragments. These are indicated by question marks. In late passage CHEF/16–2 cells, the predominate change was loss of the X chromosome long arm. Trisomy for chromosome 11, a deleted chromosome 2, and a 1 to 5 translocation were also found. Although the CHEF/16–2 cells were still predominantly diploid, *not one cell out of 24 analyzed was euploid.*

For the tumor cells, marker chromosomes are listed in columns, with the left-hand column being the most frequent marker for a given tumor. Table 5 contains the karyotypes for tumor 1. The most frequent marker is the X deletion found in the parental 16–2 cells. This change plus the frequent appearance of an isochromosome formed from the short arms of chromosome 10 suggest early divergence of karyotype for this tumor. Three of these cells contain 4q trisomies,

TABLE 3. *Individual karyotypes for CHEF 18–1 (late passage)*

						Number of chromosomes in each group							
1	2	X	Y	4	5	6	7	8	9	10	11	Marker	Total
2	2	1	1	2	2	2	2	2	2	2	2		22
2	2	1	1	2	2	2	2	2	2	2	2		22
2	2	1	1	2	2	2	2	2	2	2	2		22
2	2	1	1	2	2	2	2	2	2	2	2	7q	23
2	2	1	1	2	2	2	2	2	2	2	2		22
2	2	1	1	2	2	2	2	2	2	2	2		22
2	2	1	1	2	2	2	2	2	2	2	2	?	23
2	2	1	1	2	2	2	2	2	2	2	2	?	23
2	2	1	1	2	2	2	2	2	2	2	2		22
2	2	1	1	2	2	2	2	2	2	2	2		22

TABLE 4. *Individual karyotypes for CHEF 16–2 (late passage)*

						Number of chromosomes in each group								
1	2	X	Y	4	5	6	7	8	9	10	11	Marker		Total
2	2		1	2	2	2	2	2	2	2	2	Xp		22
2	2		1	2	2	2	2	2	2	2	3	Xp		23
2	2	1	1	2	2	2	2	2	2	2	3			23
2	2		1	2	2	2	2	2	2	2	2	Xp		22
2	2		1	2	2	2	2	2	2	2	2	Xp		22
2	2		1	2	2	2	2	2	2	2	2	Xp		22
2	1		1	2	2	2	2	2	2	2	2	Xp	2p+	22
2	2		1	2	2	2	2	2	2	2	2	Xp		22
1	2	1	1	2	1	2	2	2	2	2	3	t(1p5q)	t(1q5p)	23
2	2		1	2	2	2	2	2	2	2	2	Xp		22

while a fourth cell is monosomic for the chromosome 4 long arms. Seven of the cells contained small unidentifiable fragments, one cell contained five, another contained three. Other less common changes involved chromosomes 1,5, and 6.

Table 6 defines the changes in tumor 2. The most frequent change in these cells was trisomy for chromosome 11. The X long arm deletion was the most common marker, but six of these cells contained intact X chromosomes. Less common were changes in chromosomes 1,2,8,9, and 10, and a ring of unknown origin. Figure 14 illustrates a typical karyotype from tumor 2.

The most frequent change in tumor 3 (Table 7) involves chromosome 4. Five of the cells are trisomic for 4q, one is trisomic for 4p, and one is monosomic for 4q. In some cells, the changes in chromosome 4 involve an Xp4q translocation, while in others the p or q arms of chromosome 4 are single units. Other infrequent changes involve unidentifiable markers and chromosomes 2,10, and 11.

Table 8 defines the common karyotypes of tumor 4, the tumor that was

TABLE 5. *Individual karyotypes for CHEF 16–1 T1(32012)*

Number of chromosomes in each group

1	2	X	Y	4	5	6	7	8	9	10	11			Marker		Total
2	2	2	1	2	2	2	2	2	3	1	2	Xp	i(10p)	4q		24
2	2	2	1	2	2	2	2	2	2	2	1		i(10p)	4q		22
1	2	2	1	2	2	2	2	2	1	2	1					25
2	2	2	1	2	1	1	2	2	2	2	1	Xp	i(10p)	1q	?(5)	22
2	2	2	1	2	1	2	2	2	2	2	2	Xp	i(10p)	dic(5,6)		24
2	2	2	1	2	2	2	2	2	2	2	2	Xp	i(10p)	5q	?	23
2	2	2		2	3	2	2	2	3	2	1	Xp	i(10p)	4q	?	25
2	2	2	1	2	2	2	2	2	1	1	2	Xp			?	23
2	2	2	1	1	2	2	2	2	1	2	1	Xp		4p	?(3)	23
2	2	2	1	2	2	2	2	2	2	2	1	Xp	i(10p)		?	23

TABLE 6. *Individual karyotypes for CHEF 16-2 T2(135021)*

				Number of chromosomes in each group									
1	2	X	Y	4	5	6	7	8	9	10	11	Marker	Total
2	1	1	1	2	2	2	2	2	1	1	3	2p	21
2	1	1	1	2	2	2	2	2	1	2	3	2q—	22
2	2		1	2	2	2	2	2	1	2	3	Xp 9q	23
2	2	1	1	2	2	2	2	2	2	1	3		22
2	1	1	1	2	2	2	2	2	2	2	2	2p—	22
2	2		1	2	2	2	2	2	2	2	3	Xp 1q	23
2	2		1	2	2	2	2	2	3	2	2	Xp	23
2	2	1	1	2	2	2	2	2	2	2	3		23
2	2		1	2	2	2	2	2	2	2	3	Xp r(?)	24
2	2	1		2	2	2	2	2	2	2	3	Yq—	23

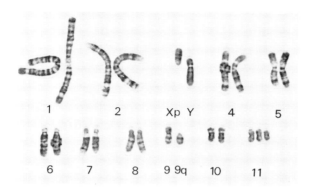

1   2   Xp Y   4   5

6   7   8   9 9q   10   11

FIG. 14. Karyotype from a CHEF 16–2 T2 cell. The abnormal chromosomes are in their proper numerical positions. They include an Xp chromosome found in 16–2 parental cells, a 9q chromosome found in two tumor lines, and a trisomy 11 found in four tumor lines as well as in the tumorigenic and nontumorigenic parental lines.

elevated in chromosome number and in cellular DNA content compared with the others. The most frequent marker in this tumor was a long arm of chromosome 9. Less frequent markers involved chromosomes 1,4,7 and unidentifiable chromosomes, which in this tumor consisted of large (No. 2-sized) homogeneously staining chromosomes. The most pronounced feature of these karyotypes was the variability of numbers of copies of normal chromosomes. In this regard, these cells resembled those found in the CHO-38 heteroploid line previously discussed. Because of these changes each cell had a unique karyotype. If the cells are considered to be tetraploids, then some chromosomes (9 and 11) were in excess, while most chromosomes had less than four copies. This bias toward increased numbers of small chromosomes accounts for the near tetraploid chromosome number, with only a 78% increase in DNA content found in these cells.

Every cell analyzed in tumor 5 contained a chromosome 4 rearrangement (Table 9). One cell was trisomic for the short arms of chromosome 4, while the others contained one normal chromosome 4 and one chromosome 4 broken into separate arms. Other markers included changes in chromosome 8 and X and an unidentifiable chromosome. The fact that one chromosome 8 was missing in all cells might suggest that the 4q markers contained chromosome 8 centromeres. Figure 15 illustrates a typical karyotype from tumor 5.

A summary of the degree of cell-to-cell variability in karyotype is shown in Table 10. If all cells analyzed had the same karyotype, the variability index would equal 0.1; if all cells had unique karyotypes, the index would equal 1.0. The table clearly indicates that all of the tumors and late passage 16–1 cells were highly variable and can, therefore, be referred to as heteroploid by our definition. Another way of attempting to draw generalizations from the tumor chromosome data is to group the rearrangements into chromosome types. In this context, out of 50 cells, 22 involved chromosome 4, 18 involved the X chromosome, 12 involved unidentifiable chromosomes, 7 involved chromosome 10, 6 involved chromosome 9, and other changes occurred with a frequency

TABLE 7. Individual karyotypes for CHEF 16–2 T3(43021)

	Number of chromosomes in each group											Marker	Total
1	2	X	Y	4	5	6	7	8	9	10	11		
1	2		1	2	2	2	2	2	2	2	2	4p · ?(2)	23
2	2		1	1	2	2	2	2	2	1	2	Xp · ?(2)	22
2	2	1	1	2	2	2	2	2	2	2	2	4q · ?	23
2	2		1	2	2	2	2	2	2	2	2		22
2	2	1	1	2	2	2	2	2	2	2	2	4q	23
2	2	1	1	1	2	2	2	2	2	2	2	4q	23
2	2		1	2	2	2	2	2	2	2	2	t(Xp4q)	22
2	2	1	1	2	2	2	2	2	2	1	2		21
2	1		1	2	2	2	2	2	2	2	2	t(Xp4q)	21
2	2	1	1	2	2	2	2	2	2	2	3	t(Xp4q)	23

TABLE 8. *Individual karyotypes for CHEF 16–2 T4(35023)*

				Number of chromosomes in each group								Marker	Total
1	2	X	Y	4	5	6	7	8	9	10	11		
3	4	2	1	4	3	3	3	4	5	2	5	9q	40
3	3	2	2	4	2	4	2	4	5	2	6	9q	40
2	3	1	1	3	2	2	2	3	4	1	4		28
4	4	2	2	4	3	3	2	4	4	2	6	9q i(7q) ?	43
3	3	2	2	3	3	3	3	4	5	1	6		38
2	4	2	2	3	3	4	2	4	4	1	6	4p 4q 1p	41
2	2	1	2	3	2	2	2	3	2	1	3		25
3	3	2	1	3	3	3	3	4	3	2	5	?	35
3	3	2	1	4	3	3	3	2	3	2	5	9q	36
3	3	1	2	3	3	3	3	3	3	3	5	9q ?(2) 7p+	38

TABLE 9. Individual karyotypes for CHEF 16–2 T5(43023)

				Number of chromosomes in each group								Marker			Total
1	2	X	Y	4	5	6	7	8	9	10	11				
2	2	1	1	1	2	2	2	1	2	2	2	4p	4q+		22
2	2	1	1	1	2	2	2		2	2	2	4p	4q+	8q+	22
2	3	1	1	1	3	2	3		2	2	3	4p	4q+	8q+	26
2	2	1	1	1	2	2	1	1	2	2	2	4p			20
2	2	2	1	1	2	2	2	1	2	2	1	4p	4q+	?	23
2	2	1	1	1	2	2	2	1	2	2	2	4p	4q+		22
2	2	1	1	1	2	2	2	1	2	2	2	4p	4q+		22
2	2		1	2	2	2	2	1	2	2	2	4p		Xp	22
2	2	1	1	1	3	1	2	1	2	2	2	4p	4q+		22
2	2	1	1	1	2	2	2	1	2	2	2	4p	4q+		22

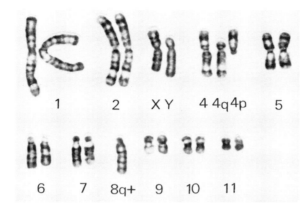

1   2   X Y   4 4q 4p   5

6   7   8q+   9   10   11

FIG. 15. Karyotype from a CHEF 16–2 T5. The abnormal chromosomes are placed in their proper numerical positions. They include a chromosome 4 broken into 4q and 4p arms and an 8q+ chromosome composed of one chromosome 8 and a portion of the q arm of a second chromosome 8.

of less than 10%. The most frequent changes can be further broken down into trisomic or monosomic changes. In these categories, 13 cells were trisomic and 4 cells were monosomic for parts of chromosome 4, 12 cells were monosomic for part of the X chromosome, 7 cells were trisomic and 2 cells were monosomic for parts of chromosome 10, and 1 cell was monosomic for parts of chromosome 9. The CHEF/16–2 cells in late passage had 8 of 10 cells with X chromosome rearrangements, and 8 of 10 were monosomic for part of the X chromosome.

An additional attempt was made to quantitate the karyotype comparisons of CHEF/16–2 and 18–1. Chromosomes were isolated from metaphase cells from each clone and analyzed on a flow cytometer (L. Cram, D. Arndt-Jovin, B. Grimwade, and T. Jovin, manuscript in preparation). The resulting flow karyotypes are shown in Figure 16. Chromosomes from each clone were analyzed two times and these four distributions are superimposed on each other in Figure 16. The total number of chromosomes analyzed for each analysis was greater than 50,000. Following receipt, the cells were subcultured approximately four

TABLE 10. *Karyotype variability indices**
for CHEF 16–2, CHEF 18–1, and CHEF 16–2T
cells

16–2 (Early Passage)	0.2
18–1 (Early Passage)	0.2
16–2 (Late Passage)	0.6
18–1 (Late Passage)	0.3
16–2 T1(32012)	0.8
16–2 T2(35021)	1.0
16–2 T3(43021)	0.9
16–2 T4(35023)	1.0
16–2 T5(43023)	0.6

* Karyotype variability index = number of different karyotypes/total karyotypes.

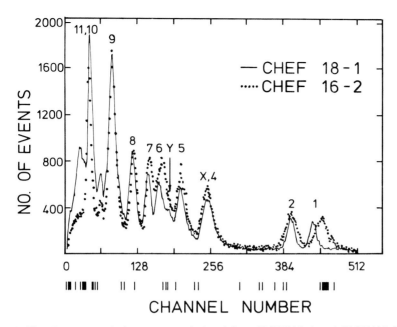

FIG. 16. Flow karyotype of chromosomes isolated from CHEF/18–1 and CHEF/16–2 cells. Isolated chromosomes were stained with 10 μM Hoechst-33342 and analyzed for total fluorescence. Two separate measurements were made for each cell line. The two corresponding distributions were averaged on a channel-by-channel basis. The resulting 16–2 and 18–1 distributions were normalized to the same total number of events and the No. 2 chromosomes were normalized to have the same intensity (see text).

times before chromosome isolation. At this time the clones were predominately euploid. In general, the flow karyotypes were in good agreement. The major differences were in numbers of chromosomes in the peaks for chromosomes 4 to 7 and in peak location for chromosome 1. The good agreement in position between peaks 4 to 7 suggests that the numerical differences were due to monosomies in CHEF/18–1 or trisomies in CHEF/16–2 for the chromosomes in these peaks. However, the karyotype data do not provide supportive evidence for this interpretation and the differences may be due to selective chromosome recovery following isolation. The peak location differences for chromosome 1 may be more significant. The ratio between chromosomes 1 and 2 from arm length measurements was 1.37 (Hsu and Zenzes 1964); the ratio from thymidine incorporation measurements was 1.21 (Stubblefield et al. 1975). The ratios between chromosomes 1 and 2 from the flow distributions were 1.25 for CHEF/16–2 and 1.07 for CHEF/18–1. This suggests that the chromosome 1 in CHEF/ 18–1 may be the abnormal one because it has a lower than normal DNA content. A careful morphological analysis of the No. 1 chromosomes in each line failed to reveal any consistent differences between them to account for the flow measurements. However, in some cells from clone 18–1 one No. 1 homologue had a light staining p terminal band. The frequency of these cells varied from 1% in early passage to 40% in late passage.

DISCUSSION

The studies of cellular DNA content in heteroploid cells suggest that while karyotype stability has gone awry, the maintenance of DNA per cell is stringent and comparable to that of euploid cells from which they were derived. Atkin recognized and discussed the relatively tight DNA distributions found in 119 malignant tumors. He believed that there was some small but real increase in cell-to-cell DNA content variability, and that chromosome count data were exaggerated due to the presence of broken cells (Atkin 1970). The development of flow cytometers and the resulting improvement in CV measurements as described in this chapter indicates that there is no difference at all in cell-to-cell DNA content variability between euploid and rapidly growing heteroploid populations. When these data are compared with karyotype analyses of heteroploid cells, which indicate high levels of variability (Miller et al. 1971, Kraemer et al. 1972, Deaven et al. 1975), the two types of data appear to be highly inconsistent. Because the flow data could not be adequately explained by preexisting concepts, Kraemer et al. (1972) presented an alternative model to explain the loss of karyotype control in heteroploid cell lines. This concept suggested that chromosomes may not always exist as separate entities, that during some portion of interphase much of the cellular DNA is continuous, and that heteroploidy represents the aberrant packaging of chromosomes, which then appear at metaphase as rearrangements. This model is predicated on the existence of latent kinetochore and telomere regions that under heteroploid conditions may be expressed. Unfortunately, since the time of its description no data have become available that could lead to the proof or disproof of this model.

The classical explanation for cell-to-cell chromosome number variability is gain or loss due primarily to nondisjunction or other mitotic errors. Our results suggest that the variability found in HeLa cells and in heteroploid Chinese hamster cells is not due to elevated rates of nondisjunction. The results illustrated in Figures 9a and 9b indicate that heteroploid cells that gain or lose chromosomes are rapidly lost from the population. These studies are not all inclusive and Colcemid-induced nondisjunction may be more deleterious to cells than spontaneous nondisjunction. Nevertheless, the tight CVs found in heteroploid cells leave insufficient room for cellular gains or losses of DNA in proportion to chromosome number variabilities.

The limited data on arm length measurements in HeLa cells also suggest that DNA content per cell is more stringently controlled than karyotype. Karyotype analyses in HeLa and CHO-38 cells indicate that there is some bias in favor of variability of chromosome number among the smaller chromosomes (Deaven et al. 1975, Deaven 1976). This would be an expected result of constant DNA and chromosome number variability and has been reported previously

(Atkin and Baker 1966); however, these results are in contrast to those of Minkler et al. (1970), as well as others, who report number variability in all chromosome size groups.

An alternative way of explaining the DNA-chromosome number paradox would be to invoke some type of cellular selection for a given amount of DNA. In this context, cellular evolution would involve selections for improved gene balance as well as selection for a given and constant amount of DNA. The optimal amounts of DNA in this construction would be multiples of the haploid genome where highly evolved heteroploid and most normal eukaryotic cells tend to cluster. The basis of this DNA selection process could be the nuclear membrane. Comings (1968) has developed a case for an orderly spatial organization of nuclear components with interrelationships between these components and the nuclear membrane. Shaw and Chen (1974) have speculated on the consequences of errors in this organization, and on the relationship such rearrangements may have to the onset of malignancy. Most events would be lethal while others would initiate a series of changes that would be selected on the basis of both gene and nuclear structural balance. For example, most trisomies, either in vivo or in vitro, are only marginally stable, eventually giving rise to additional chromosome changes and then progressing to heteroploidy. The studies of Terzi (1972) may provide some supportive evidence for this construction. Terzi observed ploidy changes in Chinese hamster cells (Wg) following treatment with 1,500 rads of X rays. The surviving cells went through a series of changes in modal chromosome number but eventually most of the clones returned to a chromosome number of 22. Terzi believed that changes from the parental chromosome number were disfavored and lost by selection. He suggested that the mitotic apparatus has "structural restrictions" that act as a control over chromosome number. This may be so, but an alternative explanation for his data is that the cells were selected for near diploid amounts of DNA, regardless of chromosome number. A few of his clones actually stabilized at a modal number of 21 while other clones appeared with 22 chromosomes, then diminished and disappeared, only to give rise to new clones with a modal number of 22, but with a different karyotype. This behavior could be ascribed to selection for DNA content rather than chromosome number. In addition, Chinese hamster lines may stabilize at 20 (V79), 21 (CHO), or 22 (M3–1) chromosomes. A feature common to all of these lines is near diploid DNA values. If heteroploid cells were selected only for favorable gene balance we would expect to find many random combinations of chromosomes and, therefore, DNA per cell in these populations. The cell-to-cell assortments of chromosomes found in these lines could come from known mitotic abnormalities. For example, multipolar mitoses are commonly found in a variety of tumor tissues and in some heteroploid cells in vitro (Oksala and Therman 1974). In one study a clear correlation was established between increasing frequencies of multipolar mitoses and tissue change from normal through dysplasia (Scarpelli and von Haam 1957). Endoreduplicated cells, common in many malignant tumors, have been found to form tripolar mitoses at frequencies as high as 50% (Schmid 1966). After multipolar divisions, daughter nuclei are especially prone to fuse. These processes could permit an enormous amount of chromosome reassortment. If the additional constraint of favorable DNA content per nucleus is added, cell populations with constant DNA but variable numbers or copies of chromosomes would rapidly predominate in the population.

The chromosome changes found in the tumors induced by CHEF/16–2 cells fail to support the Boveri hypothesis, at least at the level of visible chromosome alteration. While the alterations are in some sense nonrandom, many cells, including all cells analyzed for tumor 2, did not show evidence of the most common rearrangement found (altered chromosome 4). Furthermore, in looking for evidence of chromosome imbalance, the most common trisomy or monosomy was found in only 13 of 50 cells (trisomy for parts of chromosome 4).

The most striking results to date from Chinese hamster material are those of Bloch-Shtacher and Sachs (1977). These investigators reported clear-cut increases in amount of chromosome 3 material (chromosome 4 in the standard Hsu-Zenzes idiogram) in tumors induced by SV40 and methylcholanthrene. Modal chromosome numbers in these tumors ranged from 26 and 27 to 38 and 39. No chromosome alterations other than single p or q arms of chromosome 4 were found. In contrast to these data, Zuna and Lehman (1977) analyzed the karyotypes of five SV40 transformed clones and tumors derived from three of these clones. The tumors were in the near tetraploid chromosome number range, contained numerous additions or deletions of normal chromosomes, several markers, and a nonrandom frequency of altered No. 1 chromosomes. In an attempt to further identify the chromosome(s) associated with SV40 transformation and malignancy, Lehman and Trevor (1979) selected a pseudodiploid SV40-transformed clone and carried it through three successive passages in nude mice. The clone maintained a modal number of 22 chromosomes through the three passages and was characterized karyotypically by changes in chromosomes 1 and 2, loss of the X chromosome, and loss of chromosomes 5, 6, and 11. Many cells also contained two marker chromosomes, one of which appears to be similar to the large, dull-staining chromosome found in tumor 4 in the present study. Chinese hamster cell lines transformed with 1-methylguanine were analyzed for karyotype changes by Trewyn et al. (1979). Three of these lines were frequently trisomic for the 4q arm. Two of the lines were found to be tumorigenic while the third was not. This observation indicates that excess amounts of chromosome 4 do not necessarily provide malignant potential to Chinese hamster cells.

The largest study of the chromosomes in primary Chinese hamster tumors used the Schmidt-Ruppin strain of Rous sarcoma virus as the inducing agent (Kato 1968). This work was done without banding, but the Chinese hamster karyotype is sufficiently distinctive to permit a fair amount of identification without bands. Our experience with CHO cells, however, suggests that these data should be interpreted with some caution (Deaven and Petersen 1973). Kato studied 42 individual tumors and found 28 different karyotypes in 68 stem and side lines. About half of the stem lines were normal diploids. The most common alterations were additions of chromosomes 5,6, and 10. Kato found a "possible correlation between trisomy 6 and the fibrosarcoma type of tumor."

Our data and the above comparative studies all demonstrate some degree of nonrandom changes. However, these changes involve different chromosomes in different tumors induced by the same agent (our data), nonrandom changes that may or may not be tumorigenic (Trewyn et al. 1979), and nonrandom changes in the midst of random rearrangements (our data, Zuna and Lehman 1977, Lehman and Trevor 1979). Recent reports give good support to the specific chromosome change hypothesis, at least in myeloproliferative disorders (see Mark 1977 for review). These specific changes seem associated with the type of target cell rather than the etiologic agent (Ohno et al. 1979). If this conclusion is correct, perhaps the fibroblast target cell could respond with a variety of chromosome changes and then be selected on the basis of other advantages. The available data on Chinese hamster cells suggest that this may be the case. The general conclusion from all of these data is that chromosome instability per se is in some way related to tumorigenic potential. A similar conclusion has been reached by German (1972) by considering the simultaneous increase in chromosome instability and cancer incidence in patients with rare genetic disorders.

What is the nature of this initial change? Perhaps a nonlethal chromosome break or addition alters the nuclear order. Or possibly it is a more subtle genome-perturbing event such as an insertion element that disrupts control mechanisms (McClintock 1956). In this context, the nature of the chromosome 1 differences found by flow karyotyping the CHEF/16–2 and 18–1 parental lines may be of particular interest.

SUMMARY

We have discussed evidence that heteroploid cells are karyotypically diverse but retain high levels of control over DNA content per cell. We have suggested that these cells evolve under selection parameters involving specific gene content *and* DNA content as dictated by nuclear topological features. Additional data on chromosome changes in primary tumors of the Chinese hamster are presented and compared with other available data. We conclude that all of these studies are drawn together by a common characteristic—chromosome instability.

ACKNOWLEDGMENTS

This work was performed under the auspices of the United States Department of Energy. We wish to thank Arlene Nock and Evelyn Campbell for technical assistance, and Juliemarie Grilly and Jo Ann Brown for assistance in manuscript preparation.

REFERENCES

Atkin, N. B. 1969. Perimodal variation of DNA values of normal and malignant cells. Acta. Cytol. 13:270–273.

Atkin, N. B. 1970. Cytogenetic studies on human tumors and premalignant lesions: The emergence of aneuploid cell lines and their relationship to the process of malignant transformation in man, in Genetic Concepts and Neoplasia (The University of Texas System Cancer Center 23rd Annual Symposium on Fundamental Cancer Research). Williams and Wilkins Co., Baltimore, pp. 36–56.

Atkin, N. B. and M. C. Baker. 1966. Chromosome abnormalities as primary events in human malignant disease: Evidence from marker chromosomes. J. Natl. Cancer Inst. 36:539–557.

Bader, S. 1959. A cytochemical study of the stem cell concept in specimens of human ovarian tumor. J. Biophys. Biochem. Cytol. 5:217–229.

Barlogie, B., W. Gohde, and B. Drewinko. 1979. Flow cytometric analysis of DNA content for ploidy determination in human solid tumors. J. Histochem. Cytochem. 27:505–507.

Bergsma, D. (Ed.) 1966. Standardization in human cytogenetics, Birth Defects 2(2):21.

Bloch-Shtacher, N., and L. Sachs. 1977. Identification of a chromosome that controls malignancy in Chinese hamster cells. J. Cell Physiol. 93:205–212.

Boveri, T. 1914. Zur frage der entwicklung maligner Tumoren, Gustav Fischer, Jena.

Chang, S. 1954. Continuous subcultivation of epithelial-like cells from normal human tissues. Proc. Soc. Exp. Biol. Med. 87:440–443.

Chu, E. H. Y., and N. H. Giles. 1958. Comparative chromosomal studies on mammalian cells in culture. I. The HeLa strain and its mutant clonal derivatives. J. Natl. Cancer Inst. 20:383–401.

Comings, D. E. 1968. The rationale for an ordered arrangement of chromatin in the interphase nucleus. Am. J. Hum. Genet. 20:440–468.

Cox, D. M., and T. T. Puck. 1969. Chromosomal nondisjunctions: The action of Colcemid on Chinese hamsters cells in vitro. Cytogenetics 8:158–169.

Deaven, L. L. 1976. Cellular DNA content and chromosome number in heteroploid cell populations, in The Automation of Uterine Cancer Cytology (G. L. Wied, G. F. Bahr and P. H. Bartels, eds.), Tutorials of Cytology, Chicago, pp. 304–310.

Deaven, L. L., and D. F. Petersen. 1973. The chromosomes of CHO, an aneuploid Chinese hamster cell line: G-band, C-band, and autoradiographic analysis. Chromosoma 41:129–144.

Deaven, L. L., and D. F. Petersen. 1974. Measurements of mammalian cellular DNA and its localization in chromosomes. Methods Cell Biol. 8:179–204.

Deaven, L. L., P. C. Sanders, J. L. Grilly, P. M. Kraemer, and D. F. Petersen. 1975. Chromosome G-banding and DNA constancy in aneuploid cell populations, in Mammalian Cells: Probes and Problems (C. R. Richmond, D. F. Petersen, P. F. Mullaney and E. C. Anderson, eds.), ERDA Symposium Series CONF-731007. National Technical Information Service, Springfield, Virginia, pp. 212–227.

Eagle, H. 1955. Propagation in a fluid medium of a human epidermoid carcinoma, strain KB. Proc. Soc. Exp. Biol. Med. 89:362–364.

German, J. 1972. Genes which increase chromosomal instability in somatic cells and predispose to cancer. Prog. Med. Genet. 8:61–101.

Gey, G. O., W. D. Coffman, and M. T. Bubicek. 1952. Tissue culture studies of the proliferative capacity of cervical and normal epithelium. Cancer Res. 12:264–265.

Hauschka, T. S., and A. Levan. 1958. Cytologic and functional characterization of single cell clones isolated from the Krebs-2 and Ehrlich ascites tumors. J. Natl. Cancer Inst. 21:77–135.

Howell, A. N., and R. Sager. 1978. Tumorigenicity and its suppression in cybrids of mouse and Chinese hamster cell lines. Proc. Natl. Acad. Sci. USA 75:2358–2362.

Hsu, T. C. 1952. Mammalian chromosomes in vitro. I. The karyotype of man. J. Hered. 43:167–172.

Hsu, T. C. 1961. Chromosomal evolution in cell populations. Int. Rev. Cytol. 12:69–161.

Hsu, T. C., and P. S. Moorehead. 1957. Mammalian chromosomes in vitro. VII. Heteroploidy in human cell strains. J. Natl. Cancer Inst. 18:463–471.

Hsu, T. C., and M. T. Zenzes. 1964. Mammalian chromosomes in vitro. XVII. Idiogram of the Chinese hamster. J. Natl. Cancer Inst. 32:857–869.

Kato, R. 1968. The chromosomes of forty-two primary Rous sarcomas of the Chinese hamster. Hereditas 59:63–119.

Kato, H., and T. H. Yosida. 1970. Nondisjunction of chromosomes in a synchronized cell population initiated by reversal of Colcemid inhibition. Exp. Cell Res. 60:459–464.

Klein, G. 1951. Comparative studies of mouse tumors with respect to their capacity for growth as "ascites tumor" and their average nucleic acid content per cell. Exp. Cell Res. 2:518–573.

Kraemer, P. M., L. L. Deaven, H. A. Crissman, and M. A. Van Dilla. 1972. DNA constancy despite variability in chromosome number. Advances in Cell and Molecular Biology, 2:47–108.

Lehman, J. M., and K. Trevor. 1979. Karyology and tumorigenicity of a simian virus 40-transformed Chinese hamster cell clone. J. Cell Physiol. 98:443–450.

Makino, S. 1952. Cytological studies on cancer. III. The characteristics and individuality of chromosomes in tumor cells of the Yoshida sarcoma which contribute to the growth of the tumor. Gann 43:17–34.

Mark, J. 1977. Chromosomal abnormalities and their specificity in human neoplasms: An assessment of recent observations by banding techniques. Adv. Cancer Res. 24:165–222.

McClintock, B. 1956. Controlling elements and the gene. Cold Spring Harbor Symp. Quant. Biol. 21:197–216.

Miller, O. J., D. A. Miller, P. W. Allderdice, V. G. Dev, and M. S. Grewal. 1971. Quinacrine fluorescent karyotypes of human diploid and heteroploid cell lines. Cytogenetics 10:338–346.

Minkler, J. L., J. W. Gofman, and R. K. Tandy. 1970. A specific common chromosomal pathway for the origin of human malignancy. II. Evaluation of long-term human hazards of potential environmental carcinogens. Adv. Biol. Med. Phys. 13:107–141.

Ohno, S., M. Babonits, F. Wiener, J. Spira, G. Klein, and M. Potter. 1979. Nonrandom chromosome changes involving the Ig gene-carrying chromosomes 12 and 6 in pristane-induced mouse plasmacytomas. Cell 18:1001–1007.

Oksala, T., and E. Therman. 1974. Mitotic abnormalities and cancer, *in* Chromosomes and Cancer, J. German ed., John Wiley & Sons, New York, pp. 240–261.

Sager, R. and P. E. Kovac. 1978. Genetic analysis of tumorigenesis: I. Expression of tumor-forming ability in hamster hybrid cell lines. Somatic Cell Genetics 4:375–392.

Scarpelli, D. G., and E. von Haam. 1957. A study of mitosis in cervical epithelium during experimental inflammation and carcinogenesis. Cancer Res. 17:880–884.

Schmid, W. 1966. Multipolar spindles after endoreduplication. Exp. Cell Res. 42:201–204.

Shaw, M. W., and T. R. Chen. 1974. The application of banding techniques to tumor chromosomes, *in* Chromosomes and Cancer, J. German ed., John Wiley & Sons, New York, pp. 135–149.

Stubblefield, E., S. Cram, and L. Deaven. 1975. Flow microfluorometric analysis of isolated Chinese hamster chromosomes. Exp. Cell Res. 94:464–468.

Terzi, M. 1972. On the selection for the modal chromosome number in Chinese hamster cells. J. Cell. Physiol. 80:359–366.

Tjio, J. H., and A. Levan. 1956. The chromosome numbers of man. Heriditas 42:1–6.

Trewyn, R. W., S. J. Kerr, and J. M. Lehman. 1979. Karyotype and tumorigenicity of 1-methylguanine-transformed Chinese hamster cells. J. Natl. Cancer Inst. 62:633–638.

Wells, R. S., E. W. Campbell, D. Swartzendruber, L. M. Holland, and P. M. Kraemer. 1980. Tumorgenicity of mouse cell lines tested by a new assay. J. Natl. Cancer Inst. (in press).

Zuna, R. E. and J. M. Lehman. 1977. Heterogeneity of karyotype and growth potential in simian virus 40-transformed Chinese hamster cell clones. J. Natl. Cancer Inst. 58:1463–1471.

Genetics of Cancer

Genes, Chromosomes, and Neoplasia, edited by
Frances E. Arrighi, Potu N. Rao, and Elton Stubblefield.
Raven Press, New York © 1981.

Human Cancer Genes

Alfred G. Knudson, Jr.

*The Institute for Cancer Research, The Fox Chase Cancer Center,
Philadelphia, Pennsylvania 19111*

In 1914 Boveri enunciated the hypothesis that genetic change in somatic cells was the initiating event in the origin of cancer. His hypothesis reflected his own observations on the relationship between developmental anomalies and specific karyotypic abnormalities in sea urchins. The term "somatic mutation" was first used in connection with cancer by Tyzzer (1916) as a result of his observations on genetic stability and change of host response to transplantable tumors in mice. The subsequent discovery that X rays could produce both mutation and cancer gave further support to the somatic mutation hypothesis, as does the more recent finding of correlation between mutagenicity and carcinogenicity of chemicals. There is now considerable evidence that many cancers can be interpreted as being initiated by spontaneous or induced somatic mutations.

What I wish to discuss here is not whether somatic mutation can "cause" cancer. I wish to assume that at least some cancers arise in this fashion and to pass on to the problem of the identity and function of the genes whose mutations mediate the disease. Here we have an opportunity to learn about fundamental defects in cancer cells.

CANCER CAN RESULT FROM SOMATIC MUTATION AT SPECIFIC GENE SITES

If spontaneous or induced mutations in somatic cells constitute a major class of events that initiate the process of carcinogenesis, then we are interested not only in the inciting agents and their molecular mechanisms of action but also in the targets. Where do the critical mutations occur? Is there a specific gene site for a specific cancer? What is the normal function of a gene whose mutation leads to cancer?

Early events in clonal evolution of tumor cells are the most clearly nonrandom, and it is only these with which we are concerned here. The first, and still most specific, change known is that of the Philadelphia (Ph[1]) chromosome, which results from a translocation between chromosome 22 and another chromosome, usually No. 9 (Rowley 1973). Nearly all cases of chronic myelocytic leukemia (CML) show this change, and for most of the course of the disease it is the only aberration observed in the leukemic cells. For most patients there is no known cause of their disease. For some there is strong suspicion that ionizing radiation played an etiologic role, whether from atomic fallout or X irradiation. For others there is the possibility of chemical induction. Most cases, of course, are not associated with a known environmental agent and are attributable to "spontaneous" chromosomal aberrations. Published reports on two families indicate that a specific predisposition to CML can be dominantly inherited (Weiner 1965, Hirschhorn 1968, Tokuhata et al. 1968). Some cells in such persons demonstrated no visible cytogenetic change, although Ph[1] leukemic cells were present.

A similar relationship seems to exist between Burkitt's lymphoma and chromosome 14 (Klein 1979). The tumor cells of most cases of this disease show a translocation between chromosomes 8 and 14. The same abnormality has been found in both the African and the American forms of this lymphoma and in the B-cell form of acute lymphocytic leukemia. The abnormality is independent of infection with the Epstein-Barr virus. Other aberrations of chromosome 14 have been associated with other lymphoreticular neoplasms, and it may be that chromosome 14 contains major genes for the regulation of B lymphocyte responsiveness. On the other hand, reports of cases of translocation of chromosome 8 to other chromosomes suggest that No. 8 may be the critical chromosome.

Extensive cytogenetic investigations in acute leukemia have demonstrated several nonrandom aberrations there too. The most consistent change seems to be a translocation between chromosomes 15 and 17 in many cases of acute promyelocytic leukemia (Rowley et al. 1977, Van Den Berghe et al. 1979). However, in about one half of the cases of acute leukemia, both lymphocytic and nonlymphocytic, the karyotypes are normal, and important genetic abnormalities may be subtle and not yet recognized.

For some leukemias and lymphomas there does seem to be some specificity of gene site, although that site appears to be different for different subclasses of these neoplasias. Unfortunately, we do not know whether the abnormalities precede or follow the appearance of a neoplastic clone. If they are initiating, and if each subclass of human cancer is associated with a specific gene site, there must be at least 100–200 human cancer genes.

CANCER CAN RESULT FROM GERMINAL MUTATION AT SPECIFIC GENE SITES

Although there are a few recessively inherited human conditions, such as xeroderma pigmentosum and ataxia-telangiectasia, that predispose to cancer, the more typical familial cancers are dominantly inherited. These include polyposis coli, the family cancer syndromes, and retinoblastoma. Virtually every human cancer exists in a dominantly heritable form as well as a nonhereditary form (Knudson et al. 1973). The fraction of such heritable cases may be rather high, as with retinoblastoma (about 40%), or very low, as with cervical cancer (probably less than 1%). The total number of such dominant cancer genes is not known, but at least 30 are listed in McKusick's catalog of human genes (McKusick 1978). If each cancer is associated with one or a few such genes, we may conclude again that there are at least 100–200 of them in the human genome.

The main reason we do not know the number of such genes is that for most of them there is no invariant sign of their presence in a carrier. For a few, such as polyposis coli, there is always, or nearly always, clinical indication of the presence of the gene. For most, as with retinoblastoma, only tumors betray the gene. If many gene carriers escape tumors, or develop them late in life, the presence of the gene will be in doubt. The highly penetrant genes that are expressed early in life, such as that for retinoblastoma, are few in number.

Another complication is that more than one gene may produce a particular tumor. At least two genes are known for familial pheochromocytoma, one associated with medullary carcinoma of the thyroid and one not (Knudson and Strong 1972a). The former association itself may be caused by two different genes, since we do not know whether Sipple's disease and the mucosal neuroma syndrome are produced by mutations at one or two gene loci. For colon cancer there are also several different genes, the best known being polyposis coli and the family cancer syndrome first reported by Warthin (1913). Each of these may be caused by more than one gene; for example, it is not known whether Gardner's syndrome is allelic with, or separate from, polyposis coli.

The existence of heritable neoplasia with considerable tissue specificity suggests that the heritable form of each cancer is caused by mutation at one or a few genetic loci. For three tumors, we are fortunate to have some cytogenetic evidence that this is the case. In one instance the evidence is confined to a single family in which renal carcinoma was found in ten members, bilaterally in six (Cohen et al. 1979). In the same family a constitutional cytogenetic abnormality, a translocation between chromosomes 3 and 8, was also segregating. No person with a tumor had a normal karyotype. The three persons with the translocation but without tumors are still young adults. The authors calculated a probability of 0.87 that translocation carriers would develop at least one renal carcinoma by the age of 60 years, whereas only one person per thousand in the general population develops renal carcinoma by that age (Knudson 1979). Presumably, a "renal carcinoma gene" is located on either chromosome 3 or 8.

No other pedigree of this type has been reported. In one recent report, however, aniridia and Wilms' tumor have been found together in three members (two half-sibs and a maternal aunt) of one family and associated with an unbalanced translocation between chromosome 11 and 2 in one (Yunis and Ramsay 1980). The unaffected mother and an unaffected half-sib were found to be carriers of the corresponding balanced translocation. The chromosomal imbalance involved an interstitial deletion in the short arm of chromosome 11, the same site affected in isolated cases of aniridia and Wilms' tumor (Riccardi et al. 1978, Francke et al. 1979). Evidently, genes for aniridia and Wilms' tumor are located close to each other, as predicted earlier (Knudson and Strong 1972b), and deletion simultaneously eliminates both genes. Penetrance is apparently complete for aniridia but discordance for Wilms' tumor in a pair of identical twins shows that penetrance for the tumor is incomplete (Francke et al. 1979).

We cannot say whether all hereditary cases of Wilms' tumor involve genetic abnormality at this same site. It is possible, of course, that in familial cases without aniridia there is a submicroscopic alteration of the Wilms' tumor gene with no change at the aniridia locus. But there may also be another gene that predisposes to Wilms' tumor, especially in view of the association sometimes noted with hemihypertrophy and with the Beckwith-Wiedemann syndrome. So far there are no reports of karyotypic abnormality in such cases.

The tumor that has been most thoroughly studied cytogenetically is retinoblastoma. About 40% of cases can be attributed to a dominantly heritable mutation that imposes a risk of tumor about 100,000 times as high as the risk for a child who does not carry this mutation (Knudson 1976). In a few cases all of the somatic cells of a patient show a deletion in the long arm of chromosome 13 ($13q^-$). A comparative study of these cases shows that one band, 13q14, is affected in all. Yunis and Ramsay (1978) have applied a more refined banding method to show that a specific subband seems to be the critical site at which a retinoblastoma gene is located. The deletion cases resemble hereditary cases in their age-specific incidences and bilaterality (Knudson et al. 1976), and it is likely that application of this new method will reveal that some inherited cases with apparently normal karyotypes have small deletions.

Cytogenetic observation plays a major role in chromosomal localization of human cancer genes because there are no known gene products such as are necessary for the elegant somatic cell hybridizaton studies that have localized so many human genes. However, it might be possible to demonstrate genetic linkage between a human cancer gene and another gene whose chromosomal location has been determined by such methods. Such an analysis would require numerous pedigrees unless the genetic linkage were very close. It would also require that there not be very many different genes for the tumor in question. In fact, such an analysis has been carried out for one human tumor. King et al. (1980) have shown linkage between a gene for breast cancer and the locus for glutamyl-pyruvate transaminase, a locus known to be on chromosome 10. This is an extremely important finding because it establishes clearly that there is a major gene for this important tumor and identifies its location as well. We now hope to discover whether some uniform aberration of chromosome 10 is found in breast cancer cells, whether from hereditary or nonhereditary cases.

SOMATIC AND HERITABLE MUTATIONS OCCUR AT THE SAME CANCER GENE SITES

There is very limited evidence bearing on the relationship between sites altered in hereditary and nonhereditary forms of the same cancer. It has already been noted that in two pedigrees of familial CML the same cytogenetic abnormality (Ph[1]) was found in leukemic cells as is usually found in leukemic cells in nonhereditary cases. There are no published studies to indicate whether the nonhereditary forms of renal carcinoma and Wilms' tumor may be associated with cytogenetic change at the same loci as are affected in hereditary cases. For retinoblastoma there is some evidence that the same abnormality can occur constitutionally or just in tumor cells. One study, done before banding techniques were available, has shown deletion in a D group chromosome, perhaps No. 13, in tumors of four of five individuals whose karyotypes were normal in lymphocytes (Hashem and Khalifa 1975). In two instances the tumor was unilateral and the cases may be presumed to be nonhereditary. However, in the other instances the case was hereditary. Balaban-Malenbaum et al. (Balaban-Malenbaum, Gilbert, Nichols, Hill, Shields, and Meadows, unpublished observations) confirmed with banding techniques that this can happen; in one instance a bilateral case with normal lymphocytes showed a 13q⁻ aberration in tumor cells. This case is very much like the familial cases of CML with Ph[1] leukemic cells.

Although not conclusive, these results are consistent with the hypothesis that the same gene sites are affected in both hereditary and nonhereditary forms of the same tumor. The results do demonstrate that the specific abnormality seen in constitutional deletion cases must be exerting its influence directly in the tumor target tissue rather than indirectly through some other tissue. The deletion site evidently includes a specific human cancer gene. In cases without microscopically visible aberration we assume that submicroscopic changes, such as small deletions and point mutations, are present.

CANCER GENES ARE DIFFERENTIATION GENES

These cancer genes show considerable tissue specificity in their expression, even when more than one tumor results. The polyposis coli gene leads to colon cancer, not to lung or breast cancer. In this and some other instances there is a histologic abnormality, such as hyperplasia, hamartoma, or benign tumor, suggesting some interference with normal development even in the absence of cancer. Careful investigation may reveal such changes in other cases too. Thus, in Wilms' tumor it has been found that various anomalies of development are found in the kidneys under conditions in which a germinal mutation is likely, i.e., with bilateral tumor (Bove and McAdams 1976). In a recent survey Jackson et al. (1979) have shown that C-cell hyperplasia of the thyroid medulla is associated with the hereditary form of medullary carcinoma of the thyroid, but not with the nonhereditary form. It seems likely that the normal alleles of these human cancer genes are important for tissue differentiation, as has been suggested for those effecting neoplasia in the nervous system (Knudson and Meadows 1978).

Even though these dominantly inherited cancer genes cause specific cancers with high probability, a tumor is a rare outcome at the level of the cell. Thus, in a child with the retinoblastoma gene, only three or four retinoblasts out of millions will be converted to tumor cells; malignant transformation is a rare event. A second event seems to be necessary for retinoblastoma production. This would be true for nonhereditary cases as well, except that *both* events would occur in retinoblasts, i.e., in somatic cells. Dr. Hethcote and I (1978) constructed a mathematical model that relates the cellular processes of proliferation, differentiation, and mutation to the age-specific incidences for the tumor in gene carriers and normal children, and demonstrated that the observed incidences could be related to a two-step process. This model can be applied to any cell line that ultimately differentiates and stops dividing, but it cannot be applied to tumors that arise in renewal tissues.

This two-step model has been developed into a general model for all cancers (Moolgavkar and Venzon 1979, Moolgavkar and Knudson, unpublished observation).

By modeling the growth, differentiation, and mutation of stem cells in renewal tissues like skin and intestine, Moolgavkar developed a single mathematical expression that relates these events to incidences in populations for any cancers, including childhood tumors, which emerge as special cases in which normal stem cell proliferation ceases. The model gives a good description of the epidemiology of breast cancer (Moolgavkar et al., 1980). In this model the overall probability that cancer will emerge depends upon the number of cells that pass through the two steps, while the shape of the incidence curve depends upon growth and differentiation, or birth and death processes, in the target tissue.

In this model, benign tumors, hyperplasias, or developmentally anomalous cells that lie on the pathway to cancer are considered to be clones of intermediate cells. Thus, for example, polyps of the colon, C-cell hyperplasia of the thyroid, neurofibromas, and the microscopic renal anomalies associated with Wilms' tumor are considered to be preneoplastic lesions consisting of intermediate cells. Any agent that interferes with differentiation could lead to increased proliferation of the intermediate cells and thus increase the probability of the second event leading to cancer. It may be that some agents that increase cancer risk, but do not cause mutation, operate by this mechanism.

Studies of affected females who are heterozygous at the X-chromosomal locus for the enzyme glucose-6-phosphate dehydrogenase may be helpful in testing the hypothesis about intermediate cells. The neurofibromas seen in patients with von Recklinghausen's disease express both alleles in heterozygotes and must therefore be polyclonal in origin (Fialkow et al. 1971). We view these as "one-hit" lesions and predict that this behavior will be demonstrated by analogous lesions in other hereditary cancers (Knudson and Meadows, 1980). On the other hand the "two-hit" tumors should be monoclonal. The expression of a single allele has already been found for both medullary carcinoma of the thyroid and pheochromocytoma in Sipple's disease, another heritable disorder that affects neural crest derivatives (Baylin et al. 1978). We agree that in Sipple's disease the precursor lesions, C-cell hyperplasia and adrenal medullary hyperplasia, arise from cells with one "hit" (Jackson et al. 1979), and speculate that they are analogous to hereditary neurofibromas and are also of polyclonal origin. On the other hand, neurofibromas can arise by somatic mutation in persons without von Recklinghausen's disease and should then be monoclonal, as is indeed the case (Fialkow 1977).

The nature of a second hit is unknown, although it could be a genetic event too. Mutation at some second gene is possible. If the second gene were the one corresponding to the first, in the homologous chromosome, then the origin of cancer could be said to involve the production of clones defective for a recessive gene. Under that circumstance, cancer cells could also be produced by somatic recombination in cells heterozygous for a cancer gene. The second event could be the result of either mutation or recombination. Furthermore, any agent that could increase the probability of recombination would be carcinogenic without being mutagenic. Such an agent might be the genetic condition known as Bloom's syndrome, in which homologous chromosome exchange is known to occur in somatic cells at an increased rate and which predisposes victims to many different kinds of cancer (Festa et al. 1979). An important corollary of the recessive cancer gene hypothesis is that cancer cells at the time of origin are cells with a single gene abnormality. Explication of the function of such a single gene should provide the understanding of the fundamental defect in a cancer cell.

HUMAN CANCER GENES CAN BE CHARACTERIZED AND ISOLATED

We may hypothesize that the genes that are the sites of germinal and somatic cancer mutations can interfere with differentiation in heterozygous individuals and cells and cause cancer in homozygous cells. The characterization and isolation of such a gene should illuminate the fundamental defect of a cancer.

Two approaches to understanding these cancer genes are conceivable; one involves delineation of phenotypic abnormality, the other, genotypic abnormality. Characterization of phenotypic abnormality depends upon finding a somatic cell that can express it, a requirement that might not be easily met for genes with considerable tissue specificity. Still, the solution may be attainable in two instances. Fibroblasts from patients with the hereditary form of retinoblastoma are more radiosensitive in vitro than are those from patients with the nonhereditary form (Weichselbaum et al. 1978). Unfortunately, there are presently no clues regarding the mechanism. When fibroblasts from patients with polyposis coli are treated with the chemical promoter 12-O-tetradecanoyl phorbol-13-acetate, they are capable of producing tumors after injection into the anterior chambers of the eyes of nude mice (Kopelovich et al., 1979). These findings suggest that the cancer mutation can be studied in cell lines in vitro.

Isolation of these cancer genes has not been attempted yet, because the prospect of finding the appropriate gene in the absence of knowledge of its product seems small. Still, there are some factors whose careful consideration might lead to the gene. One is that chromosomes can be sorted with increasing accuracy.

In fact it happens that chromosome 13 is obtained in the purest form of all chromosomes by flow cytometry and sorting (Carrano et al. 1979). Consider then the possibility that three variants of a particular chromosome could be obtained: (1) normal, (2) those deleted for a cancer gene, and (3) those altered in submicroscopic fashion, as by intragenic deletion or base change. Could the technology of molecular genetics be applied to successful isolation? It is noteworthy that polymorphic genes with identifiable products are closely linked to both the retinoblastoma and the Wilms' tumor genes, esterase D in the former and the β-globin gene in the latter (Sparkes et al. 1980, Gusella et al. 1979).

SUMMARY

The evidence available from the study of somatic and germinal mutations in human cancer supports the hypothesis that cancer results from mutation at one of 100–200 human cancer genes. The normal alleles of these genes seem to be important for specific tissue differentiation, since constitutional heterozygotes often show developmental aberrations selectively in the target tissue. It may be that cancer develops when a heterozygous cell becomes homozygously defective, as by somatic mutation or somatic recombination. The genes that predispose so strongly to specific cancers promise to shed light on the nature of the fundamental defects of cancer cells. The characterization and isolation of cancer genes may be possible.

ACKNOWLEDGMENT

Supported in part by an appropriation from the Commonwealth of Pennsylvania and by research grant CA-06927 from the National Cancer Institute.

REFERENCES

Baylin, S. B., S. H. Hsu, D. S. Gann, R. C. Smallridge, and S. A. Wells. 1978. Inherited medullary thyroid carcinoma: a final monoclonal mutation in one of multiple clones of susceptible cells. Science 199:429–431.

Bove, K. E., and A. J. McAdams. 1976. The nephroblastomatosis complex and its relationship to Wilms' tumor: a clinico-pathologic treatise, in Perspectives in Pediatric Pathology, vol. 3. Year Book Medical Publishers, Chicago, pp. 185–223.

Boveri, T. 1914. Zur Frage der Entstehung Maligner Tumoren. Gustav Fischer, Jena, 64 pp.

Carrano, A. V., J. W. Gray, R. G. Langlois, K. J. Burkhart-Schultz, and M. A. Van Dilla. 1979. Measurement and purification of human chromosomes by flow cytometry and sorting. Proc. Natl. Acad. Sci. USA 76:1382–1384.

Cohen, A. J., F. P. Li, S. Berg, D. J. Marchetto, S. Tsai, S. C. Jacobs, and R. S. Brown. 1979. Hereditary renal-cell carcinoma associated with a chromosomal translocation. N. Engl. J. Med. 301:592–595.

Festa, R. S., A. T. Meadows, and R. A. Boshes. 1979. Leukemia in a black child with Bloom's syndrome: somatic recombination as a possible mechanism for neoplasia. Cancer 44:1507–1510.

Fialkow, P. J. 1977. Clonal origin and stem cell evolution of human tumors, in Genetics of Human Cancer, J. J. Mulvihill, R. W. Miller, and J. F. Fraumeni, Jr., eds. Raven Press, New York, pp. 439–453.

Fialkow, P. J., R. W. Sagebiel, S. M. Gartler, and D. L. Rimoin. 1971. Multiple cell origin of hereditary neurofibromas. N. Engl. J. Med. 284:298–300.

Francke, U., L. B. Holmes, L. Atkins, and V. M. Riccardi, 1979. Aniridia-Wilms' tumor association: evidence for specific deletion of 11p13. Cytogenet. Cell Genet. 24:185–192.

Gusella, J., A. Varsanyi-Breiner, F. T. Kao, C. Jones, T. T. Puck, C. Keys, S. Orkin, and D. Housman. 1979. Precise localization of human β-globin gene complex on chromosome 11. Proc. Natl. Acad. Sci. USA 76:5239–5243.

Hashem, N., and S. Khalifa. 1975. Retinoblastoma: a model of hereditary fragile chromosomal regions. Hum. Hered. 25:35–49.

Hethcote, H. W., and A. G. Knudson. 1978. A model for the incidence of embryonal cancers: application to retinoblastoma. Proc. Natl. Acad. Sci. USA 75:2453–2457.

Hirschhorn, K. 1968. Cytogenetic alterations in leukemia, in Perspectives in Leukemia (A Presentation of the Leukemia Society of America, Inc., December, 1966), W. Dameshek and R. M. Dutcher, eds. Grune & Stratton, New York, pp. 113–120.

Jackson, C. E., M. A. Block, K. A. Greenawald, and A. H. Tashjian. 1979. The two-mutational-event theory in medullary thyroid cancer. Am. J. Hum. Genet. 31:704–710.

King, M. C., R. C. P. Go, R. C. Elston, H. T. Lynch, and N. L. Petrakis. 1980. Allele increasing susceptibility to human breast cancer may be linked to the glutamate-pyruvate transaminase locus. Science 208:406–408.

Klein, G. 1979. Lymphoma development in mice and humans: diversity of initiation is followed by convergent cytogenetic evolution. Proc. Natl. Acad. Sci. USA 76:2442–2446.

Knudson, A. G. 1976. Genetics and the etiology of childhood cancer. Pediatr. Res. 10:513–517.

Knudson, A. G. 1979. Persons at high risk of cancer. N. Engl. J. Med. 301:606–607.

Knudson, A. G., and A. T. Meadows. 1978. Developmental genetics of neural tumors in man, in Cell Differentiation and Neoplasia (The University of Texas System Cancer Center 30th Annual Symposium on Fundamental Cancer Research, 1977). Raven Press, New York, pp. 83–92.

Knudson, A. G., and A. T. Meadows. 1980. Regression of neuroblastoma IV-S: a genetic hypothesis. N. Engl. J. Med. 302:1254–1256.

Knudson, A. G., A. T. Meadows, W. W. Nichols, and R. Hill. 1976. Chromosomal deletion and retinoblastoma. N. Engl. J. Med. 295:1120–1123.

Knudson, A. G., and L. C. Strong. 1972a. Mutation and cancer: neuroblastoma and pheochromocytoma. Am. J. Hum. Genet. 24:514–532.

Knudson, A. G., and L. C. Strong. 1972b. Mutation and cancer: a model for Wilms' tumor of the kidney. J. Natl. Cancer Inst. 48:313–324.

Knudson, A. G., L. C. Strong, and D. E. Anderson. 1973. Heredity and cancer in man. Prog. Med. Genet. 9:113–158.

Kopelovich, L., N. E. Bias, and L. Helson. 1979. Tumour promoter alone induces neoplastic transformation of fibroblasts from humans genetically predisposed to cancer. Nature 282:619–621.

McKusick, V. A. 1978. Mendelian Inheritance in Man, Ed. 5. The Johns Hopkins University Press, Baltimore, 975 pp.

Moolgavkar, S. H., N. E. Day, and R. G. Stevens, 1980. Two-stage model for carcinogenesis: epidemiology of breast cancer in females. J. Natl. Cancer Inst. 65:559–569.

Moolgavkar, S. H., and D. J. Venzon. 1979. Two-event model for carcinogenesis: incidence curves for childhood and adult tumors. Math. Biosci. 47:55–77.

Riccardi, V. M., E. Sujansky, A. C. Smith, and U. Francke. 1978. Chromosomal imbalance in the aniridia—Wilms' tumor association: 11p interstitial deletion. Pediatrics 61:604–610.

Rowley, J. D. 1973. A new consistent chromosomal abnormality in chronic myelogenous leukemia identified by quinacrine fluorescence and Giemsa staining. Nature (London) 243:290–293.

Rowley, J. D., H. M. Golomb, and C. Dougherty. 1977. 15/17 Translocation, a consistent chromosomal change in acute promyelocytic leukemia. Lancet 1:549–550.

Sparkes, R. S., M. C. Sparkes, M. G. Wilson, J. W. Towner, W. Benedict, A. L. Murphree, and J. J. Yunis. 1980. Regional assignment of genes for human esterase D and retinoblastoma to chromosome band 13q14. Science 208:1042–1044.

Tokuhata, G. K., C. L. Neely, and D. L. Williams. 1968. Chronic myelocytic leukemia in identical twins and a sibling. Blood 31:216–225.

Tyzzer, E. E. 1916. Tumor immunity. J. Cancer Res. 1:125–156.

Van Den Berghe, H., A. Louwagie, A. Broeckaert-Van Orshoven, G. David, R. Verwilghen, J. L. Michaux, and G. Sokal. 1979. Chromosome abnormalities in acute promyelocytic leukemia (APL). Cancer 43:558–562.

Warthin, A. S. 1913. Heredity with reference to carcinoma: as shown by the study of the cases examined in the pathological laboratory of the University of Michigan, 1895–1913. Arch. Intern. Med. 12:546–555.

Weichselbaum, R. R., J. Nove, and J. B. Little. 1978. X-ray sensitivity of diploid fibroblasts from patients with hereditary or sporadic retinoblastoma. Proc. Natl. Acad. Sci. USA 75:3962–3964.

Weiner, L. 1965. A family with high incidence leukemia and unique Ph[1] chromosome findings. Blood 26:871.

Yunis, J. J., and N. Ramsay. 1978. Retinoblastoma and subband deletion of chromosome 13. Am. J. Dis. Child. 132:161–163.

Yunis, J. J., and N. K. C. Ramsay. 1980. Familial occurrence of the aniridia-Wilms' tumor syndrome with deletion 11p13-14.1. J. Pediatr. 96:1027–1030.

Genes, Chromosomes, and Neoplasia, edited by
Frances E. Arrighi, Potu N. Rao, and Elton Stubblefield.
Raven Press, New York © 1981.

Genetic-Environmental Interactions in Human Cancer

Louise C. Strong

Department of Medical Genetics, The University of Texas System Cancer Center, M. D. Anderson Hospital and Tumor Institute, Houston, Texas 77030

As reviewed in the previous chapter by Knudson (1981, see pages 453 to 462, this volume), family studies (Anderson 1978), chromosomal studies (Francke et al. 1979, Cohen et al. 1979), and prospective studies of progeny of survivors of cancer (Lynch et al. 1977, Schappert-Kimmijser et al. 1966) have indicated that primary genetic factors account for at least some significant fraction of human tumors. While the overall fraction of human cancer attributable to major genes may be small, study of the development of tumors in gene carriers may provide a unique opportunity to identify environmental factors that may affect cancer development in general and to study mechanisms of carcinogenesis.

There are rare opportunities to study populations in which genetic and environmental factors are easily defined. My intent is to outline the various approaches that have been used in studying genetic and environmental interactions in cancer and to examine their implications.

DISORDERS OF MUTAGENESIS AND CARCINOGENESSIS

Certain rare autosomal recessive diseases are associated with very high cancer risks apparently related to genetic (chromosomal) instability (see review by Ray and German 1981, pages 351 to 378 this volume). The in vitro findings of high mutation rates in these conditions associated with high cancer rates in vivo strongly support the association between mutagenesis and carcinogenesis. The mutation rates and the chromosomal instability in vitro may be dramatically increased by exposure to specific mutagenic or carcinogenic agents; in vivo, there are very high cancer rates and, at least for xeroderma pigmentosum, an obvious correlation between ultraviolet exposure and tumor development.

In ataxia-telangiectasia, there appears to be a defect in repair of gamma radiation–induced DNA damage as indexed by increased sensitivity to cell killing by radiation, a severe acute toxic reaction to radiation therapy, and an increased risk of cancer, frequently of the radiosensitive hematopoietic system (for review, see Strong 1977a).

In Bloom's syndrome, increased numbers of sister chromatid exchanges (Chaganti et al. 1974), considered by some to be an index of mutagenesis, and a high frequency of quadriradials, indicative of homologous chromosome exchange, have been observed in vitro. The disorder is otherwise characterized by an increased risk of various cancers, such that all patients who have reached the age of 35 years have developed some cancer (German et al. 1977). The observations of patches of hyperpigmented and hypopigmented skin in a patient with Bloom's syndrome and leukemia suggested to Festa et al. (1979) that somatic recombination had occurred at a high rate, resulting in clones of cells homozygous for a given gene affecting pigmentation. Extrapolation of those observations to the etiology of the leukemia suggested to the authors that homozygosity achieved by somatic recombination for a gene predisposing to leukemia or other cancer might provide a mechanism for the high cancer rate.

Patients with Fanconi's anemia suffer pancytopenia and often develop acute myelomonocytic leukemia or squamous cell carcinoma, disorders often associated with myelosuppression. In addition, hepatoma and hepatocellular carcinoma following relatively short, low-dose androgen treatments, or macronodular cirrhosis with hemochromatosis secondary to multiple transfusions, have been observed. In vitro study of cells from patients with Fanconi's anemia reveals chromosomal instability and increased sensitivity to certain mutagens; in addition there may be a defect in repair of gamma radiation–induced DNA damage, or some defect in the repair of cross-linking of DNA (for review, see Strong 1977a).

CANCER RISK IN SURVIVORS OF EMBRYONAL TUMORS RELATED TO AUTOSOMAL DOMINANT PREDISPOSING GENES AND RADIATION THERAPY

Individuals with autosomal dominant cancer-predisposing disorders (as demonstrated by many mechanisms, including family studies, chromosomal studies, or prospective studies of familial cancer patients) have been shown to have a risk of cancer of a specific tissue type approaching 100% over the indicated years at risk (Schappert-Kimmijser et al. 1966, Lynch et al. 1977, Cohen et al. 1979, Anderson 1978). While for the individual and the overall target tissue this risk implies a gene with very high penetrance or likelihood of a given phenotype (cancer), probably millions of target cells carrying the predisposing genes are at risk and only occasionally does one develop tumor. The small risk of tumor per target cell implies that some other rare event(s) must occur in a predisposed cell to produce tumor. That such a rare event is a discrete event occurring in a single cell is supported by the demonstration with biochemical and genetic markers of a clonal origin for tumors in at least one heritable tumor syndrome (Baylin et al. 1978).

Given, then, that a single gene may dramatically increase the risk of cancer for a specific target tissue or set of tissues, but that the gene alone is not sufficient to cause tumor in every cell, and some other rare discrete event(s) must occur, I will attempt to review the available data regarding the factors that may affect that (those) event(s).

The approach that we have taken has been to study the cancer risk in survivors of childhood cancer, some of whom have a demonstrable heritable tumor (e.g., bilateral retinoblastoma, nevoid basal cell carcinoma syndrome, or familial Wilms' tumor or other childhood tumors), and some of whom have been treated with known carcinogens including radiation or chemotherapy, or both, at known times and doses. Although the numbers of patients and of second tumors are small, some definite patterns have emerged.

Nevoid Basal Cell Carcinoma Syndrome (NBCCS)

The NBCCS is an autosomal dominant disorder characterized by multiple skeletal anomalies, dentigerous keratocysts, ectopic calcifications, and palmar pits, in addition to a propensity to develop multiple basal cell carcinomas, ovarian fibromas, and occasionally medulloblastoma or other tumors. The distribution of basal cell carcinomas in NBCCS patients is similar to that of basal cell carcinomas in the general population, presumably reflecting the role of ultraviolet exposures, e.g., basal cell carcinomas are most frequent in sun-exposed areas, and are rare in black patients. However, the onset and frequency of basal cell carcinomas are unique in that patients with NBCCS develop tumors at an early age (an average age at diagnosis of 15 years) and may develop hundreds of primary tumors (Strong 1977a).

We have previously reported three patients with NBCCS and medulloblastoma treated by radiation therapy (Strong 1977a). Each developed a similar and characteristic distribution of nevi that histologically were basal cell carcinomas, i.e., along the margins of the radiation field, within 1 to 3 years after radiation treatment. Among more than 30 long-term survivors of brain tumors or acute leukemias treated with cranial and or spinal radiation of comparable doses during early childhood at M. D. Anderson Hospital, none have developed basal cell carcinomas in 3 to 25 years of follow-up. However, similar lesions within the irradiated areas have occurred soon after radiation in every child reported with NBCCS who received radiation treatment, including one patient treated by radiation therapy for thymic enlargement (Strong 1977a,b, Cutler et al. 1979, Hawkins et al. 1979, Schweisguth et al. 1968, Southwick and Schwartz 1979, Scharnagel and Pack 1949). Many of these children had an extended family history of NBCCS, implying that they did not necessarily have a new or more severe mutation predisposing to medulloblastoma and earlier than usual onset of basal cell carcinomas.

Follow-up of the three previously reported patients with NBCCS over 8 to 15 years following radiation (Strong 1977a) revealed that after the initial burst of basal cell carcinomas in the radiation-treated areas over a 1- to 3-year period, few new lesions developed in areas not further exposed to ionizing or ultraviolet radiation. Basal cell carcinomas in sun-exposed areas, however, became increasingly numerous and severe.

Ovarian fibromas are frequent in postpubertal females with NBCCS. However, the only two female survivors of NBCCS and medulloblastoma, treated with spinal radiation, have developed ovarian fibromas and fibrosarcomas at the ages of 4 and 8 years, respectively. No other pediatric patient treated with cranial or spinal radiation at M. D. Anderson Hospital has developed ovarian fibromas or fibrosarcomas, and no other reported NBCCS patient has developed ovarian fibromas before puberty (Burket and Rauh, 1976).

The effect of radiation therapy in enhancing the frequency of the specific tumors to which the NBCCS patients are uniquely predisposed and the relatively quick tumor development after radiation therapy suggest the following:

1. The carcinogenic effect of radiation, at least within 8 to 15 years, is tissue specific for those tissues predisposed to cancer in the NBCCS. That the specificity of the effect is related to the genetic background and not the characteristics of radiation in general is evidenced by the absence of such effect in other patients similarly treated. Both ultraviolet and ionizing radiation appear to enhance tumor development.

2. The effect of exposure to ionizing radiation seems to be limited in time, with a short latent period and early peak time for tumor development; whether there will be a different level of response after the usual 20-year latent period is unknown, but at 10-year follow-up no dramatic new tumor development has been noted.

3. The primary genetic defect must be very different from that of xeroderma pigmentosum (XP). In XP, hundreds of skin tumors develop in sun-exposed areas, but the carcinogenic effect is specific for ultraviolet radiation and not for ionizing radiation. Furthermore, the tumors that are induced by ultraviolet radiation are not only basal cell carcinomas but also include squamous cell carcinomas and melanomas. The major carcinogenic effects in XP are consistent with a defect in repair of ultraviolet-induced DNA damage's causing an increased mutation rate, which in turn causes a nonspecific increased cancer rate in exposed tissues.

By contrast, in the NBCCS both ultraviolet and ionizing radiation are effective in inducing skin tumors, but the tumors are almost exclusively basal cell carcinomas. These findings suggest that the genetic defect may lie within a specifically predisposed target tissue, and that perhaps exposure to any carcinogen enhances that predisposition.

4. It has been suggested that cancer arises in a multistage process, involving two to seven stages (Whittemore 1977). If the development of basal cell carcinoma is due to a multistage process, and if in the NBCCS the first stage is inherited, then the radiation effects, including the multiplicity of basal cell carcinomas, the short latent period, and the relative brevity of the effect suggest

that the number of stages is small and that radiation may induce the second or final stage.

Retinoblastoma

Retinoblastoma is unique among childhood cancers in that it has long been noted to occur more often than expected in close relatives of patients who have had the disease. The relatively favorable prognosis associated with retinoblastoma has permitted many patients to survive and reproduce. Study of their progeny has demonstrated that patients with bilateral retinoblastoma, and a small fraction (10–15%) of patients with unilateral retinoblastoma, have an autosomal dominant condition with high gene penetrance. Other unilateral retinoblastoma patients probably have a nonheritable tumor. Rare patients with retinoblastoma have an identifiable genetic defect, a chromosome deletion (Knudson et al. 1976). The early age at diagnosis and frequency of multifocal tumors in patients with heritable retinoblastoma, as contrasted to those with nonheritable retinoblastoma, suggested to Knudson (Knudson 1971, Knudson et al. 1975) that tumor development fit a two-stage model in which the first event might be a germinal or somatic mutation, the second a stochastic event occurring with the expected frequency of a somatic mutation.

The retinal tumors occur so early in life and so predictably in gene carriers that the role of environmental factors is difficult to determine, although variations from the expected tumor distributions according to a stochastic process may be in part environmental. However, a more obvious opportunity to observe environmental effects on a uniquely predisposed patient population involves study of the development of second malignant neoplasms.

Patients with bilateral or heritable retinoblastoma have a high risk of second malignant neoplasms, most often osteosarcoma (Sagerman et al. 1969, Jensen and Miller 1971, Kitchen and Ellsworth 1974, Abramson et al. 1976, 1979). Abramson et al. (1979) noted that spontaneous second malignant neoplasms occurred in 14% of patients with bilateral retinoblastoma who received no radiation therapy and were followed over 1 to 32 years. However, radiation-related second malignant neoplasms have been observed in nearly one third of patients treated with 11,000 to 15,000 rads kilovoltage radiation and followed for 25 to 35 years (Sagerman et al. 1969). The most common spontaneous or radiation-related second malignant neoplasm has been osteosarcoma, a tumor also observed in close relatives of retinoblastoma patients (Gordon 1974). For spontaneous osteosarcoma (osteosarcoma arising in the areas not receiving radiation therapy), the reported risk is increased 500-fold to 1 in 200 persons over 40 years (Kitchen and Ellsworth 1974); for radiation-related osteosarcoma, the risk determined by life-table methods in patients followed for up to 35 years indicates a dose-related risk greater than 1 in 10 for patients treated with more than 6,000 rads (Sagerman et al. 1969).

The usual peak latent period for radiation-related neoplasms in survivors of childhood cancer is greater than 15 years after radiation (Li et al. 1975). In retinoblastoma survivors, the peak time to tumor development after radiation is 5 to 6 years (Sagerman et al. 1969, Strong 1977b). As most retinoblastoma patients are diagnosed and treated before the age of 2 years, the age peak for radiation-related osteosarcoma and other second malignant neoplasms is long before the peak age for spontaneous osteosarcoma (Glass and Fraumeni 1970). The highest rate of second malignant neoplasms occurs during an age at which spontaneous cancer rates are low, and the most frequent expected spontaneous tumors would be leukemias or brain tumors (Cutler and Young 1975). Thus, the second neoplasms demonstrate tumor specificity that cannot be explained by age or radiation characteristics.

These findings in survivors of heritable retinoblastoma, e.g., the high spontaneous risk of second malignant neoplasms, the high risk of radiation-related second malignant neoplasms, the relative tissue specificity for osteosarcoma in spontaneous and radiation-related tumors, the dose-response relationship for the radiation-related cancer risk, and the short latent period for radiation-induced tumors, demonstrate that radiation may be enhancing the natural tumor predisposition in retinoblastoma gene carriers, increasing the risk of tumors to which the individual is predisposed. The short latent period and tissue specificity for the radiation-induced cancers are consistent with radiation's providing the final cellular event, presumably a mutation, in a predisposed target tissue.

Nove et al. (1979) investigated the in vitro effect of radiation on tissues from retinoblastoma patients and demonstrated that some patients with heritable retinoblastoma seem to have an increased sensitivity to cell killing by radiation. However, it is not clear whether these findings are due to the "retinoblastoma gene," adjacent linked genes, or some other phenomenon, or whether the "sensitivity" is specific for ionizing radiation or clinically correlated with the occurrence of radiation-related second neoplasms in some subset of retinoblastoma patients.

Survivors of Other Embryonal Tumors

Although it has been suggested that some small fraction of other embryonal tumors is heritable, for no other tumor types are there such readily identifiable "markers" for the heritable fraction as the anomalies of the NBCCS or the bilaterality of heritable retinoblastoma. However, for other embryonal cancers there are some data that suggest that radiation-related second malignant neoplasms occur most often in the small fraction of patients with some evidence of a genetic syndrome.

The characteristics of Wilms' tumor patients with radiation-induced second malignant neoplasms reported by Meadows et al. (1978) contrasts with characteristics of an unselected series of Wilms' tumor patients (Pendergrass 1976) (Table 1). The high frequency of patients in the former series who had evidence of a genetic predisposing background suggests that those genetic conditions are asso-

TABLE 1. *Wilms' tumor (WT): Characteristics of patients with radiation (XRT)-related second malignant neoplasms (SMN)*

Characteristic	Observed : NWTSG[+]		Observed : XRT-related SMN[*]	
	No.	%	No.	%
Total cases	547	—	20	—
Sib with WT	2	0.4	2	10
Sib with other childhood cancer	0	—	1	5
Neurofibromatosis	0	—	2	10
Bilateral WT	22	4.0	1	5

[+] NWTSG, National Wilms' Tumor Study Group; reported by Pendergrass (1976).
[*] Patients with SMN reported by Meadows et al. 1978, and updated, Meadows, personal communication, 1979.

ciated with an increased risk of a radiation-related second malignant neoplasm.

Case reports and reviews of survivors of childhood cancer who develop radiation-related second malignant neoplasms further suggest that rare genetic disorders expected to represent less than 1% of childhood tumors, including von Recklinghausen's neurofibromatosis, breast cancer-sarcoma familial cancer syndrome, and familial childhood cancers, are associated with a high frequency of radiation-related second malignant neoplasms (McKeen et al. 1978, Reimer et al. 1977, Li and Fraumeni 1969a, 1969b, 1975, Meadows et al. 1978). The observations are consistent with the model developed for NBCCS and retinoblastoma, that radiation-related second malignant neoplasms are dose related, nonrandom among persons exposed to comparable radiation doses, and most often of the specific tissue type to which the individuals are predisposed.

Considerable data are accumulating in the literature concerning the risk of second malignant neoplasms in patients treated with chemotherapy. However, none of these reports has provided information on familial or congenital conditions, and to date the information from survivors of childhood cancers has not been sufficient to determine whether specific genetic subgroups of patients are more susceptible to the carcinogenic effects of chemotherapy.

The study of genetic and environmental factors and of the risk of radiation-related second malignant neoplasms in childhood cancer survivors has provided some new insights into mechanisms of carcinogenesis, and the results have practical implications for identifying those patients at greatest risk of the carcinogenic effects of radiation. However, many questions remain to be answered, including whether the genetic predisposition is related to radiation effects or includes increased susceptibility to many or most environmental carcinogens; whether identifiable genetic subgroups are more susceptible to the carcinogenic effects of chemotherapy; and whether programs of prevention or early detection can alter the natural course of the condition.

The above studies involve study of the cancer risk in a population of patients for whom at least some information on the genetic background and environmental carcinogen exposures are available. Such populations are rare. However, other approaches to the study of genetic and environmental interactions in human cancer have involved study of environmental factors affecting cancer risk in persons with apparent genetic syndromes and review of genetic (most often familial) factors in patients with tumors apparently related to environmental carcinogens.

IMMUNODEFICIENCY AND CANCER: DUNCAN'S DISEASE OR AUTOSOMAL RECESSIVE SEX-LINKED LYMPHOPROLIFERATIVE DISEASE

Patients with heritable and acquired immunodeficiency disorders have long been noted to have an increased cancer risk. The risks seem to be specific for selected tumor types. Study of at least one heritable immunodeficiency syndrome has suggested that the oncogenic "trigger" may be a common environmental factor, the Epstein-Barr virus (EBV).

In 1974, Purtilo et al. reported a family in which several young male members of a large kindred had died of lymphoproliferative disease following exposure to EBV. Subsequent studies (Purtilo et al. 1978) have revealed an X-linked recessive lymphoproliferative syndrome in which an inherited immunodeficiency to control of EBV-induced B cell proliferation in males results in fatal infectious mononucleosis, immunoblastic sarcoma, plasmacytoma, American Burkitt's lymphoma, histiocytic lymphoma, or agammaglobulinemia. Study of families with this rare condition has revealed that various types of clinical immunodysfunction may be related to interaction between a single gene defect and exposure to an almost ubiquitous virus. The complex immune interactions that lead to the observed phenotypes provide a unique model for study of immunoregulatory mechanisms and B cell neoplasia in general.

FAMILIAL CANCER AGGREGATIONS AND ENVIRONMENTAL FACTORS

The above review has focused on relatively rare cancers for which genetic and environmental factors may be apparent. However, probably among common cancers as well, genetically determined factors affect susceptibility to environmental carcinogens. A classic example involves the interaction of familial risk and smoking-related risk of lung cancer as described by Tokuhata (1976). Study of relatives of lung cancer patients and relatives of controls indicated that a positive familial history of lung cancer in the absence of smoking in the relatives of patients or controls was associated with a fourfold increased risk of lung cancer, smoking in the absence of a positive familial history was associated with a fivefold increased risk of lung cancer, but a positive familial history

TABLE 2. *Selected examples of reports of genetic and environmental factors in common cancers*

Indication for study	Genetic factor	Environmental factor	Reference
Familial bladder cancer	Metabolic defect in tryptophan metabolism	Tryptophan metabolites	Leklem and Brown 1976
Familial bladder cancer	Autosomal dominant heritable predisposition to early onset bladder cancer	Industrial exposures, cigarette smoking	Purtilo et al. 1979
Familial BK mole/melanoma	Autosomal dominant heritable syndrome	Ultraviolet radiation (sun lamp)	Greene and Fraumeni 1979
Familial Hodgkin's disease	Familial aggregation	Woodworking occupation; Exposure to chemical preservatives	Greene et al. 1978
Leukemia in benzene workers	Familial aggregation of affected workers	Occupational exposure to benzene	Aksoy et al. 1976
Bladder cancer among aniline dye workers	Familial aggregation of affected workers	Occupational exposure to aniline dye	Purtilo et al. 1978
High-risk geographic region for nasopharyngeal cancer	Immunogenetic determinants, histocompatibility-linked genes (A2-Sin 2 linkage disequilibrium)	Asian Chinese lifestyle, possibly including fume exposure	Simons et al. 1977

and smoking were associated with a 14-fold increased risk. More recent studies by Cohen et al. (1977a, 1977b, Cohen 1978) have suggested that familial (genetic?) factors and smoking independently but perhaps also synergistically predispose to chronic obstructive lung disease, a precursor to lung cancer.

Familial aggregations of cancer do not define a "genetic" factor; however, observations of familial cancer aggregations may provide etiologic clues related to genetic or environmental factors. Tumors occurring in multiple family members, distributed throughout a single germ line, at relatively early ages for the specific tumors, and at multiple primary sites are findings most suggestive of a major heritable factor. However, more subtle genetic mechanisms likely also affect cancer risk and may be linked to metabolism of specific agents. Recording of environmental exposures in reports of "familial" cancer and of familial or other genetic characteristics in reports of "environmental" cancers is urged. Examples of selected reports are summarized in Table 2.

CONCLUSION

I have attempted to review the data on genetic and environmental interactions in human cancer that seem to contribute to the understanding of carcinogenesis in general. These data strongly support a multistage model for at least some cancers. That at least one stage involves acquired or inherited mutation is supported by the correlation between in vitro mutation rates and in vivo cancer rates in patients with disorders of genetic instability, by the autosomal dominant heritable tumors, and by the relationship of chromosomal deletion or rearrangement to cancer. That the number of discrete cellular changes is greater than 1, but still a small number, is supported by the rare development of tumor in tissues carrying an autosomal dominant, tissue-specific mutation and by the enhancement of tumor development by a brief exposure to a known carcinogen, radiation, within a short time after the exposure. The consistent finding that radiation-induced tumors are tissue specific for the unique genetic predisposition in persons with different autosomal dominant, cancer-predisposing mutations suggests that those mutations are not specific for radiation-induced defects, but for tissue susceptibility. These findings raise concern that such individuals may be susceptible to other environmental carcinogens.

The nature of the changes induced by radiation in the predisposed tissues is not clear. However, Festa et al. (1979) suggested an appealing hypothesis, consistent with data from syndromes of genetic instability and autosomal dominant diseases. According to that hypothesis, the first stage in carcinogenesis is a stable mutation, somatic or inherited, and the second an event of somatic recombination at the same genetic locus. The recombination event would produce a cell homozygous recessive for the cancer-predisposing mutation. If this model is correct, individuals with an autosomal dominant heritable tumor predisposition might be uniquely susceptible to carcinogens that increase somatic recombination.

ACKNOWLEDGMENTS

This investigation was supported by Grant Number RR5511-17 awarded by the National Institutes of Health; CA2559 by the National Cancer Institute; and 6-235 from the National March of Dimes.

REFERENCES

Abramson, D. H., R. M. Ellsworth, and L. E. Zimmerman. 1976. Nonocular cancer in retinoblastoma survivors. Trans. Am. Acad. Ophthalmol. Otolaryngol. 81:OP454–OP457.

Abramson, D. H., H. J. Ronner, and R. M. Ellsworth. 1979. Second tumors in nonirradiated bilateral retinoblastoma. Am. J. Ophthalmol. 87:624–627.

Aksoy, M., S. Erdem, G. Erdogan, and G. Dincol. 1976. Combination of genetic factors and chronic exposure to benzene in the aetiology of leukaemia. Hum. Hered. 26:149–153.

Anderson, D. E. 1978. Familial cancer and cancer families. Semin. Oncol. 5:11–16.

Baylin, S. B., S. H. Hus, and D. S. Gann. 1978. Inherited medullary thyroid carcinoma: A final monoclonal mutation in one of multiple clones of susceptible cells. Science 199:429–430.

Burket, R. C., and J. L. Rauh. 1976. Gorlin's syndrome: Ovarian fibromas at adolescence. Obstet. Gynecol. 47:43s–46s.

Chaganti, R. S. K., S. Schonberg, and J. German. 1974. A manyfold increase in sister chromatid exchanges in Bloom's syndrome lymphocytes. Proc. Natl. Acad. Sci. USA 71:4508–4512.

Cohen, A. J., F. P. Li, S. Berg, D. J. Marchetto, S. Tsai, S. C. Jacobs, and R. S. Brown. 1979. Hereditary renal-cell carcinoma associated with a chromosomal translocation. N. Engl. J. Med. 301:592–595.

Cohen, B. H. 1978. Is pulmonary dysfunction the common denominator for the multiple effects of cigarette smoking? Lancet 2:1024–1027.

Cohen, B. H., W. C. Ball, Jr., S. Brashears, E. L. Diamond, P. Kreiss, D. A. Levy, H. A. Menkes, S. Permutt, and M. S. Tockman. 1977a. Risk factors in chronic obstructive pulmonary disease (COPD). Am. J. Epidemiol. 105:223–232.

Cohen, B. H., C. G. Graves, D. A. Levy, S. Permutt, M. S. Tockman, E. L. Diamond, P. Kreiss, H. A. Menkes, and S. Quaskey. 1977b. A common familial component in lung cancer and chronic obstructive pulmonary disease. Lancet 2:523–526.

Cutler, S. J., and J. L. Young, Jr. 1975. Third National Cancer Survey: Incidence Data, NCI Monogr. 41, DHEW Publ. No. (NIH) 75–787. National Cancer Institute, Bethesda, Maryland.

Cutler, T. P., C. A. Holden, and D. M. MacDonald. 1979. Multiple naevoid basal cell carcinoma syndrome (Gorlin's syndrome). Clin. Exp. Dermatol. 4:373–379.

Festa, R. S., A. T. Meadows, and R. A. Boshes. 1979. Leukemia in a black child with Bloom's syndrome. Cancer 44:1507–1510.

Francke, U., L. B. Holmes, L. Atkins, and V. M. Riccardi. 1979. Aniridia-Wilms' tumor association: Evidence for specific deletion of 11p13. Cytogenet. Cell Genet. 24:185–192.

German, J., D. Bloom, and E. Passarge. 1977. Bloom's syndrome. V. Surveillance for cancer in affected families. Clin. Genet. 12:162–168.

Glass, A. G., and J. F. Fraumeni, Jr. 1970. Epidemiology of bone cancer in children. J. Natl. Cancer Inst. 44:187–199.

Gordon, H. 1974. Family studies in retinoblastoma. Birth Defects 10:185–190.

Greene, M. H., L. A. Brinton, J. F. Fraumeni, and R. D'Amico. 1978. Familial and sporadic Hodgkin's disease associated with occupational wood exposure. Lancet 2:626–627.

Greene, M. H., and J. F. Fraumeni, Jr. 1979. The hereditary variant of malignant melanoma, *in* Human Malignant Melanoma, W. H. Clark, Jr., L. I. Goldman, and M. J. Mastrangelo, eds. Grune & Stratton, Inc., New York, pp. 139–166.

Hawkins, J. C., III, H. J. Hoffman, and L. E. Becker. 1979. Multiple nevoid basal-cell carcinoma syndrome (Gorlin's syndrome): Possible confusion with metastatic medulloblastoma. J. Neurosurg. 50:100–102.

Jensen, R. D., and R. W. Miller. 1971. Retinoblastoma: Epidemiologic characteristics. N. Engl. J. Med. 285:307–311.

Kitchin, F. D., and R. M. Ellsworth. 1974. Pleiotropic effects of the gene for retinoblastoma. J. Med. Genet. 11:244–246.

Knudson, A. G., Jr. 1971. Mutation and cancer: A statistical study of retinoblastoma. Proc. Natl. Acad. Sci. USA 68:820–823.

Knudson, A. G., Jr. 1981. Human cancer genes, *in* Genes, Chromosomes, and Neoplasia (The University of Texas System Cancer Center 33rd Annual Symposium on Fundamental Cancer Research), F. E. Arrighi, P. N. Rao, and E. Stubblefield, eds. Raven Press, New York, pp. 453–462.

Knudson, A. G., Jr., H. W. Hethcote, and B. W. Brown. 1975. Mutation and childhood cancer: A probabilistic model for the incidence of retinoblastoma. Proc. Natl. Acad. Sci. USA 72:5116–5120.

Knudson, A. G., Jr., A. T. Meadows, W. W. Nichols, and R. Hill. 1976. Chromosomal deletion and retinoblastoma. N. Engl. J. Med. 295:1120–1123.

Leklem, J. E., and R. R. Brown. 1976. Abnormal tryptophan metabolism in a family with a history of bladder cancer. J. Natl. Cancer Inst. 56:1101–1104.

Li, F. P., J. R. Cassady, and N. Jaffe. 1975. Risk of second tumors in survivors of childhood cancer. Cancer 35:1230–1235.

Li, F. P., and J. F. Fraumeni, Jr. 1969a. Soft-tissue sarcomas, breast cancer, and other neoplasms. A familial syndrome? Ann. Intern. Med. 71:747–749.

Li, F. P., and J. F. Fraumeni, Jr. 1969b. Rhabdomyosarcoma in children: Epidemiologic study and identification of a familial cancer syndrome. J. Natl. Cancer Inst. 43:1365–1373.

Li, F. P. and J. F. Fraumeni, Jr. 1975. Familial breast cancer, soft-tissue sarcomas, and other neoplasms. Ann. Intern. Med. 83:833–834.

Lynch, H. T., R. E. Harris, P. M. Lynch, H. A. Guirgis, J. F. Lynch and W. A. Bardawil. 1977. Role of heredity in multiple primary cancer. Cancer 40:1849–1854.

McKeen, E. A., J. Bodurtha, A. T. Meadows, E. C. Douglass, and J. J. Mulvihill. 1978. Rhabdomyosarcoma complicating multiple neurofibromatosis. J. Pediatr. 93:992–993.

Meadows, A. T., G. J. D'Angio, A. E. Evans, N. Jaffe, O. Schweisguth, and J. van Eys. 1978. Spontaneous and treatment-related second malignant neoplasms in children, *in* Tumors of Early Life in Man and Animals, L. Severi, ed. Perugia Quadrennial International Conferences on Cancer, Monteluce, Italy, pp. 45–59.

Nove, J., J. B. Little, R. R. Weichselbaum, W. W. Nichols, and E. Hoffman. 1979. Retinoblastoma, chromosome 13, and in vitro cellular radiosensitivity. Cytogenet. Cell Genet. 24:176–184.

Pendergrass, T. W. 1976. Congenital anomalies in children with Wilms' tumor. Cancer 37:403–409.

Purtilo, D. T., B. McCarthy, J. P. S. Yang, G. H. Friedell, and the Worcester Urology Group. 1979. Familial urinary bladder cancer. Semin. Oncol. 6:254–256.

Purtilo, D. T., C. K. Cassel, and J. P. S. Yang. 1974. Fatal infectious mononucleosis in familial lymphohistiocytosis. N. Engl. J. Med. 291:736.

Purtilo, D. T., L. Paquin, and T. Gindhart. 1978. Genetics of neoplasia–Impact of ecogenetics on oncogenesis. Am. J. Pathol. 91:609–687.

Ray, J. H., and J. German. 1981. The chromosome changes in Bloom's syndrome, ataxia-telangiectasia, and Fanconi's anemia, *in* Genes, Chromosomes, and Neoplasia (The University of Texas System Cancer Center 33rd Annual Symposium on Fundamental Cancer Research), F. E. Arrighi, P. N. Rao, and E. Stubblefield, eds. Raven Press, New York, pp. 351–378.

Reimer, R. R., J. F. Fraumeni, Jr., R. Reddick, and E. L. Moorhead II. 1977. Breast carcinoma following radiotherapy of metastatic Wilms' tumor. Cancer 40:1450–1452.

Sagerman, R. H., J. R. Cassady, P. Tretter, and R. M. Ellsworth. 1969. Radiation induced neoplasia following external beam therapy for children with retinoblastoma. Am. J. Roentgenol. Radium Ther. Nucl. Med. 105:529–535.

Schappert-Kimmijser, J., G. D. Hemmes, and R. Nijland. 1966. The heredity of retinoblastoma. Ophthalmologica 151:197–213.

Scharnagel, I. M., and G. T. Pack. 1949. Multiple basal cell epitheliomas in a five-year-old child. Am. J. Dis. Child. 77:647–651.

Schweisguth, O., R. Gerard-Marchant, and J. Lemerle. 1968. Naevomatose baso-cellulaire association a un rhabdomyosarcome congenital. Arch. Franc. Ped. 25:1083–1093.

Simons, M. J., G. B. Wee, D. Singh, S. Dharmalingham, N. K. Yong, J. C. W. Chau, J. H. C. Ho, N. E. Day, and G. De-The. 1977. Immunogenetic aspects of nasopharyngeal carcinoma.

V. Confirmation of a Chinese-related HLA profile (A2, Singapore 2) associated with an increased risk in Chinese for nasopharyngeal carcinoma. J. Natl. Cancer Inst. 47:147–151.

Southwick, G. J., and R. A. Schwartz. 1979. The basal cell nevus syndrome. Cancer 44:2294–2305.

Strong, L. C. 1977a. Genetic and environmental interactions. Cancer 40:1861–1866.

Strong, L. C. 1977b. Theories of pathogenesis: Mutation and cancer, *in* The Genetics of Human Cancer, J. J. Mulvihill, R. W. Miller, and J. F. Fraumeni, Jr., eds. Raven Press, New York, pp. 401–414.

Tokuhata, G. K. 1976. Cancer of the lung: Host and environmental interaction, *in* Cancer Genetics, H. T. Lynch, ed. Charles C. Thomas, Springfield, Illinois, pp. 213–232.

Whittemore, A. S. 1977. The age distribution of human cancer for carcinogenic exposures of varying intensity. Am. J. Epidemiol. 106:418–432.

Genes, Chromosomes, and Neoplasia, edited by
Frances E. Arrighi, Potu N. Rao, and Elton Stubblefield.
Raven Press, New York © 1981.

A Population-Based Assessment of Familial Cancer Risk in Utah Mormon Genealogies

M. Skolnick, D. T. Bishop, D. Carmelli, E. Gardner, R. Hadley,
S. Hasstedt, J. R. Hill, S. Hunt, J. L. Lyon,* C. R. Smart, †
and R. R. Williams‡

*Department of Medical Biophysics and Computing, *Department of Family and Community Medicine, †Department of Surgery, and ‡Department of Internal Medicine, LDS Hospital and University of Utah Medical Center, Salt Lake City, Utah 84143*

The last century has ushered in an era of modern medical care. Consequently, many infectious diseases and other afflictions of industrialization and poor sanitation, which accompanied urbanization, have been eradicated. As the expectation of life continues to increase, we are left with new major causes of illness and death that must be understood and eventually controlled.

Many late-age onset diseases, such as cancer, used to be rare but have become more common with a shift in life expectancy. For some of them, it is known that life style can affect relative risks, but for others there is no clear insult such as exposure to a particular high-risk occupation, drug, or diet. This leads us to search for underlying genetic predispositions that singly, or in combination with other genes or nongenetic factors, may produce a disease. To date, genetic analysis has led to the description of many large kindreds with genetic predispositions. However, because of the lack of a reference population and the difficulty of correcting for ascertainment, the relative importance of these families in the total burden of cancer is unclear. In this paper, we begin the assessment of genetic factors in a large, well-defined population, the Utah descendants of the Mormon pioneers.

Two major types of risk factors for cancer depend on familial data. The first are family formation variables (age at first delivery, onset and end of the reproductive period, number of children, and age at last delivery), which affect the risk of breast cancer (Stasjewski 1971, Trichopoulos et al. 1972, MacMahon et al. 1973, Stavraky 1974, Henderson et al. 1974, Tulinius et al. 1978, Hunt et al. 1980). The second is familial history, which increases risk for cancer at some sites (Knudson et al. 1973, Anderson 1975). In particular, the excess of breast cancer in certain families and the corresponding increased risk in relatives have been reported (Lynch et al. 1976a). Further, pairs of cancer sites have been associated and reported as aggregating in families (e.g., breast and ovarian cancers [Lynch et al. 1976b]). For most major sites, the nature of the familial factors has not previously been determined. However, for breast cancer there is a growing body of evidence that familiality reflects the presence of major genes segregating in families (Hill et al. 1978, Bishop and Gardner 1980, King and Bishop 1980, King et al. 1980). Genetic modes of transmission of a predisposition to cancer have been established for a variety of the less common neoplasms such as Gardner's syndrome (polyp colon cancer, Gardner 1951, 1962), heritable retinoblastoma (Knudson 1971), neuroblastoma (Knudson and Strong 1972a), and Wilms' tumor of the kidney (Knudson and Strong 1972b). These rare cancers of genetic predisposition are more easily ascertained in pedigrees than are genetically induced common cancers. This is due to the relatively large probability that a common sporadic cancer will mimic genetic transmission by chance in a small number of pedigrees drawn from a large population. Thus, the definition of gene-environment interactions for common sites can proceed with proper ascertainment only after studies of familial clusters in defined populations have been conducted.

In this chapter we present an analysis of familial aggregation of various cancer sites in Utah Mormon genealogies and an analysis of family formation variables in the same population. We then discuss the interaction of the two types of risk factors in a large, well-studied breast cancer pedigree (Gardner 1980), where the risk of breast cancer shows the effects of both a genetic predisposition and a late age at first delivery.

THE UTAH POPULATION

The Utah descendants of the Mormon pioneers have several characteristics that make them an ideal study population for familial and genetic risk factors. First of these is large family size, which was greater than eight children per couple for much of the last century (Skolnick et al. 1978, Mineau et al. 1979). Second, because of their cultural emphasis on genealogy, an immense resource has been created to aid any person interested in ancestral research (Bean et al. 1980). From these data, housed in the Genealogical Society of Utah, we gathered family group sheets (Figure 1) on all families that contained at least one individual who was born or had died in Utah or along the pioneer trail. Data on the 1.2 million individuals listed on these sheets have beem computerized and the individuals linked into a genealogy representing the majority of the Utah descendants of the Mormon pioneers (Skolnick 1977, Skolnick et al. 1979).

The cancer cases for the population studies come from the Utah Cancer Registry. It began in 1952 in some hospitals and become statewide in 1966. Since all cancers diagnosed in the state are recorded, this file represents a completely ascertained data source. This means that we see families with low incidence of cancer as well as with high. Since, for several major sites, the Utah Mormons have a lower cancer incidence than the United States as a whole, genetically predisposed families would be expected to stand out more from background noise (sporadic cancers) than in other populations. The familial

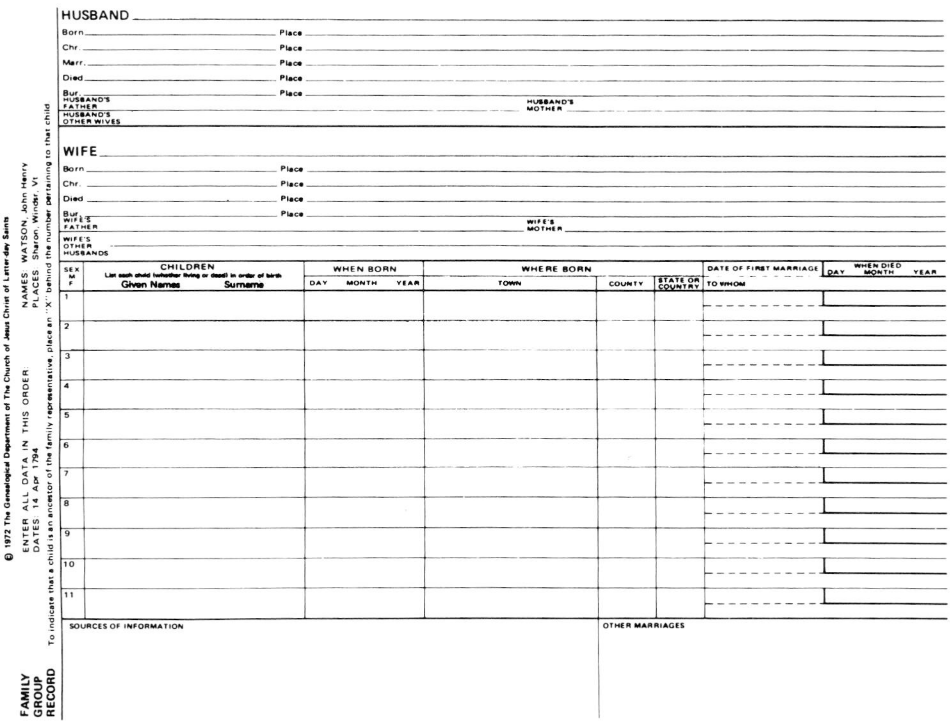
HUSBAND
Born — Place
Chr. — Place
Marr. — Place
Died — Place
Bur. — Place
HUSBAND'S FATHER — HUSBAND'S MOTHER
HUSBAND'S OTHER WIVES

WIFE
Born — Place
Chr. — Place
Died — Place
Bur. — Place
WIFE'S FATHER — WIFE'S MOTHER
WIFE'S OTHER HUSBANDS

SEX M F	CHILDREN List each child (whether living or dead) in order of birth — Given Names / Surname	WHEN BORN — DAY / MONTH / YEAR	WHERE BORN — TOWN	COUNTY	STATE OR COUNTRY	DATE OF FIRST MARRIAGE — TO WHOM	WHEN DIED — DAY / MONTH / YEAR
1							
2							
3							
4							
5							
6							
7							
8							
9							
10							
11							

SOURCES OF INFORMATION — OTHER MARRIAGES

© 1972 The Genealogical Department of The Church of Jesus Christ of Latter-day Saints
ENTER ALL DATA IN THIS ORDER: NAMES: WATSON, John Henry PLACES: Sharon, Windsr, Vt DATES: 14 Apr 1794
To indicate that a child is an ancestor of the family representative, place an "X" behind the number pertaining to that child
FAMILY GROUP RECORD

FIG. 1. Family group sheet.

relationships and demographic information are extracted from the genealogy file. The number of cancer cases by site and sex that are linked to the genealogy are given in Table 1.

STRATEGY OF ANALYSIS

The study of family formation risk factors extracts variables of interest from the genealogy file for women with breast cancer and their matched controls. For each variable, a log-linear model is fit to the data to determine which are the most important variables. Logistic regression is done on the entire sample and within substrata of the sample to determine the strength of association and degree of interaction of each variable. Relative odds for the extreme divisions of each variable adjusted for the remaining variables are given by Mantel-Haenszel contingency table analysis.

Several statistical problems arise in cancer aggregation studies. Foremost is the lack of genetic independence among members of any pedigree. Since a significant fraction of the Utah Mormon population is descended from some 10,000

TABLE 1. *Frequencies of cancer groups by sex*

General site	Sites included	ICDA codes	Number in tumor file	
			Male	Female
Lip	Lip, tongue, salivary glands, mouth	140.0–149.9	517	167
Digestive	Esophagus, stomach	150.0–151.9	315	179
Intestine	Small intestine	152.0–152.9	37	28
Intestinal tract	Tract, other digestive organs	159.0–159.9	3	2
Colon	Colon, rectum	153.0–154.9	902	813
Peritoneum	Retroperitoneum, peritoneum	158.0–158.9	24	24
Pancreas	Liver, bile ducts, gallbladder, pancreas	155.0–157.9	272	222
Lung	Nose, larynx, trachea, lung	160.0–165.9	864	136
Blood	Blood, bone marrow, leukemia, spleen, hematopoietic system	169.0–169.9	415	266
Bone	Bones, connective, subcutaneous soft tissue	170.0–172.9	277	238
Breast	Female breast	174.0–174.9	-0-	2,025
Uterus	Uterus, cervix, placenta, corpus uteri	179.0–182.9	-0-	829
Female genital	Vagina, clitoris, genital tract	184.0–184.9	-0-	54
Ovary	Ovary, fallopian tube	183.0–183.9	-0-	340
Prostate	Prostate, testis	185.0–187.9	2,034	-0-
Urinary	Urinary bladder, kidney	188.0–189.9	605	207
CNS	Eye, lacrimal gland, brain, CNS	190.0–192.9	236	203
Thyroid	Thyroid, other endocrine glands	193.0–194.9	102	193
Lymph	Lymph nodes, Hodgkin's disease	196.0–196.9	338	228

pioneers, complex genetic relationships exist between contemporary Mormons. The failure of the observations to satisfy statistical independence complicates the use of standard statistical results, which rely on statistical independence.

Two methods of looking at familiality of cancer are used. The first, called the genealogical index, compares the mean kinship coefficient of cases to the mean for matched controls. The cases are individuals with cancer at some site. The controls are matched at random from the genealogy, and the kinship is determined using the genealogy. The second method compares the number of sibs or cousins concordant for cancer at certain sites to the expected numbers.

Potential Biases in Cancer Family Assessment

Several sources of bias must be considered when assessing familial risks for cancer. A difference in the number of ancestors could conceivably exist between the genealogy members with cancer and those in the genealogy at large, which would lead to bias toward excess kinship in one group or the other. Studies of the number of ancestors found for cases and controls failed, however, to demonstrate any significant differences between the two groups.

Similarly, the cases found in the genealogy could be different in some way from the Utah Cancer Registry. Initially, one important characteristic, age at onset for breast cancer, was examined, and no significant difference was found (Hill 1980a).

Familial awareness of the occurrence of a cancer in a line of descent might influence relatives of a patient towards more intensive screening than they would otherwise seek. This might be more probable for kindreds living in an urban environment (whose members are in close contact with one another and who patronize centralized medical facilities) than for those in rural environments. Also, as urban families probably have greater access to cancer screening clinics and treatment centers, a familial bias might be introduced into the Utah Cancer Registry records.

FAMILY FORMATION

There are currently 6,363 women in the Utah Cancer Registry with breast cancer, and 2,025 (31%) of them are linked to the genealogy file. After excluding the women who were unmarried, apparently nulliparous, or without known birth dates for themselves or their children, 1,119 women (18%) were left to analyze. For each case, four controls were matched by exact year of birth (YOB) and place of birth (Utah or not Utah) from the genealogy file. For each of these women the following data were extracted: number of children (NC), age at first delivery (AFD), age at last delivery (ALD), and lengths of reproductive period (ALD − AFD).

Significant differences ($p < 0.05$) between cases and controls were found for AFD (24.0 for cases, 23.1 for controls), ALD (34.9,33.9), and NC (4.4,4.5). The crude relative odds for the association for a late AFD with breast cancer (Table 2) shows a strong trend of increasing risk with a later AFD. Various adjustments were used to assess the independence of AFD from the other family formation variables (Table 3). When the relative odds was adjusted by ALD using the Mantel-Haenszel method, it was lower but still significant. Adjustment for NC had no effect, while the combination of ALD and NC decreased the odds slightly. Table 2 also shows a consistent trend for the association of a late ALD with breast cancer. NC had a large confounding effect on ALD,

TABLE 2. *The association with breast cancer of age at first delivery (AFD), age at last delivery (ALD), and number of children (NC)*

Variable		Crude relative odds	N Cases	N Controls
AFD:	<21	1.0	249	1,275
	21–22	1.2	248	1,069
	23–25	1.3*	286	1,123
	26–29	1.6†	203	658
	30–34	1.9†	93	248
	35–49	2.0†	40	103
ALD:	<25	1.0	57	339
	25–29	1.2	146	756
	30–34	1.5*	306	1,197
	35–39	1.7†	371	1,303
	40–44	1.6*	208	774
	45+	1.7‡	31	107
NC:	1	1.3	66	267
	2	1.5*	172	616
	3	1.3	201	844
	4	1.5*	232	815
	5	1.3	157	653
	6	1.3	120	485
	7	1.4	82	318
	8+	1.0	89	478

Reference groups: AFD, <21; ALD, <25; NC, 8+.
* p <0.01.
† p <0.001.
‡ p <0.05.

TABLE 3. *Effects of adjustment variables on the association of AFD, ALD, and NC with breast cancer*

Exposure variable	Adjustment variables	Mantel-Haenszel relative odds
AFD	None	1.7*
	ALD	1.5*
	NC	1.7*
	ALD, NC	1.3†
ALD	None	1.5*
	AFD	1.3†
	NC	2.3*
	AFD, NC	2.2*
NC	None	1.3
	AFD	1.0
	ALD	1.7*
	AFD, ALD	1.1

Exposed vs. unexposed for each variable: AFD, 26+ vs. <21; ALD, 40+ vs. <30; NC, 1.2 vs. 7+.
* p <0.001, chi-square significance level.
† p <0.05, chi-square significance level.

and the relative odds increased after adjustment for NC and AFD (Table 3). The number of children appeared to be weakly associated with breast cancer, although no definite trend was apparent.

A saturated log-linear model was fit to the data and reduced to six terms (Table 4). The interaction of YOB, AFD, and ALD was very significant, as were the main effects of AFD and ALD individually. Removal of the marginally significant NC and AFD effects and the interaction of AFD and ALD reduced the goodness of fit, but not significantly.

By an additive multivariate regression model, the coefficients for ALD and NC were significant, while AFD was borderline (Table 5). The large ALD effect came mostly from the late age at diagnosis (65 years and over) groups, as the AFD and ALD coefficients were the same when the age at diagnosis was 40–64. The interaction effects were modeled by different variable combinations, with AFD × ALD/NC having the largest effect, although it was not significant.

The relative odds and standardized coefficients for AFD specific for age at breast cancer diagnosis are given in Table 6. The relative odds were adjusted for ALD, and the logistic coefficients were adjusted for ALD and NC. The largest association for AFD is in the earlier age at diagnosis groups and for ALD in the later age at diagnosis groups.

TABLE 4. *Approximate Z-statistics for breast cancer data fitted by a log-linear model*

Terms	Model			
	I	II	III	IV
YOB a	0.01	−0.18	0.21	—
NC b	−0.38	−0.01	0.01	—
ALD c	2.47*	3.17†	3.32‡	3.42‡
AFD d	2.23*	3.54‡	4.20‡	4.60‡
ab	0.09	−0.06	—	—
ac	−0.86	−1.21	−1.11	—
ad	−0.39	−0.16	—	—
bc	−1.31	−1.40	−1.74	−1.61
bd	−0.82	−0.86	—	—
cd	−0.36	−0.66	−1.57	−1.87
abc	−1.56	−2.01*	−2.00*	−1.97*
abd	−0.53	—	—	—
acd	−1.07	−2.22*	−2.34*	−2.54*
bcd	−0.28	—	—	—
Goodness of fit (p)	1.00	.78	.92	.95

Z statistics have an approximate standard normal distribution and dashes indicate terms not included in model. Division points for variables: Birth year (YOB)(a) = (<1900,1900+), NC(b) = (1–3,4+), AFD(c) = (<21,26+), ALD(d) = (<35,35+). abcd term lost due to zero cell in contingency table.

* p <0.05.
† p <0.01.
‡ p <0.001.

TABLE 5. Logistic regression coefficients and standard errors

| | Coefficients | | |
| | Univariate | Multivariate additive model | Multivariate with interaction term |
Variables			
AFD	0.20*	0.08†	0.13‡
ALD	0.16*	0.23*	0.23*
NC	−0.07‡	−0.16§	−0.20*
AFD × ALD/NC	0.11*	—	−0.08
	Standard errors		
AFD	0.031	0.043	0.054
ALD	0.034	0.052	0.052
NC	0.034	0.053	0.059
AFD × ALD/NC	0.053	—	0.055

* $p < 0.001$.
† $p < 0.10$.
‡ $p < 0.05$.
§ $p < 0.01$.

TABLE 6. The association of AFD with breast cancer specific for age at diagnosis

Age at diagnosis	Mantel-Haenszel relative odds	Crude relative odds	Standardized logistic coefficient
25–39	1.82	2.40	0.43*
40–54	2.23†	2.30†	0.18‡
55–64	1.76§	2.02†	0.18‡
65–74	0.86	1.02	0.11
75–99	1.25	1.40	0.05
All	1.47†	1.70†	0.08*

Mantel-Haenszel relative odds adjusted for ALD. Logistic coefficient adjusted for ALD and NC.
* $p < 0.10$.
† $p < 0.001$.
‡ $p < 0.05$.
§ $p < 0.01$.

GENEALOGICAL INDEX

The genealogical index calculates Malécot coefficients of kinship (coefficient de parenté, Malécot 1948) for all possible pairwise relationships among a group of individuals of interest. These coefficients were calculated in integral powers of ½ (Mendelian segregation) and express the probability that randomly selected homologous chromosomes from two individuals are identical by descent from a common ancestor. The calculation can be simplified through the use of genetic path diagrams (Wright 1921); it becomes necessary only to count the number of birth events along all possible pathways between two individuals.

The genealogical index computer program (Bishop and Hasstedt 1980) calculates the distribution of genetic relationships and the mean kinship coefficient. In the kinship distribution of Figure 2, the horizontal axis plots kinship exponents (the power to which ½ is raised for the corresponding coefficient) 1 to 13, while the vertical axis plots the number of pairwise relationships found corresponding to each kinship exponent. The basic strategy pursued in this study involves the comparison of the kinship distributions of the major cancer sites to the kinship distributions of their matched controls.

Selection of Controls

Matched controls are chosen from the genealogy for each individual with cancer. These controls are selected at random from five-year birth cohorts, since the genetic structure of the Mormon population changes markedly with time. Cohorts born in Utah or elsewhere were found to differ in the number and degree of pairwise relationships. Further discussion of biases in the control population has been reported elsewhere (Hill 1980a).

Comparisons are made between the distributions and means of the kinship coefficients for cases and controls. Since samples of controls exhibit variation, it is possible to improve the estimate of mean and variance for the controls by repeated sampling. This provides an empirical significance level for comparisons.

Results of a Kinship Survey of Major Cancer Sites

Kinship control distributions have been estimated by repeated sampling for the following cancer sites: prostate, breast, colon, cervix/uterus, hematopoietic and reticuloendothelial system, lung/bronchus, central nervous system, rectum/anus, and stomach. The results of the genealogical index analyses of these sites are represented in two ways in Figure 2, A–H and Figure 3. One type of plot shows the actual distributions of the pairwise relationships over the first thirteen degrees of kinship. Statistical analysis of the number of ancestors found for each set of cases and their controls does not demonstrate any significant difference.

Table 7 contains tabulations of means of the number of related pairs found for (1) the cases and (2) the control distributions. It also contains mean kinship coefficients for cases and controls. With the single exception of the prostate cases, no significant difference can be found between the number of related pairs for cases and controls for any of the sites analyzed. The information from these kinship coefficients is therefore mainly contained within the differences in shape between the kinship distributions for cases and controls rather than in the number of relationships (prostate cancer excepted). Thus, in Figures 2A

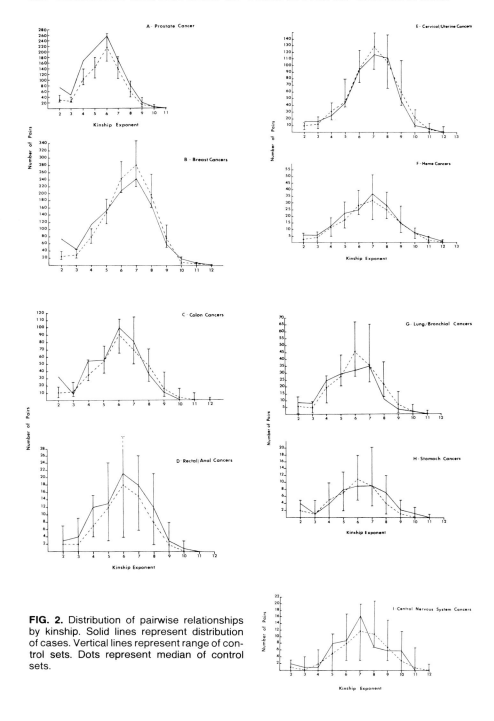

FIG. 2. Distribution of pairwise relationships by kinship. Solid lines represent distribution of cases. Vertical lines represent range of control sets. Dots represent median of control sets.

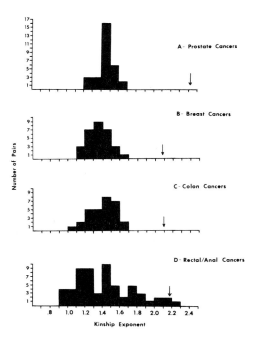

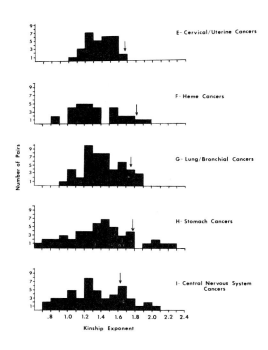

FIG. 3. Mean kinship distributions of reiterated control sets for selected cancer sites. Arrows indicate mean kinship of cases.

through 2I, it is the excess of cases over controls in the first five kinship categories that accounts for the differences between the mean kinships. The sites in Table 7 are ranked by an empirical significance index that is defined as the difference in the mean kinship coefficient of cases and controls normalized by the standard error of the control distribution.

Comparison of the mean kinship distributions in Figure 3 yields a clear distinction between the various sites. The mean kinships of prostate, breast, and colon cancers (ranking first, second, and third, respectively) lie entirely outside their control distributions, suggesting strongly that the observed genetic relationships among these cancer sets did not occur by chance alone. Prostate cancer clearly exhibits the greatest excess kinship of all the major sites surveyed so far. It also exhibits a distinct excess of related pairs, unlike the remaining sites of Table 7. The colon cancers actually have a higher mean kinship than the breast cancers (2.14×10^{-5} and 2.08×10^{-5}, respectively). Their reversal when ranked by the empirical significance index is due to the larger standard error for the colon cancer controls because of their smaller sample size. Rectal/anal cancers rank fourth, showing an increased mean kinship over random controls of uncertain significance. These were followed by cervical/uterine, heme, lung/bronchial, stomach and central nervous system cancers, all of which show considerably lower mean kinship.

In summary, this study of the comparative kinship of major cancer sites has identified prostate, breast, colon, and probably rectal/anal cancers as being outstanding in their tendency to aggregate in families as measured by excess kinship in a population-based genealogy.

A Kinship Survey of Pairwise Coaggregation of Major Cancer Sites

The genealogical index technique can also calculate kinship relationships *between* two sets of individuals of interest by computing the coefficient of kinship between each possible pair of individuals, one from each set. A major goal of this study is the ranking of pairs of sites by their tendency to coaggregate among *different* individuals within the same pedigree. Table 8 lists the results of pairwise coaggregation ranked by the empirical significance index.

Calculation of the empirical significance index requires estimation of both mean and variance of the control populations. Analysis of single-site aggregation demonstrated that the best predictor of control population variance is the number of related pairs found, rather than the number of individuals in a set or the total number of related and unrelated pairs (Hill 1980b). This follows from the fact that, for this population with relatively recent founders, the number of related pairs is dependent on the age structure, which in turn influences the amount of remote kinship that exists. A more detailed description of the method of control population variance will be given elsewhere (Hill 1980b).

Table 7 lists the average value of mean kinship found by repeated sampling of the matched control populations for nine major cancer sites. While there

TABLE 7. Survey of numbers of related pairs for major cancer sites and matched control populations*

Site	Cases No. related pairs (r)	Controls			$\frac{r - \mu}{\sigma}$	Mean kinship cases $\times 10^{-5}$	Average mean kinship controls $\times 10^{-5}$	Empirical significance index
		No. random samples	Mean (μ) no. related pairs					
Prostate 1,981 cases	1,025	30	762.8		4.13	2.42	1.45	9.08
Breast 2,025 cases	1,110	30	1,121.7		−0.14	2.08	1.36	5.83
Colon 1,240 cases	377	30	346.1		0.81	2.15	1.40	4.97
Rectum/anus 565 cases	88	60	70.4		1.50	2.14	1.44	2.14
Cervix/uterus 1,241 cases	502	30	519.7		−0.32	1.68	1.37	1.95
Heme 714 cases	166	30	150.0		0.74	1.81	1.33	1.64
Lung/bronchus 859 cases	154	50	171.5		−0.66	1.66	1.43	1.39
Stomach 459 cases	45	50	42.8		0.28	1.78	1.40	0.92
CNS 392 cases	56	50	57.0		−0.09	1.63	1.35	0.88

* Sample size is equal to number of cases for each site.
CNS, central nervous system.

TABLE 8. *Survey of pairwise aggregations of major cancer sites ranked by empirical significance levels*

Paired sites	No. nonzero pairs	Mean kinship $\times 10^{-5}$	Ranking by mean kinship	Empirical significance index
1 GI vs colon	1,457	2.025	6	9.054
2 GI vs rectal	695	2.060	5	6.721
3 GI vs prostate	2,163	1.707	17	5.910
4 Prostate vs breast	2,113	1.681	20	5.454
5 Colon vs rectum	362	2.120	2	5.301
6 Prostate vs colon	1,127	1.745	15	4.835
7 Female reproductive vs uterus	1,323	1.652	25	4.060
8 Prostate vs CNS	468	1.861	9	4.055
9 Prostate vs heme	777	1.729	16	3.905
10 Female reproductive vs ovary	373	1.867	8	3.672
11 GI vs stomach	473	1.806	12	3.659
12 GI vs female reproductive	2,050	1.549	36	3.397
13 Prostate vs pancreas	336	1.799	13	3.053
14 Prostate vs female reproductive	1,731	1.537	40	2.984
15 Prostate vs uterus	1,310	1.830	10	2.936
16 Heme vs CNS	193	1.924	7	2.935
17 Stomach vs pancreas	76	2.252	1	2.858
18 Prostate vs rectum	537	1.674	23	2.828
19 GI vs uterus	1,534	1.526	42	2.680
20 Prostate vs stomach	387	1.706	18	2.632
21 Breast vs ovary	426	1.681	21	2.578
22 Stomach vs ovary	89	2.078	4	2.512
23 Stomach vs colon	237	1.787	14	2.507
24 GI vs breast	2,171	1.483	52	2.467
25 Breast vs pancreas	367	1.659	24	2.250
26 Ovary vs pancreas	65	2.112	3	2.245
27 Female reproductive vs rectum	494	1.612	27	2.236
28 Female reproductive vs breast	1,821	1.477	53	2.197
29 GI vs ovary	472	1.608	28	2.157
30 Colon vs uterus	769	1.536	41	2.038
31 Colon vs breast	1,166	1.495	49	2.002
32 Colon vs female reproductive	1,044	1.501	47	1.968
33 Stomach vs rectum	116	1.810	11	1.848
34 Uterus vs heme	564	1.547	38	1.848
35 Breast vs lung	875	1.501	48	1.811
36 Breast vs rectum	580	1.538	39	1.798
37 Breast vs stomach	436	1.569	32	1.791
38 GI vs pancreas	358	1.593	30	1.786
39 Female reproductive vs heme	725	1.513	45	1.768
40 Female reproductive vs pancreas	296	1.605	29	1.700
41 Female reproductive vs lung	755	1.502	46	1.698
42 Female reproductive vs stomach	368	1.570	31	1.656
43 Breast vs uterus	1,392	1.451	56	1.617
44 Lung vs stomach	159	1.682	19	1.593
45 Uterus vs rectum	370	1.553	35	1.546
46 Prostate vs ovary	381	1.518	43	1.330
47 Ovary vs uterus	279	1.549	37	1.323
48 Ovary vs rectum	112	1.676	22	1.317
49 Colon vs heme	476	1.493	50	1.292
50 Breast vs CNS	472	1.492	51	1.279
51 Breast vs heme	822	1.452	55	1.274

TABLE 8 (continued)

Paired sites	No. nonzero pairs	Mean kinship $\times 10^{-5}$	Ranking by mean kinship	Empirical significance index
52 Lung vs pancreas	129	1.613	26	1.160
53 Lung vs uterus	576	1.458	54	1.125
54 Colon vs pancreas	172	1.555	34	1.070
55 Lung vs ovary	156	1.557	33	1.029
56 Uterus vs pancreas	227	1.516	44	1.021
57 Lung vs GI	851	1.393	61	0.703
58 Lung vs prostate	727	1.396	58	0.680
59 Colon vs ovary	251	1.444	57	0.673
60 Lung vs heme	349	1.392	62	0.451
61 CNS vs uterus	292	1.381	64	0.337
62 CNS vs colon	275	1.381	64	0.337
63 Heme vs ovary	143	1.395	60	0.303
64 Uterus vs stomach	272	1.373	65	0.289
65 CNS vs pancreas	74	1.396	59	0.222
66 Heme v rectum	199	1.329	66	0.030
67 CNS vs stomach	76	1.319	67	−0.012
68 Colon vs lung	431	1.319	68	−0.029
69 GI vs CNS	506	1.298	70	−0.195
70 GI vs heme	861	1.303	69	−0.202
71 Pancreas vs rectum	88	1.254	71	−0.228
72 Pancreas vs heme	126	1.254	72	−0.273
73 CNS vs rectum	121	1.234	73	−0.345
74 CNS vs lung	166	1.215	74	−0.490
75 Rectum vs lung	215	1.208	75	−0.592
76 CNS vs ovary	81	1.007	77	−1.003
77 Heme vs stomach	153	0.980	78	−1.493
78 CNS vs female reproductive	381	1.073	76	−1.705

CNS, central nervous system; Heme, hematopoietic system.

are some differences in the average kinship for different control populations, the variance is small (1.325×10^{-13}), suggesting a substantial genetic homogeneity within the genealogy. The control population mean kinship is estimated by repeated sampling of the matched control population for a single pairwise comparison of breast and colon cancer. This mean kinship value, 1.323×10^{-5}, is used as an estimate of the mean kinship of the 78 control populations. It should be recognized that individuals who were diagnosed with both paired cancer sites were excluded from this survey, so the kinship indices are a measure of coaggregation of paired sites among *different* members within the same families. Reference to the kinship distributions that have been sampled from the genealogy and plotted in Figure 3 suggests that the assumptions underlying the calculation of empirical significance are reasonable, as there is small variance among the means of the distributions, while dispersion of the distributions increases with decreasing number of related pairs.

The rankings listed in Table 8 appear to confirm and extend the results of the survey of single major sites. With two exceptions (No. 8, prostate versus CNS and No. 16, heme versus CNS), all of the paired sites that appear in the top 34 rankings are either reproductive, breast, or digestive (gastrointestinal and pancreatic) cancers. Further, the major sites that ranked lowest in the single-site analysis, lung/bronchial, heme, and CNS cancers, predominate in the lowest rankings of paired sites. For example, 83% of the paired sites that include prostate cancer and 75% of those that include breast cancer appear in the first half of the ranking order, while only 17% of the CNS pairings and 8% of the lung/bronchial pairings are in the upper half of the rankings.

Examples of considerable discrepancy between the two ranking schemes presented in columns 4 and 5 of Table 8 can readily be found. For example, the pairing of breast and prostate cancers ranks fourth by empirical significance but 20th by mean kinship alone. The difference is due to the large number of related pairs (2,113) found between the prostate and breast cancer sets, which gives greater confidence in the difference between the mean kinship found between prostate and breast cancers (1.681×10^{-5}) and the estimated mean kinship of the control population (1.323×10^{-5}). This example should be contrasted with No. 26, ovarian and pancreatic cancers, which has a high mean kinship (2.112×10^{-5}) calculated from a small number (65) of related pairs. Based on mean kinship alone, this pairing would rank third rather than 26th. However, ranking by mean kinship ignores sampling biases that are accounted for in the empirical significance index. Thus, it is far more likely for the ovarian and pancreatic aggregations, which yield a mean kinship of 2.122×10^{-5}, to have occurred by chance alone than for the prostate and breast aggregations, with a lower mean kinship of 1.681×10^{-5}, to have occurred by chance.

A conclusion from these initial familial studies in the Utah Mormon population is that reproductive, breast, and digestive cancers both aggregate and coaggregate to a roughly similar excessive extent. Further, subsets of a particular organ tract tend to coaggregate with each other. For example, the reproductive cancers tend to be roughly as related to the digestive cancers as they are to each other. The subsets of the digestive cancers (pancreas, stomach, colon, and rectal) appear to coaggregate mutually in different individuals within the same families. A similar pattern appears for breast and reproductive cancers, especially for prostate. The fact that this pattern of coaggregation is fairly consistent among and between most of the breast, reproductive, and digestive cancers seems to suggest a common familial epidemiology for these cancers, with only a few exceptions. On the other hand, lung/bronchial and CNS cancers tend to dominate the least familial rankings, suggesting a lesser tendency to aggregate in families or to coaggregate with other sites. The hematopoietic cancers appear to fall in between these two groups, suggesting a moderate tendency to coaggregate with the digestive, breast, and reproductive sites. Interesting exceptions to the above generalizations appear with the eighth (CNS and prostate) and 16th (heme and CNS) rankings. Even though CNS and heme cancers do not appear to coaggregate appreciably with the reproductive and digestive cancers, they appear to coaggregate with each other.

FAMILIAL AGGREGATION

The cancer sites were divided into 19 categories, which were based on ICD codes. Table 1 lists these classifications together with the actual number of cases for each category in the tumor file. Also included is a synonym for each group, which is used in the following tables. For each of the relationships studied and each pair of the cancer classifications, the number of pairs of individuals with the appropriate relatedness and tumors was calculated. The various relationships contained the following numbers of pairs:

sister-sister	285 pairs
brother-brother	387 pairs
brother-sister	566 pairs
aunt-niece	375 pairs
uncle-nephew	389 pairs
aunt-nephew or uncle-niece	799 pairs
female cousins	774 pairs
male cousins	1,052 pairs
opposite sex cousins	1,621 pairs

The largest counts for any one cancer site were obtained for male cousins who both had prostate cancer (143 pairs) and brothers with prostate cancer (68 pairs). There were 51 pairs of sisters with breast cancer and 159 pairs of oppostite sex cousins in which the male had prostate cancer and the female, breast cancer. These counts constitute the observed values in the analysis.

The null hypothesis we are testing against is that for each relationship the cancers are randomly distributed, with frequencies given by Table 9. Under this hypothesis, an expected frequency of each pair of cancers can be calculated. As an example, for opposite sex sibs in which the male had colon cancer and the female had uterine cancer, the expected number of such pairs is as follows:

$$566 \times (902/6{,}941) \times (829/6{,}099) = 10.0$$

where 566 is the number of opposite-sex sib pairs, 902 is the number of colon cancers found in a total of 6,941 males, 829 is the number of uterine cancers found in 6,099 females, and 10.0 is the expected number of pairs.

A χ^2 was then calculated for each category where

$$\chi^2 = (\text{observed} - \text{expected})^2/\text{expected}.$$

Thus, in the above

$$\chi^2 = (19 - 10.0)^2/10.0 = 7.2.$$

Also, a relative risk (RR) defined by

RR = observed/expected (= 1.90 for this example) was calculated. Table 9

TABLE 9. *Cancer sites showing increased or decreased pairwise clustering using expected values calculated from the tumor file*

Relation	Site 1	Site 2	Observed	Expected	Relative risk	χ^2
Sister-	breast	breast	51	29.7	1.72	14.5
sister	uterus	uterus	12	5.3	2.26	7.4
Brother-	prostate	prostate	68	33.2	2.05	35.3
brother	colon	colon	13	6.5	2.00	5.4
Brother-	(male)	(female)				
sister	colon	uterus	19	10.0	1.90	7.2
	prostate	pancreas	13	6.0	2.17	6.9
	colon	colon	18	9.8	1.84	6.0
Female	breast	bone	9	19.5	0.46	5.1
cousins	uterus	uterus	23	14.3	1.61	4.7
	ovary	colon	4	11.5	0.35	4.3
	uterus	breast	84	68.0	1.24	3.6
Male	prostate	prostate	143	90.3	1.55	30.1
cousins	colon	colon	28	17.8	1.57	5.3
	lung	lung	26	16.3	1.60	5.2
	prostate	digestive	44	31.3	1.40	4.8
	colon	digestive	22	13.9	1.58	4.2
	prostate	colon	98	80.1	1.22	3.8
Opposite	(male)	(female)				
sex	colon	uterus	47	28.6	1.64	11.1
cousins	colon	CNS	15	7.0	2.14	8.0
	colon	lip	13	5.8	2.24	7.9
	prostate	lymph	8	17.8	0.45	4.8
Aunt-	uterus	uterus	17	6.9	2.46	13.2
niece						
Uncle-	CNS	prostate	16	7.8	2.05	7.7
nephew						
Opposite	(male)	(female)				
sex	prostate	uterus	60	31.8	1.89	24.1
third-	colon	breast	16	33.5	0.48	8.7
degree	prostate	bone	16	9.1	1.76	4.4
relatives	urinary	uterus	16	9.5	1.68	3.8

contains the relationship and sites that had a χ^2 value of more than 3.8, and an expected value of at least 5.0. This χ^2 value would correspond to the 95th percentile for a single χ^2 test, but in this case because of the number of tests performed, the significance level was lower.

In same sex sibs, only pairs with the same sites (breast and uterus, colon and prostate) satisfy the requirements for the χ^2 statistic and expected value. Of these sites, three (uterus, prostate, and colon) also appear for same sex cousins, in each case with decreased RR and χ^2 value. It is interesting that in each case where a site is paired with itself, the RR is greater than 1.0. This suggests that these sites may be overrepresented in the preceding analysis.

The tumor file was searched to find all pairs of individuals with a given relationship. Each of these individuals was then placed in one of the 19 cancer categories based on their tumor site, and a 19 × 19 table of the frequency of each of the pairwise cancer categories was formed. The marginal frequency of each category was then computed and compared by a χ^2 test to the corresponding frequency obtained from the tumor file. Table 10 records the results of this test together with a list of sites that were increased or decreased from expected when taken individually. This latter test was performed using a comparison of binomial proportions.

Table 11 repeats the analysis from Table 10, but the expected values are based on the marginal pairwise frequencies. Effectively, this treats the analysis as a contingency table. In this case, there are fewer pairwise relationships with a χ^2 of greater than 3.8 and expected value greater than 5.0. However, five of the six associations with a relative risk greater than 1.0 also appear in Table 9. The only new association (lung-digestive in brothers) has 11 observed cases with an expected value of 5.3.

TABLE 10. *Unusual occurrence of a site in a familial cluster by different degrees of relationship*

Degree of relationship	χ^2	Sig.	Deviation of binomial proportion from expected [N(x)>1.96]	
			Pos.	Neg.
Sister-sister	25.16	.025	breast uterus	CNS bone
Brother-brother	39.19	.005	prostate CNS	pancreas
Brother-sister (M)	23.01	.025	none	blood
(F)	22.40	.05	pancreas uterus	thyroid
Female-female cousins	38.30	.005	uterus breast lung	bone CNS
Male-male cousins	50.44	.005	colon prostate	bone lymph
Male-female cousins (M)	15.54	NS	colon	
(F)	43.52	.005	uterus	bone, thyroid ovary, lymph
Aunt-niece	14.68	NS	uterus	
Uncle-nephew	26.34	.01	bone	blood
Uncle-niece (M) and	25.00	.01	bone prostate	blood
aunt-nephew (F)	29.97	.005	uterus	breast

TABLE 11. *Cancer sites showing increased or decreased pairwise clustering in the contingency table*

Relationship	First site	Second site	Obs.	Exp.	Relative risk	χ^2
Male cousins	lung	lung	26	14.4	1.64	8.6
Male cousins	prostate	lung	62	82.7	0.75	4.9
Male cousins	prostate	prostate	143	118.8	1.20	4.7
Male cousin-female cousin	colon	breast	59	77.7	0.76	4.3
Male cousin-female cousin	colon	CNS	15	8.7	1.72	3.9
Sisters	breast	breast	51	38.0	1.34	4.1
Brothers	lung	digestive	11	5.3	2.08	5.2
Brothers	prostate	prostate	68	53.2	1.28	3.8

Obs., observed; Exp., expected.

FAMILY FORMATION AND GENETIC RISK IN KINDRED 107

Classical epidemiological studies have identified a number of risk factors to be predictive of various types of cancer. In most of these studies, genetic factors have not been taken into account, and thus the strength of associations established may be confounded by the "noise" produced by the unstudied genetic variations among study subjects. On the other hand, genetic studies suffer from environmental variation that make isolating the strictly genetic components difficult and results hard to reproduce. The interaction of gene and environmental factors has long been recognized as realistic, but few studies have been designed to investigate the interactions.

One of the greatest attributes of the Utah Mormon genealogy is the capability to investigate gene-environment interactions. From an epidemiological viewpoint, the genealogy may be analyzed as a population-based study, but at the same time we have longitudinal demographic and structural information on the relationships among cases and controls.

A major study is at present being done on a large pedigree, Kindred 107. Currently, pedigree members are being assayed for biochemical markers, fluorescent banding patterns, and sister chromatid exchange rates, as well as being asked to answer a questionnaire on medical history and dietary habits. This study was started in 1947 when a university student mentioned to E. J. Gardner that two of his great aunts had died of breast cancer. Kindred 107 has been examined for family formation variables and genetic risk factors (Hill et al. 1978, Bishop and Gardner 1980, Gardner 1980), thus allowing the interactive factors to be evaluated.

At present, Kindred 107 represents about 1,400 individuals plus an additional generation of children. There are 29 breast cancers (including breast cancers in two males), 17 benign breast tumors, and 36 other cancers and tumors. Pedigree analysis supports the hypothesis of a dominant major gene that predisposes to breast cancer. The analysis suggests that this allele has incomplete penetrance, even for females, such that at least 15% of all females with the abnormal genotype will not express cancer (Bishop and Gardner 1980). It is speculated that this partial penetrance shows that the susceptibility allele is in some cases related to other cancer sites or to benign tumors, and in other cases that there is no expression of the gene. A number of individuals at high risk of carrying the allele did have other forms of cancer, but there appears to be no pattern for an alternate site.

The risk factors associated with family formation have been examined, and age at first delivery was found to be associated with breast cancer in the Mormon genealogy as a whole. For a woman whose first delivery was after age 25, breast cancer was 1.7 times more frequent than for a woman with a first delivery before 21 years of age. This frequency is in agreement with those on other populations (MacMahon et al. 1973). In Kindred 107, the relative odds were four to one. If the gene(s) were penetrant, one would expect no association with age at first delivery. However, the relative odds were more than doubled in Kindred 107, suggesting that a late age at first delivery enhances the genetic pathway, aiding in tumor initiation or growth.

Thus, an increase in the association of an environmental variable with a disease in a related sample as compared to an independent sample would indicate environmental effects may augment the genetic effect. They may act along the same pathway and be cumulative in effect, resulting in earlier onset of the cancer, increased severity, or a greater penetrance of the gene.

Figure 4 shows the mean age at first delivery in Kindred 107 for women with breast cancer, breast tumors, other cancers, and other tumors, as well as for normal individuals. These values plotted by year of birth compared with

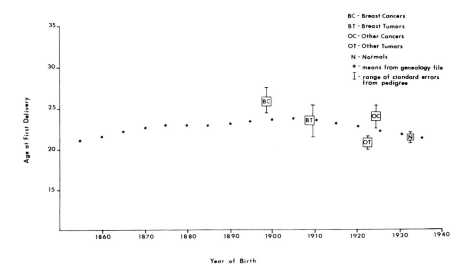

FIG. 4. Age at first delivery by disease status for Kindred 107.

the means for the genealogy as a whole support the gene-environment interaction seen in Kindred 107. The figure also shows the advantage of being able to compare pedigree values with a standard population. The combined use of genetic and family formation risk factors provides more knowledge on which to base future studies of breast cancer etiology and provides more information for genetic counseling.

DISCUSSION

This article reports on a preliminary analysis of the Utah Mormon genealogy file linked to the Utah Cancer Registry. The combination of a genealogical file and cancer records provides the basis for a large-scale investigation of the genetic and familial aspects of cancers of all sites. It also allows us to explore gene-environment interactions. There are a number of advantages to this structure over those in previous epidemiological studies. First, many studies have had no reference population. The Utah Mormon genealogy provides us with such a control population. Second, there is no limit to the specific sites which can be studied other than that imposed by the Utah Cancer Registry. Third, there is total ascertainment of the population, which removes the problem of a bias induced by the study of a small sample set. Ascertainment of cancer-prone families can be accomplished under well-defined rules that allow correction factors to be applied.

The data base provides a major epidemiological resource that can be used at every stage in an analysis. Initially, gross effects such as the relationship between age at first pregnancy and breast cancer can be identified. Clusters of cancer sites in relatives for breast, prostate, and colon were found by two different methods. Analysis of pairs of sites confirmed that GI and reproductive cancers are the most important. Then pedigrees such as Kindred 107, which may be informative for genetic analysis, can be selected under a given criterion. Pedigree analysis refines within individual pedigrees gross effects that are indicated by analysis of the genealogy. Finally, the identification of other potential effects within the individual pedigrees returns the emphasis from the pedigrees to the population data base for further study.

ACKNOWLEDGMENTS

This research was supported by grants CA-16573 and GM-27192 from the National Institutes of Health.

REFERENCES

Anderson, D. E. 1975. Familial susceptibility, in Persons at High Risk of Cancer: An Approach to Cancer Etiology and Control, J. F. Fraumeni, ed. Academic Press, New York, pp. 39–54.

Bean, L. L., G. P. Mineau, K. A. Lynch, and J. D. Willigan. 1980. The Genealogical Society of Utah as a data resource for historical demography. Popul. Index (in press).

Bishop, D. T., and E. J. Gardner. 1980. Analysis of the genetic predisposition to cancer in individual pedigrees, *in* Banbury Report 4: Cancer Incidence in Defined Populations, J. Cairns, J. L. Lyon, and M. Skolnick, eds. Cold Spring Harbor Laboratory, Cold Spring Harbor, New York, pp. 389–408.

Bishop, D. T., and S. J. Hasstedt. 1980. The genealogical index. Technical Report No. 17, Department of Medical Biophysics and Computing, University of Utah, Salt Lake City.

Gardner, E. J. 1951. A genetic and clinical study of intestinal polyposis, a predisposing factor for carcinoma of the colon and rectum. Am. J. Hum. Genet. 3:167–176.

Gardner, E. J. 1962. Follow-up study of a family group exhibiting dominant inheritance for a syndrome including intestinal polyps, osteomas, fibromas, and epidermal cysts. Am. J. Hum. Genet. 14:376–390.

Gardner, E. J. 1980. Prevention and cure for hereditary cancers, *in* Banbury Report 4: Cancer Incidence in Defined Populations, J. Cairns, J. L. Lyon, and M. Skolnick, eds. Cold Spring Harbor Laboratory, Cold Spring Harbor, New York, pp. 365–377.

Henderson, B., E. D. Powell, I. Rosario, C. Keys, R. Hanisch, M. Young, J. Casagrande, V. Gerkins, and M. C. Pike. 1974. An epidemiologic study of breast cancer. J. Natl. Cancer Inst. 53:609–614.

Hill, J. R. 1980a. A survey of cancer sites by kinship in the Utah Mormon population, *in* Banbury Report 4: Cancer Incidence in Defined Populations, J. Cairns, J. L. Lyon, and M. Skolnick, eds. Cold Spring Harbor Laboratory, Cold Spring Harbor, New York.

Hill, J. R. 1980b. A kinship survey of cancer in the Utah Mormon population. Ph.D. dissertation, University of Utah, Salt Lake City.

Hill, J. R., D. Carmelli, E. J. Gardner, and M. Skolnick. 1978. Likelihood analysis of breast cancer predisposition in a Mormon pedigree, *in* Genetic Epidemiology, N. E. Morton and C. S. Chung, eds. Academic Press, New York, pp. 247–253.

Hunt, S. C., R. R. Williams, M. Skolnick, J. L. Lyon, and C. R. Smart. 1980. Breast cancer and reproductive history from genealogical data. J. Natl. Cancer Inst. 64:1047–1053.

King, M. C., and D. T. Bishop. 1980. Genetic epidemiology of breast cancer in Mormon kindreds, *in* Banbury Report 4: Cancer Incidence in Defined Populations, J. Cairns, J. L. Lyon, and M. Skolnick, eds. Cold Spring Harbor Laboratory, Cold Spring Harbor, New York, pp. 379–385.

King, M. C., N. L. Petrakis, R. C. P. Go, R. C. Elston, and H. T. Lynch. 1980. An allele increasing susceptibility to human breast cancer may be linked to the GPT locus. Science 208:406–408.

Knudson, A. G. 1971. Mutation and cancer: Statistical study of retinoblastoma. Proc. Natl. Acad. Sci. USA 68:820–823.

Knudson, A. G., and L. C. Strong. 1972a. Mutation and cancer: Neuroblastoma and pheochromocytoma. Am. J. Hum. Genet. 24:514–532.

Knudson, A. G., and L. C. Strong. 1972b. Mutation and cancer: A model for Wilms' tumor of the kidney. J. Natl. Cancer Inst. 48:313–324.

Knudson, A. G., L. C. Strong, and D. E. Anderson. 1973. Heredity and cancer in man. Prog. Med. Genet. 9:113–156.

Lynch, H. T., A. J. Krush, R. J. Thomas, and J. Lynch. 1976a. Cancer family syndrome, *in* Cancer Genetics, H. T. Lynch, ed. Charles C Thomas, Springfield, Illinois, pp. 355–388.

Lynch, H. T., G. M. Mulcahy, P. Lynch, H. Guirgis, F. Brodkey, J. Lynch, K. Maloney, and L. Rankin. 1976b. Genetic factors in breast cancer: A survey. Pathol. Annu. 11:77–101.

MacMahon, B., P. Cole, and J. Brown. 1973. Etiology of human breast cancer: A review. J. Natl. Cancer Inst. 50:21–42.

Malécot, G. 1948. Les mathématiques de l'heredité. Masson et Cie., Paris.

Mineau, G. P., L. L. Bean, and M. Skolnick. 1979. Mormon demographic history. II. The family life cycle and natural fertility. Popul. Stud. 33:429–446.

Skolnick, M. 1977. Prospects for population oncogenetics, *in* Genetics of Human Cancer, J. J. Mulvihill, R. W. Miller, and J. F. Fraumeni, eds. Raven Press, New York, pp. 19–25.

Skolnick, M., L. L. Bean, D. May, V. Arbon, K. de Nevers, and P. Cartwright. 1978. Mormon demographic history. I. Nuptiality and fertility of once-married couples. Popul. Stud. 32:5–19.

Skolnick, M., L. L. Bean, S. M. Dintelman, and G. P. Mineau. 1979. A computerized family history data base system. Sociol. Soc. Res. 63:506–527.

Staszewski, J. 1971. Age at menarche and breast cancer. J. Natl. Cancer Inst. 47:935–940.

Stavraky, K., and S. Emmon. 1974. Breast cancer in premenopausal and postmenopausal women. J. Natl. Cancer Inst. 53:647–654.

Trichopoulos, D., B. MacMahon, and P. Cole. 1972. Menopause and breast cancer risk. J. Natl. Cancer Inst. 48:605–613.

Tulinius, H., N. Day, G. Johannesson, O. Bjarnason, and M. Gonzales. 1978. Reproductive factors and risk for breast cancer in Iceland. Int. J. Cancer 21:724–730.

Wright, S. 1921. Correlation and causation. J. Agricul. Res. 20:557–585.

Genes, Chromosomes, and Neoplasia, edited by
Frances E. Arrighi, Potu N. Rao, and Elton Stubblefield.
Raven Press, New York © 1981.

Cancer Control through Genetics

John J. Mulvihill

Clinical Epidemiology Branch, National Cancer Institute, Bethesda, Maryland 20205

Cancer is a genetic disease, at least at the cellular level. If the cancer cell, however defined, did not pass on its secret for escaping the rules for good behavior of normal cells, then no one would die of cancer. In a sense, my title, "Cancer Control through Genetics," is the theme of the entire volume. Mostly, genetic control in this monograph refers to how genes interact among themselves or with other agents to direct the cancer process from the initial mutation or chromosomal abnormality to cell differentiation or dedifferentiation and death through loss of normal control. If, on exposure to Epstein-Barr virus, chromosome 14 breaks at q13 and translocates to the terminus of the other chromosome 14, as it does in various lymphomas (McCaw et al. 1975), the proximity of the break point to the gene locus for nucleoside phosphorylase invites speculation on the possible role of the enzyme in controlling the expression of malignancy (Mulvihill 1980). The future excitement lies in exploiting such control, in isolating cancer genes, and in developing new rationales for interfering with carcinogenesis.

But even now, much can be done to control or prevent cancer through knowledge of the genetic determinants of human neoplasia. Despite the cliché that 80% to 90% of cancer is caused by environmental factors (Higginson and Muir 1979), some people get cancer because of their genes (Mulvihill et al. 1977a, Schimke 1978). Applying knowledge of the many genetic, familial, and congenital conditions that predispose to cancer can speed prevention and control in three ways. First, appreciation of inborn factors improves diagnoses and possibly prevents cancer in individuals or relatives of already affected patients through wisely applied screening, surveillance, protection from exogenous factors, and sometimes prophylactic surgery. Second, the sustained effort of medical geneticists and their professional and political allies to control genetic disease, mostly of children, contain lessons for oncologists and their colleagues concerned with preventing cancer, mostly of adults. Finally, apart from these biopolitical approaches, ultimate control of cancer by understanding chemical and cellular mechanisms will be speeded by research in cancer genetics.

These prevention methods and strategies may eventually change cancer rates. Medical geneticists have been criticized for studying clinical rarities, but have as often shown the scientific advantage of deriving principles by starting with nature's exceptions.

CONTROL BY IDENTIFYING PERSONS AT HIGH RISK

Environmental Versus Genetic Risk Factors

Tobacco use is a known and large environmental risk factor in lung cancer (Hoover 1978). For example, a 55-year-old white male smoker has 15 times the risk of lung cancer that nonsmokers have. Intuition says that is a large multiplier, one that might make prudent men avoid such a risk by not smoking. Why, then, has heavy tobacco use continued? Perhaps one reason is that the absolute risk is still too small; it is 15 times the age-specific incidence of lung cancer in nonsmokers, which approximates 28/100,000 white males per year. [The calculation assumes a total age-specific incidence of 186/100,000/year (Cutler and Young 1975) and that 27.5% of white males never smoked (Advisory Committee to the Surgeon General 1979).] The product is an absolute annual risk of about 1 in 250 or 0.4%. In counseling prospective parents about the recurrence of severe congenital anomalies, such as myelomeningocele or congenital heart defects, much larger risks, 3% to 4%, are often not high enough to cause parents to change plans for additional children (Carter et al. 1971, Leonard et al. 1972). This comparison in how human beings perceive risk leads to the question: Is smoking for recreation any easier to change than sex for procreation? Reproductive plans do not change until the absolute risk for a defective child is 10% or more.

So, for cancer prevention that requires an individual to alter life style, epidemiologists must identify huge risk factors that people can take personally. The best known environmental carcinogens do not have such risks, but genetic cancers do. Knudson et al. (1975) calculated that three retinoblastomas arise in every individual with the retinoblastoma gene, but only one in 30,000 persons without it; dividing these rates gives a relative risk of 90,000. The frequency of unilateral cases that eventually manifest multifocality, that is, hereditary disease, is large enough and the desire to preserve sight great enough that ophthalmologic surveillance is standard practice after an infant has enucleation for unilateral retinoblastoma.

In 1964, Wilms' tumor was associated with two rare birth defects: sporadic, bilateral aniridia (absence of the colored iris of the eye) and hemihypertrophy (overgrowth of one limb or part of the body) (Miller et al. 1964). The clinical association defined certain patients that deserved surveillance, but the yield was small and the best screening test involved X irradiation. Now, screening can be done by ultrasound and, more importantly, an even higher risk subgroup has been found—namely, those children with aniridia, growth and mental retardation, and a constitutional karyotype of deletion of the short arm of chromosome 11 (11p−) (Riccardi et al. 1978).

A recent clue to pathogenesis and an opportunity for prevention come from a report of eight adults with renal cell carcinoma in three generations of a family (Cohen et al. 1979). Because we routinely karyotype cancer family members, S. Tsai, in our contract-supported laboratory, discovered a 3;8 translocation chromosome in all available tumor patients and deduced its presence in others. Screening for kidney cancer revealed asymptomatic cancers in three relatives, all with the translocation. Despite study of additional patients with sporadic and familial renal cell carcinoma, this family remains a unique example of an adult tumor as the only manifestation of a constitutional chromosomal defect. The probability of kidney cancer in translocation carriers was estimated at 87%.

Breast cancer, twice as frequent as retinoblastoma, Wilms' tumor, and renal cell carcinoma combined (Cutler and Young 1975), provides additional contrasts in the magnitude of risk factors. With reason, much effort is made to clarify and quantify the risk of breast cancer in women who take estrogens. The wide use of conjugated estrogens for menopausal symptoms and of pure steroids, as in the birth control pill, justifies such studies, especially in the face of unclear benefits or satisfactory alternatives for birth control. To date, risk ratios seem small, rarely exceeding 2.0 (MacMahon 1979). Doubling the already low annual incidence yields such a low estimate that women or their doctors seem unwilling to change drug habits.

At first, similar low risk estimates, about threefold, were obtained for family history of breast cancer; but, subgroups with much higher risks were delineated, like the sisters of already affected young women and their mothers that have a 39-fold increased risk (Anderson 1976, 1977).

Counseling to Control Breast Cancer

Such high risk estimates move some women to consider scrupulous surveillance or even surgery to prevent breast cancer (Mulvihill et al. 1977b, Lynch et al. 1978). Using Anderson's figures on familial risk, we reported 13 women and have counseled some 19 more who were referred or sought counseling for breast cancer risk, usually because of their family history. In clinic, we pick the pedigree type defined by Anderson (1977) that best describes our patient, the consultand, as a sister or daughter in a mother, sister, or second-degree pedigree. The family risk factor, which is specific for three age groups (under 40 years, 40–59, and 60 and over), is multiplied by the other risk factors to get a summary relative risk estimate. This figure can be multiplied by the age-specific incidence (which properly should be for populations without the risk factors present in the consultand). Multiplying by five gives a five-year probability estimate, 18%, or 1 in 5.6, in the example (Table 1). The benefits and risks of medical surveillance and prophylactic mastectomy are presented with as little prejudice as possible. We try to distinguish medical fact from opinion and to bolster a woman's decision.

How the risk estimate is received and used must be complex. Several women,

TABLE 1. *Sample risk calculation for breast cancer in a 30-year-old woman from breast cancer family*

Factor	Patient
Family (young sister in mother-type pedigree)	39.5
×	×
Late age at first term pregnancy	2
×	×
Fibrocystic disease	2
=	=
Relative risk	158
×	×
Age-specific annual incidence	22.5/100,000
×	×
Five years	5
=	=
Probability of breast cancer in next five years	0.18 or 1/5.6

who had a high risk and understood it, had not, on follow-up, started a program of medical surveillance. Usually the choice is made before our counseling and, surprisingly, bears little relationship to our estimate, which is admittedly imperfect and based on many arguable assumptions.

An impediment to improving risk estimates is the imperfect knowledge of breast cancer genetics. Great hope now lies in the report of genetic linkage analysis in 13 breast cancer families (King et al. 1980). A statistically significant probability of genetic linkage, a lod score of 2.4, was found between breast cancer phenotype and the known single-gene marker, glutamate-pyruvate transaminase, which is assigned to the short arm of chromosome 10 (10p) (McKusick 1978).

With this etiologic insight, women at high risk may be identified for further research on pathogenesis and interaction with environmental factors, like drugs and diet, as well as clinical counseling with a goal of breast cancer prevention.

CONTROL BY PRIMARY PREVENTION

As discussed so far, the identification of individuals to be screened for early cancer is secondary prevention; that is, although cancer has occurred, death from cancer may be prevented or delayed. For primary prevention, there are two logical options: blocking the host-environment interaction that leads to cancer or removing the organ that is predisposed to cancer.

Surgery to Prevent Genetic Cancer

Prophylactic surgery has enough morbidity that it must be reserved for the highest risk patients. In general, these will be persons predisposed to cancer because of some genetic trait, either Mendelian (single gene) or chromosomal. For example, despite wide notice of diet as a cause of colon cancer (Higginson and Muir 1979), the single most potent cause is the mutant gene for multiple polyposis of the colon. The risk for cancer is considerable: by age 40 years, about half of all patients with multiple polyposis have large bowel cancer; by age 70, nearly all (Veale 1965). Currently, patients with the mutant gene have prophylactic colectomy by age 20 years to prevent cancer. If the political purveyors of prevention wish to begin efforts to prevent some colon cancers, they should expedite public and medical education about this considerable and controllable risk factor. Surgeons who remove a colon because of this genetic disease incur an obligation for alerting the family to the hereditary nature of the disorder.

Polyps are sufficient clinical markers of the mutant gene, but in vitro phenotypes are being studied to gain insight into the origins of colon cancer occurring, as it usually does, without polyposis. Traits deserving exploration by independent laboratories include tetraploidy in epithelial cultures of skin biopsies of Gardner's syndrome patients (Danes and Alm 1979), and, in fibroblast cultures, low serum requirement for growth, lack of contact inhibition, increased plasminogen activator, easy transformation by Kirsten murine sarcoma virus, deformed actin cable matrix, and unusual sensitivity to the tumor promoter tetradecanoylphorbol-13-acetate (Kopelovich et al. 1977, 1979).

Similar clinical use and research leads are seen in the syndromes of dysgenetic gonads (Figure 1). Malignancy in these disorders seems confined to the patients

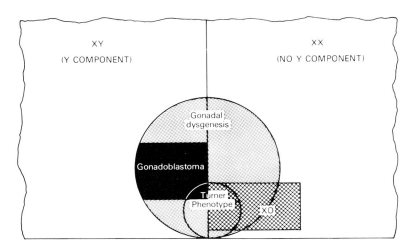

FIG. 1. The Y chromosome predisposes to malignancy in gonadal dysgenesis. The fragmented rectangle consists of the universe of males with a Y chromosome and females without it. Circles straddling the interface represent persons with ambiguous genitalia and gonadal dysgenesis with or without the classic Turner's phenotype of short stature, webbed neck, and primary amenorrhea. Those in circles with a simple 45,X karyotype have little risk of gonadal malignancy. Those with a Y chromosome, even when seen only in the germ cells, have a high frequency of gonadoblastoma. (Adapted from Mulvihill et al. 1975).

TABLE 2. *Cancer prevention by prophylactic surgery*

Congenital or genetic trait	Surgery
Cryptorchism	orchiopexy
Polyposes coli	colectomy
Familial colon cancer	colectomy
Multiple mucosal neuroma syndrome	thyroidectomy
Familial breast cancer	mastectomy
Familial ovarian cancer	oophorectomy
Gonadal dysgenesis	gonadectomy

with gonadal dysgenesis with or without the Turner's phenotype who have a Y component either completely or as a minor mosaicism, even confined to the germ line. The research question is, why? The relevance to cancer prevention lies in the obligation of clinicians to be as sure as possible by clinical and cytogenetic means that a patient with gonadal dysgenesis does not have a Y chromosome.

In summary, surgery has always been the first treatment of most cancers; it also seems to have a role in preventing cancer in a few congenital and genetic disorders that predispose nonvital organs to malignancy (Table 2). When a clear marker of the genotype is lacking, as in site-specific familial aggregation of breast, colon, and ovarian cancer, surgery is a more controversial approach to cancer control than it is in autosomal dominant traits, like the genetic polyposes and multiple mucosal neuroma syndrome. The list is short, but long enough to establish a principle.

Blocking Host-Environment Interactions

A global warning to the public to follow a certain diet or avoid certain exposures seems to go unheeded. The population at risk must be sharply defined. To illustrate, sunlight is the commonest known cause of cancer in the U.S.; but, paradoxically or perversely, most people look forward to sun exposure and congratulate each other on a healthy tan. However, when parents whose child with xeroderma pigmentosum died of melanoma were told to protect their affected infant twin boys from ultraviolet light, they did so and prevented cancer (Lynch et al. 1977).

This area of genetic-environmental interaction has been termed ecogenetics (Mulvihill 1976). By analogy to pharmacogenetics, a better known word, which means the study of genetic variations in response to drugs, ecogenetics is the study of genetic variations in response to environmental agents in general. Some 13 rare genetic traits predispose affected individuals to cancer or other untoward outcome on exposure to certain physical or chemical agents (Table 3). Except for the XO Turner's syndrome, the traits are determined by single mutant genes. In the future, other examples of ecogenetics in human oncology may emerge,

TABLE 3. *Genetic-environmental interactions (ecogenetics) in human malignancy (Mulvihill 1980)*

Environmental agent	Genetic trait	Tumor or outcome
Ionizing radiation	ataxia-telangiectasia with lymphona	radiation toxicity
	retinoblastoma	sarcoma
	nevoid basal cell carcinoma syndrome	basal cell carcinoma
Ultraviolet radiation	xeroderma pigmentosum	skin cancer, melanoma
	cutaneous albinism	skin cancer
	? B-K mole syndrome	melanoma
Stilbestrol	XO Turner's syndrome	adenosquamous endometrial carcinoma
Androgen	Fanconi's pancytopenia	hepatoma
Iron	hemochromatosis	hepatocellular carcinoma
Tyrosine	tyrosinemia	hepatocellular carcinoma
Epstein-Barr virus	Purtilo X-linked lympho-proliferative syndrome	Burkitt's and other lymphomas
	HLA-A Sin 2	nasopharyngeal carcinoma
Papillomavirus, type 5	Epidermodysplasia verruciformis	skin cancer

and genetic variations in carcinogen metabolism may be clarified. Drugs may even be given to interfere with host-carcinogen interactions and to control expression of human cancer.

In short, current knowledge of cancer genetics could be better applied to controlling cancer through primary and secondary prevention, prophylactic surgery, blocking host-environment interactions, and surveillance with early detection.

BIOPOLITICAL STRATEGY OF PREVENTION THROUGH GENETICS

The science of genetics sets forth principles and examples of cancer prevention. The experience of applying medical genetics to the control of congenital and hereditary diseases suggests biopolitical strategies for cancer prevention through genetics.

In general, the current belief that knowledge of etiology leads to prevention is too simple. It is usually necessary, but not at all sufficient. Etiology is something to know; prevention is something to do. Many other steps are needed; a possible sequence could be from knowledge of etiology, through professional and public education and consensus building, to motivation for legislation and enforcement, if prevention is an action by society, or to individual counseling and provision of health services, if prevention is to be a personal action.

A major lesson from medical genetics is to base any efforts on a firm scientific foundation of etiologic knowledge. Mendel's laws give firmer estimates of recurrence risk than empiric epidemiologic data can. If a child has cystic fibrosis, an autosomal recessive trait, the recurrence risk is 25%. Techniques for counting chromosomes as devised by Dr. Hsu and others for identifying one from the other explain many cases of multiple congenital anomalies, like Down's syndrome, and facilitate prevention through prenatal diagnosis (Hsu 1979). Knowledge of metabolic pathways gird efforts in primary prevention through prenatal diagnosis and in secondary prevention by blocking postnatal expression of the mutant gene with dietary or drug management. In very few areas is the scientific knowledge of human cancer as extensive as it is in genetic diseases.

Because of limited funding, geneticists had to start small; perhaps cancer prevention would likewise benefit from a small budget. Geneticists picked a rare disease in an ethnic group at higher risk than others—Tay-Sachs disease in Ashkenazi Jews—and effected a tremendously successful prevention program (Kaback 1977). Part of the success was in achieving intellectual and emotional consensus with the target population. Failure to pass this hurdle in sickle cell screening evoked cries of racial discrimination, and screening efforts had to be dropped (Murray et al. 1980). The same lack of consensus threatens efforts at cancer prevention. Severe congenital defects of the central nervous system may be detected by alpha-fetoprotein serology coupled with confirmatory amniocentesis; but the merits of making the techniques part of general obstetric practice were debated in the United Kingdom because of doubts that enough personnel were available to handle the estimated load (Harris 1980). As new knowledge emerged, it became obvious that geneticists could use the same health care setting—the newborn nursery, in this case—to prevent complications of several inborn errors of metabolism.

Finally, geneticists have turned to educators and experts in health care behavior to incorporate into the design of programs to prevent genetic disease the ability to measure the impact of such interventions (Lubs and de la Cruz 1977). The findings from such studies emphasize that knowledge of etiology is a necessary but insufficient condition for effective disease prevention.

CONCLUSION

Popularizers of the possible environmental causes of cancer perpetuate the misconception that such cancers are preventable; they may not be. Geneticists likewise struggle with the converse misconception that genetic diseases are hopeless; they are, in fact, treatable and preventable.

Genetic factors are conspicuous in the origins of many human cancers, whether they are chromosomal, single gene, or multifactorial. (The last category, ecogenetics, underscores the host-environment interactions needed to produce cancer.) Finally, the biopolitics of preventing genetic disease has many lessons for oncologists who are now extending the cancer war on the front called prevention.

REFERENCES

Advisory Committee to the Surgeon General of the Public Health Service. 1979. Smoking and Health: A Report of the Surgeon General. DHEW Publication No. (PHS) 79–50066, p. A–10.

Anderson, D. E. 1976. Genetic predisposition to breast cancer. Recent Results Cancer Res. 57:10–20.

Anderson, D. E. 1977. Breast cancer in families. Cancer 40:1855–1860.

Carter, C. O., J. A. F. Roberts, K. A. Evans, and A. R. Buck. 1971. Genetic clinic: A follow-up. Lancet 1:281–285.

Cohen, A. J., F. P. Li, S. Berg, D. J. Marchetto, S. Tsai, S. C. Jacobs, and R. S. Brown. 1979. Hereditary renal-cell carcinoma associated with a chromosomal translocation. N. Engl. J. Med. 301:592–595.

Cutler, S. J., and J. L. Young, Jr., eds. 1975. Third National Cancer Survey: Incidence Data. Natl. Cancer Inst. Monogr. 41, DHEW Publ. No. (NIH) 75–787. U.S. Government Printing Office, Washington, D.C., 454 pp.

Danes, B. S., and T. Alm. 1979. In vitro studies on adenomatosis of the colon and rectum. J. Med. Genet. 16:417–422.

Harris, R. 1980. Maternal serum alphafetoprotein in pregnancy and the prevention of birth defect. Br. Med. J. 280:1199–1202.

Higginson, J., and C. S. Muir. 1979. Environmental carcinogenesis: Misconceptions and limitations to cancer control. J. Natl. Cancer Inst. 63:1291–1298.

Hoover, R. 1978. Epidemiology: Tobacco and geographic pathology, in Pathogenesis and Therapy of Lung Cancer, C. C. Harris, ed. Marcel Dekker, Inc., New York, pp. 3–24.

Hsu, T. C. 1979. Human and Mammalian Cytogenetics: An Historical Perspective. Springer-Verlag, New York, 186 pp.

Kaback, M. M., ed. 1977. Tay-Sachs Disease: Screening and Prevention. Alan R. Liss, Inc., New York, 433 pp.

King, M-C, R. C. P. Go, R. C. Elston, H. T. Lynch, and N. L. Petrakis. 1980. Allele increasing susceptibility to human breast cancer may be linked to the glutamate-pyruvate transaminase locus. Science 208:406–408.

Knudson, A. G., Jr., H. W. Hethcote, and B. W. Brown. 1975. Mutation and childhood cancer: A probabilistic model for the incidence of retinoblastoma. Proc. Natl. Acad. Sci. USA 72:5116–5120.

Kopelovich, L., N. E. Bias, and L. Helson. 1979. Tumour promoter alone induces neoplastic transformation of fibroblasts from humans genetically predisposed to cancer. Nature 282:619–621.

Kopelovich, L., S. Conlon, and R. Pollack. 1977. Defective organization of actin in cultured skin fibroblasts from patients with inherited adenocarcinoma. Proc. Natl. Acad. Sci. USA 74:3019–3022.

Leonard, C. O., G. A. Chase, and B. Childs. 1972. Genetic counseling: A consumers' view. N. Engl. J. Med. 287:433–439.

Lubs, H. A., and F. de la Cruz, eds. 1977. Genetic Counseling. Raven Press, New York, 541 pp.

Lynch, H. T., B. C. Frichot, III, and J. F. Lynch. 1977. Cancer control in xeroderma pigmentosum. Arch. Dermatol. 113:193–195.

Lynch, H. T., R. E. Harris, C. H. Organ, Jr., and J. F. Lynch. 1978. Management of familial breast cancer. Arch. Surg. 113:1053–1067.

MacMahon, B. 1979. Oestrogens in the genesis of endometrial and breast cancer, in The Regulation of Fertility: Evaluation and Perspectives (INSERM Symposium Series, Vol. 83), Editions INSERM, Paris, pp. 81–92.

McCaw, B. K., F. Hecht, D. G. Harnden, and R. L. Teplitz. 1975. Somatic rearrangement of chromosome 14 in human lymphocytes. Proc. Natl. Acad. Sci. USA 72:2071–2075.

McKusick, V. A. 1978. Mendelian Inheritance in Man, 5th ed. Johns Hopkins University Press, Baltimore.

Miller, R. W., J. F. Fraumeni, Jr., and M. D. Manning. 1964. Association of Wilms's tumor with aniridia, hemihypertrophy and other congenital malformations. N. Engl. J. Med. 270:922–927.

Mulvihill, J. J. 1976. Genetic factors in pulmonary neoplasms. Birth Defects 12:99–111.

Mulvihill, J. J. 1980. Clinical observations of ecogenetics in human cancer. Ann. Intern. Med. 92:809–813.

Mulvihill, J. J., R. W. Miller, and J. F. Fraumeni, Jr., eds. 1977a. Genetics of Human Cancer. Raven Press, New York, 519 pp.

Mulvihill, J. J., A. W. Safyer, and J. K. Bening. 1977b. Counseling for familial breast cancer: Role of prophylactic mastectomy. Am. J. Hum. Genet. 29:80A.

Mulvihill, J. J., W. M. Wade, and R. W. Miller. 1975. Gonadoblastoma in dysgenetic gonads with a Y chromosome. Lancet 1:863.

Murray, R. F., Jr., N. Chamberlain, J. Fletcher, E. Hopkins, R. Jackson, P. A. King, and T. M. Powledge. 1980. Special consideration for minority participation in prenatal diagnosis. J.A.M.A. 243:1254–1256.

Riccardi, V. M., E. Sujansky, A. C. Smith, and U. Francke. 1978. Chromosomal imbalance in the aniridia-Wilms' tumor association: 11p interstitial deletion. Pediatrics 61:604–610.

Schimke, R. N. 1978. Genetics and Cancer in Man. Churchill Livingstone, New York, 108 pp.

Veale, A. M. O. 1965. Intestinal polyposis. Eugenics Laboratory Memoirs, series 40. Cambridge University Press, London.

Genes, Chromosomes, and Neoplasia, edited by
Frances E. Arrighi, Potu N. Rao, and Elton Stubblefield.
Raven Press, New York © 1981.

Genes, Chromosomes, and Neoplasia: An Overview

H. J. Evans

Medical Research Council, Clinical and Population Cytogenetics Unit, Western General Hospital, Edinburgh, United Kingdom

The topic of this monograph is very broad, encompassing the range from molecule to man; it is also of course of considerable importance from both the fundamental viewpoint of the basic scientist and the more practical viewpoint of the clinical scientist, so this symposium monograph has something for everybody. We have been privileged to read 27 excellent chapters, each of which provides some excitement for some readers and all of which provide some measure of stimulation for everyone.

It would be entirely presumptious of me to attempt to summarize each of the topics so lucidly presented by the authors, but when each of us reads each finds that there are one, or many, points made that strike a chord—or sometimes a dischord—with his own thoughts, and in most cases he learns of new facts and new fancies, and sees old facts in a new light. In some instances, for example in the case of DNA-mediated gene transfer discussed in this book by Drs. Ruddle, Siminovich, and Sweet, an important topic has been covered fairly exhaustively and so I will say no more about it. What I am going to do therefore is to pick up only some of the points that I found particularly interesting; to introduce a few items that were not covered, and which I think may be relevant to our discussion; and to embellish or to question some of the themes that have been discussed in relation to genes, chromosomes, and neoplasia.

CHROMATIN AND CHROMOSOME STRUCTURE AND THE QUESTION OF SCAFFOLDS

A chromosome as we see it at metaphase consists of about equal parts by weight of DNA, basic proteins, and acidic proteins. And, as we have read, in man for example the DNA in each nucleus is equivalent in total length to a DNA duplex some 2 m long. This means that even the smallest of our chromosomes has a couple of centimeters of DNA. How this DNA is packaged with its proteins to give a chromosome a couple of microns long still remains something of a mystery to me. We now have a good understanding of the association between DNA and the basic histones at the nucleosome level and good evidence for a coiled coil, or solenoid, structure giving a 100 Å fiber, but the packaging and folding at higher levels—and the involvement of a protein core (Stubblefield and Wray 1971) or scaffold (Paulson and Laemmli 1977)—are still matters of debate.

Classical genetics has long established a predetermined and fixed linear arrangement of genetic information along the length of the chromosome, and cytogenetics has further shown, albeit at a coarser level, a characteristic linear pattern of bands. In man, the average number of genes on a chromosome is probably somewhere around 2,000, but the average number of resolvable chromosome bands is less than 20, two orders of magnitude less. The amount of DNA in a human nucleus is around 3×10^9 nucleotide pairs, so that with roughly 300 metaphase bands per genome, each band must contain an average of around 10^7 nucleotide pairs. That many nucleotide pairs, let me remind you, would be enough to code for 5,000 different mRNA transcripts, each originally 2 kilobases (kb) long. The first point that I want to emphasize then is that *our cytological techniques of defining breakage points, or sites of rearrangements, are exceedingly coarse in relation to the location or involvement of specific genes.*

Now we see no obvious structural features of the nucleoprotein fibers of chromosomes that might indicate how a predetermined arrangement of genetic material in the chromosome is maintained. However, I think that the evidence that the chromosome may have an acidic protein core structure, or scaffold, is becoming more and more credible.

The demonstration of an acidic protein scaffold in dehistonized human chromosomes by Laemmli and Paulson was discussed by Dr. Stubblefield and has been repeated in various laboratories including my own. Moreover, my colleague, Peter Jeppesen, has shown that a protein scaffold is also present in isolated Chinese hamster chromosomes in which the DNA is removed by DNase digestion (Jeppesen et al. 1978). Most of us are familiar with the fact that in meiotic chromosomes we can silver stain acidic protein structures that run the whole length of the chromosomes and that correspond to the lateral elements of the synaptonemal complex. Howell and Hsu (1979), using the same kind of technique applied to Indian muntjac somatic cell chromosomes, have shown that there also exists a silver-staining core protein structure that runs the length of the chromatids. Moreover, this structure is not destroyed when the chromosomes are exposed to DNase or RNase digestion, but disappears when the chromosomes are exposed to trypsin.

An important question to ask is "how is the DNA of the chromosome anchored to the scaffold?" If we subject chromosomes to DNase digestion, we can remove all the DNA except for the approximately 0.1% that is firmly bound to the acidic protein core. When this core DNA is removed, it consists of nearly uniform short fragments. My colleagues have recently cloned a large number of these fragments, and these are currently being sequenced, so we hope to have information shortly on at least the base and sequence composition of these DNAs.

Staying for the moment with the problem of chromosome structure, I should like to refer briefly to two very relevant recent publications that have not been presented in this monograph, and introduce the topic of Z-DNA.

A NEW FORM OF DNA STRUCTURE—Z-DNA

Although fiber diffraction studies on DNA show the presence of at least three different molecular forms, the structure in solution had always been believed to be a regular B form. Occasionally, suggestions have been made that some DNAs, particularly those with unusual base sequences, might adopt quite different conformations, but the evidence has never been conclusive (Crick et al. 1979, Arnott 1979). Recently, however, Wang et al. (1979), on the basis of studies on the crystal structure of the hexanucleotide d(CpGpCpGpCpGp), have shown that this type of deoxypolymer, with a simple alternating purine-pyrimidine sequence, crystallizes as a left-handed double helical molecule with Watson-Crick base pairs and an antiparallel organization of the sugar phosphate chain. This so-called Z-DNA therefore differs significantly from the standard right-handed B-DNA. The left-handed stacking in the hexanucleotide DNA fragments is also apparently observed in longer polymers of deoxynucleotides (Arnott et al. 1980), and there is some evidence that the Z-DNA conformation is not restricted to poly(dG-dC)·poly(dG-dC), since Arnott and co-workers report a similar Z form in fibers containing poly(dG-dT)·poly(dA-dC).

Arnott et al. (1980) interestingly speculate that a segment of left-handed DNA may act as a store of negative winding and could compensate for the unwinding of positive B-DNA when this is melted to give access for DNA synthesis, transcription, etc. An important difference between B-DNA and Z-DNA, perhaps more relevant to the topic of carcinogenesis, is that in the latter there is a vastly increased accessibility of the guanine N7 and C8 atoms. There are many carcinogenic agents, e.g., the alkylating agents, that are relatively specific for the O6, N7, or C8 sites in guanine. The Z-DNA structure would therefore increase the reactivity of guanine considerably. It should also be noted that the GC base pair forms part of the outer surface of Z-DNA so that the C5 position of cytosine is more accessible to any enzyme than it is in B-DNA. CpG sequences are relatively infrequent in eukaryotic DNA, but they are often highly methylated on cytosine C5, especially in inactive genes.

Does the conformation of Z-DNA act as a signal for methylation? Wang et al. (1979) point out that if one looks through biologically active sequences, one can see that occasionally nature preserves alternating sequences of guanine and cytosine residues. For example, there is a highly conserved segment in rodent parvoviruses near the origin of replication that contains a total of 20 alternating G and C residues in two segments. Furthermore, a segment of the histidine D gene of *Salmonella* has a mutational hot spot that occurs at the site of eight alternating GC residues. The importance of this Z form of DNA in altering the susceptibility of chromatin to attack by carcinogens or in providing a means of facilitating gene inactivation are interesting speculations. It seemed tempting to me therefore to consider the possibility that the highly methylated inactive rDNAs discussed by O. J. Miller might be switched to the Z-form.

Before I turn to gene action, there is one other facet of chromosome structure in the context of this symposium that we have not referred to—jumping genes.

JUMPING GENES

In recent years, bacterial geneticists have described a variety of mobile DNA segments, including plasmids, which can insert themselves into the genome but also detach themselves completely and multiply independently in the cell; insertion sequences (IS), which are short stretches of DNA around 800–1,400 nucleotides long that appear to be able to insert themselves almost anywhere in the bacterial genome; transposons, which are elements that can hop about on the genome in the same way as IS, but carry other genes with them; and of course viruses. The transposons usually have insertion sequences on either end of them running in opposite directions, so one may consist of a coding piece of DNA with an inverted repeat at each end.

One feature of the bacterial insertion sequences that move from place to place is that they may alter the pattern of expression of adjacent genes, and so, in terms of effect, they directly parallel Barbara McClintock's transposable controlling elements in maize (McClintock 1957). A similar indication of transposable genes yielding instability and high mutation frequencies was originally uncovered by Green in his studies on mutable loci on *Drosophila,* and in particular on those elements operating at the white locus (Green 1973, 1976). Indeed, there is now considerable evidence for the mobility of certain genes around the genome in the fruit fly, so that we should not be surprised if roving bits of DNA exist in the mammalian genome.

The DNA copies of the RNA retroviruses discussed so elegantly by Drs. Bishop and Collett in this symposium monograph are essentially lengths of DNA bounded or flanked by inverted repeats, and these integrate into the DNA of the avian host genome at up to 1,000 or more different sites. Dr. Bishop asked whether these retroviruses could be mutators. I believe that they may well be. Moreover, many mammalian DNA viruses with linear DNA genomes, e.g., the adenovirus and herpes genomes, are essentially pieces of DNA enclosed within inverted repeats, and these also insert themselves into the host genome at a variety of sites.

In discussing jumping genes in the context of chromosome structure and neoplasia, it is certainly relevant to refer to the recent intriguing studies in *Drosophila* on what are referred to as dispersed repeated gene families. Finnegan et al. (1978) have recently described two repeated gene families that are widely scattered over the *Drosophila* genome. The genes that characterize these two families have been cloned and are called the *copia* and *412* genes, and each family contains 30–40 copies of identical or nearly identical genes that are not linked in tandem arrays, but are scattered throughout the genome. The number and chromosomal location of the two dispersed repeated gene families differ between *Drosophila* strains and indeed between individuals. However, the structures of the elements themselves, as determined by the patterns of restriction enzyme cleavage sites, are very closely conserved (Strobel et al. 1979). The large variation in the distribution of these elements among individual flies from the same laboratory stock implies that there is a high rate of transposition around the genome and that the number of potential insertion sites is large, numbering at least several hundred. Each of the genes codes for an abudant mRNA (the *copia* gene codes for 3% of the cytoplasmic polyadenylated RNA probe) that is present in a variety of cell types, so the genes are actively transcribed, although it is not known whether their mRNAs are translated to give useful proteins.

It is generally considered that transcribed genes that are repeated tandemly behave as blocks, or single domains, for transcription so that tandem arrays provide a means for rapid synthesis of one kind or message. Dispersed repeated genes, however, will each have different flanking gene sequences and each may be considered to reside in a separate domain. Dispersal therefore may allow for different frequencies of transcription of the same gene associated with different blocks of genes in the genome that are either transcribed at different times or in different cells in the developing and functioning organism. It seems not unlikely that dispersed repeated genes are not confined to the fruit fly, and indeed dispersed repeated DNA sequences with elements closely similar to *412* and *copia* gene families have recently been found in the yeast *Saccharomyces cerevisiae* (Cameron et al. 1979). Like its *Drosophila* counterpart, the yeast element is about 5 kb long, codes for an abundant poly(A)-containing RNA, and is bordered by direct terminal repeats of around 0.25 kb. The yeast elements also appear to be capable of transposition and are located in different places in different yeast strains.

Both the *Drosophila* and yeast elements are similar in structure to prokaryotic transposable elements and also to the integrated DNA genomes of RNA viruses of the retrovirus group. We might expect that elements such as *412* and *copia* that change their location within the genome may also affect the expression of other genes with which they are brought into contact, and it is possible that these sorts of genes are responsible for the mutator effects observed in *Drosophila* and in maize. Moreover, it seems highly likely that mobile dispersed repeated gene families must be present in the genomes of vertebrates, including mammals. Do we have any evidence for their existence? Although the tools necessary for demonstrating their presence are available, we have, as yet, no direct evidence for this kind of jumping gene in man. I am tempted to speculate, however, that some of the inherited genetic instabilities of the sort discussed in this book by Dr. Ray, e.g., Bloom's syndrome, could well be a reflection of the presence of a roving DNA insert at important sites in the genome.

GENE EXPRESSION AND CHROMOSOME STRUCTURE

I would like to make three comments under the heading of chromosome structure in relation to gene expression.

1. There is some evidence that DNA segments that are being continuously transcribed, for example ribosomal DNA, may not have any associated nucleosomes with them. This also appears to be true for transcribing regions of *Xenopus* lampbrush chromosomes (Scheer 1978). For active *continuous* transcription, therefore, the nucleosomal structure of the chromosome may be dispensed with.

2. Dr. Weintraub and his colleagues (Weisbrod et al. 1980) elegant DNase I digestion experiments showed that the DNAs of transcriptionally inactive chick embryonic or adult globin genes and the ovalbumin genes are more resistant to digestion than the DNAs of these genes when they are being transcribed, and the nucleosome conformation is different in these two states. Moreover, a change in conformation is evident *before* the actual activation of a gene, and this change is due to an association of the DNA with the high-mobility group proteins HMG14 and HMG17. This work provides some of the first indications of a role for the HMG proteins in altering chromosome structure for transcription, but we have some way to go before we can fully understand the kinds of changes in structure that may be necessary for, or associated with, controlled transcription.

3. A structural finding associated with transcriptional control is contained in the recent report by Van der Ploeg et al. (1980) on the DNA composition of the $\gamma\delta\beta$-globin gene complex on chromosome 11 in γ-β-thalassemia. These authors have demonstrated that a deletion of the γ- and δ-genes blocks the transcription of the β-gene, although the β-gene is present together with a large amount of flanking DNA. This is perhaps the first clear demonstration of a position effect at the molecular level in man and is very important when we consider the kinds of chromosome structural changes that we see associated with human neoplastic states.

SPLICING AND GENE EXPRESSION

The discovery that the RNA transcripts of coding loci in eukaryotes contain noncoding inserts, or introns, has raised the very important question of how the cell manages to excise the noncoding regions to give a viable functional mRNA. It is now clear from work like the elegant studies described by Dr. Ting on the chick ovalbumin and ovomucoid genes in this volume, that the whole of the coding sequence of a gene, including the introns, are transcribed to give the heterogeneous RNA in the nucleus, and that this is then spliced down to give the specific messenger for each transcribed locus. Several groups have examined the sequences around these spliced junctions in hn RNA and have defined a "consensus sequence" from which any one sequence diverges by only a few nucleotides. This is generally thought to be the preferred sequence at which splicing may occur. Recently, Lerner et al. (1980) had presented the intriguing hypothesis, and an equally intriguing set of data, suggesting that the small RNAs found in the nuclei of eukaryotic cells (sn RNA), or rather the discrete ribonucleic protein particles that they produce in association with a set of seven polypeptides (sn RNP), may function as the splicing machinery.

A question that we still have no answer to is why there is so much intron DNA. Does it assist in chromosomal pairing and preventing mismatching? I would not have thought so. Perhaps a more plausible suggestion is that having a number of short, discontinuous structural sequence runs reduces the severity of any introduced frameshift errors. Introns might also represent degenerated transposable sequences, and their continued presence might also provide innocuous homes for jumping genes and viruses, but these are merely a few of the less outrageous suggestions.

Selective control at the RNA splicing level has been suggested to account for the switching that occurs in the production of immunoglobulins. However, Rabbitts et al. (1980) have recently shown, by using cloned cDNA plasmid probes and mouse cells producing different immunoglobulin chains, that there is good evidence for a deletion of gene sequences involved in the switch from one type of IgG to another. In this work it was shown that the DNA of an IgG 2b secreting cell line lacks the $C_{\gamma1}$ gene and that IgG 2a secreting cells lose the $C_{\gamma1}$ and $C_{\gamma2b}$ genes. Switching due to gene deletion is, of course, irreversible, and one wonders how widespread such a phenomenon might be for other loci in mammalian cells. Furthermore, *we may ask whether this process of programmed deletion does in its own right increase the risk of unwanted genetic change at the deletion site that could be associated with neoplastic transformation?* This is a question I shall return to later.

METHYLATION AND GENE EXPRESSION

The significance of methylation is not entirely unclear, although Dr. Miller presented convincing evidence that methylation may be involved in preventing transcription of amplified mammalian ribosomal genes and Dr. Silverstein that methylation prevents re-expression of the *TK* gene, although methylation of the virus genome does not prevent viral transformation. In bacteria, it has been suggested that methylation may provide a marker (and a "preservative") for parental DNA, allowing for a one-way correction of mismatched bases to daughter DNA following replication, and prior to methylation, of the new strand. But it turns out that 5-methyl cytosine residues are hot spots for mutagenesis in *Escherichia coli,* due to deamination of 5-MC to thymine (Coulondre et al. 1978). 5-Methyl cytosine is present in mammalian genomes, and particularly so in simple sequence repetitive DNAs, but the function and importance of methylation is obscure here. For example, in the large intron of the rabbit β-globin gene, restriction enzyme analysis shows that a CCGG site is methylated at the internal C in about 50% of the gene copies in most somatic tissues, in 100% in sperm, and in about 80% in brain, but it is not methylated at all in a rabbit cell line (Waalwijk and Flavell 1978). Again, for example, the somatic rDNA in *Xenopus* is heavily methylated (Bird and Southern 1978), but in the sea urchin it is not (Bird et al. 1979). Moreover, in over half of the sea urchin genome, the DNA is unmethylated, which would argue against a suggested role of methylation in both protecting parental DNA strands and as a means of preventing DNA-protein binding and blocking polymerase action and hence transcription. Methylation does not therefore appear to be a universal mechanism for blocking transcription.

THE *Sarc* GENE AND GENE EXPRESSION

I do not intend to say very much about the impressive papers by Drs. Bishop and Collett, for the work on the avian retroviruses is well known and remarkable in showing that a single viral gene (the oncogene) coding for a single protein is responsible for inducing malignant transformation; that this gene, or something very similar, is present as a normal constituent of the host genome; that the virus oncogene–coded and equivalent host gene–coded proteins are associated with the plasma membrane of the cell; and that these viruses all appear to transform by arresting differentiation in a proliferating tissue. The defective retrovirus that causes a mouse leukemia, the Abelson murine leukemia virus, also produces a single protein, P120, which turns out to have almost identical properties to the Rous sarcoma virus pp60src (Witte et al. 1980), and oncogene equivalents are widespread throughout vertebrates. How relevant are they to the induction of human neoplasms, and more particularly of human leukemias, is an intriguing question, as is the suggestion that they may be related to cancer genes in "cancer families" in man. It seems that many of these viruses arrest differentiation and produce polyclonal rather than monoclonal neoplasms, and, despite their obvious importance, it may be that the avian lymphoid leukosis virus, which does not have an oncogene and produces a monoclonal neoplasm, may turn out to be a better human model.

GENE AMPLIFICATION

Gene amplification is fashionable, exciting, and important and was given a pretty fair run in this book. I think that we all agree that: (1) double minutes (DMs) are more common than we had hitherto believed, and may be present in a wide range of malignant cells; (2) they really are secondary phenomena of selection and are not primary or essential changes for neoplastic transformation; (3) DMs represent amplified genes plus amplified associated DNA; (4) DMs are unstable and may be lost in the absence of selection and their stable counterparts are the homogenous staining regions (HSR) of Biedler and Spengler (1976) and the C-bandless chromosomes, or CMs, described by Dr. Levan in this volume.

How do these amplified regions arise? DMs could be produced by a rolling circle replication or by a replication bubble with multiple initiations of replication at one site (the onion skin analogue) similar to that proposed for DNA-puffs in polytene chromosomes of *Rhyncosciara.* In vitro amplification follows stepwise selection, and so a build-up of copies due to chromatid exchange (Bostock et al. 1979) also cannot be ruled out as a mechanism.

How widespread is this phenomenon of amplification? Has it been observed in relation to genes other than those specifying dihydrofolate reductase and cells other than those derived from malignancies or subjected to step-wise selection procedures in vitro? Answers to these questions are beginning to emerge. We have already read that there is some evidence for a 100-fold amplification of a locus coding for aspartic transcarbamylase without any associated and obvious HSR or DMs.

CONSTITUTIONAL GENETIC CHANGES AND CANCER

DNA Repair and Chromosomal Instability Syndromes

We have for many years now become accustomed to the fact that DNA is not an inviolate rigid molecule, but undergoes change very frequently and that the cell has enzymes not only for replicating DNA, but for repairing DNA lesions, for recombining DNA molecules, and for modifying DNA. Inefficiencies or failures on the part of these enzyme systems are paramount in failures to maintain the integrity of the genome. Where such failures are preprogrammed by inherited genetic information, the consequences may be drastic. We have been particularly concerned with the association between inherited defects in DNA repair and the very high incidence of cancer. I shall return to this shortly, but here I might point out that a number of conditions in man that involve neurological degeneration also appear to be associated with a defect in DNA repair. For example, severe forms of xeroderma may be associated with severe mental retardation and neurological complications. Similarly, ataxia-telangiectasia is accompanied by cerebellar ataxia and progressive neurological deterioration, and neurological complications are also associated with Cockayne's syndrome and with some examples of Bloom's syndrome. More recently, two groups have reported on cells obtained from patients suffering from Huntington's chorea, which is a dominant autosomal disease producing progressive mental deterioration, and these cells appear to be more sensitive than normal to killing by ionizing radiation (Arlett and Muriel 1979, Moshell et al. 1980). As yet we have no information on any association between Huntington's chorea and early cancer, but we should look.

In considering these human syndromes, which appear to involve a deficiency or abnormality in DNA repair processes, it is worthwhile pointing out that although many of these conditions appear to be relatively rare, the frequency of heterozygotes must be quite high; for instance in the case of ataxia-telangiectasia an incidence of 1 in 40,000 would suggest that perhaps 1% of the population are heterozygotes. Considering the known half dozen or so different syndromes that are associated with DNA repair defects in man (Table 1) and indeed if we expect that over 100 such genes might be associated with DNA repair and recombination processes, it may well be that most of us are heterozygous at one or more loci involved in DNA repair and that there will be a range in efficiency for DNA repair throughout the population. Moreover, it seems likely that many of these mutations involving DNA-housekeeping genes would be homozygous lethals and that the half dozen or so known DNA-repair-defect syndromes represent only a small fraction of individuals with a diminished DNA repair capacity. Individuals who are constitutional heterozygotes for a repair defect will, because of spontaneous somatic mutations, produce over their lifetimes large numbers of cell homozygous for a repair defect. With these kinds of thoughts in mind, a number of groups have been looking at cancer incidence in such heterozygotes.

Some years ago, Swift (1971) showed that heterozygotes carrying the recessive gene for Fanconi's anemia are more cancer prone than average and later showed a similar increased incidence of cancer in heterozygotes of ataxia-telangiectasia (Swift et al. 1976). More recently, Swift and Chase (1979) have reported on the incidence of cancer in 31 families each having an index case with xeroderma pigmentosum. As in the earlier studies on families with Fanconi's anemia and ataxia-telangiectasia, differences between the cancer incidence in blood relatives and spouse controls were sought after, and were seen to be more striking in people living in the Southern states or working outside. In the Fanconi studies it was remarked that if Fanconi heterozygotes are three times more likely to die from neoplasms than those who are not blood relatives, then they could comprise around 5% of all patients dying from acute leukemias and around 1% of all patients dying from all cancers including leukemias. The same calcula-

TABLE 1. *Estimates of the heterozygote frequency for the autosomal recessive syndromes associated with cancer and leukemia* *

Syndromes	Incidence (approx.)	Estimated heterozygote frequency (%)
Ataxia-telangiectasia	1 in 40,000	1
Bloom's	In Ashkenazi Jews	0.5 (?)
Chediak-Higashi	?	?
Fanconi's anemia	1 in 360,000	0.3
Werner	1–22 in 10^6	0.2–0.9
Xeroderma pigmentosum	1–4 in 10^6	0.2–0.4

* Data taken from Swift (1975).

tions on the data on ataxia-telangiectasia (AT) heterozygotes by Swift et al. (1976) suggest that AT heterozygotes may account for 5% of all patients dying from cancer before the age of 45. I have no wish to comment on the validity of these calculations, but I would like to put in a strong plea for the development of reliable methods for identifying heterozygotes for these inherited conditions predisposing to cancer, since they could be of enormous use in screening for individuals at increased risk.

Constitutional Chromosome Anomalies and Cancer

In addition to the chromosomal instability and DNA repair-deficient syndromes, there are a number of heritable constitutional chromosome anomalies that predispose to cancer. We are familiar with, for example, the increased incidence of leukemia in patients with Down's syndrome, trisomy 21; the presence of a deletion of 13q (13q14) in some, but not in all, patients with bilateral retinoblastoma; and the presence of a deletion of part of 11p (11p13) in patients with aniridia with Wilms' tumor. In this monograph, Dr. Knudson has discussed these overt cytogenetic anomalies that predispose to cancer and both he and Dr. Mulvihill have referred to the recent remarkable report by Cohen et al. (1979) on a family with a hereditary renal cell carcinoma associated with a balanced translocation involving chromosomes 3 and 8 t(3:8)(p21;q24). The translocation was present in all five live members of the family who had had renal cancer and could be assigned to three of five other, deceased members of the family who had had renal cancer. No family member with renal cancer had shown a normal karyotype.

Those of you who have been following the saga of the involvement of chromosome 8 in the leukemias and lymphomas will prick up your ears on reading this report on the renal cell carcinoma, particularly since the region of chromosome 8 involved in translocation in the renal carcinoma story (8q24) is the same as in the lymphomas and acute myeloid leukemias that we will refer to in a moment. Dr. Knudson suggested that perhaps a gene predisposing to renal cell carcinoma is located at the site of the translocation, but I suggest that the association of the 8q24 region with other types of malignancies may not be fortuitous, but rather reflects the presence of a locus at or near this region that is concerned with the control of cell cycling or growth control. I shall also return to this subject.

ACQUIRED GENETIC CHANGES AND CANCER

Because of the refractory nature of solid tumors for cytogenetic study, most of our information on acquired chromosomal changes and cancer comes from studies on reticuloendothelial neoplasms. Drs. Rowley, Sandberg, and Mitelman have each contributed significantly to this field, and in this volume review recent developements. At one time I was completely bamboozled by the plethora of chromosome markers that turn up consistently, or inconsistently, in reticuloen-dothelial neoplasms and feared that we should never make any sense of this. My pessimistic mood is now being slowly replaced by some optimism, particu-larly when we bear in mind three problems:

1. The first of these is simply that in studying chromosomes in proliferating cells of bone marrow or blood in leukemia patients, a constant problem has been whether the cells examined were of malignant origin or were normal cells. Some headway has been made in this area, particularly with regard to B cell leukemias, and interesting findings are beginning to emerge.

2. If, and this is but a supposition, an important genetic or epigenetic change that is required for malignant transformation is the loss or suppression of a single locus or a single chromosome site (assuming prior, or constitutional, loss or inactivation of the homologous site) then *in many cases such a loss will not be detectable by standard cytogenetic methods*. However, in a proportion of cases gross changes may involve a locus important in malignant transforma-tion, so that *we might expect some degree of consistency of cytogenetic change in some, but only some, cases*.

3. The proliferative capacity of malignant cells and, in contrast to embryonal cells, their lack of a requirement for maintaining a wide armory of information for differentiation, allow for a considerable evolution of karyotype during tumor cell development. We have long recognized, therefore, that many chromosome changes are secondary to the development of malignancy. Although they may be important and of interest in their own right, they are not informative with regard to the *origin* of transformation. We need therefore to clearly distinguish between any chromosomal change involved in neoplastic transformation itself and any other changes that facilitate growth—under varying conditions in differ-ent tumors and different locations—and which offer advantages in terms of *proliferation rate* and *viability*.

With these three points in mind a number of developments of interest have recently emerged.

Malignant Cells with Aberrations Involving Chromosomes Containing Ig Genes

An intriguing correlation has recently turned up between chromosomes known to possess the immunoglobulin heavy chain gene in mouse and man and the involvement of these chromosomes in structural changes in malignant cells pro-ducing Ig. Translocations involving chromosome 14 have been described in a variety of human lymphomas, B-cell neoplasms, and plasmacytomas, and re-cently Croce et al. (1979) have shown that the gene for human immunoglobulin heavy chains is located on this particular chromosome. Whether this association is of any importance at all is of course not known, but it is interesting to note that Ohno et al. (1979) have recently reported a remarkable nonrandom chromosome change involving the Ig gene–carrying chromosomes 12 and 6 in mouse plasmacytomas. In this work, seven primary and early passage plasmacytomas induced in BALB/c mice were shown in all cases to contain a deleted chromosome 15 with a translocation of the distal part of that chromosome onto either chromosome 12 or chromosome 6. It is interesting to note here that T-cell leukemias in the mouse show a regular trisomy of chromosome 15, whereas all the plasmacytomas have this consistent translocation of the distal part of chromosome 15 to either chromosome 6 or 12. In the mouse, chromosome 12 is known to carry the heavy chain and chromosome 6 the light chain determinants, and whether the initial breakage events that ultimately give rise to translocations are breakage events involved in normal processing of Ig chains is not known. This may, however in the light of my earlier comments, be an interesting possibility. What can be said from this type of study is that mouse T-cell leukemias, whether induced by a variety of viruses, X rays, or various chemical carcinogens, all show trisomy 15, and the duplication of chromosome 15 is taken to relate directly to the neoplastic behavior of the precursor cell, the prothymocyte.

Chromosome Abnormalities in Luekemic B Lymphocytes

Chromosome analyses reveal abnormalities in many cases of lymphoma, non-lymphocytic leukemia, and acute lymphoblastic leukemia. However, most studies on chronic lymphocytic leukemia (CLL) have shown normal karyotypes, although some reports describe certain abnormalities. In these studies a continuing difficulty has been to distinguish between normal and leukemic cells. Polyclonal B-cell mitogens and Epstein-Barr virus can induce transformation and immunoglobulin synthesis in lymphocytes from patients with CLL. A number of groups have recently used methods to specifically identify the leukemic B lymphocytes and have revealed a number of interesting chromosome anomalies (Gahrton et al. 1979, 1980, Hurley et al. 1980). I am not going to detail or discuss these findings, but I simply want to draw attention to the possibilities that are opened by this approach.

Burkitt's Lymphoma and the Involvement of Chromosome 8

Burkitt's lymphomas, both African and North American, have associated with them a specific chromosome abnormality involving a translocation between chromosomes 8 and 14, t(8q−;14q+). In general, the emphasis has been upon the importance of the presence of the 14q+ element in the same way as the alteration in the 22q has been implicated as the major factor in the 9/22 translocation observed in chronic myeloid leukemia. Recently, however, there have been at least three reports of Burkitt's lymphomas in which no 14q+ was present, but in which a translocation involving chromosome 8 was evident (Miyoshi et al. 1979, Van den Berghe et al. 1979, Berger et al. 1979). The region of chromosome 8 involved (q24) is common to the site involved in the 8/14 translocations, and the reports raise a question of whether the rearrangement of chromosome 8 is just as important as, or more important than, the rearrangement of 14q in Burkitt's lymphoma.

We have already commented on the fact that a rearrangement involving 8q24 is present in a family showing inheritance of renal cell carcinoma. We can add to this a number of reports describing translocations involving chromosomes 8 and 21 in patients with acute myeloid leukemia (e.g., Trujillo et al. 1979, Prigogina et al. 1979). Although human gene mappers have recently allocated a number of loci to chromosome 8 and an interferon receptor site to chromosome 21, we have no information on the gene content of the terminal region of 8q, and I look forward to any information on this region with interest.

CONCLUDING COMMENTS

From some of the comments that I have already made, it will be evident that I incline to the view that there will be various domains in different parts of the genome that are responsible for the genetic control of different differentiation pathways and different cell functions, and that each domain may possess one or more loci directly or indirectly involved in controlling—suppressing, maintaining, or initiating—mitotic cycling. From this view, mutations, or indeed epigenetic changes, at such sites may be responsible for initiating and maintaining a neoplastic state. Some of the sites where gross chromosomal changes occur with varying degrees of consistency with particular neoplastic states, such as the terminal regions of chromosome 22q in CML and 8q in lymphomas and AML and the intercalary regions of 14q and 17q in lymphomas and leukemias, 13q in retinoblastoma, and 11p in Wilms' tumor with aniridia, could all represent such sites. In enumerating these I am reminded of some of the data acquired by Dr. Harold Klinger and colleagues (Klinger et al. 1978) in experiments in which malignant mouse A9 cells were hybridized to human cells. It transpires that the presence of but one of a number of human chromosomes suppresses the malignant nature of the hybrid cell, and these include chromosomes 8, 11, 13, and 17.

I think we all acknowledge the very considerable clinical benefits that cytogenetics and cell biology have provided in diagnosis and prognosis, particularly in the leukemias, and chromosome analysis has a continuing and important role to play in this area. However, I have emphasized that only a proportion of genetic changes causally associated with neoplastic transformation will be identifiable cytologically as chromosome structural changes. We must not therefore expect that everything will turn out to have the high degree of concordance of the translocation of chromosome 22 and CML, or the involvement of chromosomes 8 or 14, or both, and the lymphomas. In saying this, I end by making two pleas. The first and obvious is for the extension of our cytogenetic studies to solid tumors. The second, which is not really contradictory, is for spending perhaps a little less time looking down the microscope and a little more time looking at gels to try to characterize those changes that are not so evident at the cytological level, but that we presume must exist at the molecular level. In this latter context, the kinds of comparisons that may be informative are those between the organization of DNA around the translocation points in a given type of tumor cell with a characteristic translocation compared with those chromosomes in a similar tumor cell with no translocation and with those chromosomes in normal cells. Similarly, in those tumors that are sometimes associated with specific deletions of 11p or 13q, we must examine the molecular structure of these chromosomes to characterize what is deleted and then examine those same tumors in which there is no *overt* chromosome aberration to see if we can detect common deleted sequences. For this kind of work one hope will be that we can use the fluorescence-activated chromosome sorter (Lebo et al. 1979) to isolate in a pure state, and in quantity, the chromosomes of interest. Another is that we have the appropriate molecular probes to identify large areas of the genome. Only in these ways will we be able to identify the critical loci involved in neoplastic change and evaluate the tempting speculations of a causal association between neoplasia and deletion of, for example, the Ig locus and the nucleoside phosphorylase locus on chromosome 14, the TK locus on chromosome 17, and the interferon receptor locus on chromosome 21.

REFERENCES

Arlett, C. F., and W. J. Muriel. 1979. Radiosensitivity in Huntington's chorea cell strains: A possible preclinical diagnosis. Heredity 42:276.

Arnott, S. 1979. Is DNA really a double helix? Nature 278:780–781.

Arnott, S., R. Chandrasekaran, D. L. Birdsall, A. G. W. Leslie, and R. L. Ratliff. 1980. Left-handed DNA helices. Nature 283:743–745.

Berger, R., A. Bernheim, H.-J. Weh, G. Flandrin, M.-T. Daniel, J.-C. Brouet, and N. Colbert. 1979. A new translocation in Burkitt's tumor cells. Hum. Genet. 53:111–112.

Biedler, J. L., and B. A. Spengler. 1976. Metaphase chromosome anomaly: Association with drug resistance and cell-specific products. Science 191:185–187.

Bird, A. P., and E. M. Southern. 1978. Use of restriction enzymes to study eukaryotic DNA methylation: I. The methylation pattern in ribosomal DNA from *Xenopus laevis*. J. Mol. Biol. 118:27–47.

Bird, A. P., M. H. Taggart, and B. A. Smith. 1979. Methylated and unmethylated DNA compartments in the sea urchin genome. Cell 17:889–901.

Bostock, C. J., E. M. Clark, N. G. L. Harding, P. M. Mounts, C. Tyler-Smith, V. van Heyningen, and P. M. B. Walker. 1979. The development of resistance to methotrexate in a mouse melanoma cell line. Chromosoma 74:153–177.

Cameron, J. R., E. Y. Loh, and R. W. Davis. 1979. Evidence for transposition of dispersed repetitive DNA families in yeast. Cell 16:739–751.

Cohen, A. J., F. P. Li, S. Berg, D. J. Marchetto, S. Tsai, S. C. Jacobs, and R. S. Brown. 1979. Hereditary renal cell carcinoma associated with a chromosomal translocation. N. Engl. J. Med. 301:592–595.

Coulondre, C., J. H. Miller, P. J. Farabaugh, and W. Gilbert. 1978. Molecular basis of base substitution hotspots in *Escherichia coli*. Nature 274:775–780.

Crick, F. H. C., J. C. Wang, and W. R. Bauer. 1979. Is DNA really a double helix? J. Mol. Biol. 129:449–461.

Croce, C. M., M. Shanden, J. Martinis, L. Cicurel, G. G. D'ancona, T. W. Dolby, and H. Koprowski. 1979. Chromosomal location of the genes for human immunoglobulin heavy chains. Proc. Natl. Acad. Sci. USA 76:3416–3419.

Finnegan, D. J., G. M. Rubin, M. W. Young, and D. S. Hogness. 1978. Repeated gene families in *Drosophila melanogaster*. Cold Spring Harbor Symp. Quant. Biol. 42:1053–1063.

Gahrton, G., L. Zech, K.-H. Robert, and A. G. Bird. 1979. Mitogenic stimulation of leukemia cells by Epstein-Barr virus. N. Engl. J. Med. 301:438–439.

Gahrton, G., K.-H. Robert, K. Friberg, L. Zech, and A. G. Bird. 1980. Extra chromosome 12 in chronic lymphocytic leukaemia. Lancet 1:146–147.

Green, M. M. 1973. Some observations and comments on mutable and mutator genes in Drosophila. Genetics Suppl. 73:187–194.

Green, M. M. 1976. Mutable and mutator loci, *in* The Genetics and Biology of *Drosophila*, vol. 1b, M. Ashburner and E. Novitski, eds. Academic Press, New York, pp. 929–946.

Howell, W. M., and T. C. Hsu. 1979. Chromosome core structure revealed by silver staining. Chromosoma 73:61–66.

Hurley, J. N., S. M. Fu, H. G. Kunkel, R. S. K Chaganti, and J. German. 1980. Chromosome abnormalities of leukaemic B lymphocytes in chronic lymphocytic leukaemia. Nature 283:76–78.

Jeppesen, P. G. N., A. T. Bankier, and L. Sanders. 1978. Non-histone proteins and the structure of metaphase chromosomes. Exp. Cell Res. 115:293–302.

Klinger, H. P., A. S. Baim, C. K. Eun, T. B. Shaws, and F. H. Ruddle. 1978. Human chromosomes which affect tumorigenicity in hybrids of diploid human with heteroploid human or rodent cells. Cytogenet. Cell Genet. 22:245–249.

Lebo, R. V., A. V. Carrano, K. Burkhart-Schultz, A. M. Dozy, L. C. Yu, and Y. W. Kan. 1979. Assignment of human β-, γ- and δ-globin genes to the short arm of chromosome 11 by chromosome sorting and DNA restriction enzyme analysis. Proc. Natl. Acad. Sci. USA 76:5804–5808.

Lerner, M. R., J. A. Boyle, S. M. Mount, S. L. Wolin, and J. A. Steitz. 1980. Are snRNPs involved in splicing? Nature 283:220–224.

McClintock, B. 1957. Controlling elements in the gene. Cold Spring Harbor. Symp. Quant. Biol. 21:197–216.

Miyoshi, I., S. Hiraki, I. Kimura, K. Kijamoto, and J. Sato. 1979. 2/8 translocation in a Japanese Burkitt's lymphoma. Experientia 35:742.

Moshell, A. N., S. F. Barrett, R. E. Tarone, and J. H. Robbins. 1980. Radiosensitivity in Huntington's disease: Implications for pathogenesis and presymptomatic diagnosis. Lancet 1:9–11.

Ohno, S., M. Babonits, F. Wiener, J. Spira, G. Klein, and M. Potter. 1979. Nonrandom chromosome changes involving the Ig gene-carrying chromosomes 12 and 6 in pristane-induced mouse plasmacytomas. Cell 18:1001–1007.

Paulson, J. R., and U. K. Laemmli. 1977. The structure of histone-depleted metaphase chromosomes. Cell 12:817–828.

Prigogina, E. L., E. W. Fleischman, G. P. Puchkova, O. E. Kulagina, S. A. Majakova, S. A. Balakirev, M. A. Frenkel, N. V. Khvatova, and I. S. Peterson. 1979. Chromosomes in acute leukemia. Hum. Genet. 53:5–16.

Rabbitts, T. H., A. Forster, W. Dunnick, and D. L. Bentley. 1980. The role of gene deletion in the immunoglobulin heavy chain switch. Nature 283:351–356.

Scheer, U. 1978. Changes of nucleosome frequency in nucleolar and non-nucleolar chromatin as a function of transcription: An electron microscopic study. Cell 13:535–549.

Strobel, E., P. Dunsmuir, and G. M. Rubin. 1979. Polymorphisms in the chromosomal locations of elements of the 412, copia and 297 dispersed repeated gene families in *Drosophila.* Cell 17:429–439.

Stubblefield, E., and W. Wray. 1971. Architecture of Chinese hamster metaphase chromosomes. Chromosoma 32:262–294.

Swift, M. 1971. Fanconi's anaemia in the genetics of neoplasia. Nature 230:370–373.

Swift, M., L. Sholman, M. Perry, and C. Chase. 1976. Malignant neoplasms in the families of patients with ataxia telangiectasia. Cancer Res. 36:209–215.

Swift, M., and C. Chase. 1979. Cancer in families with xeroderma pigmentosum. J. Natl. Cancer Inst. 62:1415–1421.

Trujillo, J. M., A. Cork, M. J. Ahearn, E. L. Youness, and K. B. McCredie. 1979. Hematologic and cytologic characterization of 8/21 translocation acute granulocytic leukemia. Blood 53:695–706.

Van Den Berghe, H., C. Parloir, S. Gosseye, V. Englebienne, G. Cornu, and G. Sokal. 1979. Variant translocation in Burkitt lymphoma. Cancer Genet. Cytogenet. 1:9–14.

Van der Ploeg, L. H. T., A. Konings, M. Oort, D. Roos, L. Bernini, and R. A. Flavell. 1980. γ-β-Thalassaemia studies showing that deletion of the γ- and δ-genes influences β-globin gene expression in man. Nature 283:637–642.

Waalwijk, C. and R. A. Flavell. 1978. DNA methylation at a CCGG sequence in the large intron of the rabbit β-globin gene: Tissue-specific variations. Nuclei Acid Res. 5:4631–4641.

Wang, A. H. J., G. J. Quigley, F. J. Kolpak, J. L. Crawford, J. H. van Book, G. van der Marel, and A. Rich. 1979. Molecular structure of a left-handed double helical DNA fragment at atomic resolution. Nature 282:680–686.

Weisbrod, S., M. Groudine, and H. Weintraub. 1980. Interaction of HMG 14 and 17 with actively transcribed genes. Cell 19:289–301.

Witte, O. N., A. Dasgupta, and D. Baltimore. 1980. Abelson murine leukaemia virus protein is phosphorylated in vitro to form phosphotyrosine. Nature 283:826–831.

Author Index

529

Subject Index